NEW FRONTIERS IN HIGH-ENERGY PHYSICS

Studies in the Natural Sciences

A Series from the Center for Theoretical Studies

University of Miami, Coral Gables, Florida

Recent Volumes in this Series

Volume 5 – FUNDAMENTAL THEORIES IN PHYSICS
Edited by Stephan L. Mintz, Laurence Mittag, Susan M. Widmayer, Chui-Shuen Hui, and Joseph F. Malerba

Volume 6 – PROGRESS IN THE NEUROSCIENCES AND RELATED FIELDS
Edited by Behram Kursunoglu, Stephan L. Mintz, Susan M. Widmayer, Chui-Shuen Hui, Joseph Hubbard, and Joseph F. Malerba

Volume 7 – TOPICS IN ENERGY AND RESOURCES
Edited by Behram Kursunoglu, Stephan L. Mintz, Susan M. Widmayer, Chui-Shuen Hui, Joseph Hubbard, Joseph Malerba, and George Soukup

Volume 8 – PROGRESS IN LASERS AND LASER FUSION
Edited by Behram Kursunoglu, Arnold Perlmutter, Susan M. Widmayer, Uri Bernstein, Joseph Hubbard, Christian Le Monnier de Gouville, Laurence Mittag, Donald Pettengill, George Soukup, and M. Y. Wang

Volume 9 – THEORIES AND EXPERIMENTS IN HIGH-ENERGY PHYSICS
Edited by Behram Kursunoglu, Arnold Perlmutter, Susan M. Widmayer, Uri Bernstein, Joseph Hubbard, Christian Le Monnier de Gouville, Laurence Mittag, Donald Pettengill, George Soukup, and M. Y. Wang

Volume 10 – NEW PATHWAYS IN HIGH-ENERGY PHYSICS I
Magnetic Charge and Other Fundamental Approaches
Edited by Arnold Perlmutter

Volume 11 – NEW PATHWAYS IN HIGH-ENERGY PHYSICS II
New Particles–Theories and Experiments
Edited by Arnold Perlmutter

Volume 12 – DEEPER PATHWAYS IN HIGH-ENERGY PHYSICS
Edited by Behram Kursunoglu, Arnold Perlmutter, Linda F. Scott, Mou-Shan Chen, Joseph Hubbard, Michel Mille, and Mario Rasetti

Volume 13 – THE SIGNIFICANCE OF NONLINEARITY IN THE NATURAL SCIENCES
Edited by Behram Kursunoglu, Arnold Perlmutter, Linda F. Scott, Mou-Shan Chen, Joseph Hubbard, Michel Mille, and Mario Rasetti

Volume 14 – NEW FRONTIERS IN HIGH-ENERGY PHYSICS
Edited by Behram Kursunoglu, Arnold Perlmutter, Linda F. Scott, Osman Kadiroglu, Jerzy Nowakowski, and Frank Krausz

ORBIS SCIENTIAE

NEW FRONTIERS IN HIGH-ENERGY PHYSICS

Chairman
Behram Kursunoglu

Editors
Arnold Perlmutter
Linda F. Scott

Scientific Secretaries
Osman Kadiroglu
Jerzy Nowakowski
Frank Krausz

Center for Theoretical Studies
University of Miami
Coral Gables, Florida

PLENUM PRESS • NEW YORK AND LONDON

Library of Congress Cataloging in Publication Data

Orbis Scientiae, University of Miami, 1978.
New frontiers in high-energy physics.

(Studies in the natural sciences; v. 14)
"Held by the Center for Theoretical Studies, University of Miami, Coral Gables, Florida, January 16–19, 1978.
Includes index.
1. Particles (Nuclear physics) – Congresses. I. Kursunoğlu, Behram, 1922- II. Perlmutter, Arnold, 1928- III. Scott, Linda F. IV. Miami, University of, Coral Gables, Fla. Center for Theoretical Studies. V. Title. VI. Title: High-energy physics. VII. Series.
QC793.07 1978 539.7'21 78-15767
ISBN 0-306-40037-5

Proceedings of Orbis Scientiae 1978 on New Frontiers in High-Energy Physics
held by the Center for Theoretical Studies, University of Miami,
Coral Gables, Florida, January 16–19, 1978

A Division of Plenum Publishing Corporation
227 West 17th Street, New York, N.Y. 10011

Printed in the United States of America

PREFACE

The editors are pleased to submit to the readers the state of the art in high energy physics as it appears at the beginning of 1978.

Appreciation is extended to Mrs. Helga S. Billings for her assistance with the conference and for her skillful typing of the proceedings which was done with great enthusiasm and dedication. Also, appreciation is extended to Dr. Osman Kadiroğlu for his assistance with the proceedings this year.

Orbis Scientiae 1978 received some support from the Department of Energy.

The Editors

ORBIS SCIENTIAE 1978 PARTICIPANTS

CONTENTS

NEW APPROACH TO COSMOLOGICAL THEORY

P.A.M. Dirac

Florida State University

Tallahassee, Florida

THE AGE OF THE UNIVERSE

In the 1920's after Einstein's theory had become established, people turned their attention to cosmological equations. One of the main problems then was to estimate the age of the universe. A serious difficulty soon appeared. There were two schools of thought, whose conclusions were contradictory.

On the one hand there were the people who studied the lifetimes of radioactive elements. They were led to infer an age of about 10^{10} years. On the other hand there were the people who studied the evolution of galaxies and stars clusters according to the law of gravitation of Newton or Einstein. They found that 10^{10} years was not nearly long enough. They needed at least $8x10^{10}$ years, possibly longer.

For many years there was an impasse. Eventually, a hope of reconciliation was provided by E. A. Milne. He proposed that there are really two measures of time of importance to physicists and they are not in agreement. There is atomic time, which is to be used for atomic processes governed by quantum laws, and there is mechanical time, which is to be used with the classical laws for the motion of bodies.

Milne introduced the symbol t for atomic time, τ for mechanical time. He built up an elaborate cosmological theory, and was led to the connection

$$\tau = \log t \tag{1}$$

between the two time measures. The details of Milne's theory are obscure but the conclusion (1) is probably correct. It is a reasonable relation because it means that, as one goes back to the beginning of the Universe, one makes $t \to 0$, and then $\tau \to -\infty$. The Universe had a finite beginning, a Big Bang, only when referred to atomic time. Referred to mechanical time the Universe has always existed and there has been plenty of time for the evolution of galaxies and star clusters.

The mechanical time τ is the ephermeris time of astronomers, while t is the time measured by an atomic clock. The relation (1) leads to a discrepancy between them, which it should be possible to detect with sufficiently accurate observations.

In 1929 Hubble made an estimate of the relation between the velocity of recession of a nebula and its distance from us, as a result of which he inferred (assuming the velocity of recession to be constant) that the age of the Universe is about 2×10^9 years. Hubble's method evidently refers to atomic time t, since it uses spectroscopic data and not mechanical laws. His estimate was a poor one, because of faulty statistics, and later work by the same method has increased the estimate by a factor of about 10. The most recent estimate by this method is 18×10^9 years, but there is still a good deal of uncertainty in the figure.

We now have the problem of building a cosmological theory that incorporates the two times connected by (1).

THE LARGE NUMBERS HYPOTHESIS

From data provided by atomic physics and astronomy one can

construct some large dimensionless numbers. The most important ones are: (i) the ratio of the electric to the gravitational force between an electron and a proton, e^2/Gm_em_p. It has the value about 7×10^{39}. (ii) the age of the Universe t, expressed in terms of a unit of time provided by atomic constants, say e^2/m_ec^2. It has the value about 2×10^{39}. (iii) the mass of that part of the Universe that is receding from us with a velocity $< \frac{1}{2}c$, expressed in units of the proton mass. Call it N. Its value is rather uncertain, because one does not know how much matter there is that is not seen by our telescopes, but it is probably of the order 10^{78}.

The Large Numbers Hypothesis asserts that these numbers are related by equations in which the coefficients are closed to unity. Since the number (ii) varies with the age of the Universe, the hypothesis requires that the others must also vary. We get

$$e^2/Gm_em_p :: t \quad , \tag{2}$$

$$N :: t^2 \quad . \tag{3}$$

One can formulate the relation (2) by saying that $G :: t^{-1}$ when expressed in atomic units.

To take into account the relation (3) I have been assuming continuous creation of matter. However, there are grave difficulties with this assumption. The preferred form of the assumption requires each piece of matter to be continually growing by the creation of new matter. It is hard to understand how ancient rock crystals can retain their crystalline form with such a growth. Also, from the close observations of the moon that have now been made, one can be fairly certain that the moon has not changed its size very much since its formation 4.5×10^9 years ago and cannot have grown to the extent required by the continuous creation assumption*. (See next page) In view of this evidence I believe that one must abandon the assumption of continuous creation.

One can reconcile the relation (3) with conservation of mass by assuming that the velocity of recession of a galaxy is continually decreasing, so that more and more galaxies are continually appearing with velocity of recession $< \frac{1}{2}c$. This was the picture adopted in my first paper on the subject (1938).

The law of retardation of the galaxies can easily be adjusted to fit the formula (3). A galaxy with a velocity of recession $\frac{1}{2}c$ will be at a distance about ct. There is a coefficient here whose value depends on the law of retardation, but it is of order unity and does not matter. Thus we have a mass N distributed over a sphere of radius about ct, so the average density is roughly $N/(ct)^3$. From (3) this is proportional to t^{-1}. Thus the distance of a galaxy :: $t^{1/3}$. The velocity of recession is then :: $t^{-2/3}$.

With the distance of a nebula varying in proportion to f(t), the Hubble constant is

$$H = f'(t)/f(t).$$

With a constant rate of expansion, f(t) :: t and

$$H = t^{-1}$$

with the new theory proposed have, f(t) :: $t^{1/3}$ and

$$H = \frac{1}{3} t^{-1}. \qquad (4)$$

So with the same value for the Hubble constant, the age of the Universe is reduced by a factor 3. The presently accepted value for the Hubble constant corresponds to an age of the Universe of about 18×10^9 years with a constant rate of expansion. It thus

*I am indebted to Professor S. K. Runcorn for pointing this out to me during his visit to Tallahassee.

corresponds to about 6×10^9 years with the new law of retardation of the galaxies. This is rather less than the age one usually believes, but it is not impossible. It must be remembered that this is the age as it would be measured by an atomic clock. The age by ephemeris time is something different.

A PLANET IN A CIRCULAR ORBIT

Consider a planet moving in a circular orbit (apart from a secular variation of the radius). Let r_A be the radius of the orbit in atomic units and let v be the velocity, which is dimensionless and so is the same in all units. The Newtonian law of gravitation gives

$$MG=v^2 r_A,$$

where M is the mass of the sun. Taking M to be constant and G to vary according to t^{-1}, we get

$$v^2 r_A :: t^{-1}. \qquad (5)$$

The period of the orbit, measured in atomic time, is

$$\Delta t=2\pi r_A/v.$$

This time interval is a unit of ephemeris time, so it corresponds to $\Delta\tau=1$. Thus we get

$$\Delta t/\Delta\tau=2\pi r_A/v.$$

According to (1) we have $\Delta t/\Delta\tau::t$, so

$$r_A/v :: t.$$

Combining this with (5), we get

$$r_A :: t^{1/3} \quad v :: t^{-2/3} \tag{6}$$

So these simple arguments lead to definite laws for the variation of the radius of the orbit and the velocity.

One sees that the radius of the orbit increases according to the same law as the distance of a galaxy, and the velocity in the orbit decreases according to the same law as the velocity of recession of a galaxy. These results are probably just co-incidences. I do not see any deep reason behind them.

THE CURVATURE OF THREE-DIMENSIONAL SPACE

Let us consider the three-dimensional space corresponding to a particular epoch t. If we neglect local irregularities it will be a uniform space of constant curvature (constant in space, not necessarily constant in time). The curvature is either positive, zero or negative.

The case of positive curvature can be immediately ruled out. It would require the Universe to have a finite total mass which, expressed in proton units, would give a large number that is constant, and thus contradicts the Large Numbers Hypothesis.

The case of negative curvature can also be ruled out, though not quite so directly. Introduce coordinates ξ^1, ξ^2, ξ^3 in the three-dimensional space such that they are constant for any particular galaxy. Then, corresponding to a definite $d\xi^1$, $d\xi^2$, we shall have a distance ds which varies $::t^{1/3}$. The radius of curvature R is determined by such distances and therefore also varies $::t^{1/3}$. The total mass within a sphere of radius R is R^3 times the density and is thus independent of t. It gives a constant large number, which is not allowed.

We are thus left with the case of zero curvature, or flat three-dimensional space, as the only one consistent with the Large Numbers Hypothesis. Taking Cartesian coordinates, we have in the three dimensional space

$$ds_A^{\;2} = t^{2/3}((d\xi^1)^2 + (d\xi^2)^2 + (d\xi^3)^2).$$

The suffix A is attached to ds to show that it refers to atomic units.

Let us now introduce a fourth coordinate ξ^0 such that the velocity of light referred to these coordinates is 1. Then in four-dimensional space-time

$$ds_A^{\;2} = t^{2/3}((d\xi^o)^2 - (d\xi^1)^2 - (d\xi^2)^2 - (d\xi^3)^2) \qquad (7)$$

Thus a null-line, for example, is $d\xi^0=d\xi^1, d\xi^2=0,\ d\xi^3=0$.

Corresponding to a time-like element $d\xi^0$, we have

$$ds_A = t^{1/3}\, d\xi^0.$$

This ds_A is just an element of atomic time, so we may write it dt. We get

$$d\xi^0/dt = t^{-1/3}, \qquad (8)$$

so

$$\xi^0 = 3/2\ t^{\,2/3} \qquad (9)$$

This gives us the coordinate time ξ^0 in terms of atomic time.

THE CLASSICAL EQUATIONS OF MOTION

We must now introduce some assumptions to determine how bodies move in the new theory. We know that Einstein's theory holds extremely well for describing the Solar System. Its successes must be preserved. At the same time we cannot keep the Einstein theory in its entirety, because it demands that G shall be constant.

We introduce a mechanical metric ds_M in space-time such that bodies move along geodesics with respect to this metric. It will not satisfy accurately the Einstein field equations. This metric must give the deceleration of the galaxies as well as the motion of the planets in the Solar System.

With neglect of local irregularities, ds_M must be just the previous ds_A changed by a scale factor that is a function of t only. Thus, with the same coordinates ξ as before

$$ds_M{}^2 = \eta^2((d\xi^0)^2 - (d\xi^1)^2 - (d\xi^2)^2 - (d\xi^3)^2),$$

where η is a function of t or of ξ^0. With this metric, just as with the metric ds_A, the motion of galaxies is $d\xi^1 = d\xi^2 = d\xi^3 = 0$. Also, the nulllines are the same with the metric ds_M as with the metric ds_A.

To describe the Solar System we need a generalized Schwarzchild metric. Let us pass from the Cartesian coordinates ξ^1, ξ^2, ξ^3, to the polar coordinates r, Θ, ϕ, and write ξ for ξ^0. We then take

$$ds_M{}^2 = \eta^2\{(1 - \frac{2m}{r})d\xi^2 - (1 - \frac{2m}{r})^{-1}dr^2 - r^2 d\Theta^2 - r^2 \sin^2\Theta d\phi^2\}. \tag{10}$$

Here we can allow m to vary with ξ, so that we have two functions of ξ, η and m, at our disposal. They will vary only slowly with respect to ξ, so that, with neglect of this variation, the Einstein theory holds.

We must choose η and m so as to get the correct circular orbits, satisfying (6). If z^μ denote the coordinates of a planet, the geodesic equations are

$$\frac{d^2z^\mu}{ds^2} = \Gamma^\mu_{\nu\rho} \frac{dz^\nu}{ds} \frac{dz^\rho}{ds} = 0.$$

If we refer the equations to another independent variable ξ instead of s and use a dot to denote $d/d\xi$, they become

$$\ddot{z}^{\mu} + \Gamma^{\mu}_{\nu\rho}\dot{z}^{\nu}\dot{z}^{\rho} - \dot{z}^{\mu}\ddot{s}/\dot{s} = 0. \tag{11}$$

Introduce the velocity v of the planet by the definition

$$g_{oo}\, v^2 = -\, g_{ab}\dot{z}^a\dot{z}^b \;, \; (a,b = 1,2,3). \tag{12}$$

Then

$$\dot{s}^2 = g_{\mu\nu}\dot{z}^{\mu}\dot{z}^{\nu} = g_{oo}(1-v^2).$$

Thus

$$2\dot{s}\ddot{s} = (g_{oo,o} + g_{oo,a}\dot{z}^a)(1-v^2) - 2g_{oo}\, v\dot{v}.$$

Apply equation (11) with $\mu=0$. It gives

$$\Gamma^{o}_{\nu\rho}\dot{z}^{\nu}\dot{z}^{\rho} = \ddot{s}/\dot{s} = g^{oo}\frac{1}{2}(g_{oo,o} + g_{oo,a}\dot{z}^a) - v\dot{v}/(1-v^2).$$

But

$$\Gamma^{o}_{\nu\rho}\dot{z}^{\nu}\dot{z}^{\rho} = g^{oo}\{\tfrac{1}{2} g_{oo,o} + g_{oo,a}\dot{z}^a - \frac{1}{2} g_{ab,o}\dot{z}^a\dot{z}^b\}, \tag{13}$$

so

$$\frac{1}{2} g_{oo,a}\dot{z}^a - \frac{1}{2} g_{ab,o}\dot{z}^a\dot{z}^b + g_{oo}\, v\dot{v}/(1-v^2) = 0.$$

Take the case of a circular orbit, for which $\dot{z}^1$ is of order ξ^{-1} , and neglect terms of order ξ^{-2}.

We get

$$\frac{1}{2} g_{oo,1} \dot{Z}^1 - \frac{1}{2} (g_{22,o} (\dot{Z}^2)^2 + g_{33,o} (\dot{Z}^3)^2 + g_{oo} v\dot{v}/(1-v^2) = 0, \tag{14}$$

while (12) reduces to

$$g_{oo} v^2 = -(g_{22}(\dot{Z}^2)^2 + g_{33}(\dot{Z}^3)^2). \tag{15}$$

From the $g_{\mu\nu}$ given by (10) we see that

$$g_{22,o} = 2g_{22}\dot{\eta}\eta^{-1} \qquad g_{33,o} = 2g_{33}\dot{\eta}\eta^{-1},$$

and hence (14) becomes

$$\frac{1}{2} g_{oo,1} \dot{Z}^1 + g_{oo} v^2 \dot{\eta}\eta^{-1} + g_{oo} v\dot{v}/(1-v^2) = 0.$$

Since $g_{oo}=\eta^2(1-2m/r)$, this gives, if we put $z^1=r$,

$$mr^{-2}\dot{r} + (1-2m/r)\{v^2\dot{\eta}\eta^{-1}+v\dot{v}/(1-v^2)\}.$$

For r/m small and v small, this reduces to

$$mr^{-2}\dot{r} + v^2\dot{\eta}\eta^{-1}+v\dot{v}=0. \tag{16}$$

Now apply equation (11) with $\mu=1$ and neglect terms or order ξ^{-1}. It gives

$$\Gamma^1_{\nu\rho} \dot{Z}^\nu \dot{Z}^\rho = 0$$

or

$$\Gamma^1{}_{oo} + \Gamma^1{}_{22}(\dot{Z}^2)^2 + \Gamma^1_{33}(\dot{Z}^3)^2 = 0$$

or

$$m/r^2 - r(\dot{z}^2)^2 - r\sin^2\Theta\,(\dot{z}^3)^2 = 0.$$

With the help of (15) this becomes

$$m-(1-2m/r)v^2r=0.$$

For small values of m/r it reduces to

$$m=v^2r. \tag{17}$$

Putting this result in (16), we get

$$\dot{r}r^{-1}+\dot{\eta}\eta^{-1}+\dot{v}v^{-1}=0,$$

showing that

$$r\eta v=\text{constant}. \tag{18}$$

(17) and (18) are the equations that govern circular orbits for large r. The r here is measured in the same units as T. Its connection with r_A in atomic units is thus

$$r/r_A=dT/dt = t^{-1/3},$$

from (8) with $\xi^0=T$. Thus (17) and (18) become

$$m=v^2r_At^{-1/3} \qquad r_A\eta v::t^{1/3}.$$

Thus in order to get the laws (6) for circular orbits we must take

$$m::t^{-4/3} \qquad \eta::t^{2/3},$$

or from (9)

$$m::T^{-2} \quad , \quad \eta::T. \tag{19}$$

The metric (10) is now well-defined.

EXPERIMENTAL CONFIRMATION

An important feature of the new theory is that there should be a discrepancy between ephemeris time τ, marked out by the motion of the planets, and atomic time t, marked out by an atomic clock. According to Milne's law (1)

$$dt/d\tau :: t,$$

so atomic clocks should be speeding up with respect to ephemeris time. This is an effect that should show up if one makes sufficiently accurate observations of the motions of the planets with atomic clocks.

The best chance of observing the effect is with the moon, our nearest neighbor. The motion of the moon is much disturbed by tidal effects, but if one examines the difference of the motion referred to ephemeris time and referred to atomic time, tidal effects cancel out.

The motion of the moon in ephemeris time has been observed accurately for some centuries. Since 1955 it has been observed with atomic clocks. A study of these observations has been made by Van Flandern.

Let n_A denote the angular velocity of the moon, in its orbit around the earth, referred to atomic time. Thus $n_A=v/r_A$. From (16)

$$n_A::t^{-1}.$$

Thus the angular acceleration $\dot{n}_A$ in atomic time satisfies

$$\dot{n}_A/n_A = - t^{-1}. \tag{20}$$

This cosmological effect is to be superposed on the angular acceleration produced by the tides.

The angular velocity in ephemeris time is

$$n_E = n_A \; dt/d\tau = \text{constant},$$

so in ephemeris time there is no angular acceleration. So any observed angular acceleration in ephemeris time is entirely tidal. Thus with tidal effects taken into account, equation (20) must be generalized to

$$\dot{n}_A/n_A - \dot{n}_E/n_E \doteq - t^{-1}. \tag{21}$$

Van Flandern's most recent result for the right-hand side of (21) is

$$-(7.5 \pm 2.7)10^{-11} \text{ year}^{-1}.$$

The minus sign here indicates that atomic clocks really are speeding up with respect to ephemeris time. The numerical result is not very accurate because it appears as the difference of two quantities each considerably larger than itself. It corresponds to an age for the Universe about 13×10^9 years, about twice the Hubble age when the retardation of the galaxies is taken into account.

THE COSMOLOGICAL MICROWAVE RADIATION

Microwave radiation is observed to be falling on the earth with a constancy and uniformity which show that is it not a local phenomenon but is of cosmological origin. Only a small part of the spectrum has been observed. The observations are consistent with the radiation being black-body radiation at a temperature t of about 2.8°K.

The usual explanation for this radiation is that it comes from a primordial fireball that existed in the early stages of the universe. To begin with, this fireball was strongly interacting with the matter and was roughly in temperature equilibrium with it. At a certain stage it became decoupled from the matter. Then, with the expansion of the universe, it became red-shifted and cooled down to what we now observe.

Let us see what is the connection between the rate of cooling and the expansion of the universe. Take a general law of expansion.

$$ds^2_A = t^{2n}((d\xi^0)^2 - (d\xi^1)^2 - (d\xi^3)^2) \qquad (22)$$

instead of (7). Suppose we have a distribution of particles moving freely without interaction. Thus for each particle ξ^1, ξ^2, ξ^3 increase linearly with respect to ξ^0. Take an element of phase space $dx^1\ dx^2\ dx^3\ dp_1\ dp_2\ dp_3$. Each dx is increasing $:: t^n$ and each dp varies $:: t^{-n}$, so the element of phase remains constant. Hence the number of particles per unit cell of phase space remains constant. We may apply the argument to photons and we see that, even with the general expansion law (22), black-body radiation remains black-body, and only its temperature changes.

The temperature varies according to the same law as the frequency ν of a photon,

$$T :: \nu.$$

To determine the variation of ν, we see that $\nu :: (\Delta t)^{-1}$, where Δt is the time interval between two waves. This time interval corresponds to constant $\Delta\xi^0$. Thus

$$\nu :: \Delta\xi^0/\Delta t :: t^{-n}$$

and hence

$$T :: t^{-n} .$$

With the galaxies receding at constant velocity we have n=1. Then

$$T :: t^{-1}. \tag{23}$$

This is the cooling law that is usually assumed for the microwave radiation.

With our new theory we have, corresponding to (7), $n=\frac{1}{3}$. This gives

$$T :: t^{-1/3}, \tag{24}$$

which is a much slower rate of cooling.

Let us examine the question from the point of view of the Large Numbers Hypothesis. From the present temperature $T=2.8^{\circ}K$ we can construct an energy kT which can be compared with the proton self-energy $m_p c^2$ to give a dimensionless number

$$\frac{kT}{m_p c^2} = 2.5 \times 10^{-13}.$$

We should expect this to vary roughly in proportion to $t^{-1/3}$, giving $T::t^{-1/3}$. So we get agreement with the new theory (24), but no agreement with the galaxies receding at constant velocity

(23).

With the new theory the decoupling of the primordial fireball from the matter occurred very soon after the Big Bang and not after an appreciable time, as one assumes in the usual theory.

SPIN FORCES IN LARGE-$P_\perp^2$ p-p ELASTIC SCATTERING*

A. D. Krisch

Randall Laboratory of Physics

The University of Michigan, Ann Arbor, Michigan 48109

My talk will have four parts. I will first define the various spin dependent cross sections that can be measured. I will then describe the hardware that is necessary to measure these cross section. Third, I will show some of our recent results on spin effects at large transverse momentum. Finally I will try to say how these results affect our understanding of the short range nucleon-nucleon interaction.

I. PURE-SPIN CROSS SECTIONS

Let me first define the pure 4-spin cross sections in proton-proton elastic scattering. The reaction can be defined in the center of mass system by the diagram

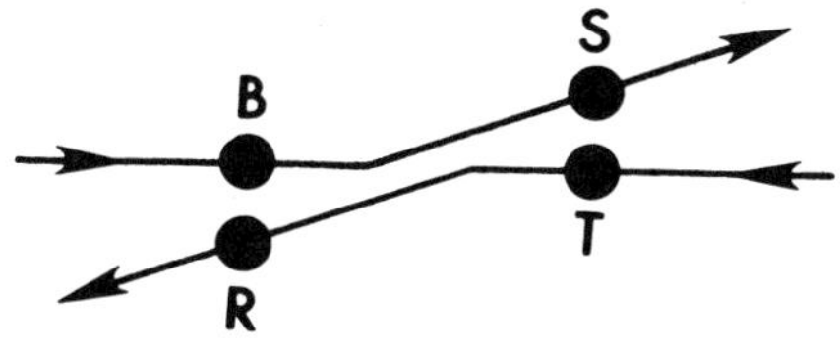

*Supported by a Research Grant from the U. S. Department of Energy

The differential cross section can be written as

$$\frac{d\sigma}{dt}(ij \rightarrow k\ell) \tag{1}$$

where the 4 indicates respectively refer to the spins of

(beam, target → scattered, recoil)

as indicated in the diagram. Each of the 4 protons' spins are measured transverse to the scattering plane along the $\vec{n}$ direction defined by

$$\vec{n} = \vec{P}_b \times \vec{P}_s \quad . \tag{2}$$

Each proton has spin 1/2 which may point either i=↑ (up) or i=↓ (down). The cross sections where the spins are measured along this $\vec{n}$ direction are called transversity cross-sections. The spins could also be measured along the direction of the momenta as in helicity cross-sections, or in other directions.

Since there are 4 protons in the reaction and each can take 2 different spin orientations there are $2^4=16$ different pure 4-spin cross sections. These pure 4-spin cross sections are the <u>fundamental</u> cross sections for proton-proton elastic scattering that one might hope to calculate directly from a theory of hadronic interactions, if we ever discover one. The spin-averaged cross section, $\langle d\sigma/dt\rangle$, which is normally measured with unpolarized beams and unpolarized targets is a complicated average of these 16 $d\sigma/dt(ij \rightarrow k\ell)$.

$$\langle d\sigma/dt\rangle = \frac{1}{4} \sum_{i,j,k,\ell=\uparrow,\downarrow} \frac{d\sigma}{dt}(ij \rightarrow k\ell) \quad . \tag{3}$$

The fundamental hadronic cross sections can not be directly obtained from the spin-averaged cross section, unless there is no spin dependence. If there is significant spin dependence then the averaging may obscure many important features of the direct hadronic interactions.

These 16 pure spin cross sections may be classified as follows:

4 non-flip such as $d\sigma/dt(\uparrow\uparrow\to\uparrow\uparrow)$

4 double-flip such as $d\sigma/dt(\uparrow\uparrow\to\downarrow\downarrow)$ (4)

8 single-flip such as $d\sigma/dt(\uparrow\downarrow\to\uparrow\uparrow)$

Not all 16 of these cross sections are independent. There is an important theorem that was first proved by A. Bohr[1]:

If parity invariance is obeyed, the all 8 single-flip transversity cross sections must be zero.

This reduces the number of $d\sigma/dt(ij\to k\ell)$ to 8. Another constraint comes from rotational invariance which implies that for identical particles

$$d\sigma/dt(\uparrow\downarrow\to\uparrow\downarrow) = d\sigma/dt(\downarrow\uparrow\to\downarrow\uparrow)$$
$$d\sigma/dt(\uparrow\downarrow\to\downarrow\uparrow) = d\sigma/dt(\downarrow\uparrow\to\uparrow\downarrow)\ . \tag{5}$$

This reduces the number of independent quantities to 6. A final constraint comes from time reversal invariance which implies that

$$d\sigma/dt(\uparrow\uparrow\to\downarrow\downarrow) = d\sigma/dt(\downarrow\downarrow\to\uparrow\uparrow)\ . \tag{6}$$

In an earlier experiment[2] we, in fact, used these relations to test parity and time reversal invariance in medium-$P_\perp^2$ hadronic interactions for the first time. We found that both invariance properties were obeyed within our ±6% precision.

Assuming that P and T are obeyed we have the following 5 independent pure-spin cross sections for proton-proton elastic scattering:

$d\sigma/dt(\uparrow\uparrow\to\uparrow\uparrow)$ up non-flip

$d\sigma/dt(\downarrow\downarrow\to\downarrow\downarrow)$ down non-flip

$d\sigma/dt(\uparrow\uparrow\to\downarrow\downarrow)$ parallel double-flip (7)

$d\sigma/dt(\uparrow\downarrow\to\uparrow\downarrow)$ anti-parallel non-flip

$d\sigma/dt(\uparrow\downarrow\to\downarrow\uparrow)$ anti-parallel double-flip .

Note that these 5 transversity cross sections are each the square of a transversity amplitude

$$d\sigma/dt(ij\to k\ell) = |f(ij\to k\ell)|^2\ . \tag{8}$$

Each transversity amplitude can be written as a linear combination of the five independent helicity amplitudes $h(\alpha\beta \to \gamma\delta)$ and vice versa.

$$f(jk\to k\ell) = \sum_{n=1}^{5} c_n \, h(\alpha\beta\to\gamma\delta) \quad . \tag{9}$$

The transversity and helicity amplitudes are schematically indicated in Fig. 1.

Before passing on to the experimental details I would like to list my colleagues who have been collaborating with me on the recent experiments:

J. R. O'Fallon.........Argonne Universities Association
L. G. Ratner and.......Argonne National Laboratory
P. F. Schultz
K. Abe, R. C. Fernow...The University of Michigan
T. A. Mulera, A. J.
Salthouse, B. Sandler
and K. M. Terwilliger
D. G. Crabb............Oxford University, Oxford
P. H. Hansen...........Niels Bohr Institute,
Copenhagen
A. Lin.................Abadan Institute of Technology
Iran

I would like to thank them for their help and advice.

II. EXPERIMENTAL HARDWARE

To study spin-spin forces in p-p elastic scattering one needs three basic facilities:

1. A polarized proton beam
2. A polarized proton target
3. A spectrometer to detect elastic scattering events

I will describe each of these in some detail.

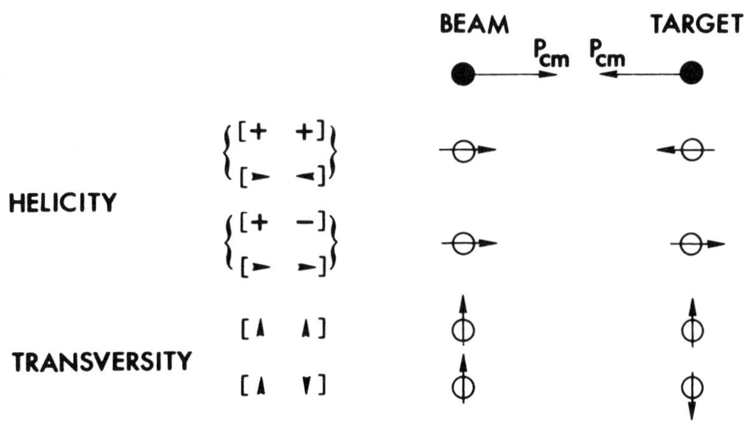

Figure 1. The spin orientations relative to the c.m. momenta for helicity and transversity cross sections.

A. Polarized Proton Beam

The polarized proton beam originates in a polarized ion source purchased from ANAC in New Zealand which now gives up to 70 μa of 20 KeV protons with a polarization of 75% ± 5%. This is an atomic beam type source which is shown in Fig. 2.

The polarized protons start as a thermal beam of hydrogen atoms which emerge from the dissociator and enter the tapered sextupole magnet. This magnet uses the Stern-Gerlach effect to focus all atoms with the electrons in one spin state $(\sigma_e, \sigma_p) = (\uparrow, \uparrow)$ and $(\uparrow, \downarrow)$; and to defocus all atoms with the electrons in the second spin state $(\downarrow, \uparrow)$ and $(\downarrow, \downarrow)$ so that they are lost from the beam. Note that the magnetic moment of the electron, μ_e, is about 670 times larger than that of the proton, μ_p, so that the Stern-Gerlach force $\vec{\nabla}(\vec{\mu}.\vec{B})$ essentially ignores the tiny contribution of μ_p. Thus emerging from the sextupole magnet is a beam of hydrogen atoms with the electron spins totally polarized but the proton spins totally unpolarized. This atomic beam passes into the intermediate field RF transition unit, which contains a uniform magnetic field, $\vec{B}$, of about 200 Gauss and an RF field with the correct frequency, ν, to induce the

$$(\uparrow,\downarrow) \rightarrow (\uparrow, \downarrow) \tag{10}$$

hyperfine transition in this 200 Gauss magnetic field,

$$h\nu = \vec{B}\cdot(\vec{\mu}_e - \vec{\mu}_p) \tag{11}$$

The RF field must have a narrow enough bandwidth to avoid the frequency

$$h\nu' = \vec{B}\cdot(\vec{\mu}_e + \vec{\mu}_p) , \tag{12}$$

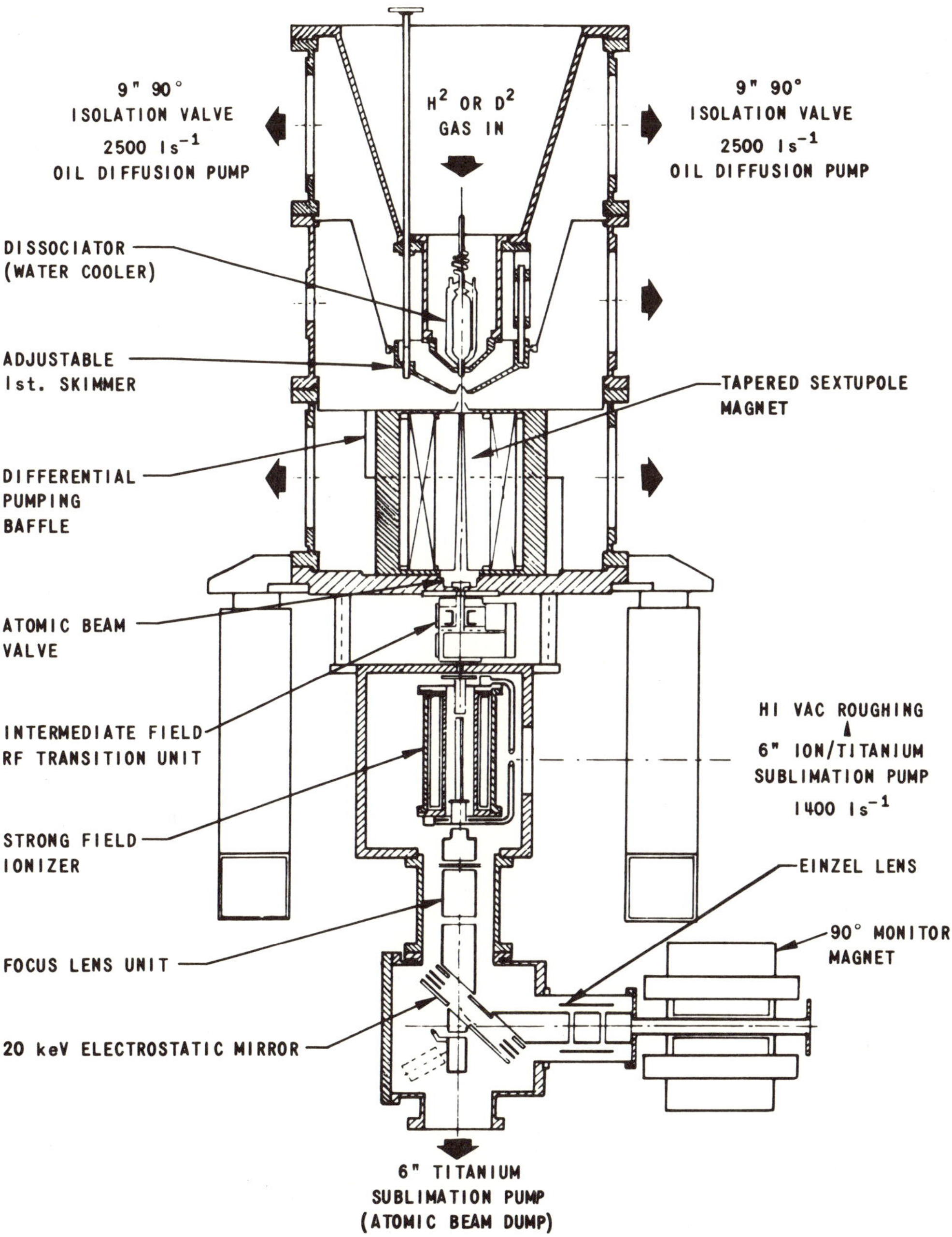

Fig. 2 Schematic of the ANAC ground state atomic beam source.

which induces the

$$(\uparrow, \uparrow) \rightarrow (\downarrow, \downarrow) \tag{13}$$

hyperfine transition. Thus emerging from the RF transition unit is a beam of hydrogen atoms with the proton spins totally polarized and the electron spins now totally unpolarized.

$$(\mu_e, \mu_p) = (\downarrow, \uparrow) \text{ and } (\uparrow, \uparrow) . \tag{14}$$

These atoms then pass into the strong field ionizer containing an electron plasma which strips the electrons from the hydrogen atoms, leaving a beam of polarized protons which are then accelerated to a few KeV by the electric field in the ionizer. The electrostatic lens and mirror then focus and accelerate the polarized proton beam and precess the spin, so that a beam of 20 KeV protons with spins transverse to the momentum emerge.

Most of the hydrogen atoms are lost in this polarizing process, and these lost atoms ruin the high vacuum which is necessary for transporting a thermal atomic beam. Thus a very elaborate vacuum pumping system is required as shown. Nevertheless the vacuum is not perfect and this reduces polarization by letting unpolarized background atoms mix with beam atoms. This lowers the source polarization from 100% to about 75%.

To obtain 70 μa E. F. Parker and his colleagues have made various source improvements including pulsing the source to take advantage of the ZGS being a pulsed accelerator.[3] The source is placed in the specially constructed pre-accelerator II dome shown in Fig. 3 where a Cockcroft-Walton column accelerates the protons to 750 KeV. The protons are then fed into the main LINAC line by a switching magnet and accelerated to 50 MeV. The polarization

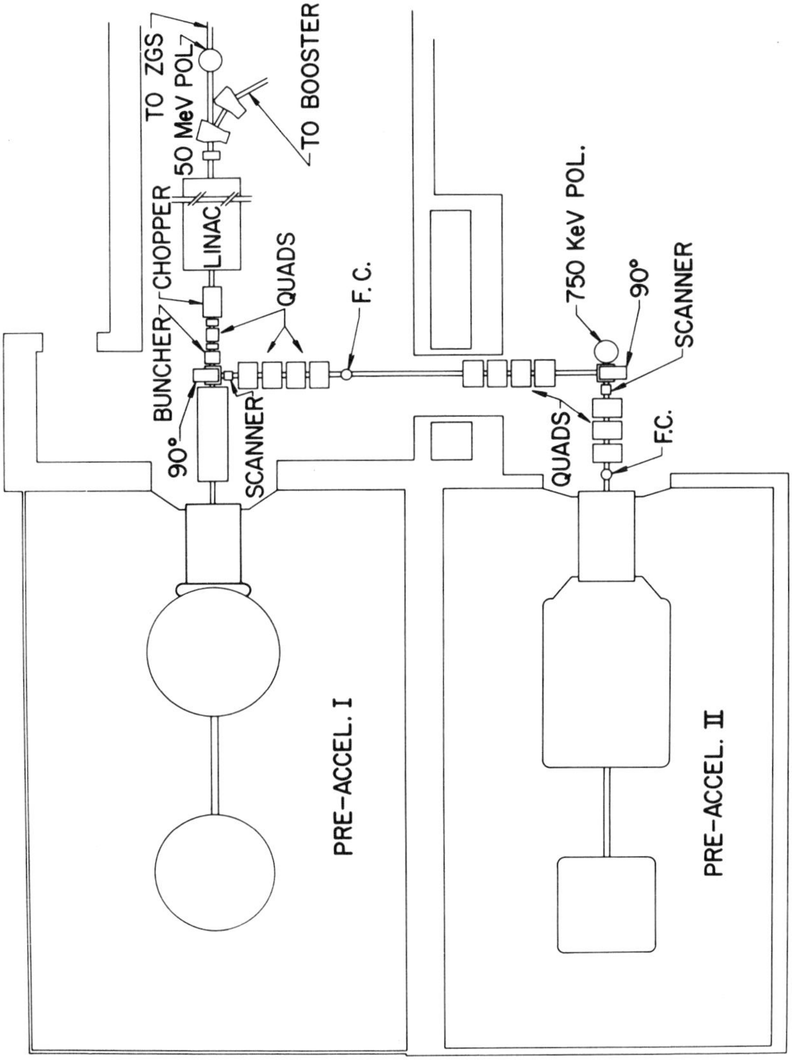

Fig. 3 Layout of the ZGS preacelerator and LINAC areas.

at 50 MeV is measured using the polarimeter shown in Fig. 4. This continuously measures the left-right asymmetry in p-Carbon elastic scattering at 55° where by cross calibration we found the analyzing power, A, to be 88% ± 5%.[4,5] At 50 MeV the beam polarization, P_B, is measured to be 75 ± 5% using the measured left-right asymmetry in the equation

$$P_B = \frac{1}{A}\left[\frac{L-R}{L+R}\right] . \qquad (15)$$

The beam is then injected into the ZGS, accelerated, and then extracted to the high energy polarimeter described below.

Maintaining the polarization of the protons during acceleration in a sychrotron is difficult, because the vertical magnetic field which keeps the protons going in circles can flip the spin. Fortunately if the proton spins are vertical, the vertical B-field will not flip them for there is no large $\vec{\mu} \times \vec{B}$ component. However, a synchrotron must have small horizontal B-field components to keep the beam focused, and these can depolarize the spin. Normally these horizontal B-fields are so weak that the depolarization per turn around a synchrotron is about 10^{-8}. There are typically 10^6 turns around a synchrotron but still the total depolarization is less than 1%.

The main problems in accelerating polarized protons in a synchrotron are "depolarizing resonances".[5,6] These occur when the Larmor precessional frequency becomes equal to the frequency with which the protons see magnetic imperfections as they circle the synchrotron. Then each proton gets a similar depolarizing perturbation every time it passes through a horizontal fringe field. These perturbations can then add coherently and rapidly depolarize the beam. In the weak focusing ZGS the "intrinsic" depolarizing resonances caused by the harmonic betatron oscillations are the most serious. The precessional frequency and the

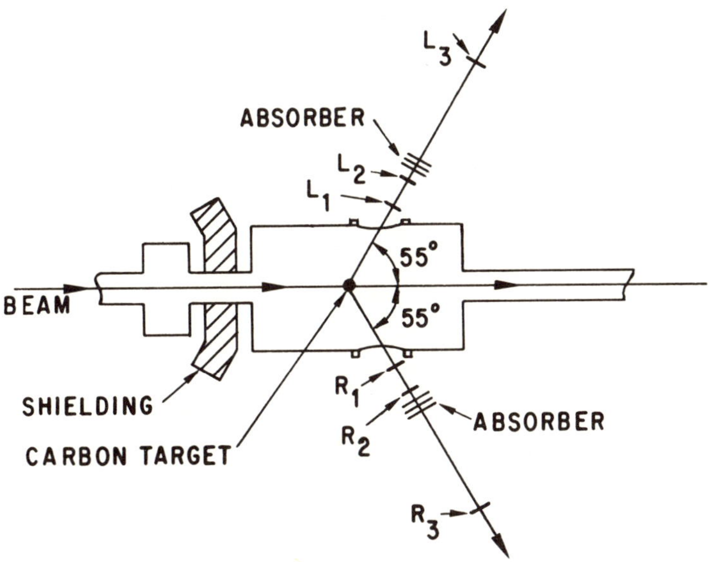

Fig. 4-Layout of the 50 MeV polarimeter, showing the two symmetric 3-counter scintillator telescopes.

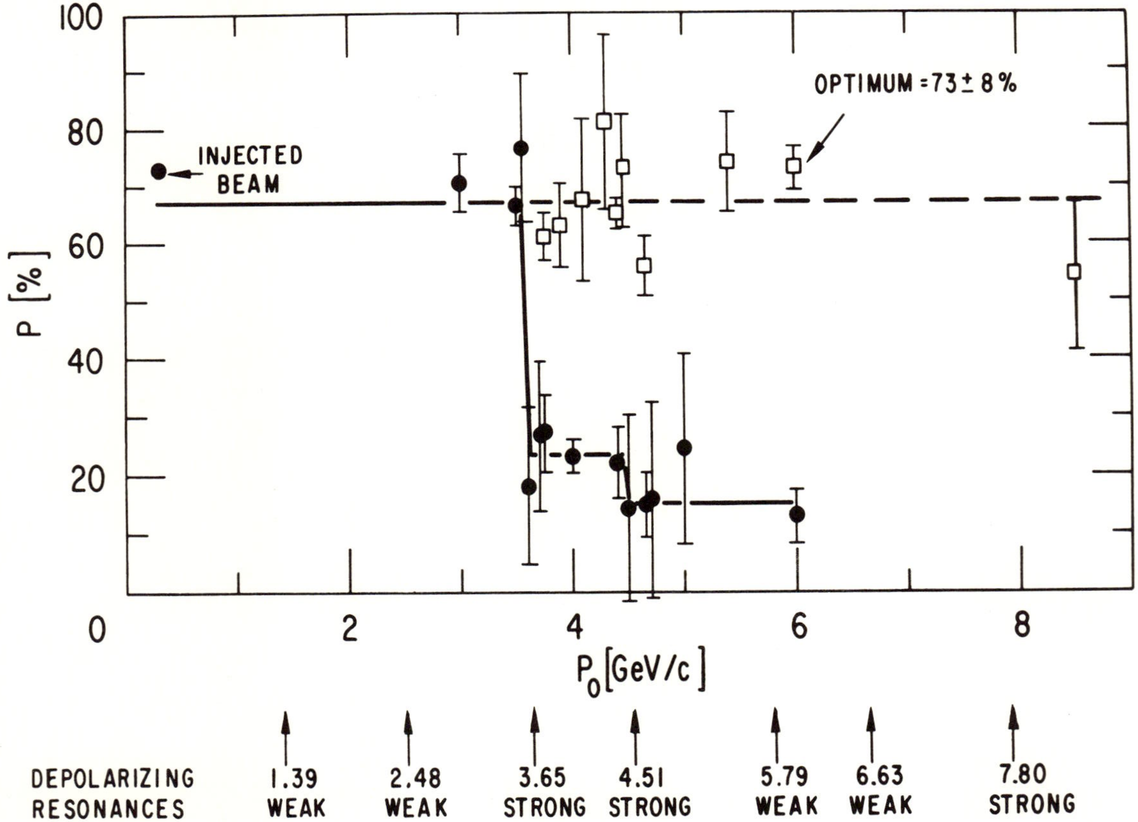

Fig. 5 Measured polarizations as a function of momentum. Open squares are with quadrupoles compensating for resonances and black dots are with quadrupoles off.

frequency of depolarizing perturbations due to vertical betatron oscillations are given by:

$$\omega_p = (\frac{g}{2} - 1)\gamma \frac{eB_o}{m}, \quad \omega_k = (k \pm \nu) \frac{eB_o}{m}. \tag{16}$$

The resonance condition occurs when ω_p equals ω_k. This occurs when the proton energy $E = m\gamma$ reaches the γ value satisfying

$$(g/2 - 1)\gamma = k \pm \nu \tag{17}$$

where ν is the number of betatron oscillations per turn around the ZGS ($\sim$.80), and k is the harmonic number (8 and 16 are the strongest). These intrinsic resonances were jumped using two pulsed quadrupoles installed in the ZGS. By pulsing these with a 20 μsec risetime, ν was rapidly changed and the beam passed through each resonance in a few turns before coherent depolariztion could occur. As shown in Fig. 5 the pulsed quadrupoles indeed overcome these resonances up to 6 GeV/c and a 75% beam polarization is maintained.

The timing and strength of these pulses was tuned by maximizing the polarization measured in the high energy polarimeter shown in Fig. 6. This measures p-p elastic scattering to the left and to the right at $P_\perp^2 = .5\ (\text{GeV/c})^2$ where the analyzing power , A, equals $.100 \pm .006$ at 6 GeV/c. The beam polarization P_B is given by

$$P_B = \frac{1}{A} \left(\frac{L-R}{L+R}\right). \tag{18}$$

Recently the ZGS staff has been using a polarimeter built at CERN which gives much better statistical precision, typically 100,000 events/pulse compared with 1000 events/pulse in the above elastic polarimeter. However, the CERN polarimeter does not have a clean elastic signal and must be calibrated frequently against

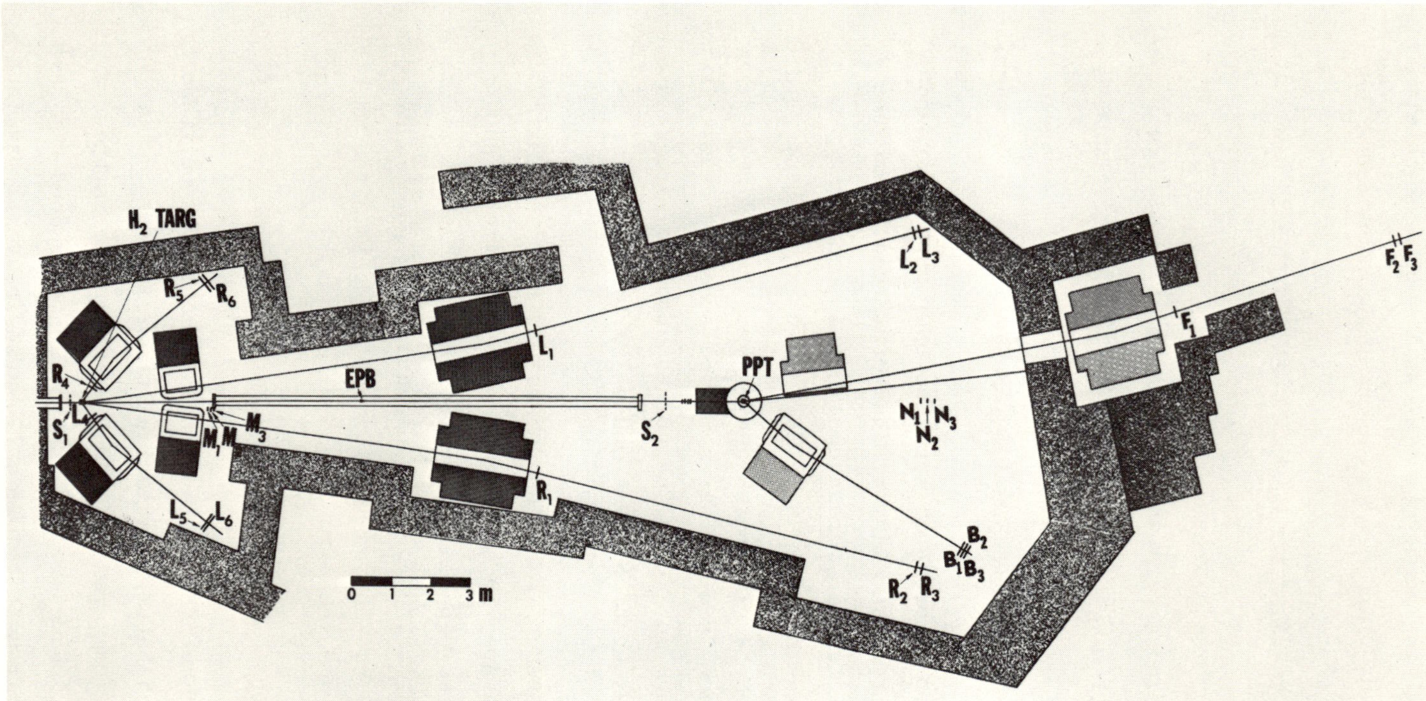

Fig. 6 Layout of the polarimeter. The polarized beam passes through the H_2 target and its polarization is measured by comparing the number of elastic events seen in the L and R spectrometers of the polarimeter.

the elastic polarimeter. Using the two polarimeters together in this way gives good statistical precision and an absolute calibration.

Above 6 GeV/c there is significant depolarization due to "imperfection" depolarizing resonances, which are caused by imperfections or misalignments in the synchrotron magnets. The kth harmonic imperfection resonance occurs when γ is given by

$$(g/2 - 1)\gamma = k \ .$$

These resonances are jumped using a pulsed horizontal magnetic field, which is tuned to compensate for the imperfection fields at each resonance energy. By carefully tuning through 10 intrinsic and 19 imperfection resonances the ZGS staff has now attained a beam polarization of 70 ± 3% at 11.75 GeV/c.[7]

B. POLARIZED PROTON TARGETS

The basic idea of a polarized proton target is to align the spins of the protons in a strong magnetic field $\vec{B}$. There is an energy $\pm\vec{\mu}_p \cdot \vec{B}$ associated with the proton's magnetic moment, μ_p, depending on its spin state. The polarization of the protons in the target is given by

$$P_T = \frac{e^{\frac{\vec{\mu}_p \cdot \vec{B}}{2kT}} - e^{\frac{-\vec{\mu}_p \cdot \vec{B}}{2kT}}}{e^{\frac{\vec{\mu}_p \cdot \vec{B}}{2kT}} + e^{\frac{-\vec{\mu}_p \cdot \vec{B}}{2kT}}} \approx 1 - e^{-\frac{\vec{\mu}_p \cdot \vec{B}}{kT}} \ . \qquad (19)$$

For a temperature, T, of 0.5°K and a magnetic field of 25,000 Gauss one obtains a polarization of about 0.3% for the protons in the target. This polarization technique is known as the "brute force method" and it clearly does not work very well, because the proton's magnetic moment is so small.

In the late 1950's a new technique called dynamic polarization was pioneered by two low temperature physicists, Abragam in France and Jeffries at Berkeley. This new technique was soon adapted for high energy physics experiments at Berkeley,[8] CERN,[9] and Argonne,[10] with Chamberlain and Borghini playing leading roles.

The basic idea of dynamic polarization is to use a magnetic field to polarize some electrons and then to transfer the electron polarization to nearby protons in the target. Since μ_e is 670 times larger than μ_p it is "easy" to get the electrons 99.5% polarized. The target material is a complex mixture of a hydrocarbon, such as ethylene glycol, $C_2H_6O_2$ (also called antifreeze), doped with some chemical such as $K_2Cr_2O_7$ which has valence electrons in one of the Cr shells that can have spin-spin interactions with the "free" protons in the hydrogen in the $C_2H_6O_2$. These Cr electrons and H protons form themselves into temporary pairs with only two possible spin states:

$$(\mu_e,\mu_p) = (\uparrow,\uparrow) \text{ and } (\uparrow,\downarrow) \tag{20}$$

since the Cr electrons are almost 100% polarized. One then feeds in microwave power to induce the hyperfine transition

$$(\uparrow,\downarrow) \rightarrow (\downarrow,\uparrow)\,. \tag{21}$$

In a B = 25,000 Gauss field this requires a frequency ν given by

$$h\nu = \vec{B}\cdot(\vec{\mu}_e - \vec{\mu}_p)\,. \tag{22}$$

This microwave power must have a narrow enough bandwidth and the B field must be uniform enough ($\sim$ 1 part in 10^4) to avoid the frequency

$$h\nu' = \vec{B}\cdot(\vec{\mu}_e + \vec{\mu}_p) \tag{23}$$

which induces the hyperfine transition

$$(\uparrow,\uparrow) \rightarrow (\downarrow,\downarrow) \tag{24}$$

and would depolarize the protons. Thus the microwave power transfers the electron's polarization to its temporary mate proton. The proton then drifts away while the electron is being repolarized by the B-field and the electron finds itself a new mate and the proton polarization process continues. Since the relaxation time of the proton's spin is 10 minutes and the electrons become repolarized in 10^{-8} seconds this complex process actually works.

Our Michigan-Argonne PPT-V polarized proton target[11] is a close copy of a CERN target[12]. The PPT is maintained at 0.5°K in a magnetic field of 25 kG and the $C_2H_6O_2$ beads doped with $K_2Cr_2O_7$ are contained in a flask 4.13 cm long by 2.9 cm in diameter. The cooling is provided by a He^3 evaporation cryostat which is pumped to 0.5°K; this He^3 cryostat is placed inside a He^4 evaporation cryostat which is pumped to 2.0°K. The free hydrogen protons are induced into a polarized state by the 70 GHz microwave power from a carcinotron tube. The proton polarization is measured using a 107 MHz NMR system with signal averaging. This NMR system is calibrated against the calculated thermal equilibrium polarization when the microwave power is off

$$P_{TE} = e^{-\frac{\vec{\mu}_p \cdot \vec{B}}{kT}} - 1 \ . \tag{25}$$

The main error in this calibration of the NMR coil comes from the measurement of the thermal equilibrium temperature which is determined by measuring the vapor pressure of the He^3 vapor in the cryostat, with a precision of ±3%.

The target polarization has been as high as P_T = 85% but the high polarized beam intensity causes radiation damage to the ethylene glycol beads which reduces the average P_T to about 65%. Maintaining even a 65% polarization requires "annealing" the PPT beads every 12 hours to remove some of the radiation damage and replacing the beads every few days. Two independent NMR coils with different diameters measure that variation of P_T with transverse position caused by the variation in radiation damage. The small coil is a straight wire along the beam axis; the large coil is a 1.0 cm diameter helix coaxial with the beam axis. With a freshly annealed PPT, the measured P_T difference between the two coils is less than 2% but after heavy irradiation is as large as 7%. We average the values of P_T measured by the two coils.

C. SPECTROMETER

The differential cross section for the elastic scattering of the polarized proton beam on the polarized proton target was measured using the double arm FB spectrometer shown in Fig. 7. This spectrometer measures both the angle and momentum of both the forward scattered (F) proton and the backward recoil (B) proton, using the 4 magnets and the 6 scintillation counters $F_1F_2F_3$ and $B_1B_2B_3$. By varying the currents in the 4 magnets and occasionally reversing the PPT magnet field we are able to cover a large $P_\perp^2$ range without moving the counters. The forward scattered proton is defined by the 15 x 20 cm (hor. x vert.) F_3 scintillation counter placed about 20 m from the PPT. The F_3 momentum bite is typically $\Delta P/P = \pm 7\%$ and the defining solid angle is typically $\Delta\Omega_{cm} \approx 10^{-3}$ sr. The other counters were overmatched to allow somewhat for various experimental difficulties such as:

a. beam size
b. beam divergence
c. magnet variations
d. multiple Coulomb scattering

The overmatching was not totally adequate to compensate for all these effects and we lost about 25% of the events. However we decided to accept this loss to keep very tight angle and momentum constraints on both arms of the spectrometer. This reduced the inelastic background which is not negligible at high $P_\perp^2$, and causes more problems in spin experiments than counting losses which are equal in all spin states. We measured the inelastic and non-hydrogen event rate in the PPT by taking runs with the $C_2H_6O_2$ beads replaced by Teflon beads which contain no hydrogen. At small $P_\perp^2$ this background rate was typically 3% but at $P_\perp^2 = 4.0\ (\text{GeV/c})^2$ it reached 16%. This was subtracted at each $P_\perp^2$. The accidentals were monitored and found to always be less than 1%.

The relative beam intensity at the PPT, I_o, was monitored by the scintillation telescopes M, N, and K, which were calibrated by foil irradiations. Our results are totally independent of the ±7% normalization error in this calibration. We monitored the size, position, and the angle of the beam at the PPT using the segmented wire ion chambers shown in Fig. 7. The beam size at the 29 mm diameter by 41 mm long PPT was about 10 mm FWHM and the beam movement was less than 1 mm. More than 97% of the beam passed through the PPT. Possible systematic error due to variations in this number was reduced to below 1% by reversing the beam spin every pulse and the target spin every 8 hours.

We obtained the four normalized elastic event rates

$$N_{ij} = E(ij)/I_o(ij) \qquad (26)$$

by simultaneously measuring the number of elastic events in the FB spectrometer, $E(ij)$, and the number of incident protons $I_o(ij)$ in each of the four initial spin states ($ij \equiv$ beam, target $= \uparrow\uparrow$, $\uparrow\downarrow$, $\downarrow\uparrow$, and $\downarrow\downarrow$). The spin-spin correlation parameter, A_{nn}, was obtained from

$$A_{nn} = \frac{N_{\uparrow\uparrow} - N_{\uparrow\downarrow} - N_{\downarrow\uparrow} + N_{\downarrow\downarrow}}{P_B P_T \Sigma N_{ij}} \quad . \qquad (27)$$

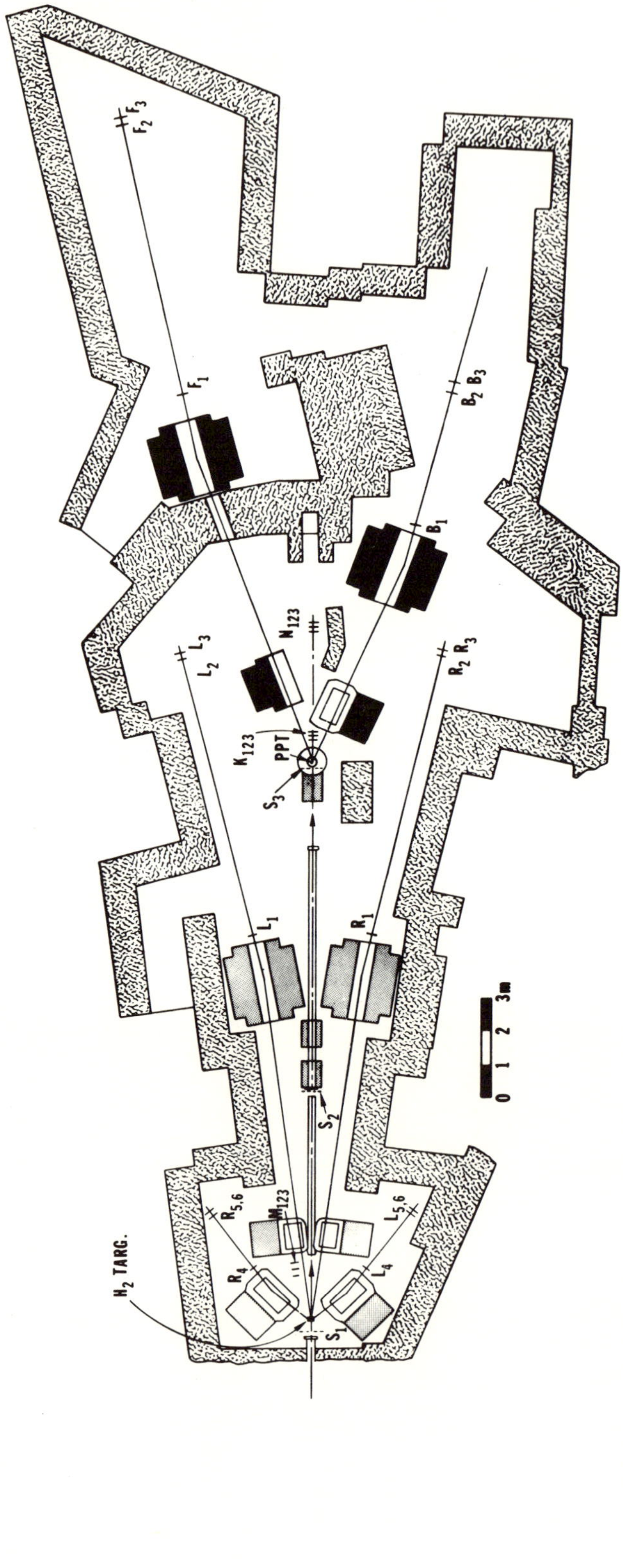

Fig. 7 Layout of the experiment. The polarized beam passes through the liquid H_2 target and its polarization is measured. The beam then scatters in the polarized proton target (PPT) and the elastic events are counted by the F and B counters. The M, N, and K counters are intensity monitors, while S_1, S_2 and S_3 monitor the beam position.

The analyzing power, A, was obtained by averaging over either the target or beam polarization

$$A_B = \frac{N_{\uparrow\uparrow} + N_{\uparrow\downarrow} - N_{\downarrow\uparrow} - N_{\downarrow\downarrow}}{P_B \Sigma N_{ij}} ,$$

$$A_T = \frac{N_{\uparrow\uparrow} - N_{\uparrow\downarrow} + N_{\downarrow\uparrow} - N_{\downarrow\downarrow}}{P_T \Sigma N_{ij}} . \tag{28}$$

The equality of A_B and A_T, required by rotational invariance, gave a consistency check which held within errors, and we averaged A_B and A_T to obtain A. We obtained the four pure two-spin cross sections, $d\sigma/dt(ij)$, from the equations

$$d\sigma/dt\ (\uparrow\uparrow) = \langle d\sigma/dt\rangle\ [1 + 2A + A_{nn}] ,$$

$$d\sigma/dt\ (\downarrow\downarrow) = \langle d\sigma/dt\rangle\ [1 - 2A + A_{nn}] , \tag{29}$$

$$d\sigma/dt\ (\uparrow\downarrow) = d\sigma/dt(\downarrow\uparrow) = \langle d\sigma/dt\rangle\ [1 - A_{nn}] ,$$

where $\langle d\sigma/dt\rangle$ is the measured spin average cross section. For our 11.75 GeV/c data this was obtained by renormalizing the large-$P_\perp^2$ 12.1 GeV/c results of Allaby et al.[13] by a factor 1.21 to agree with their later small-$P_\perp^2$ 12.0 GeV/c data,[14] which we also used. For our 90°_{cm} data we used the 90° measurements of $\langle d\sigma/dt\rangle$ by Akerlof et al.[15] and Kammerud et al.[16]

III. RESULTS

A. LARGE $P_\perp^2$ SPIN EFFECTS

First I will discuss our recent measurement of large-$P_\perp^2$ 11.75 GeV/c p-p elastic scattering in pure-initial spin states.[17] By scattering a high intensity polarized proton beam from a polarized proton target we simultaneously measured A and A_{nn} at large $P_\perp^2$.

The analyzing power, A, parameterizes the spin-orbit interaction in that it measures that part of dσ/dt which depends on whether the spins are parallel or antiparallel to the orbital angular momenta.

$$A = \frac{d\sigma/dt\ (\uparrow\uparrow) - d\sigma/dt\ (\downarrow\downarrow)}{4 < d\sigma/dt >} . \tag{30}$$

The spin-spin correlation parameter, A_{nn}, parameterizes the spin-spin forces, for it measures the difference between the spin-parallel and spin-anti-parallel cross sections.

$$A_{nn} = \frac{d\sigma/dt\ (\uparrow\uparrow) + d\sigma/dt\ (\downarrow\downarrow) - 2d\sigma/dt\ (\uparrow\downarrow)}{4 < d\sigma/dt>} . \tag{31}$$

Our data are shown in Fig. 8. Notice first the behavior of A, especially the smooth decrease at large $P_\perp^2$. At high energy the spin-orbit force does not appear to be very large in either the diffractive-type region at small $P_\perp^2$ or the diffractive-type region[18] at large $P_\perp^2$. But it does seem very important in the medium-$P_\perp^2$ region, which may be evidence that this region is not diffractive.

The spin-spin interaction has much more dramatic structure. Notice especially the rapid rise in A_{nn} starting at $P_\perp^2 = 3.6$ $(GeV/c)^2$. A_{nn} reaches a value of 30% at $P_\perp^2 = 4\ (GeV/c)^2$. This rapid rise occurs just at the start of the large $P_\perp^2$ hard scattering

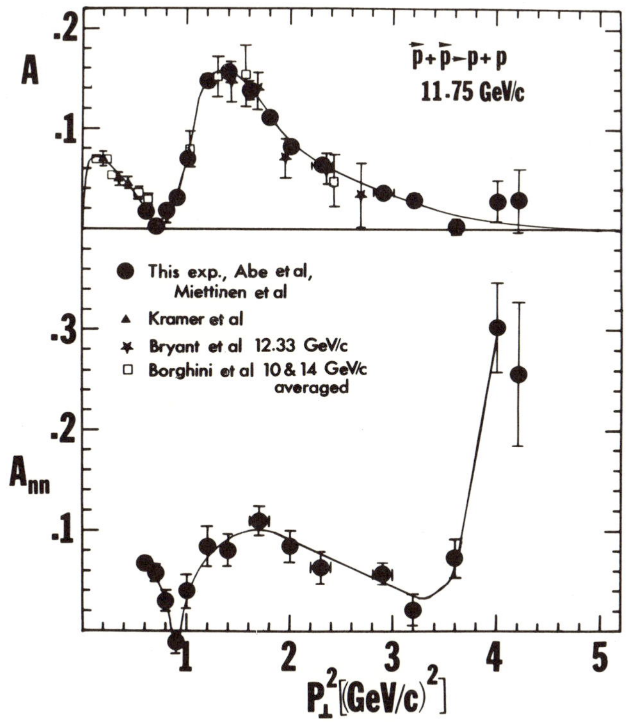

Fig. 8 Analyzing power, A, and spin-spin correlation parameter, A_{nn}, plotted against $p_\perp^2$ for p-p elastic scattering at 11.75 GeV/c.

component.

This can be seen more clearly in Figure 9, where pure-initial-spin cross sections are plotted against $\rho_\perp^2$, my scaled $-P_\perp^2$ variable,[18,19] comparing our spin data with ISR data.[20] This gives an overall picture of spin effects in high energy p-p elastic scattering. The three different $(d\sigma/dt)_{ij}$ are very close together in the diffraction peak. In the medium-$P_\perp^2$ non-diffractive region, just after the break, they move far apart and then back together again in a complicated way. In the large-$P_\perp^2$ region after the second break they split apart very rapidly until $(d\sigma/dt)_{\uparrow\uparrow}$ becomes twice as large as $(d\sigma/dt)_{\uparrow\downarrow}$.

The large A_{nn} may be associated with the break, in the sense of being caused by interference between the $\exp(-3\rho_\perp^2)$ component at medium-$P_\perp^2$ and the $\exp(-1.6\rho_\perp^2)$ component at high-$P_\perp^2$.

The large spin-spin interaction may instead be associated with the $\exp(-1.6\rho_\perp^2)$ region itself. Then the ratio $(d\sigma/dt)_{\uparrow\uparrow}/(d\sigma/dt)_{\uparrow\downarrow}$ might continue to grow with $P_\perp^2$ or reach some constant value. Looking at the different slopes in Fig. 9 it is interesting to speculate on how much the parallel scattering might dominate the anti-parallel at higher $P_\perp^2$. The maximum $P_\perp^2$ available at the ZGS polarized-proton beam is indicated by the 90° points. It is also interesting to notice that $(d\sigma/dt)_{\uparrow\uparrow}$ is pure triplet scattering while $(d\sigma/dt)_{\uparrow\downarrow}$ is a mixture of singlet and triplet. Thus at large-$P_\perp^2$ the triplet scattering dominates the singlet scattering by a very large factor, if singlet and triplet scattering are well defined at these high energies.

Notice in Fig. 9 that the $\exp(-1.6\rho_\perp^2)$ region at 11.75 GeV/c has essentially the same scaled slope as large-$P_\perp^2$ elastic scattering at $s = 2800\ (\text{GeV}/c)^2$. Thus by measuring the spin-spin forces in 11.75 GeV/c high-$P_\perp^2$ $p+p \to p+p$ we may already be directly probing the inner structure of the nucleon in a spin-sensitive way. This probing may indicate whether the inner structure can be most easily understood in terms of spinning geometrical clouds or in terms of

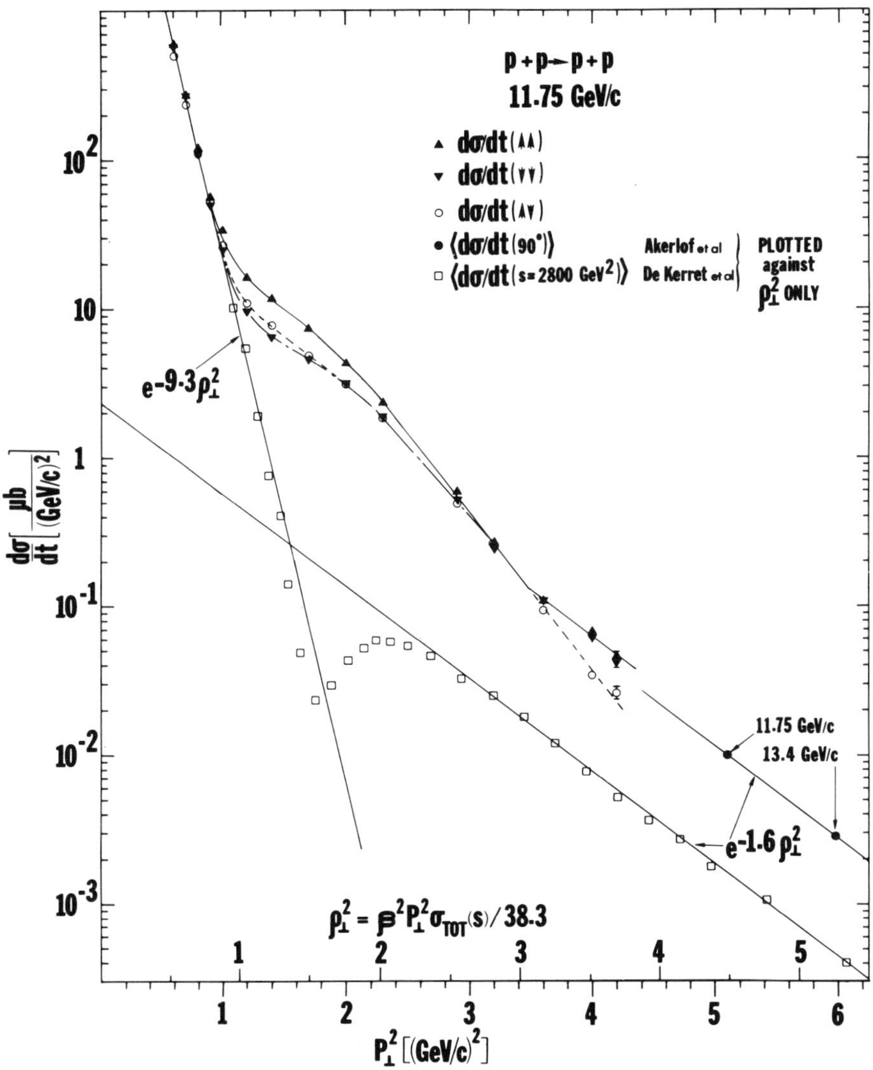

Fig. 9 Differential cross section for p-p elastic scattering in pure initial spin states at 11.75 GeV/c is plotted against $\rho_\perp^2$, the scaled $P_\perp^2$. ISR "spin averaged" elastic cross sections are also plotted.

pointlike constituent quarks with spin. Our data indicate that the spin-parallel interaction dominates the anti-parallel interaction by a factor of 2 at $P_{\perp}^2 = 4.0\ (\mathrm{GeV}/c)^2$, suggesting that the "hard" exp $(-1.6p_{\perp}^2)$ component is dominated by the spin-parallel scattering.

B. 90°_{cm} SPIN DEPENDENCE

Measuring the fixed angle energy dependence of proton-proton elastic scattering at 90°_{cm} is a sensitive way to probe the short range behavior of strong interactions.[15,16] For $p+p \to p+p$, 90°_{cm} is a special symmetry point where the pure-spin cross sections may be especially important. We recently studied the fixed-angle energy dependence of A_{nn} at 90°_{cm} using the ZGS Polarized Proton Beam and our polarized-proton target.[21] The momentum of the ZGS polarized beam was varied between 1.75 and 5.5 GeV/c. Care was taken to avoid momenta very close to known depolarizing resonances.

The spin-spin correlation parameter, A_{nn}, and the spin-orbit analyzing power A, were obtained as before from the event rates in the FB spectrometer in each of the initial spin states ($\uparrow\uparrow$, $\uparrow\downarrow$ $\downarrow\uparrow$, and $\downarrow\downarrow$) using equations 27 and 28. Because of symmetry, A at 90°_{cm} must be zero. This constraint was typically satisfied within $\pm 1.\%$. The values of A_{nn} are plotted in Fig. 10 against P_{lab}. We obtained the two independent pure two-spin cross sections at 90°_{cm} from the equations

$$(d\sigma/dt)_{\uparrow\uparrow} = (d\sigma/dt)_{\downarrow\downarrow} = \langle d\sigma/dt\rangle(1 + A_{nn}), \qquad (32)$$

$$(d\sigma/dt)_{\uparrow\downarrow} = (d\sigma/dt)_{\downarrow\uparrow} = \langle d\sigma/dt\rangle(1 - A_{nn}) .$$

The spin-average $p+p \to p+p$ cross section was taken from the 90° measurements of $\langle d\sigma/dt \rangle$.[15,16]

Recall that the spin-spin correlation parameter, A_{nn}, measures the spin-spin interaction since it parameterizes the difference

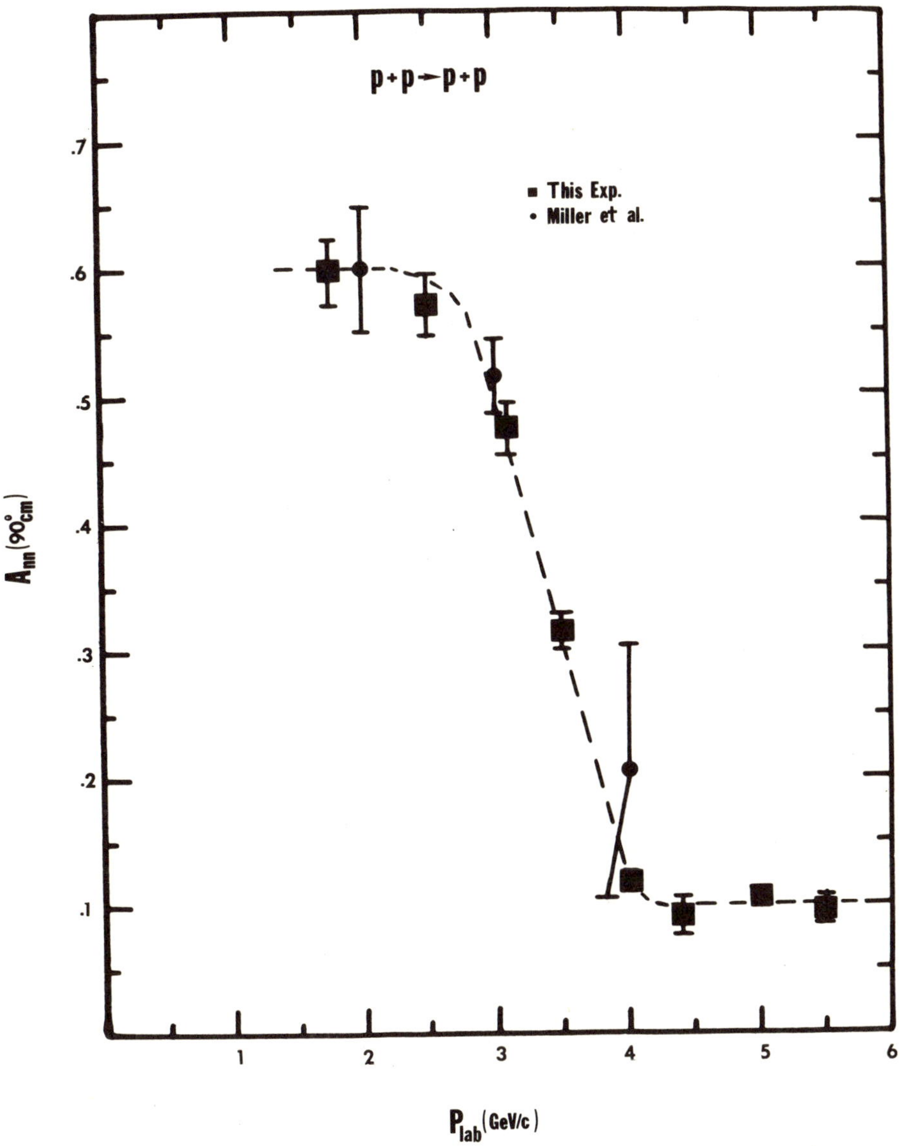

Fig. 10 The spin-spin correlation parameter, A_{nn}, for p-p elastic scattering at 90° in the center-of-mass is plotted against the incident lab momentum.

between the spin-parallel and spin-anti-parallel cross sections. In Fig. 10 the sharp change in $A_{nn}(90^\circ_{cm})$ near P_{lab} = 3.5 GeV/c is very clear. At low energy A_{nn} is approximately 60% which implies that $(d\sigma/dt)_{\uparrow\uparrow}$ is some 4 times larger than $(d\sigma/dt)_{\uparrow\downarrow}$. Throughout the range P_{lab} = 4.0 → 5.5 GeV/c, A_{nn} is equal to 10% within the 1% errors. Thus the strength of the spin-spin interaction shifts rapidly from a 60% level to a 10% level. This shift may suggest a change of the 90°_{cm} scattering mechanism for p+p→ p+p, near 3.5 GeV/c.

One can perhaps see this effect more clearly in Fig. 11, where the pure-spin cross sections are plotted against $P_\perp^2$ showing the 90°_{cm} fixed-angle energy dependence of $p_\uparrow + p_\uparrow \rightarrow p+p$. This gives an overall view of the spin dependence at 90°_{cm}, which is a special symmetry point for p-p elastic scattering. The spin-averaged 90°_{cm} cross section $\langle d\sigma/dt \rangle$ is also shown.[15,16] Notice that $(d\sigma/dt)_{\uparrow\downarrow}$ has a single $e^{-3.2P_\perp^2}$ component which extends over our entire $P_\perp^2$ range. However $(d\sigma/dt)_{\uparrow\uparrow}$ has an additional $e^{-4.4P_\perp^2}$ component in the region around $P_\perp^2 = 1(\text{GeV/c})^2$. Kammerud et al.[16] suggested the existence of such a component in p-p elastic scattering on the basis of their spin-averaged measurements at 90°_{cm}. It appears that this component only occurs in the triplet state and can therefore be seen much more clearly in our pure spin state experiment. We do not understand the theoretical reason for this triplet component.

IV. SPIN AND THE PROTON'S STRUCTURE

To understand the importance of spin in high-$P_\perp^2$ proton-proton elastic scattering it is useful to first discuss the general spin-independent properties of high-$P_\perp^2$ p-p elastic scattering. I feel that this process probes most directly and most deeply into the inner structure of the nucleon. The high available energy and luminosity allows precise and direct measurements of p-p elastic scattering over a cross section range of 10^{13} and an incident

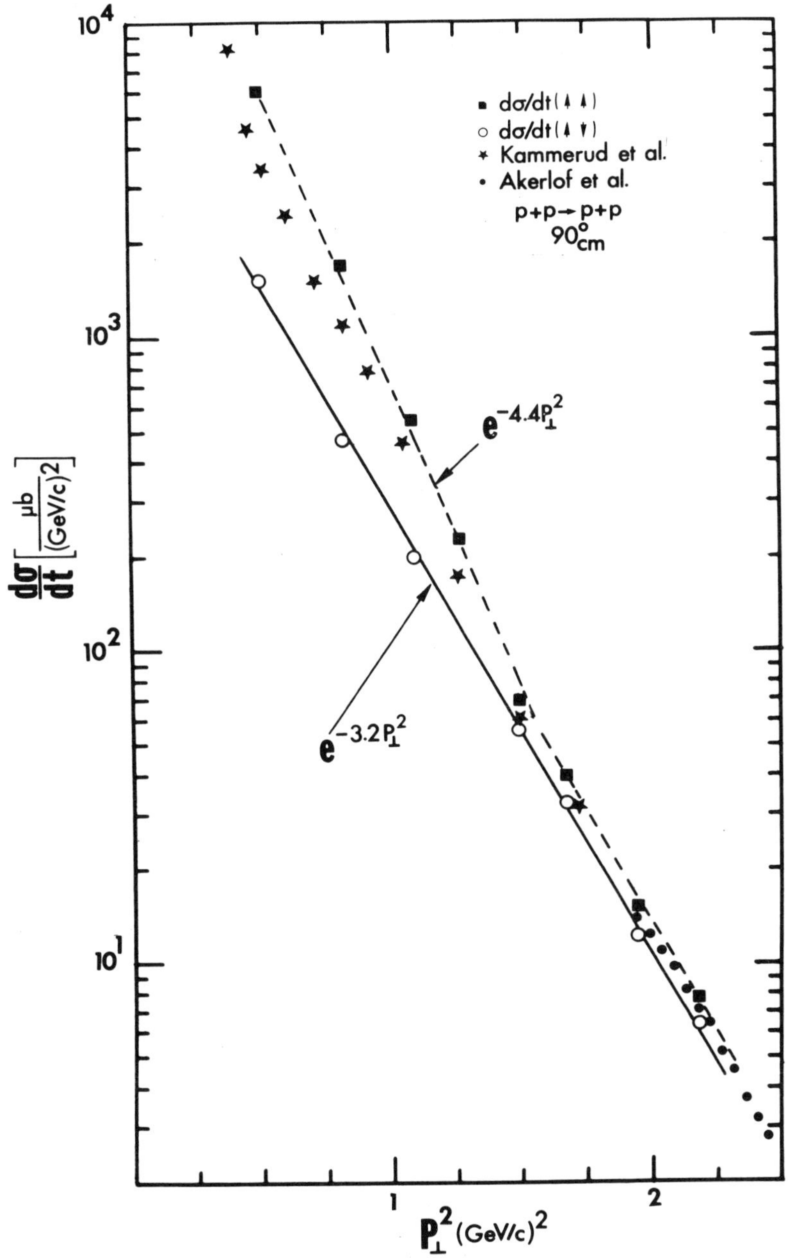

Fig. 11 The two independent pure-initial-spin differential cross sections $(d\sigma/dt)_{\uparrow\uparrow}$ and $(d\sigma/dt)_{\uparrow\downarrow}$ for p-p elastic scattering at 90° in the center-of-mass are plotted against $P_\perp^2$ as the incident P_{lab} is varied.

energy range of a few MeV to 2000 GeV. This enormous precision and range, coupled with the fundamental simplicity of elastic scattering makes this data perhaps the most severe test of any theory of strong interactions.

At small $P_\perp^2$ the cross section is large and the data is very extensive. I believe that one very important concept is now generally accepted because of all this data:

> In soft hadronic scatterings in the small-$P_\perp^2$ diffraction peak, nucleons behave as geometric objects about 1 fermi in size.

While this may now seem quite sensible, for years many theorists believed that, at the 1 fermi level, size was a totally meaningless concept and that the optical and geometric models proposed by Serber,[22] myself,[23] Van Hove,[24] Yang,[25], and others[26] were useless. I believe the ISR and Fermilab elastic measurements make it clear that the idea of a nucleon having a hadronic size of 1 fermi is a meaningful and useful concept over an enormous energy range.

Now recall that large-$P_\perp^2$ elastic scattering lets one probe very deeply into the proton. In fact, it is large $P_\perp^2$ and not high incident energy which allows probing at very small distances, since $P_\perp$ is canonically conjugate to the impact parameter b. Thus the very violent and probing "head-on" collisions at amall b can only be precisely studied at large $P_\perp^2$. Recall that

$$bP_\perp = \hbar c = .197 \text{ GeV/c-fermi} \qquad (33)$$

Thus to clearly see an object with a precision of .05 fermi requires a $P_\perp$ of 4 GeV/c or a $P_\perp^2$ of $16(\text{GeV/c})^2$. No elastic scattering event with $P_\perp = 4$ GeV/c has ever been observed because of its small cross section. But it is just such "hard scattering" events that may allow us to discover and study the inner constituents of the proton and determine if they are quarks, partons, cores, or whatever. I know of no direct way to study this problem except high-$P_\perp^2$ elastic or inelastic scattering.

When the p-p elastic differential cross section is plotted against the Mandestam variable, $-t$, it is not at all independent of energy. The slope b in an e^{bt} plot changes rapidly with energy. This change of slope with incident energy is the familiar "shrinkage" of the diffraction peak. This shrinkage indicates that either $-t$ is not a good variable or the proton size grows with energy or both. At larger $-t$ there is even more shrinkage. Thus $d\sigma/dt$ is clearly not independent of incident energy when plotted against $-t$.

Nevertheless I believe that geometric concepts are valid and important; and for 15 years I have been trying to develop a totally energy-independent geometric model which is called the Lorentz-Contracted Geometric Model and sometimes the Onion model.[19,23] It assumes that protons behave as clouds of hadronic scattering probability which are spherically symmetric except that they are Lorentz-Contracted in the direction of motion. Each proton contains several concentric Gaussian shaped clouds with different sizes. Thus in some sense the model proton looks like an onion except that the clouds overlap.

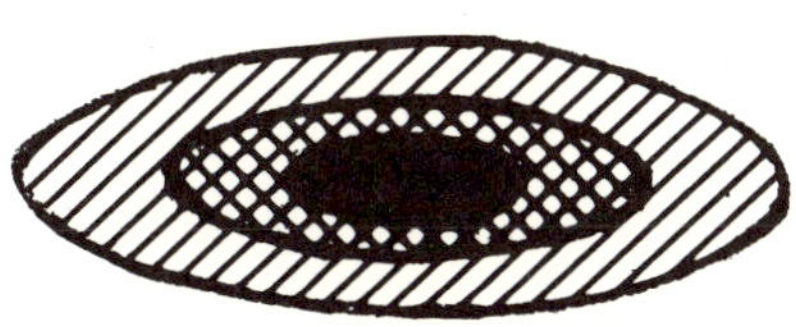

The Lorentz-Contracted proton-proton interaction probability amplitude with 3 concentric regions.

When two protons collide, the interaction probability amplitude for them to interact strongly is assumed to have the form

$$\phi = \sum_i a_i' \, e^{-\frac{1}{2}[x^2 + y^2 + z^2\gamma^2]/A_i'^2} \quad . \qquad (34)$$

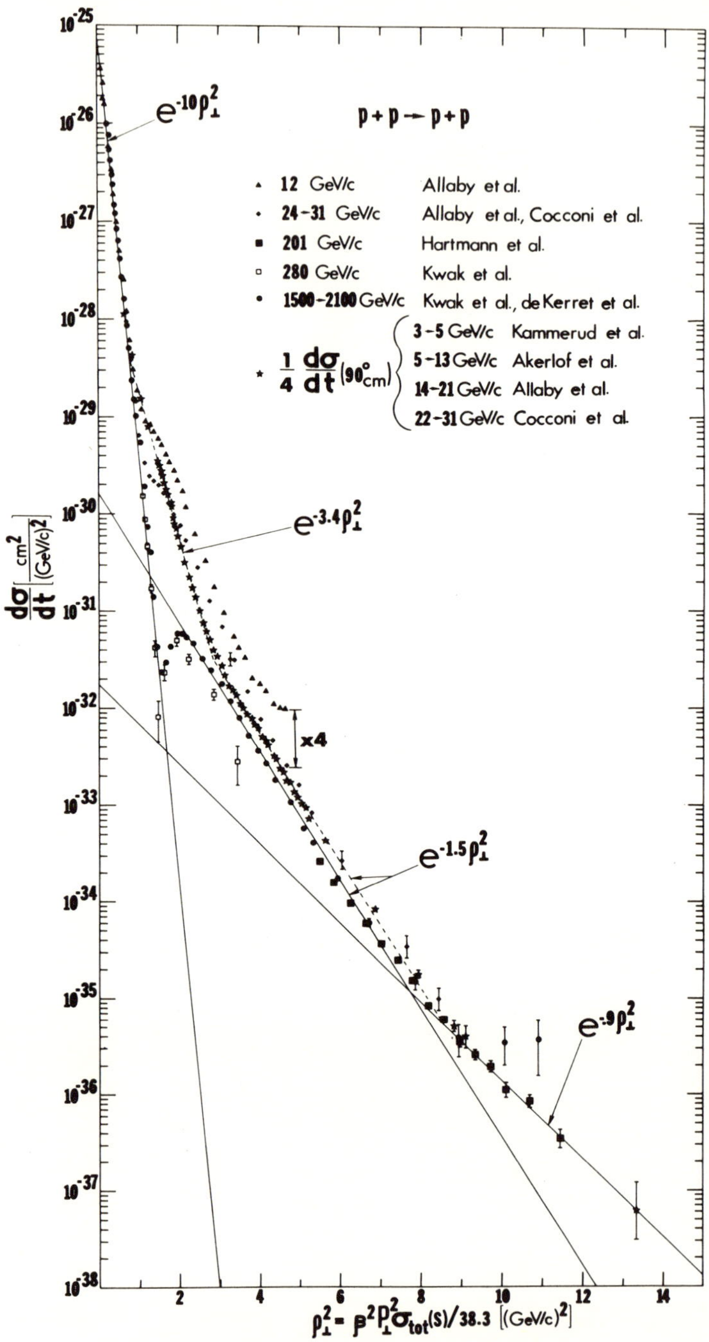

Fig. 12 Differential elastic cross section for p-p scattering is plotted against $\rho_\perp^2$.

The elastic scattering amplitude which is defined to be exactly the Fourier transform of ϕ then depends only on the variable

$$\beta^2 P_\perp^2 = tu/s \;, \tag{35}$$

where $\beta = v_{cm}/c$. This can be easily seen by Fourier transforming ϕ. The growth in $\phi_{tot}(s)$ is presumably caused by the growth in the proton size,

$$A_i'^2 = A_i^2 \, \sigma_{tot}(s)/38.3 \;. \tag{36}$$

Thus we obtain a universal variable

$$\rho_\perp^2 = \beta^2 P_\perp^2 \, \frac{\sigma_{tot}(s)}{38.3} \tag{37}$$

and the prediction that $d\sigma/dt$ should be totally independent of energy when plotted against this variable.

The only assumption required for this striking result is that the probability density is spherically symmetric except for the Lorentz contraction.

$$r^2 \to x^2 + y^2 + z^2\gamma^2 \;. \tag{38}$$

This is the basic concept of the Lorentz-contracted Geometric Model. The totally independent "Onion" model assumption is that ϕ contains several concentric Gaussian regions. This implies that the elastic cross section has the form

$$\frac{d\sigma}{dt} = \left[\sum_i a_i e^{-\frac{1}{2}A_i^2 \beta^2 P_\perp^2 \sigma_{tot}(s)/38.3} \right]^2 . \tag{39}$$

This parameterization seems to work at small $P_{\perp}^2$. Last year Peter Hansen and I compiled all p-p elastic diffraction peak data up to 2000 GeV/c and compared it with this prediction.[19] From $3 \rightarrow 2000$ GeV/c, $d\sigma/dt$ (p-p) fits a single universal function of the variable $\rho_{\perp}^2$ in the diffraction peak.

In Fig. 12 I have plotted $d\sigma/dt$ against the variable $\rho_{\perp}^2$ for p-p elastic scattering from $3 \rightarrow 2100$ GeV/c. The data spans 13 decades and includes all existing large-$P_{\perp}^2$ data. This graph clearly shows a deviation from a universal fit in the range $\rho_{\perp}^2 = 1 - 3$ $(\mathrm{GeV/c})^2$. I believe this is caused by a $e^{-3\rho_{\perp}^2}$ non-diffractive component which decreases with energy and finally sinks below the dip. Notice that the position of the dip is independent of energy in this $\rho_{\perp}^2$ plot.

At large $\rho_{\perp}^2$ and at small $\rho_{\perp}^2$ the fit appears to be excellent. As we noted earlier, in the diffraction peak all data from 3 to 2000 GeV/c fit a universal function of $\rho_{\perp}^2$ all available data extending over many decades falls on a single universal curve within about 30%.

Notice that for the 90° data I have plotted 1/4 of the measured value of $d\sigma/dt$ at 90^o_{cm}. This factor, 1/4, compensates for the constructive interference between the forward and backward scattering of the identical protons,

$$d\sigma/dt(90^o) = [f(90^o) + f(\pi-90^o)]^2 = 4\,|f(90^o)|^2 \quad . \qquad (40)$$

This particle identity effect[19] causes flattening near 90° which can clearly be seen in the 12 GeV/c data. The factor of 1/4 brings the 90° 12 GeV point exactly onto the universal curve. The particle identity factor clearly equals 1 well away from 90° but its behavior near 90° depends on the spin dependence of $d\sigma/dt$(p-p) near 90° which we are now measuring at the ZGS. Notice that the ISR[20] and Fermilab[27] data is always very far from 90° so that this, unfortunate but real, particle identity effect can be ignored.

Fig. 12 clearly demonstrated several general features of p-p elastic scattering:

1. A complicated low energy medium-$\rho_\perp^2$ component which is probably non-diffractive and disappears by 200 GeV/c.
2. An energy-independent diffraction peak with a slope of 10 $(\text{GeV/c})^{-2}$ which corresponds to the .9 fermi outer size of the proton.
3. An energy-independent large-$\rho_\perp^2$ component which may also be diffractive. Its slope of 1.5 $(\text{GeV/c})^{-2}$ corresponds to a size of 1/3 fermi, which may be the size of the proton's constituents.
4. A dip which appears to be destructive interference between the two diffractive components.
5. A possible $e^{-.9\rho_\perp^2}$ component suggested by very large $P_\perp^2$ data[20,27,28].

One can summarize the 5 points by writing the cross section as

$$\frac{d\sigma}{dt} = \left[iAe^{-\frac{1}{2}10\rho_\perp^2} + \frac{(B+ib)}{s^{\frac{1}{2}}} e^{-\frac{1}{2}3\rho_\perp^2} - iC\, e^{-\frac{1}{2}1.5\rho_\perp^2} + \ldots + \ldots \right]^2 . \tag{41}$$

Finally I will briefly discuss spin effects in nucleon-nucleon elastic scattering at large $P_\perp^2$. For many years the spin of the nucleon was considered an unfortunate complication, which hopefully would become unimportant at high energy. This does not seem to be true. The spin-spin interaction seems to be very large at high-$P_\perp^2$. In fact the spin-spin force seems to grow dramatically just at that $P_\perp^2$ where the $e^{-1.5\rho_\perp^2}$ hard-scattering component seems to become dominant. Thus these hard scattering events, which are presumably caused by the direct scattering of the protons' constituents, mostly occur when the protons' spin are parallel. What does this mean?

I believe that this large spin-spin force in high-$P_\perp^2$ elastic p-p scattering may due to the spin dependence of the direct scattering of the constituents of the proton. We do not know the exact nature of these constituents but the slope of $1.6(\mathrm{GeV/c})^{-2}$ corresponds to a size of about 1/3 fermi, which may be a measure of the size of the constituents. The large spin-spin force would then indicate that these consituents rarely scatter unless their spins are aligned.

This new fact may help us to understand the nature of the proton's constituents. For example, we observe that for elastic p-p scattering at large $P_\perp^2$

$$\frac{(d\sigma/dt)_{\uparrow\uparrow}}{(d\sigma/dt)_{\uparrow\downarrow}} = 2. \tag{42}$$

What does that imply about the ratio

$$\frac{(d\sigma/dt)_{\uparrow\uparrow}}{(d\sigma/dt)_{\uparrow\downarrow}} \tag{43}$$

for the constituents? The answer clearly depends upon the number of constituents in the proton and upon each constituent's intrinsic spin. Thus in my "Onion" model, we find that the inner 1/3 fermi "core" of each proton has a considerable spin dependence and the outer .9 fermi cloud has little spin dependence.

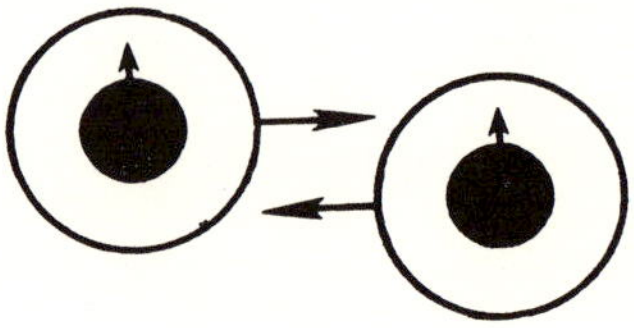

In fact we find that for the core-core collisions,

$$\frac{(d\sigma/dt)_{\uparrow\uparrow}}{(d\sigma/dt)_{\uparrow\downarrow}} \text{ (core-core)} = 2. \qquad (44)$$

If one instead considers each proton to contain 3 spin-1/2 quarks (or partons), then for the two different polarized proton cases the quark spins would look like:

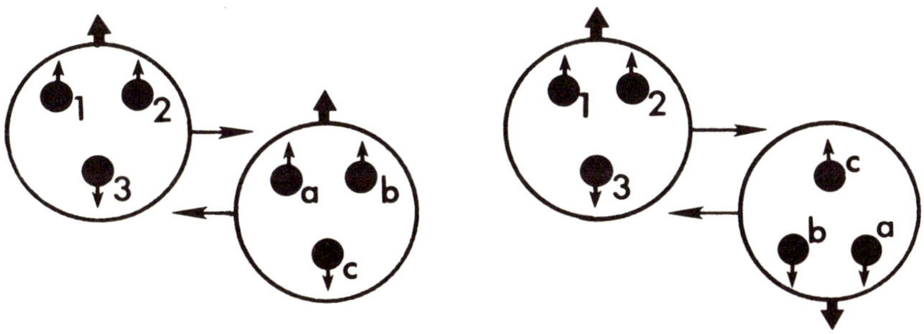

Let us assume that $\sigma_{\uparrow\uparrow} = |f_{\uparrow\uparrow}|^2$ and $\sigma_{\uparrow\downarrow} = |f_{\uparrow\downarrow}|^2$ are the quark-quark scattering cross sections and amplitudes. Then doing some simple counting implies that

$$\frac{(d\sigma/dt)_{\uparrow\uparrow}\text{ (pp)}}{(d\sigma/dt)_{\uparrow\downarrow}\text{ (pp)}} = \qquad 2 = \frac{[5f_{\uparrow\uparrow} + 4f_{\uparrow\downarrow}]^2}{[4f_{\uparrow\uparrow} + 5f_{\uparrow\downarrow}]^2} \quad \text{[coher.]} \qquad (45a)$$

$$2 = \frac{5\sigma_{\uparrow\uparrow} + 4\sigma_{\uparrow\downarrow}}{4\sigma_{\uparrow\uparrow} + 5\sigma_{\uparrow\downarrow}} \quad \text{[incoher.]} \qquad (45b)$$

depending on the coherence properties of quark-quark scattering. This gives a relation between either $f_{\uparrow\uparrow}$ and $f_{\uparrow\downarrow}$ or $\sigma_{\uparrow\uparrow}$ and $\sigma_{\uparrow\downarrow}$ for quark-quark scattering, which implies that it is difficult (45a) or impossible (45b) to explain our large spin effect in terms of 3 spin-

1/2 constituents. In a very simple-minded way this is not surprising, for if you consider a 100% polarized proton to be a polarized quark target, the maximum quark polarization is 33%. It is much easier to understand our large spin-spin interaction in terms of a single constituent.

Of course, one could assume more complex quark-quark scattering models with the different types of quarks having different scattering amplitudes. This would give other versions of equation (45). Or one could assume, as S. Brodsky has recently suggested, that we are really seeing quark exchange in large-$P_\perp^2$ p-p elastic scattering rather than quark-quark scattering. I don't yet understand if this is compatable with gluon ideas. In any case a relation like equation (44) or (45) will give a strong constraint on any theory that associates large-$P_\perp^2$ scattering experiments with the fundamental constituents of the proton. It may even suggest which of these theories is useful and which should be abandoned.

REFERENCES

1. A. Bohr, Nucl. Phys. 10, 486 (1959).
2. R. C. Fernow et al., Phys. Lett. 52B, 243 (1974).
   L. G. Ratner et al., Phys. Rev. D15, 604 (1977).
   M. Borghini et al., Phys. Rev. D17, 24 (1978).
3. E. F. Parket et al., IEEE Trans. Nuc. Sci. NS-22 1718 (1975).
4. R. M. Craig et al., 2nd International Symposium on Polarization Phenomena, Karlsruh (1965) pp. 322-323.
5. T. Khoe, R. L. Kuston, R. L. Martin, E. F. Parker, C. W. Potts, L. G. Ratner, R. E. Timm, A. D. Krisch, J. B. Roberts, J. R. O'Fallon, Particle Accel, 6, 213 (1975).
6. M. Froisart and R. Stora, Nucl. Instr. and Meth. 7; 297 (1960), D. Cohen, Rev. Sci. Inst. 33, 161 (1962).
7. L. G. Ratner et al., Private communication.
8. P. Grannis et al., Phys. Rev. 148, 1297 (1966).
9. M. Borghini et al., Phys. Lett. 24B, 77 (1967);
   M. Borghini et al., Phys. Lett. 31B, 405 (1970);
   M. Borghini et al., Phys. Lett. 36B, 501 (1971);
   M. G. Albrow et al., Nucl. Phys. B23, 445 (1970);
10. N. E. Booth et al., Phys. Rev. Lett 21, 651 (1968), and 23, 192 (1969); D. J. Sherden et al., Phys. Rev. Lett. 25, 898 (1970); J. H. Parry et al., Phys. Rev. D8, 45 (1973).
11. J. A. Bywater, R. C. Fernow, and S. W. Gray, ANL Internal Report (1975).
12. M. Borghini et al., CERN Internal Report.
13. J. V. Allaby et al., CERN Report 68-7/580 (1968).
14. J. V. Allaby et al., Nucl. Phys. B52, 316 (1973).
15. C. W. Akerlof et al., Phys. Rev. 159, 1138 (1967).
16. R. C. Kammerud et al., Phys. Rev. D4 , 1309 (1971).
17. K. Abe et al., Phys. Lett. 63B, 239 (1976).
    H. E. Miettinen et al., Phys. Rev. D16, 549 (1977).
    J. R. O'Fallon et al., Phys. Rev. Lett. 39, 733 (1977).

18. A. D. Krisch Proceedings of 1977 DPF Meeting at Argonne, AIP Conf. Proceedings (to be published).

19. A. D. Krisch, Phys. Rev. Lett. 19, 1149 (1967); Phys. Lett. 44B , 71 (1973); P. H. Hansen and A. D. Krisch Phys. Rev. D15, 3287 (1977).

20. H. DeKerret et al., Phys. Lett. 62B, 363 (1976); 68B 374 (1977).

21. A. Lin et al., Submitted in Phys. Letters B.

22. R. Serber, Phys. Rev. Lett. 10, 357 (1963); Rev. Mod. Phys. 36, 649 (1964).

23. A. D. Krisch, Phys. Rev. Lett. 11, 217 (1963); Phys. Rev. 135, B1456 (1964).

24. L. Van Hove, Nuovo Cimento 28, 798 (1963); Rev. Mod. Phys. 36, 655 (1964).

25. T. T. Wu and C. N. Yang, Phys. Rev. 137, B708 (1965); T. T. Chou and C. N. Yang, Phys. Rev. 170, 1591 (1968).

26. L. Durand and R. Lipes, Phys. Rev. Lett. 20, 637 (1968).

27. J. L. Hartmann et al., Phys. Rev. Lett. 39, 975 (1977).

28. G. Cocconi et al., Phys. Rev. 138B, 164 (1965).

ADVANCES IN THE STUDY OF SPIN EFFECTS IN NUCLEON-NUCLEON SCATTERING AT SMALL AND INTERMEDIATE MOMENTUM TRANSFERS*

Homer A. Neal

Indiana University

Bloomington, Indiana 47401

INTRODUCTION

There has been remarkable progress made in the field of high energy physics in the past few years. We have recently witnessed the discovery of new quantum numbers, the spectra of particles associated with these quantum numbers, and a much improved understanding of the role that all quantum numbers play in high energy interactions. With the wide assortment of experiments and studies being conducted in high energy physics at any one time, it is very easy for one to get the impression that the activity in the field is highly unfocused. On the one hand, it is precisely this apparent unfocused nature that has given rise to much of the excitement and richness of the field. On the other hand, as one examines the situation more closely, one finds that much of the effort in the field is indeed directed toward the solution of a few central problems. One of these problems which I will continually refer to today is that of understanding the details of the structure of the elementary particles. This problem--one of the most fascinating

*Work supported by the U.S. Department of Energy under contract E(11-1)2009, Task A.

and most difficult problems ever tackled by man--can be approached from several different angles.

For many years, it was felt that the study of polarization phenomena perhaps had little or no bearing on the perceived central problem of the constituent structure of the elementary particles. Recent developments, however, indicate that the study of spin effects may be a very powerful and effective tool for gaining information on the details of particle structure. Major technological advances such as the acceleration of polarized protons, the improvement in polarized targets, and the development of sophisticated polarimeters now make it possible for extremely complicated spin experiments to be carried out with high accuracy. Indeed, we are presently standing at the threshold of being able to extract the precise effect of a particle's spin on its interaction probability at energies of several GeV.

Because of the general complexity of polarization phenomena, even the most rudimentary theoretical understanding of the experimental observations has typically lagged behind the experimental work by a period of many years. Indeed, it is frequently the case that experimentalists will measure the polarization in a certain reaction long before the theoretical description of the differential cross-section has been satisfactorily carried out. During my talk today, however, I will point out what may be an exception to this trend. As we shall see, one of the striking discoveries in recent years has been the observation of significant polarizations in inclusive processes. I will discuss today what appears to be a successful theoretical attempt to explain the observed inclusive polarizations which, in my estimation, illustrates that polarization studies can be an effective tool in the generation and testing of production models.

In my talk, I will begin by reviewing the new developments in elastic scattering processes. This will include a survey of new developments in p-p and n-p elastic scattering cross sections, polarization, and higher order spin measurements. We will then review the existing evidence for a new dibaryon resonance and also examine recent data and theoretical analyses of inclusive baryon polarizations. Next, I will discuss a constituent model of p-p elastic polarization and then, finally, examples will be given to illustrate the status of the program to determine the scattering amplitudes.

FORMALISM

In an earlier talk at the conference the matter of notation has been taken up so I will keep my comments in this section brief. There are, in principle, sixteen distinct quantities to be measured for a spin $\frac{1}{2}$ - spin $\frac{1}{2}$ elastic scattering process since we need to know the amplitude for transitions from each of four initial spin states into each of four final spin states. It can be shown that the application of the various strong interaction invariance principles permits the problem to be completely specified in terms of only five amplitudes. The common set[1] of five independent s-channel helicity amplitudes is given below:

$$\phi_1 = \langle ++|\phi|++\rangle \qquad \text{non-flip}$$

$$\phi_2 = \langle --|\phi|--\rangle \qquad \text{double-flip}$$

$$\phi_3 = \langle +-|\phi|+-\rangle \qquad \text{non-flip}$$

$$\phi_4 = \langle +-|\phi|-+\rangle \qquad \text{double-flip}$$

$$\phi_5 = \langle ++|\phi|+-\rangle \qquad \text{single-flip}$$

In certain applications it is more convenient to deal with a set of amplitudes representing definite parity in the t channel. Such a set has been proposed by Halzen and Thomas[2] and is related to the ϕ's as shown below:

$N_o = \frac{1}{2}(\phi_1+\phi_3)$ Natural parity/non-flip

$N_1 = \phi_5$ Natural parity/single-flip

$N_2 = \frac{1}{2}(\phi_4-\phi_2)$ Natural parity/double-flip

$U_o = \frac{1}{2}(\phi_1-\phi_3)$ Unnatural parity/non-lip

$U_2 = \frac{1}{2}(\phi_4+\phi_2)$ Unnatural parity/double-flip.

A shorthand notation has been developed to designate the spin quantity measured in certain experiments. The probability of the elastic scattering process occuring with the beam, target, scattered, and recoil nucleons in specified spin states can be designated as $I(S_{beam}, S_{target}, S_{scattered}, S_{recoil})$. Now, for the various arguments, let us specify the direction of the nucleon spin by one of the following codes:

$\vec{n} \rightarrow$ normal to scattering plane

$\vec{\ell} \rightarrow$ along nucleon momentum (longitudinal vector)

$\vec{s} \rightarrow \vec{n} \times \vec{\ell}$

$o \rightarrow$ direction unspecified.

Then we can define some of the standard observables as follows:

$I_o = I(o,o;o,o)$ (the conventional differential cross section)

$I_oA = I(o,\vec{n};o,o)$ (cross section x analyzing power)

$I_oP = I(o,o;\vec{n},o)$ (cross section x polarization)

$I_oD = I(o,\vec{n};o,\vec{n})$ (cross section x depolarization)

$I_oC_{nn} \rightarrow I_oA_{nn} = I(\vec{n},\vec{n};o,o)$ (cross section x spin correlation parameter)

$I_oR = I(o,\vec{s};o,\vec{s})$ (cross section x rotation parameter) .

Any observable quantity, such as the cross section I_o, the polarization P, the depolarization D, the spin correlation parameter $C_{nn}(A_{nn})$, etc. can be readily expressed in terms of a set of helicity amplitudes. Some examples are given below.

$$I_o = \frac{d\sigma}{d\Omega} = \frac{1}{2}\{|\phi_1|^2 + |\phi_2|^2 + |\phi_3|^2 + |\phi_4|^2 + 4|\phi_5|^2\}$$

$$I_oP = \mathrm{Im}\,((\phi_5^*(\phi_1 + \phi_2 + \phi_3 - \phi_4))$$

$$I_o(1-D) = \frac{1}{2}\,(|\phi_1 - \phi_3|^2 + |\phi_2 + \phi_4|^2)$$

$$I_o(1-C_{nn}) = \frac{1}{2}(|\phi_1 - \phi_2|^2 + |\phi_3 + \phi_4|^2) \quad .$$

It is very interesting to note from the above relations how the various observables are sensitive to different amplitudes. For example, violent structure in ϕ_5 may be invisible in I_o but dominant in P.

p-p ELASTIC SCATTERING

A. Differential Cross Sections

Before we get too involved with the discussion of spin effects, it would be appropriate to consider the status of knowledge of the differential cross section in p-p elastic scattering. In Fig. 1 one can see that at low energies $d\sigma/dt$ monotonically declines with θ_{cm}. At energies near 10 GeV a secondary minimum develops near $t = -1.0\ (\mathrm{GeV}/c)^2$ and becomes more pronounced as the energy is increased. By the time we have reached ISR energies, this dip has become quite pronounced. The most recent ISR results shown in Fig. 2 illustrates that over a significant range in t beyond the first minimum the cross section is exponential in behavior, from $-t \simeq 1.5$ to $8\ (\mathrm{GeV}/c)^2$. Such a mapping of the cross section over many orders of magnitude and the spin measurements which we will discuss in later sections certainly have buried in them an enormous

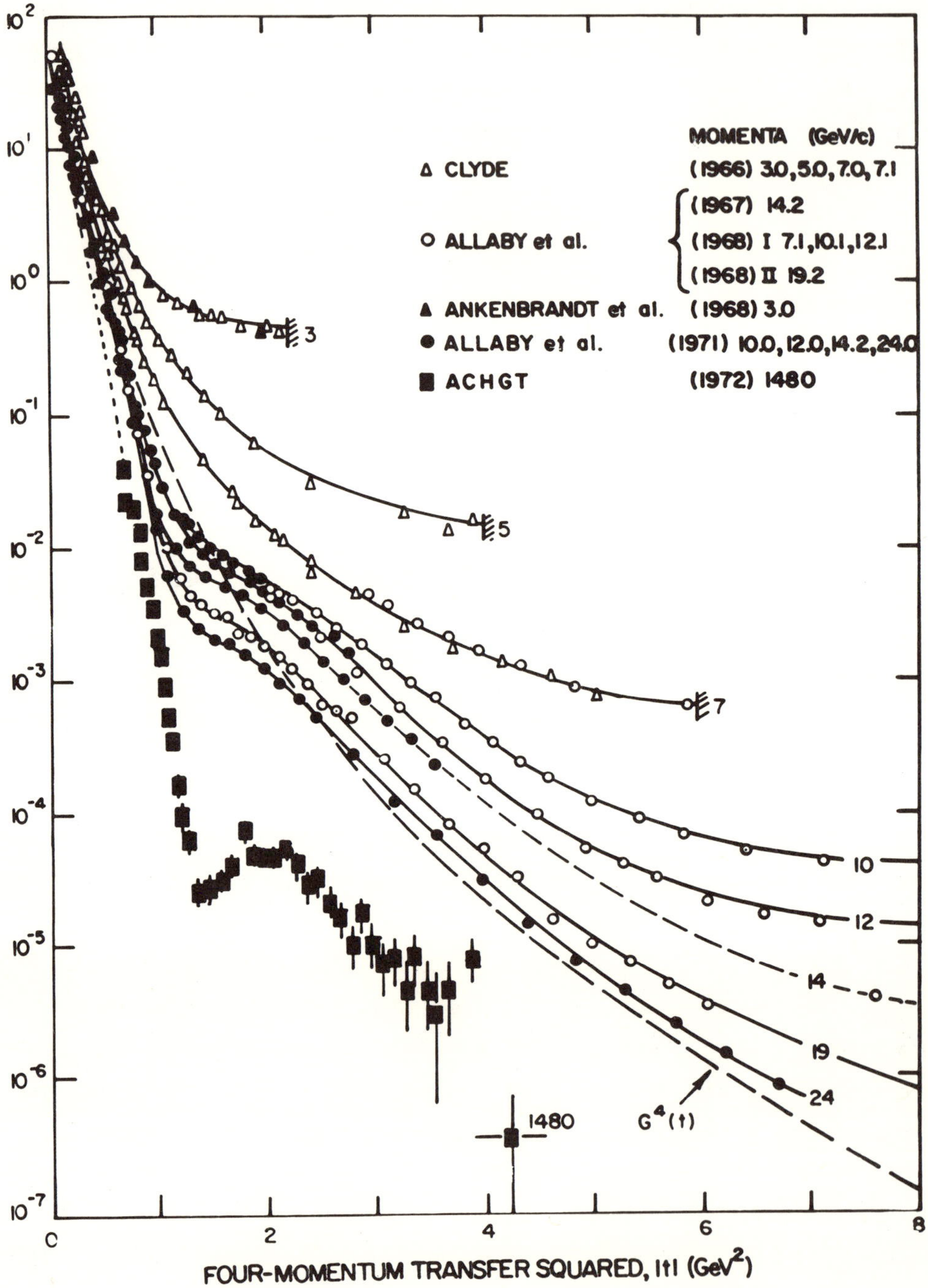

Fig. 1 Differential cross section in p-p elastic scattering (Ref. 18).

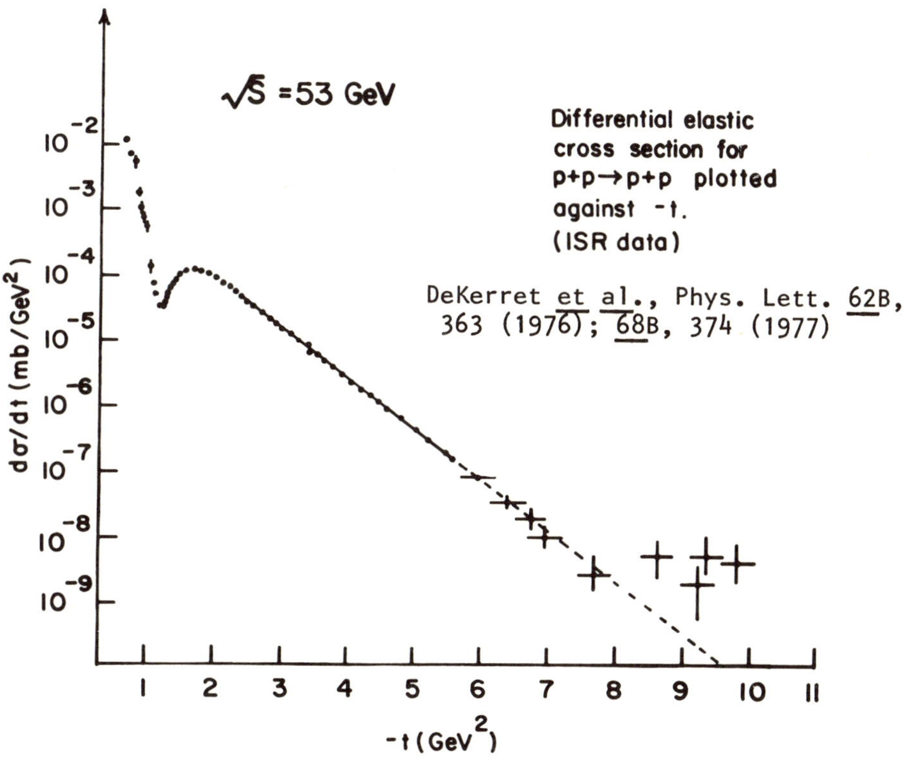

Fig. 2 Recent differential p-p cross section measurements at the ISR (Ref. 11).

amount of information on the proton--if we can only extract it.

B. Polarization in p-p Elastic Scattering in Low Energies

At energies below 1 GeV the polarization in p-p elastic scattering is about as uninteresting as it can be. With kinematic zeros at $\theta_{cm} = 0^{\circ}$ and $\theta_{cm} = 90^{\circ}$ it smoothly climbs from $P(\theta_{cm}=0)=0$ to a maximum value near $\theta_{cm} = 45^{\circ}$ and then falls smoothly to $P(\theta_{cm} = 90^{\circ}) = 0$. Of course, it is significant that finite and even large values of polarization exist at all and this point needs to be firmly kept in mind. What is missing at the low energies is the kind of structure in the polarization that we have found to be particularly pertinent to the details of the interaction. This "dullness" in P is one of the principal reasons this area of study appeared so hopeless ten years ago. But ten years ago we barely had sufficient data to explore $|t|$ values up to .6 $(GeV/c)^2$. Note that $|t|(\theta_{cm} = 90^{\circ}) \cong .4\ (GeV/c)^2$ at $P_{LAB} = 1$ (GeV/c) and that any dynamical effects in P are probably completely masked by the Pauli principle as $\theta_{cm} \to 90^{\circ}$. As the energy of study was raised from cyclotron energies into the several GeV range, we began to learn several interesting features about the p-p polarization as described below.

- In the results of Neal and Longo[3] we were able to first observe the functional form of P(t) with t sufficiently far from $t(90^{\circ})$ to detect an apparent dynamical zero in the polarization near $|t| = .7\ (GeV/c)^2$. This zero seems to be an invariant feature of p-p polarization from a few GeV to a few hundred GeV.
- At energies below 20 GeV relative maxima occur in P(t) near $|t| = .3$ and 1.7 $(GeV/c)^2$.
- In general, the magnitude of the polarization at any given t drops with increasing incident energy.

These features are evident in the plots in Fig. 3 covering the momentum region 1-12 GeV/c. With these general trends of the "low" energy data in mind, I would now like to turn to some of the details

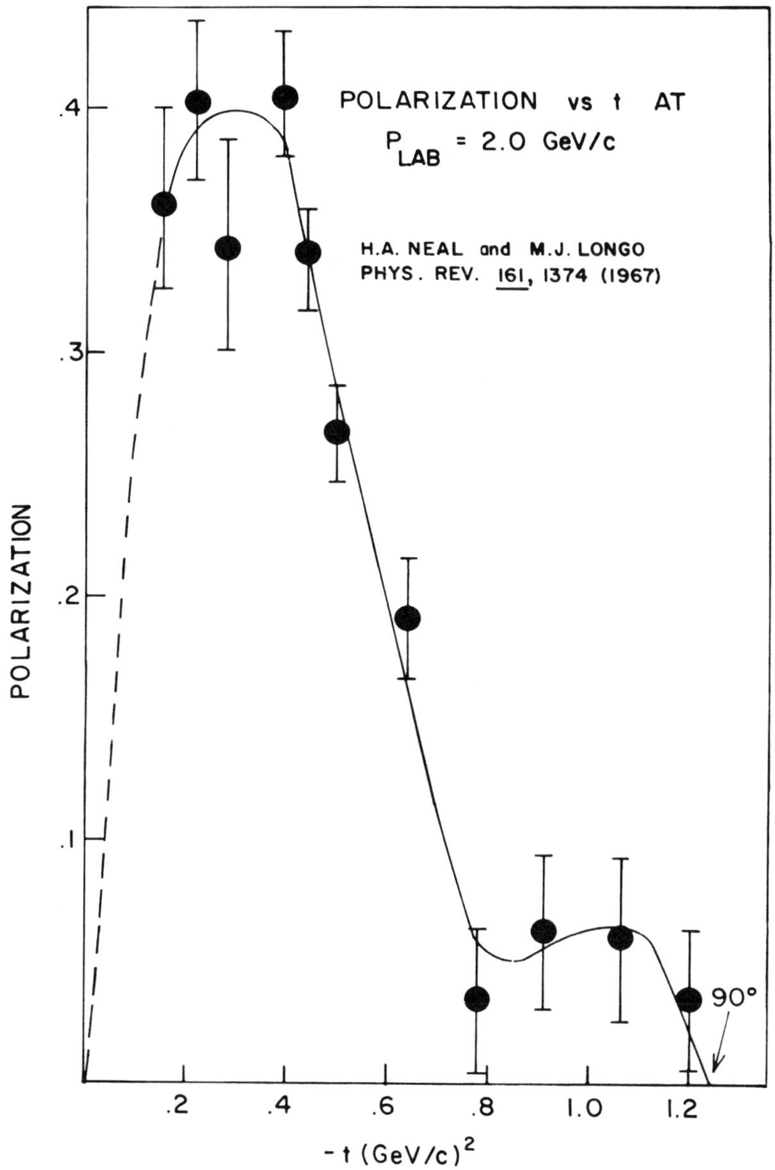

Fig. 3A Polarizations in p-p elastic scattering from 1 to 12 GeV/c.

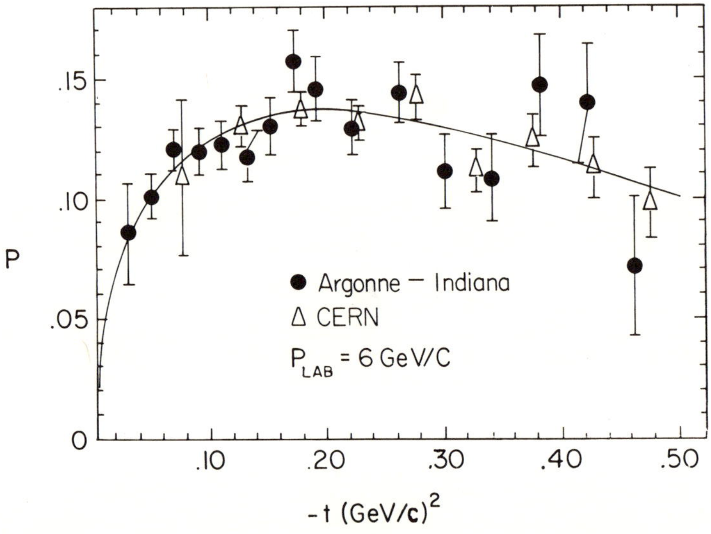
P
.15
.10
.05
0
.10
.20
.30
.40
.50
-t (GeV/c)^2
● Argonne — Indiana
Δ CERN
P_{LAB} = 6 GeV/C

Fig. 3B

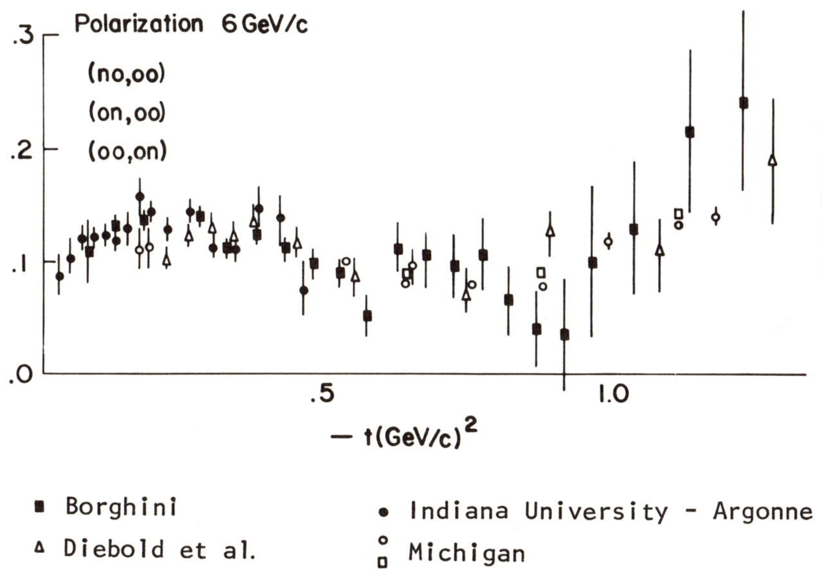
Polarization 6 GeV/c
(no,oo)
(on,oo)
(oo,on)
.3
.2
.1
.0
.5
1.0
— t(GeV/c)²
Borghini
Diebold et al.
Indiana University - Argonne
Michigan

Fig. 3C

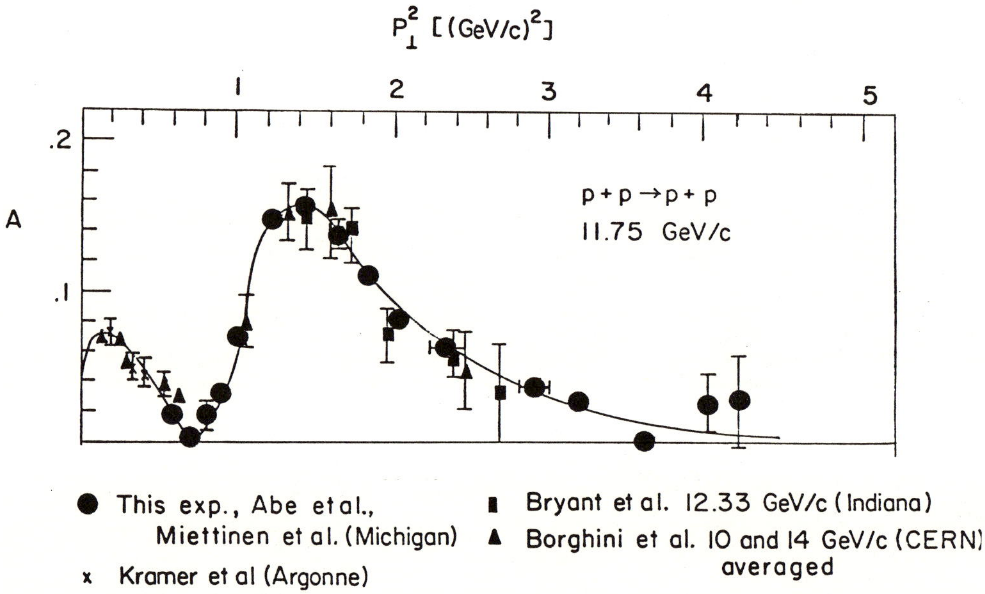

P⊥² [(GeV/c)²]
1
2
3
4
5
.2
A
.1
p + p → p + p
11.75 GeV/c
This exp., Abe et al., Miettinen et al. (Michigan)
Kramer et al (Argonne)
Bryant et al. 12.33 GeV/c (Indiana)
Borghini et al. 10 and 14 GeV/c (CERN) averaged

Fig. 3D

of the recent high energy measurements.

C. p-p Elastic Polarizations at High Energies

In this section of the talk I would like to take some time to review an experiment performed by the Indiana University group at the Internal Target Area at Fermilab to study the elastic polarization in proton-proton scattering at energies between 20 and 200 GeV/c. The measurements were confined to $|t|$ values of .3, .6, .8, and 1.0 $(GeV/c)^2$. A hydrogen gas jet was pulsed at the instant that the energy of the circulating Fermilab proton beam was at the desired value. The recoil protons resulting from the interaction of the protons of the atomic hydrogen with the circulating accelerated beam were detected in a superconducting spectrometer and analyzed in a second scattering from a carbon analyzing target.

The experimental layout is shown in Fig. 4. Elements Q1, Q2, and M are superconducting quadrupoles and dipoles. The acceptance of the spectrometer is approximately 1.4 msr. and the momentum acceptance is approximately 5%. A typical example of the missing mass distribution taken with the spectrometer is shown in Fig. 5.

Upon exiting the spectrometer the recoil protons enter a carbon polarimeter shown in Fig. 6. The incident trajectory is determined in the polarimeter by proportional chambers PC1-2. The trajectory of the proton after scattering in the carbon target C is determined by proportional chambers PC3-4. Hodoscopes HX and HY were utilized to aid in the filtering of tracks during the reconstruction. The analyzing power of the carbon target was determined in a separate calibration experiment conducted at the Argonne ZGS. Consequently, for the determination of the polarization of the incident recoil proton it is only necessary to determine the left-right scattering asymmetry, $\varepsilon = (L-R)/(L+R)$. The polarization P of the recoil protons is then simply ε/A, where A is the analyzing power determined in the calibration experiment for the particular energy of the recoil proton.

One novel feature of this experiment was the electronic

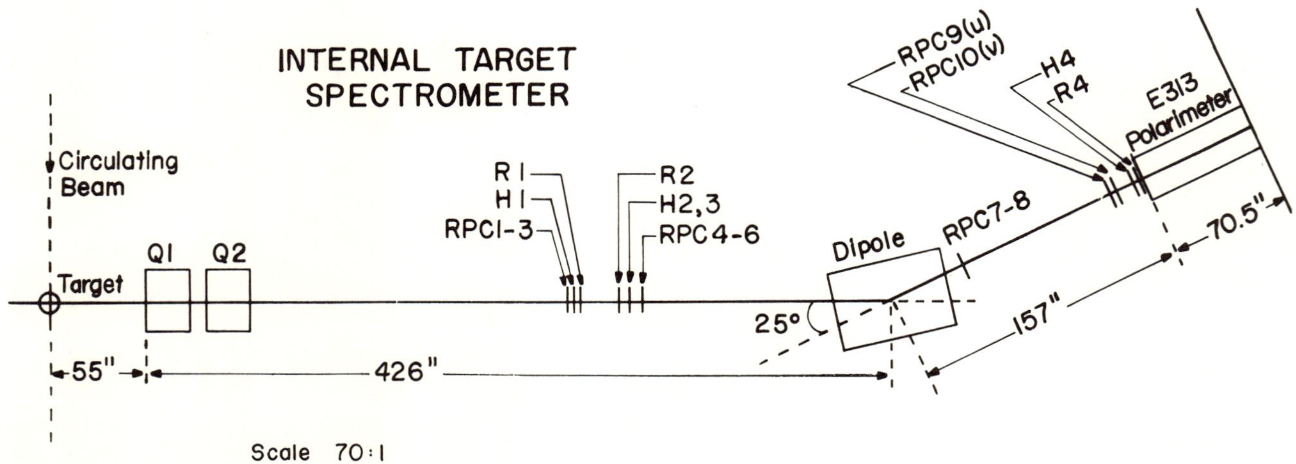

Fig. 4 Layout of Indiana University experiment E313 at Fermilab

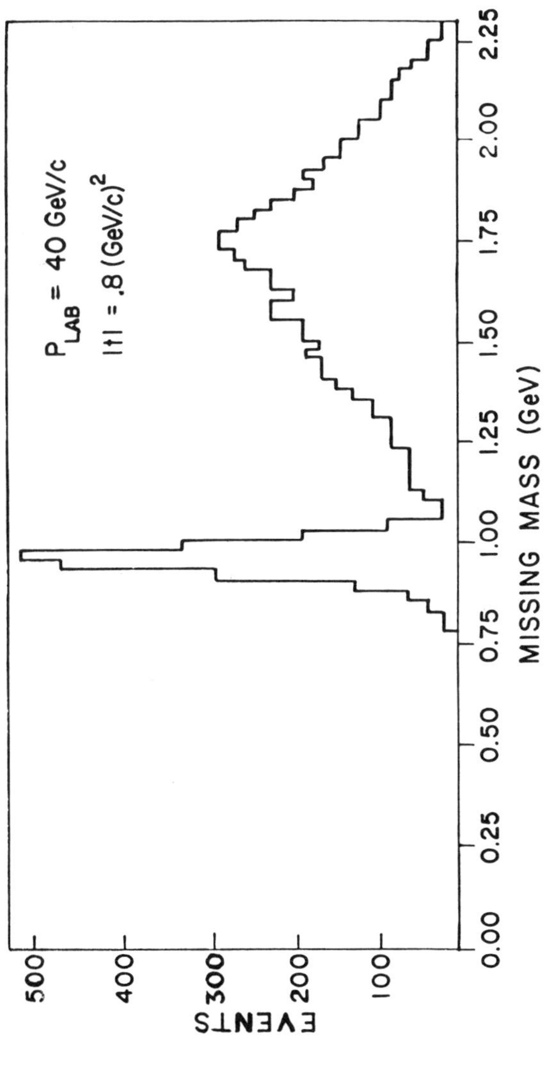

Fig. 5 Typical missing mass distribution in experiment E313.

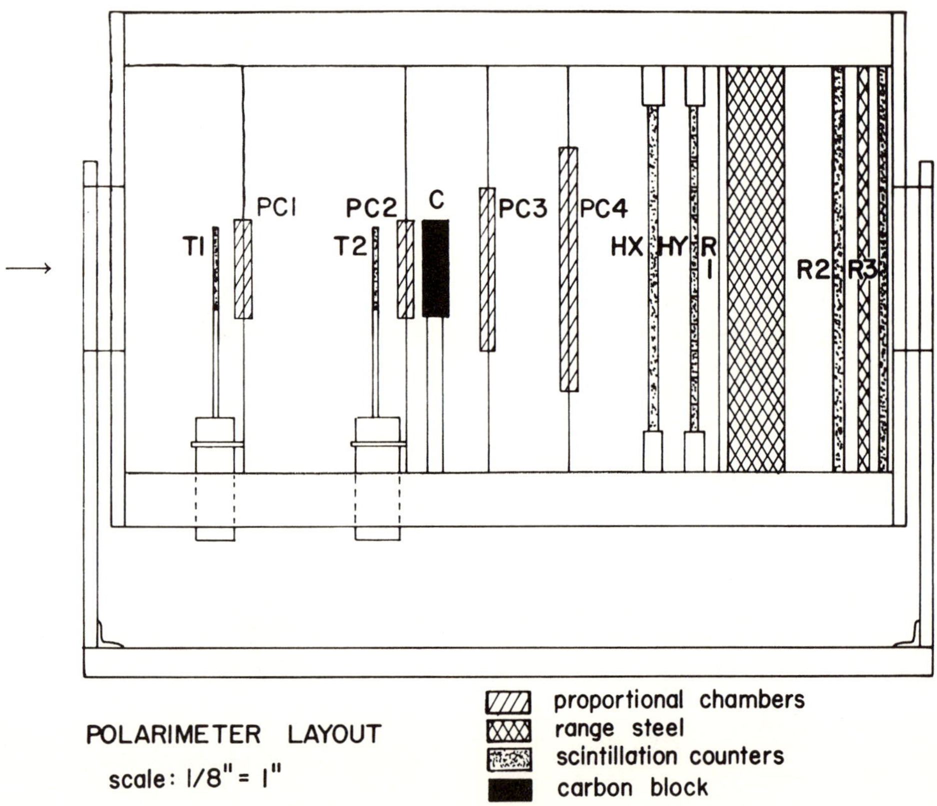

Fig. 6 Layout of carbon polarimeter used in experiment E313.

capability to reject the recording of scatters in the carbon target that were not useful for determining proton polarizations. The overwhelming majority of the protons incident on the carbon target would have preferred to pass through the carbon with the scattering of at most a few degrees. Such small angle scatters are essentially useless for polarization determination and have the effect of introducing substantial deadtime into the data-taking process. In order to avoid this, we constructed a hardware computer that examined the proportional chamber wire status after each trigger and determined if a scatter had occurred through a sufficiently large angle to be of use in the polarization determination. The effectiveness of this device is illustrated in Fig. 7 which is a scatter plot of the scattering angles in the vertical and horizontal directions. The depletion of events in the center of the plot is a direct result of the action of the polarimeter computer. Without the device this region would have been enormously saturated. Numerous tests were conducted to ensure that the polarimeter computer had no effect on the scatters at the larger angles which were used for the determination of the left-right scattering asymmetry.

Some of the tests conducted in the experiment to ensure that no significant instrumental asymmetries remained include the following. A substantial number of pion triggers were recorded and analyzed to detect any evidence of left-right scattering asymmetries. The asymmetry found in these studies was consistent with zero. Furthermore, we had the opportunity to cross-check our measured results with existing data in certain kinematic regions. The agreement found in these cases was satisfactory in all instances. We also were able to determine the up-down scattering asymmetry from the carbon target. This asymmetry, due to the fact that the p-p interaction is a strong interaction and can produce outgoing polarized protons only with the polarization vector normal to the scattering plane, is expected to be zero and was found to be consistent with zero in our data.

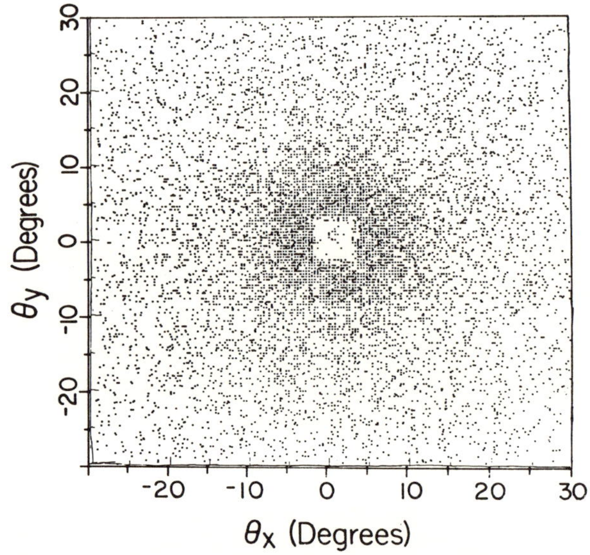

Fig. 7 Scatter plot of horizontal and vertical scattering angles in experiment E313. Depletion in center represents effect of hardwired polarimeter computer which rejects small angle scatters.

The results from the experiment[4] are shown in Fig. 8 along with published low energy data and preliminary data from experiments at FNAL and CERN. We observe in all cases a rapid decline of the polarization at fixed t as one goes to higher s values. As one goes to higher t values, there is evidence that the polarization in p-p elastic scattering becomes negative. This is an unusual phenomena for p-p elastic scattering. Indeed, from very low energies up to energies of twenty GeV, there has been no indication of the polarization in p-p elastic scattering becoming negative. This has the interesting geometrical interpretation that as one sits at a fixed center of mass angle and watches the emerging recoil proton, one observes that the outgoing proton will spin in one direction at the lower energies and will eventually spin in the opposite direction as the incident energy is increased.

In a recent paper by Pumplin and Kane,[5] the consequences were investigated of the Pomeron having quantum numbers different from the vacuum. In this paper they investigated the effects on nucleon-nucleon polarizations of the consequences of this assumption. Predictions typical of those in their paper are illustrated in Fig. 8 as dashed lines.

The possible contributions of graphs such as

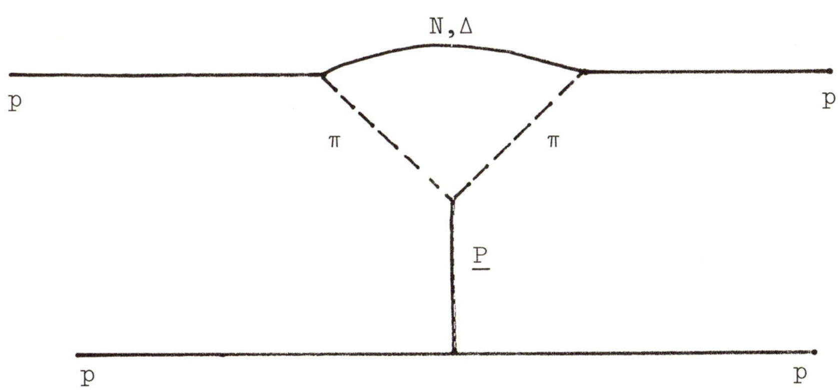

to both the flip (M_{+-}) and non-flip (M_{++}) amplitudes were evaluated. (Here the amplitude subscripts refer to the helicities of the

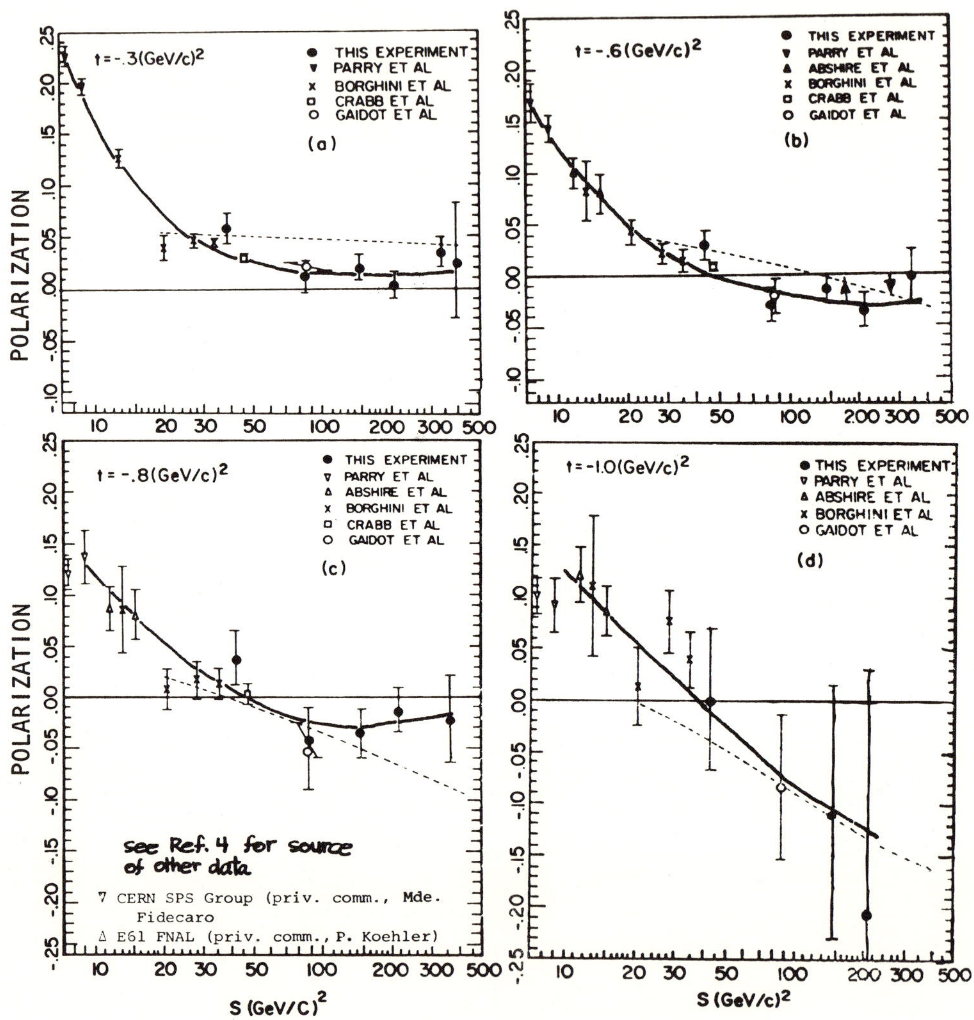

Fig. 8 Results from experiment E313.

upper vertex protons.) Estimates of the polarization were then obtained by fitting the ISR σ_{tot} and $d\sigma/dt$ data with the formula

$$M_{++} = is\ 11.5\ Re^{2.21t}\ J_1\ (R\sqrt{-t})/\sqrt{-t} + is\ 1.94\ r^2 e^{3.93t} J_0(r\sqrt{-t})\ ,$$

$$R^2 = 8.47 + .33\ (\ell ns - i\ \pi/2)\ ,$$

$$r^2 = 2.92\ (\ell ns - i\ \pi/2)\ .$$

The first term in M_{++} is central in the impact parameter and the second term is peripheral in the impact parameter and is presumably due to the two pion tail contribution. The contribution to the flip amplitude by the two pion graph would then be expected to be of the form

$$M_{+-} = isC\ 1.94\ r^2 e^{3.93t} J_1(r\sqrt{-t})\ ,$$

where C is the flip/non-flip ratio calculated from the above graph. Knowldege of M_{++} and M_{+-} permits the calculation of the polarization.

It is of obvious interest that the full implications of this model and its ability to project the general features of high energy polarizations be fully explored.

D. Polarization Due to Coulomb-Nuclear Interference

When one undertakes a study of polarization effects at high energies, it is very important that Coulomb effects are not ignored. This point has been made in a paper by Bourrely and Soffer[6]. They assumed that the non-flip helicity amplitudes received contributions only from the strong interaction and that the single-helicity flip amplitude arises purely from Coulomb effects. In this way they were able to parameterize the polarization resulting from the interference between the Coulomb amplitude and the hadronic amplitude. The maximum value of the polarization that could arise from this interference effect is shown in Fig. 9. It is clear from these

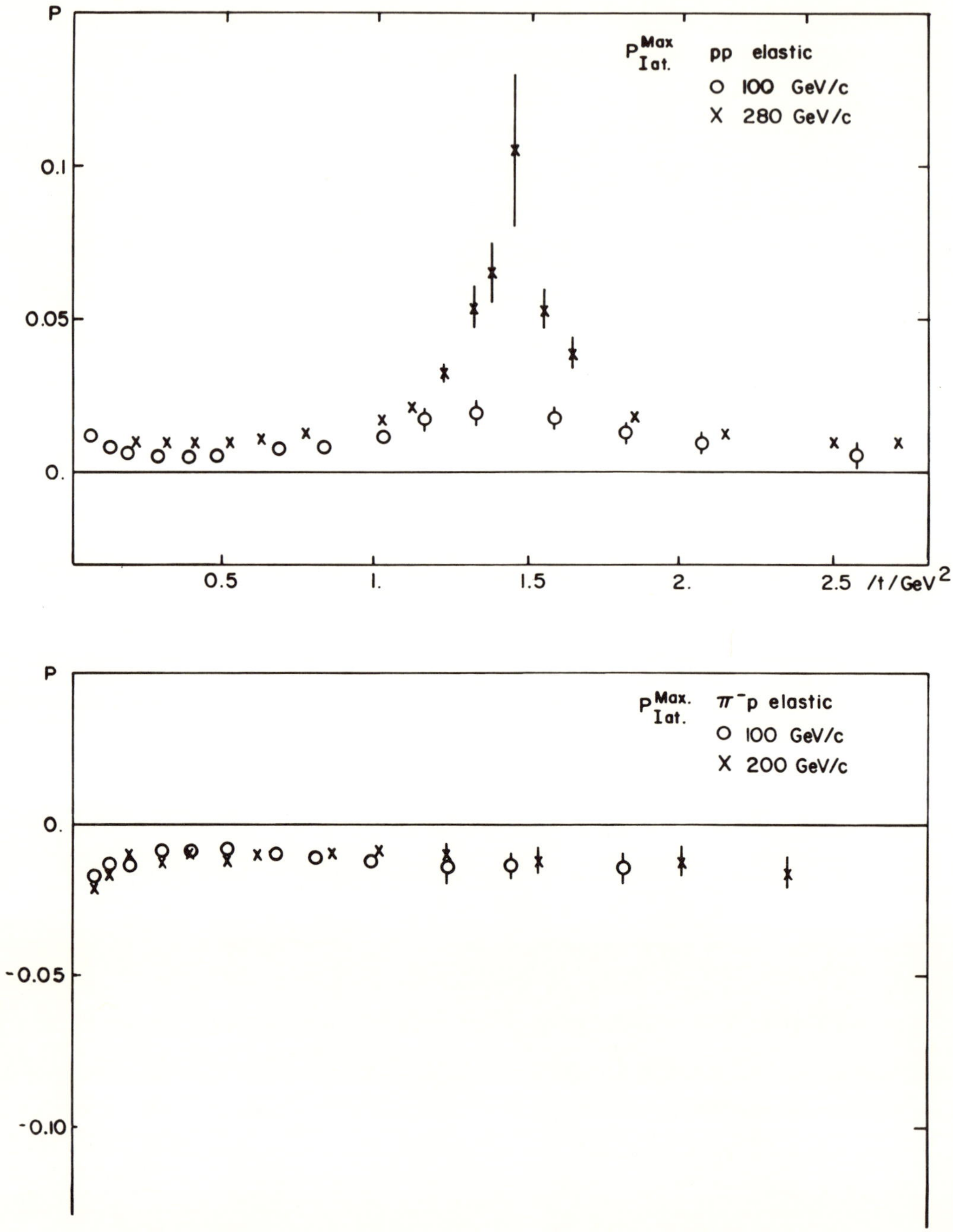

Fig. 9 Maximum polarizations vs. t arising from Coulomb interference effects (Ref. 6).

figures that the interference polarization can be a significant fraction of high energy polarizations.

DEPOLARIZATION IN p-p ELASTIC SCATTERING

I would now like to turn my attention to recent results on the depolarization parameter in p-p elastic scattering at energies above 1 GeV. Whereas the polarization parameter in nucleon-nucleon scattering measures the probability for the target particle to become polarized as a result of the elastic scattering process, the depolarization parameter is a measure of the change in the polarization of the target particle that is due to its interaction with the incident nucleon. It has been only very recently that technology has permitted the measurement of the depolarization parameter in p-p elastic scattering at energies at several GeV. The required experiment is made difficult by virtue of the fact that two spins are involved. The standard experimental technique utilizes a polarized proton target to establish the polarization of the target particle and the use of a carbon polarimeter to measure the polarization of the recoil protons. Knowledge of the target polarization, the polarization of the recoil proton, and the polarization parameter for the interaction permits the calculation of the depolarization parameter D. If the recoiling proton has essentially the same polarization as the polarization of the target proton, the depolarization proton has a value near unity.

The depolarization parameter D can be related to the helicity amplitudes mentioned earlier in the talk through the formula given below:

$$I_o\ (1-D) = \tfrac{1}{2}\ (|\phi_1 - \phi_3|^2 + |\phi_2 + \phi_4|^2).$$

It is clear from this relation that D is sensitive to a different combination of amplitudes than the polarization parameter and, in fact, can represent an independent constraint on the proton-proton system.

Next, let us examine the status of the experimental data on the measurement of the depolarization parameter in p-p elastic scattering at 6 GeV/c. The first high energy t distribution of the depolarization was measured at the Argonne ZGS a few years ago by the Indiana University group[7] (Abshire *et al.*). At that time there was also a measurement at t = .5 $(\text{GeV/c})^2$ by a Michigan-Argonne-CERN-St. Louis group[8] (Fernow *et al.*). Since that time substantially more accurate data have become available in experiments by the Michigan-Argonne-CERN-St. Louis group[9] (Ratner *et al.*) and by the Minnesota group.[10] The situation at 6 GeV/c is shown in Fig. 10. The characteristics of the data are that at small t the depolarization is indeed very close to unity. As one goes to larger t there seems to be a general decline of the polarization--reaching a value of roughly 40% for |t| in the vicinity of 1.0.

If D were actually unity it would imply that

$$\phi_1 = \phi_3$$

$$\phi_2 = \phi_4 .$$

These results are indeed expected at small t since the combinations $\phi_1 - \phi_3$ and $\phi_2 + \phi_4$ can receive contributions only from unnatural parity exchanges and such exchanges are believed, on the basis of independent evidence, to be vanishingly small for the p-p system.

Thus, the p-p depolarization parameter seems to be rather well understood at the moment and to be consistent with rather general theoretical expectations. It will, of course, be of great interest to pursue the depolarization measurements to higher momenta. In fact, such measurements have been made in the E61 experiment at Fermilab and we can look forward to seeing those results in the near future.

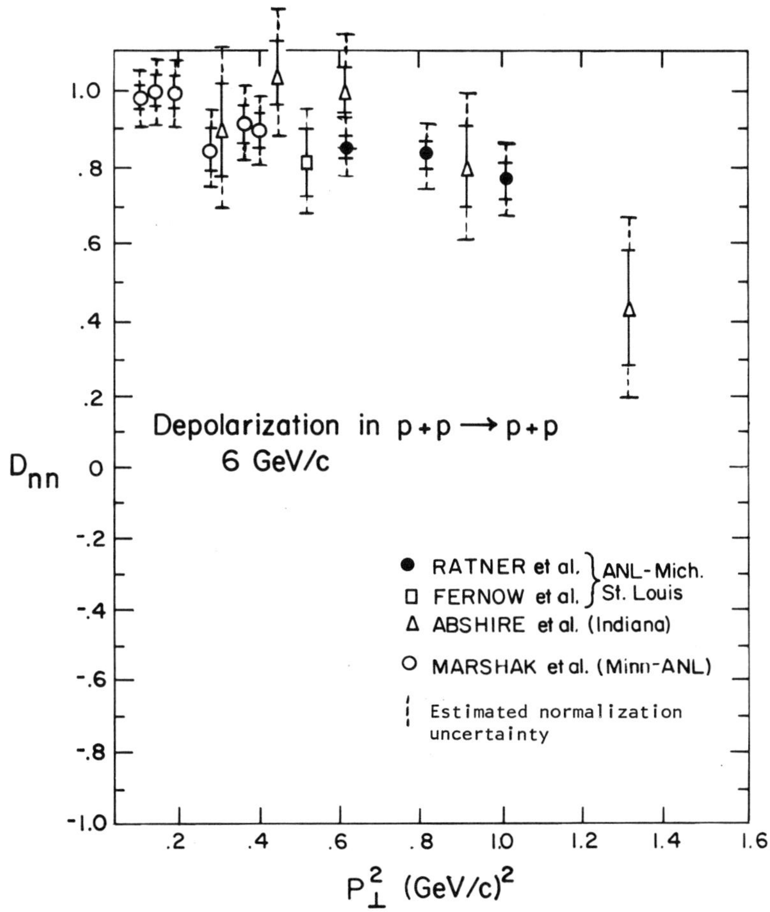

Fig. 10 Depolarization in p-p elastic scattering at 6 GeV/c (Ref. 7-10).

Higher Order Spin Measurements in p-p Elastic Scattering

Much of the progress made in the study of spin effects in p-p elastic scattering at energies below 20 GeV during the past year has been in the area of the higher order spin parameters C_{nn}, C_{ss} $[\rightarrow I(\vec{s},\vec{s};o,o)]$, and $C_{\ell\ell}$ $[\rightarrow I(\vec{\ell},\vec{\ell};o,o)]$.

Fig. 11 shows the present status of knowledge of C_{nn} from 2GeV/c to 12 GeV/c.[11,12] The variation in magnitude and shape over this range is impressive. The structure at 12 GeV/c is particularly notable and this is one of the major effects to surface in the last year--it is begging for an interpretation. How can a proton care so much (dynamically) about which direction the other's spin is pointing?

Another major advance in the study of the p-p system over the past year has been the extraction of accurate data on C_{ss} and $C_{\ell\ell}$ by the Yokosawa group at Argonne[12]. The present status of these measurements is illustrated in Fig. 12. Such measurements are absolutely essential to the process of extracting the various spin amplitudes, as will be seen in a later section.

Neutron-Proton Polarization Data

For many years the subject of neutron-proton polarization at high energies was essentially ignored and most of the effort was concentrated on studying the proton-proton system. The reason for this was simple. Neutrons are harder to produce, transport, polarize, detect, and analyze than are protons. Over the years, however, these problems have been overcome and we have just witnessed the accumulation of a rather substantial amount of high quality neutron-proton cross section data and polarization data. I will now spend some time briefly reviewing the status of these data.

The proton-proton system is in a pure isospin state of I = 1. Consequently, no amount of studying of this system can reveal any information on the I = 0 amplitudes. The n-p system, on the other hand, is a mixture of I = 1 and I = 0. Therefore, in principle, the combination of the complete n-p and p-p data will permit the isolation

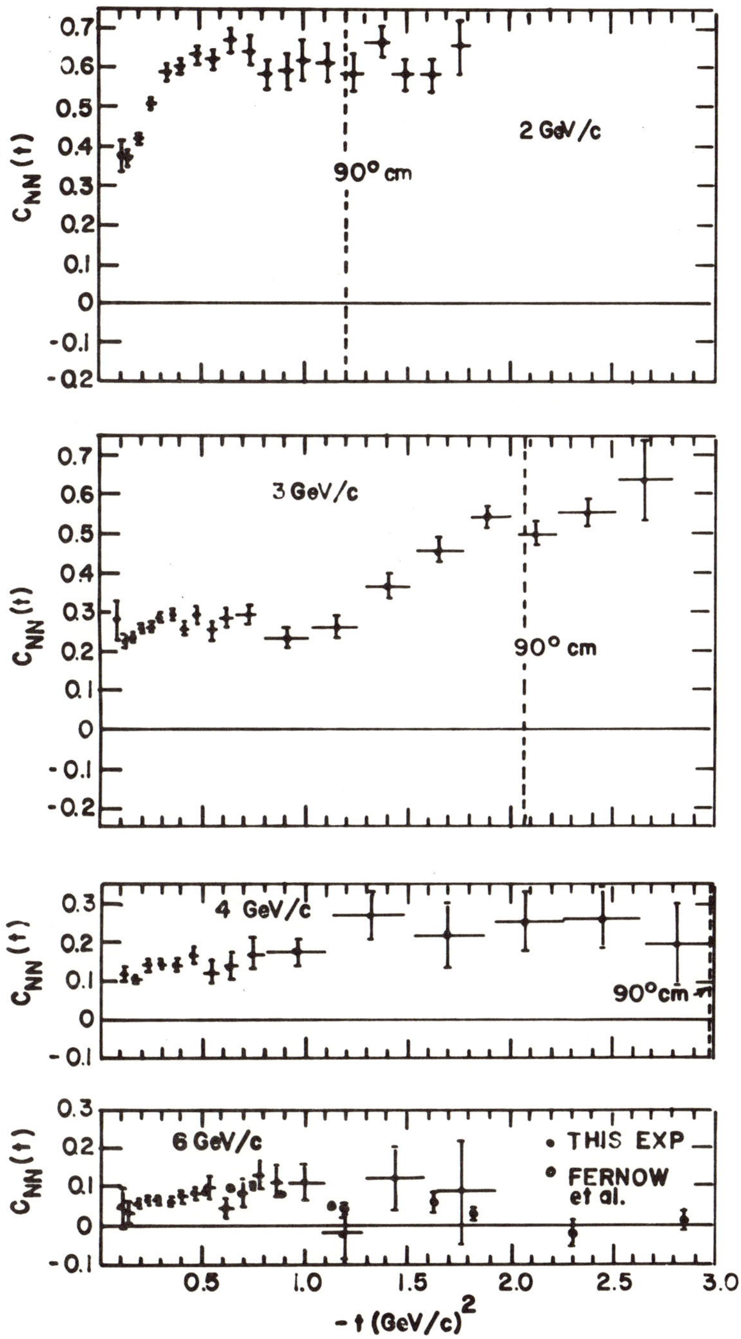

Fig 11A

Fig 11 Measurements of C_{nn} from 2 to 12 GeV/c (Refs. 11,12).

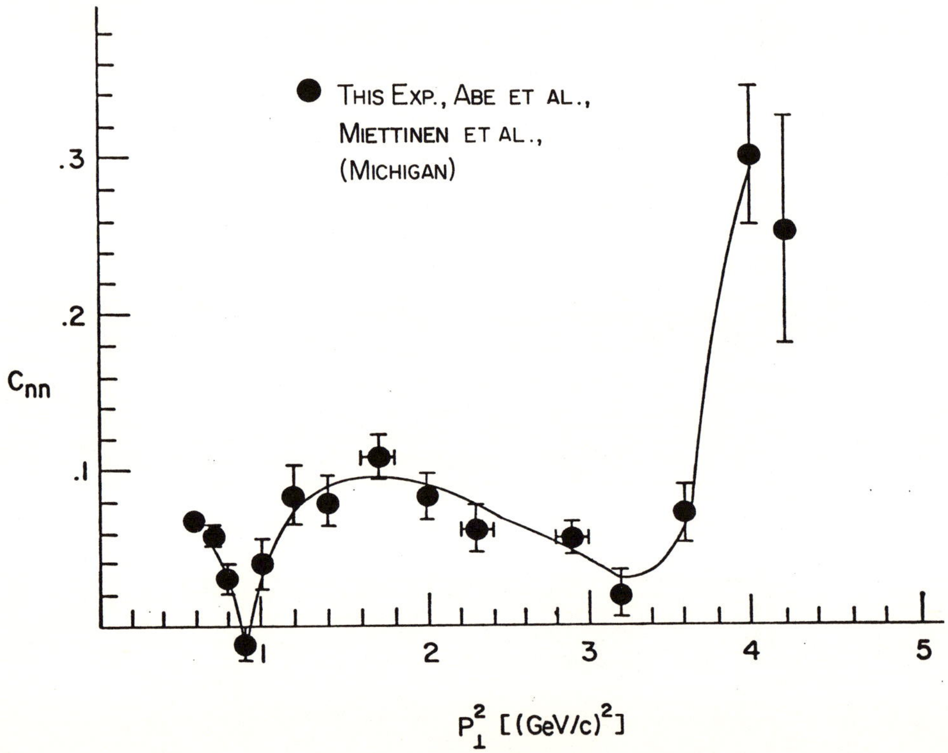
THIS EXP., ABE ET AL.,
MIETTINEN ET AL.,
(MICHIGAN)
.3
.2
.1
Cnn
1
2
3
4
5
P⊥² [(GeV/c)²]

Fig. 11B

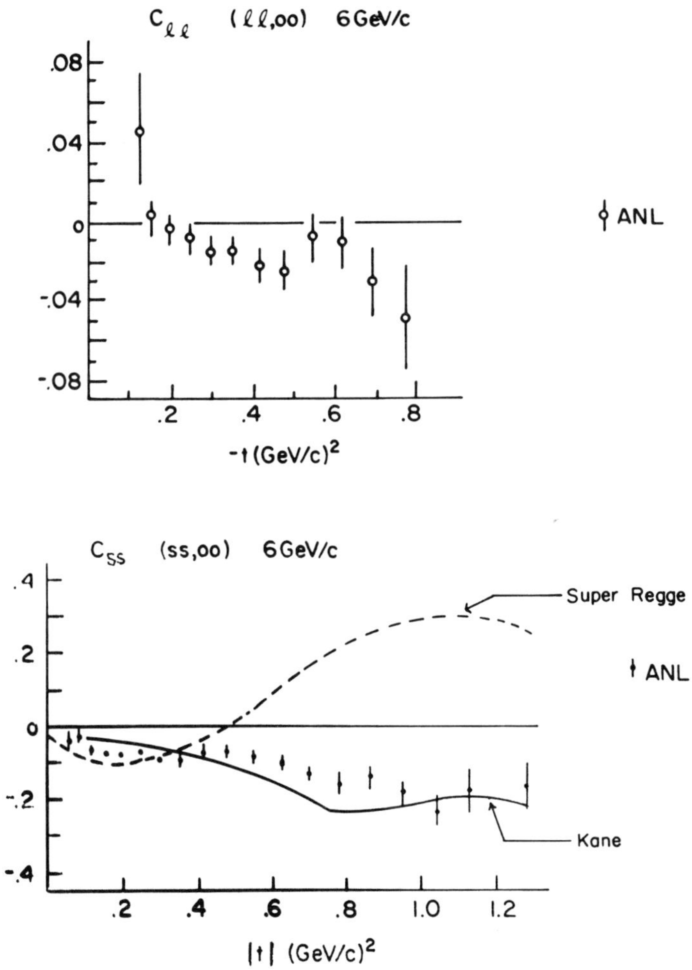

Fig. 12 Measurements of C_{ss} and $C_{\ell\ell}$ (Ref. 12).

of the I = 0 and I = 1 amplitudes. Fig. 13 shows the n-p differential cross section from 5 GeV to 12 GeV taken in an experiment at the Argonne ZGS by a University of Michigan group[13]. One observes that at each energy there is a forward diffraction peak, then a fall-off with increasing t that is somewhat less gradual, a rather flat region as one approaches large angles, and then a rising cross section as one approaches the backward peak. Note that it is meaningful to talk about the differential cross section in n-p elastic scattering beyond 90° in the center of mass since in this case we do not have the complication of identical particles as in the p-p case. Fig. 14 shows the results of measurements by this same Michigan group[14] in an experiment at Fermilab which extended the measurements to 370 GeV/c. The general features are the same: A diffraction peak, and a change in slope in the vicinity of 1 to 1.5 GeV/c. At present we are a long way from understanding all that these measurements are trying to tell us. One feature, however, that is noticeable is that because of the similarity of the slope of the forward diffraction peak in the n-p and p-p cases the radius of interaction of the p-p and n-p systems are essentially the same and are approximately one fermi. Next let us take a look at the situation with the polarizations. The Argonne Effective Mass Spectrometer group[15] has, over the past few years, studied spin effects in both p-p and n-p elastic scattering from 2 to 12 GeV/c. The incident proton beam utilized in these experiments was polarized parallel or anti-parallel to the normal of the scattering plane. The incoming beam was incident on either neutrons or protons in a deuterium target. The results from these experiments are shown in Fig. 15. One notices that at 2 GeV/c the p-p and p-n polarizations are almost the same--but not quite. As we go from 2 GeV/c to higher momenta, the p-n polarization changes structure at a rather impressive rate. At 12 GeV/c the p-p polarization remains at about .05 over the t range explored while the p-n polarization has essentially vanished over this entire region.

In an optical model one would expect the polarization for p-p

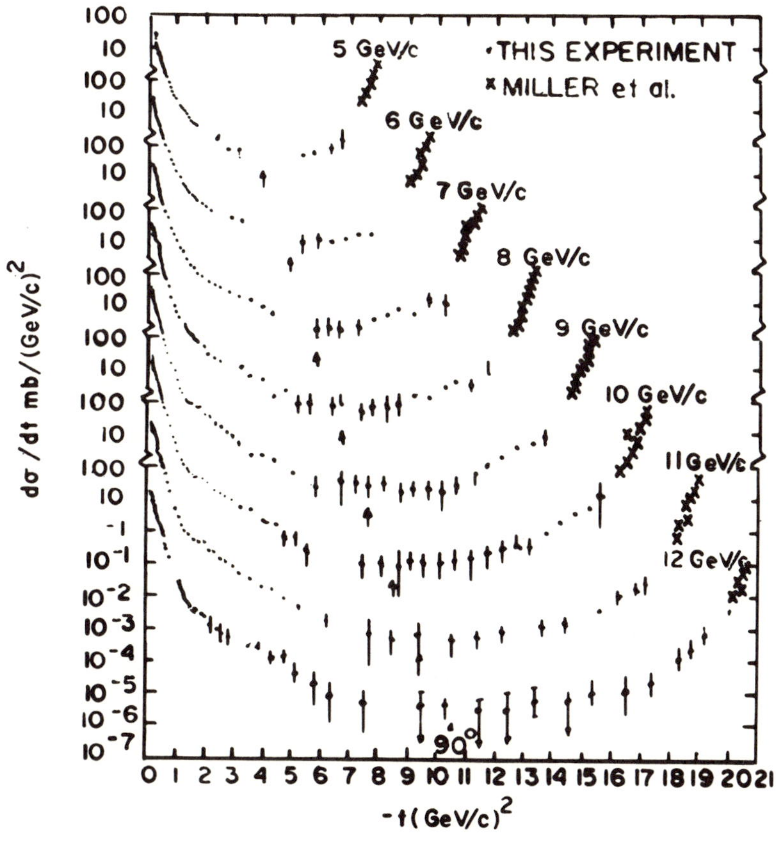

Fig. 13 n-p differential cross sections at Argonne energies (Ref. 13).

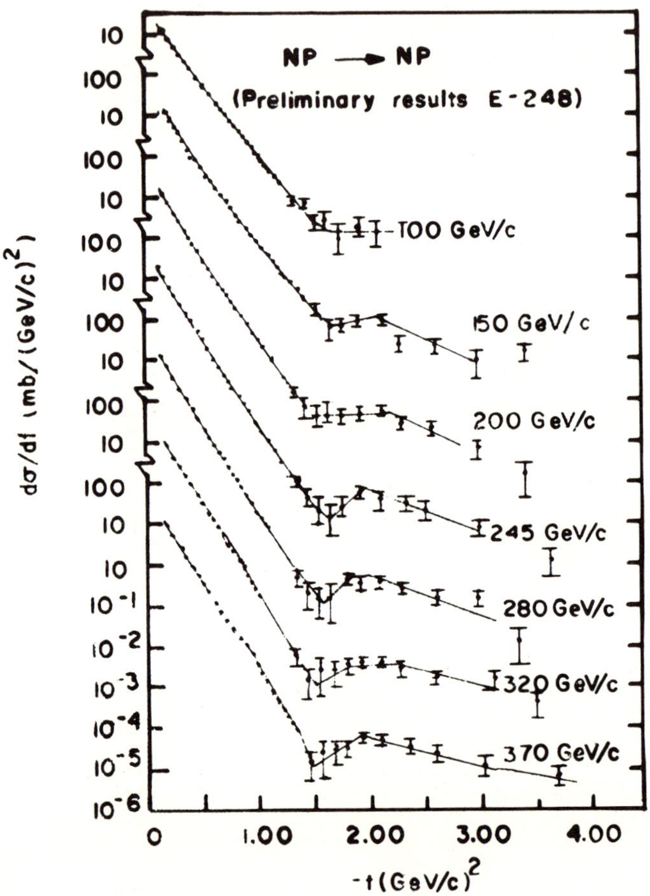

Fig. 14 n-p differential cross sections at Fermilab energies (Ref. 14).

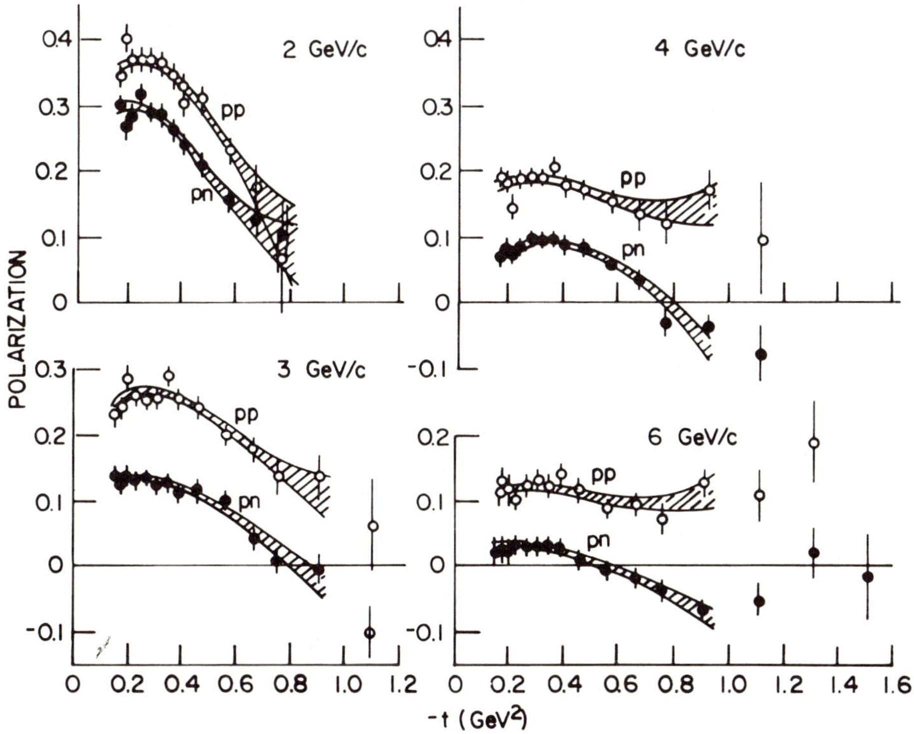

Fig. 15A p-p and n-p elastic polarizations from 2 to 12 GeV/c. (Ref. 15).

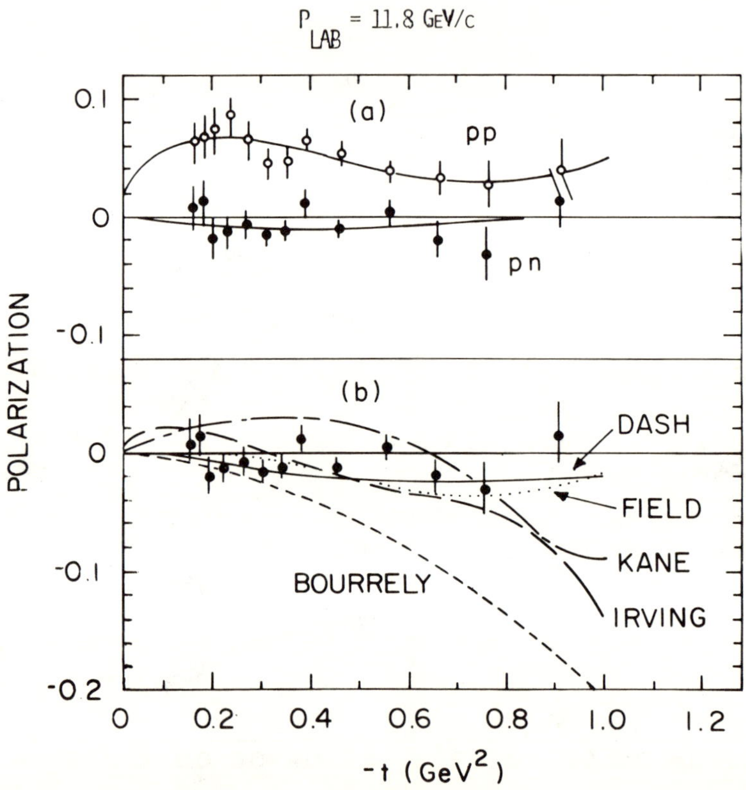
P LAB = 11.8 GeV/c
(a)
pp
pn
(b)
DASH
FIELD
KANE
IRVING
BOURRELY
POLARIZATION
0.1
0
-0.1
-0.2
0.2
0.4
0.6
0.8
1.0
1.2
-t (GeV²)

Fig. 15B

and p-n elastic scattering to be identical in shape. In a Regge model one would expect mirror symmetry just as is observed in the case of π^+p and π^-p elastic scattering. It is clear from the data presented above that neither one of these cases pertains exactly for p-p and p-n polarization. There is the general trend of moving from an optical type model at the lower energies to a Regge type picture at the higher energies.

If we designate the non-flip diffractive amplitude by N_o^o and calculate the components of the flip amplitude N_1^1 orthogonal to N_o^o as follows:

$$N_{1_\perp}^{o} = [(P\sigma)_{pp} + (P\sigma)_{pn} - \tfrac{1}{2}(P\sigma)_{cex}]/(4|N_o^o|)$$

$$N_{1_\perp}^{1} = [(P\sigma)_{pp} - (P\sigma)_{pn}]/(4|N_o^o|)\text{, where}$$

$$|N_o^o|^2 \approx (\sigma_{pp} + \sigma_{pn})/2,$$

we can display the single flip amplitudes for each of the two isospin states as a function of P_{LAB} at fixed values of t.[15] This is shown in Fig. 16. One is struck by the rapid fall-off of the I = 0 amplitude and this is the subject of a significant amount of current theoretical activity. One possible explanation of this effect requires a low lying Regge trajectory given by $\alpha = -.5 + .9t$. The rapid variation of N_1^o can not be explained as being due only to the interference of a normal Regge amplitude with a Pomeron flip contribution.

<u>The Diffraction Region: Where does it end?</u>

For many years the sharp peak in the differential cross section at small t in elastic scattering processes has been attributed to geometric diffraction effects and the corresponding region in t is called the diffraction region. The region beyond the diffraction peak has been loosely called the "large angle" region and has been assumed to be representative of constituent or exchange processes.

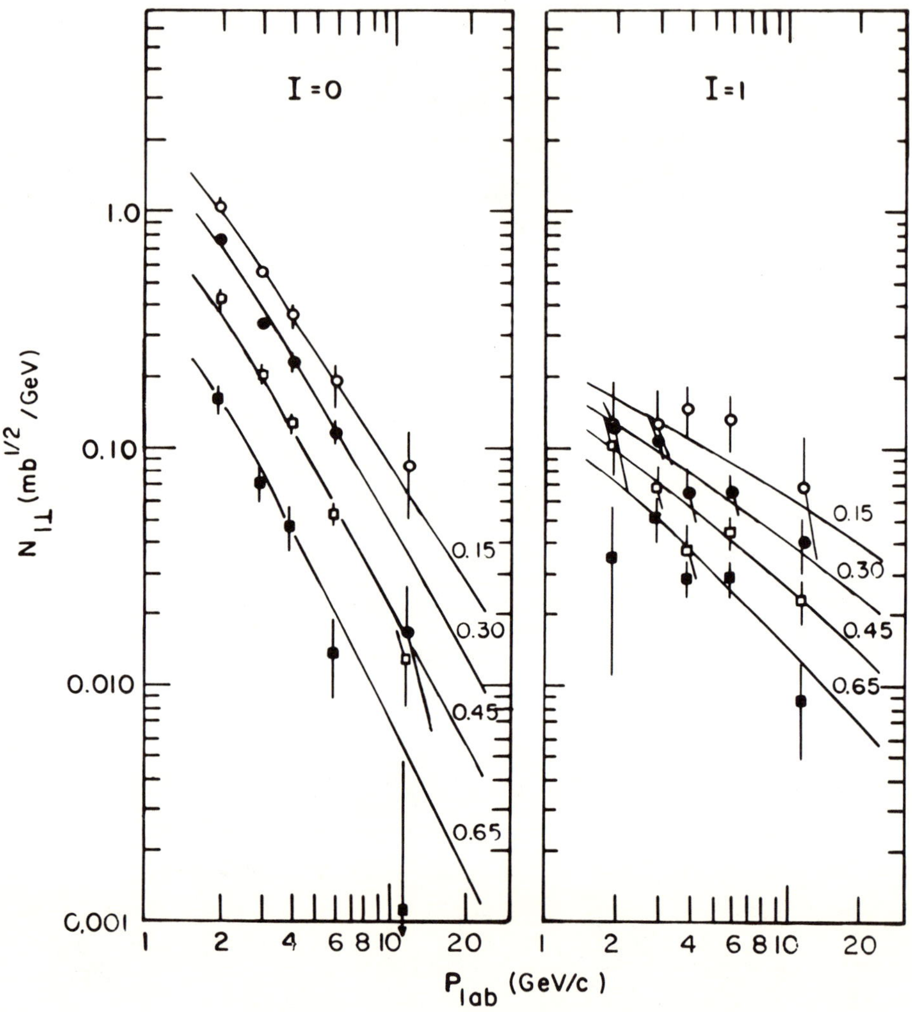

Ref. 16 Dependence of nucleon isospin spin flip amplitudes at fixed t as a function of P_{LAB} (Ref. 15).

This point of view has recently been challenged by Sukhatme.[16] Indeed, he claims that the p-p ISR data out to $|t| = 10\ (\mathrm{GeV/c})^2$ may be predominantly diffractive, whereas traditional wisdom would have the diffraction region stopping at $|t| \simeq 2.5\ (\mathrm{GeV/c})^2$. Below we will present his arguments.

The p-p elastic scattering amplitude is frequently parameterized as follows:

$$A(s,t) = P(s,t) + C(s,t) \quad ,$$

where P(s,t) represents the Pomeron (diffractive) contribution and C(s,t) is the large angle contribution. P is almost purely imaginary, it dominates at small angle and has dip structure. On the other hand, C is assumed to be smoothly varying, due to other than diffractive processes, and approximately real. It is of interest to know where $|P(s,t)| \simeq |C(s,t)|$, for this can give a rough definition of the boundary between the "diffractive" and "large angle" regions. Sukhatme utilized the phase, magnitude derivative relation to determine the value of the phase angle:

$$\phi = \tan^{-1} \frac{\mathrm{Im}A}{\mathrm{Re}A} \simeq \frac{\pi}{2} - \frac{\pi}{4} \quad \frac{d}{d(\ln s)} \ln \frac{d\sigma}{dt}$$

using the existing ISR p-p data. The result of this analyticity test is that A is still predominantly imaginary even at large $|t|$. This is then taken as evidence that the diffraction region may extend even out to $|t| = 10\ (\mathrm{GeV/c})^2$ and beyond. It is suggested that perhaps the regimes are separated by fixed θ_{cm} rather than fixed t or p_t.

Sukhatme finds that the ISR data can be accounted for in diffraction models if inelastic diffraction is taken into consideration. This discussion could be relevant to the development of polarization models for the presently accessible $|t|$ range.

Evidence of a Dibaryon Resonance

There has been evidence recently published by Auer et al.[17] in Physics Letters 67B, 113 (1977) for the existence of a resonance in the p-p system. This information is terribly exciting for at least four reasons: 1) the discovery of any resonance is certainly noteworthy, 2) the discovery of a resonance at a lower energy accelerator is particularly exciting these days since it is generally assumed that all such resonances have already been found and that the new frontiers are now at the highest energy machines, 3) there are traditional theoretical reasons why such resonances should not exist, and 4) the new evidence points up the power of utilizing spin as a tool for investigating spectroscopy.

Very good p-p total cross section data has existed in the region of a few GeV's for many years. The general shape of the cross section is illustrated in Fig. 17. As one can see, one of the most impressive characteristics of this plot is the rapid rise in the total cross section between 1 and 1.5 GeV/c. This rise has been explained by Mandelstam and others as being due to the production of the N (1238) through a single pion exchange process. The production of the higher isobars such as the N(1512) and the N(1688) appears to be substantially suppressed relative to that of the N(1238). This suppression can be understood in a model where one assumes that the dominant exchange process is that of single pion exchange. There does appear to be a small peak in the total cross section data near 3 GeV/c. This peak is consistent with N(1688) production. Also, there is some evidence for a shoulder near 2 GeV/c that might correspond to the production of the N(1512). Normally, the existence of a particle or resonance in a system reveals itself as a significant peak in the total cross section energy distribution. For example, the resonances associated with the pion-proton system at intermediate energies are evident in Fig. 18. Obviously, there is no such pronounced peaking in the proton-proton system. It is fair to say that until very recently the high energy physics community was quite

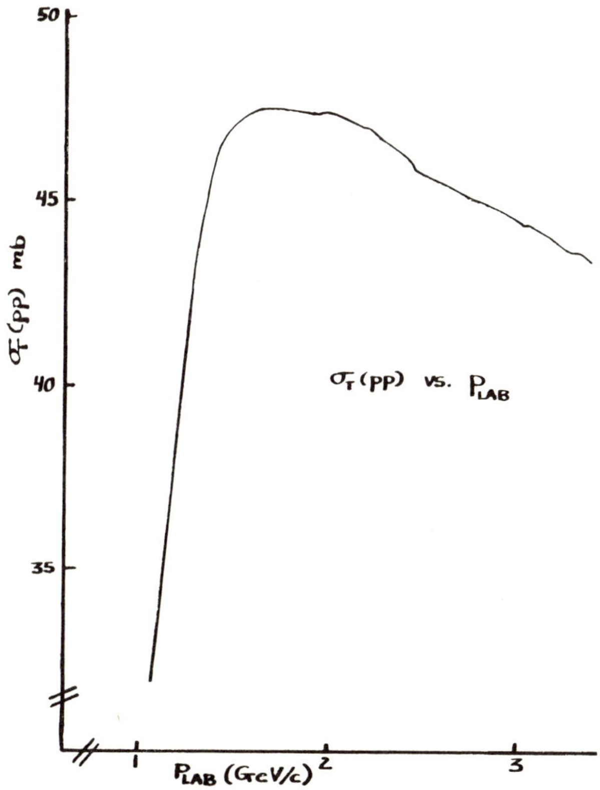

Fig. 17 Total cross section in p-p scattering (from Bugg *et al.*, Phys. Rev. **146**, 980 (1966).

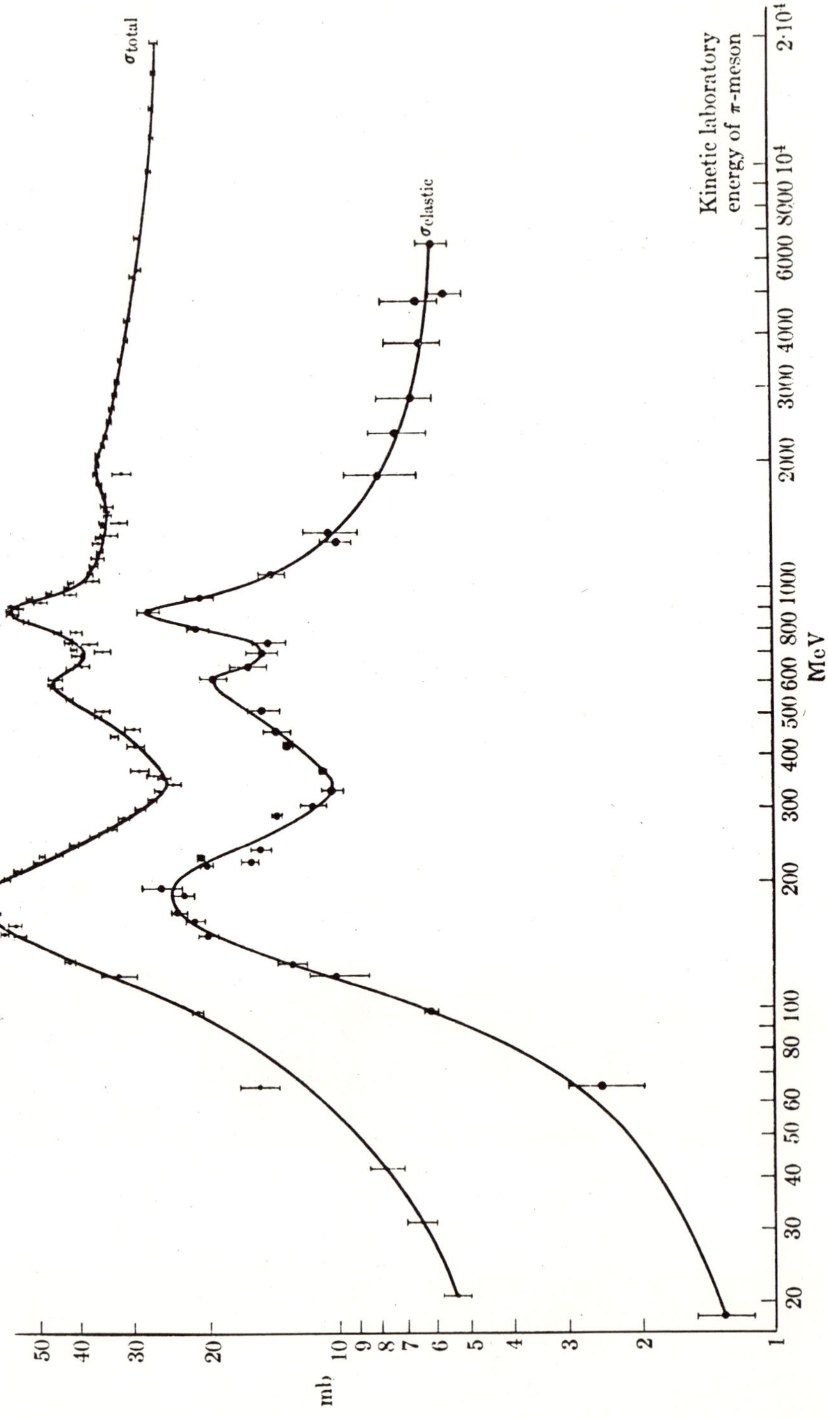

Fig. 18 Pion-proton total cross section features.

well prepared to assume that all of the structure observed in the p-p total cross section could be explained in terms of the opening of various inelastic channels. The opening of various inelastic channels is what is presumed to be responsible for keeping the total cross section at higher energies roughly constant in the face of a declining elastic total cross section.

Now let us take a look at what is new in the proton-proton system as ascertained from total cross section measurements. All of the total cross section measurements that I have referred to earlier were made without regard to the direction of the projectile or target spin. In the experiment that we will discuss, both the spin of the target particle and the incident projectile were controlled. The Argonne polarized proton beam was made incident on a polarized proton target and experimenters in the Yokosawa group measured the total cross section when the two spins were longitudinally aligned and when the two spins were longitudinally anti-parallel. The experimenters observed that the parallel spin cross section has a peak near 1.5 GeV/c while the anti-parallel spin cross section is relatively flat. Furthermore, at the position of the peak in $\sigma_{\rightrightarrows}$, it is found that $\sigma_{\rightrightarrows}$ is 50% larger than $\sigma_{\leftrightarrows}$ (Fig. 19). This difference is surprisingly large and naturally leads one to question whether or not the effect might be due to the production of a resonance state of the two incident protons. If so, the mass of the dibaryon system would be roughly 2250 MeV. This resonance would have an electric charge of two units and a baryon number of 2.

In our examination of the total cross section for unpolarized beam and target above, we observed that the rapid variation in the region of 1 to 1.5 GeV/c could be explained by the production of the N(1238). It is thus natural to inquire as to whether the observed structure in the polarized cross section could somehow be related to the N(1238). The first problem associated with this interpretation is that the N(1238) would be expected to produce an effect at a momentum lower than 1.5 GeV. Secondly, the N(1238) would be

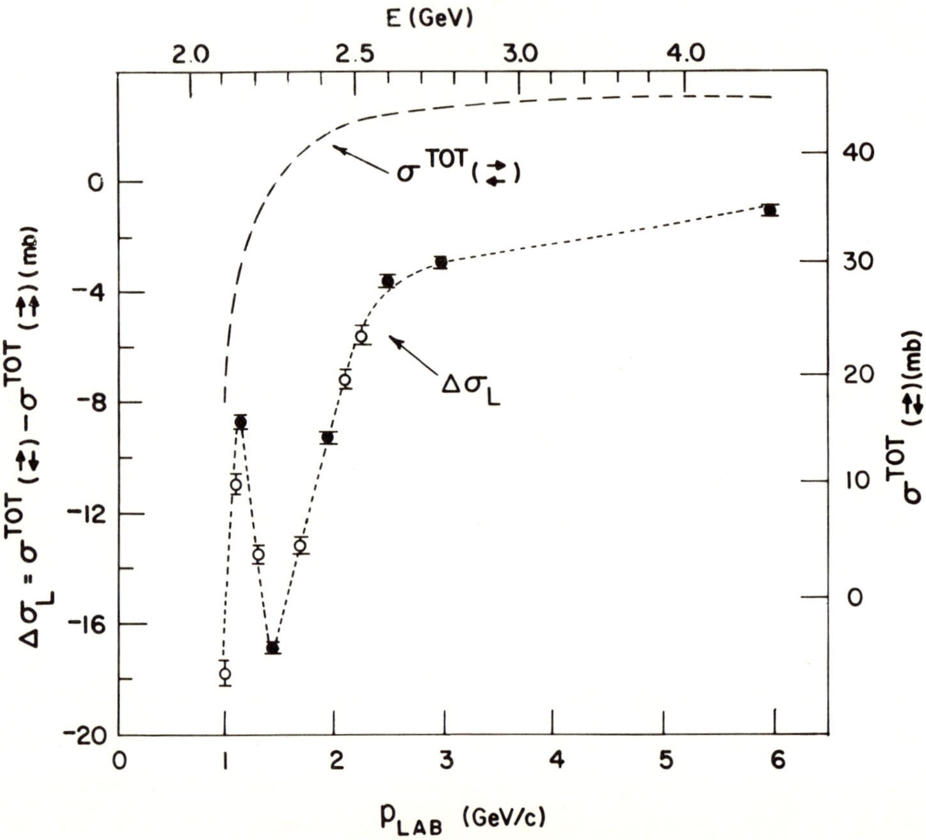

Fig. 19 Difference in longitudinal total p-p cross sections; evidence for a dibaryon resonance (Ref. 17).

expected to contribute only to the singlet spin state $^1D_2$ rather than in the triplet spin state which contributes to $\sigma_{\vec{\rightarrow}}$. Recall that it is $\sigma_{\vec{\rightarrow}}$ that shows the striking structure. By making a Legendre coefficient analysis of the differential cross sections and polarizations in p-p elastic scatterings in the neighborhood of 1 to 2 GeV/c, the authors find that their observations are consistent with a standard Breit-Wigner resonance in the $^3F_3$ state with a mass of 2250 MeV, a width of 200 MeV, and elasticity of the order of 20%, and with quantum numbers of $J^P = 3^-$.

In my opening remarks in this section I alluded to the fact that such a resonance does not fit the traditional theoretical expectations. I would like to take a few moments now to clarify that statement. The standard quark model has been amazingly successful in accounting for the large variety of particles which have been discovered. The assumed quantum numbers of the quarks are such that any known meson can be represented as a system of two of the quarks. Furthermore, any observed baryon can be characterized by a group of three of the quarks. It has long been assumed that these are the rules that govern the formation of physical elementary particles. Indeed, very ambitious experimental programs have been mounted to search for particles which did not fit into this classification. Such hypothetical particles were given the name "exotic". To data there has been no conclusive evidence of the existence of exotic elementary particles. Obviously, the existence of a diproton resonance would require an alteration in the way we view the structure of elementary particles. To make up such an object would require more than the minimal number of quarks. One possibility is, of course, the existence of quark molecules and, in fact, some models presently predict the existence of such objects.

Before we leave this topic I would like to spend a few minutes discussing the possible relationship between such a dibaryon resonance and the observed breaks in the fixed angle proton-proton differential cross sections. Several years ago in an Indiana paper by Kammerud *et al.*[18] we examined the feasibility of explaining breaks we had observed in the fixed angle cross section in terms of dibaryon resonances as a follow-up on suggestions made by Predazzi and Libby.[19] The p-p differential cross sections presented in that paper are illustrated in Fig. 20 and we can see clear evidence for some significant change in the behavior of the cross section at t values near .8, 3.0, and 6.5 GeV/c. We stated that if one assumed that the change in the slope of the fixed angle cross section was due to the crossing of the threshold for the production of a diproton resonance, the mass of the first resonances would be at 2.2 GeV/c. This is a remarkable occurrence and may point the way for explaining the structure in the fixed angle cross section. Members of the Indiana group are presently studying this question.

Λ Hyperon Inclusive Polarization Measurements

In an experiment performed by Bunce et al.[20] at Fermilab, significant polarization effects were observed in the inclusive production of Λ hyperons at 300 GeV. The experiment examined lambdas produced at production angles between 0 and 9.5 mrad in the interaction of 300 GeV protons with a Be target. The polarization of the inclusively-produced lambdas was determined from the observation of the lambda decay into protons and pions. The experimental layout is shown in Fig. 21. The proton beam enters from the left, is bent by magnet M1 onto the beryllium target T. Magnet M2 serves as a sweeping magnet to remove charged particles from the beam. V is a decay vacuum area. The pions and protons from the lambda decay are detected by proportional chambers 1-6. Magnet M3 is used to determine the charge and momentum of the decay products. By manipulating the field in magnet M1 and in other magnets not shown in the figure, the experimenters were able to choose eight production angles ranging from -2.5 mrad to +9.5 mrad.

When a lambda hyperon decays into a proton and a pion, the direction of the proton momentum relative to the spin of the lambda in its rest frame is governed by the following formula:

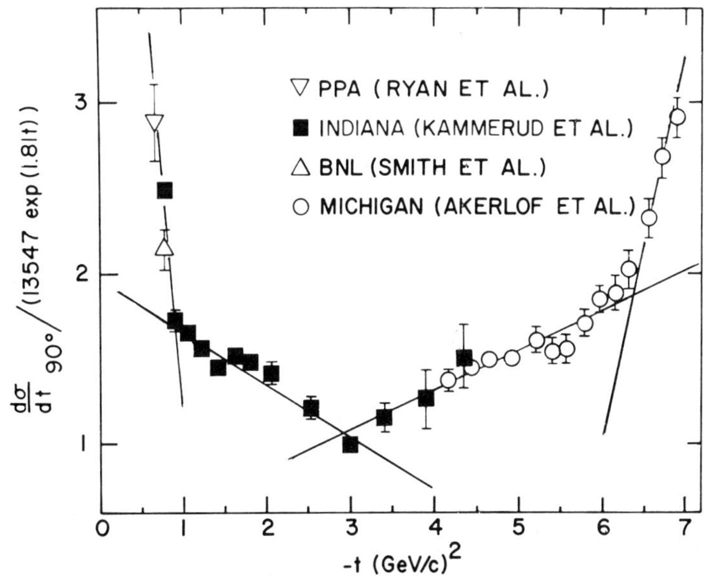

Fig. 20 Differential p-p cross section at $\theta_{cm} = 90^{\circ}$ observed in Indiana University experiment (Ref. 18,25) divided by exponential in t.

(a)

(b)

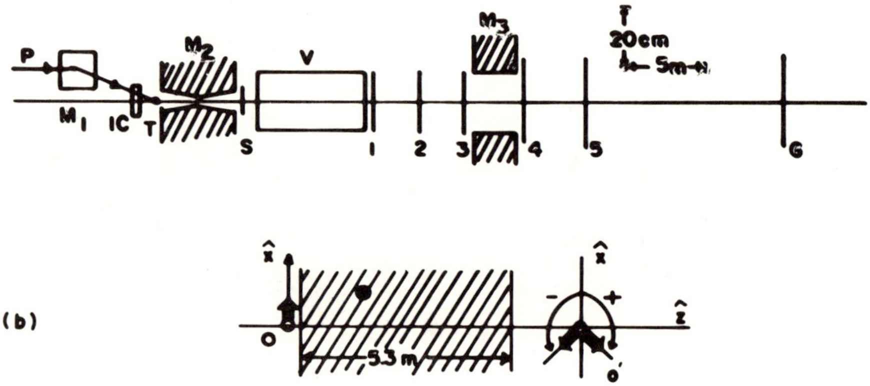

Fig. 21 Experimental layout for inclusive lambda polarization experiment (Ref. 20).

$$\frac{dN}{d\Omega} = (1 + \alpha P\cos\theta/4\pi,$$

where θ is the angle between the proton momentum and the lambda spin, P is the magnitude of the lambda polarization and $\alpha = .64 \pm .013$. Consequently, an observation of the angular distribution of the proton resulting from the lambda is sufficient to permit the determination of the polarization of the parent lambdas.

I will not go into the details of the experimental procedure of the various tests that were made to insure that all observed asymmetries were due to the physical process of lambda production and decay. The interested reader is referred to Physical Review Letters 36, 1113 (1976). The results of studies by this group are illustrated in Fig. 22 (which also includes their 24 and 400 GeV/c results), where the polarization is plotted against the transverse lambda momentum. One sees the very surprising results that, at large p_t, the polarization reaches the value of almost 30%. This result is initially surprising because it is only in rare circumstances that one observes polarization effects above 25% at any energy above a few GeV. Another feature of this result, however, creates a cause for even more excitement. Normally, inclusive processes are thought to occur as a result of many competing and random production mechanisms. It is very difficult to believe that particles can be produced inclusively with large polarization in such an enviroment. Indeed, the observation of large polarization may be a principle indicator that such production processes take place in an orderly, non-random fashion. There are several questions raised by this experiment that simply must be answered in the near future. For example, does the almost linear increase in polarization with p_t continue until the polarization saturates at unity? How important is the target Be to the results of the experiment? Would the result be the same if the experiment were conducted off of a hydrogen target? What would be the result of a similar experiment conducted on the sigma hyperons? How does the inclusively-produced lambda

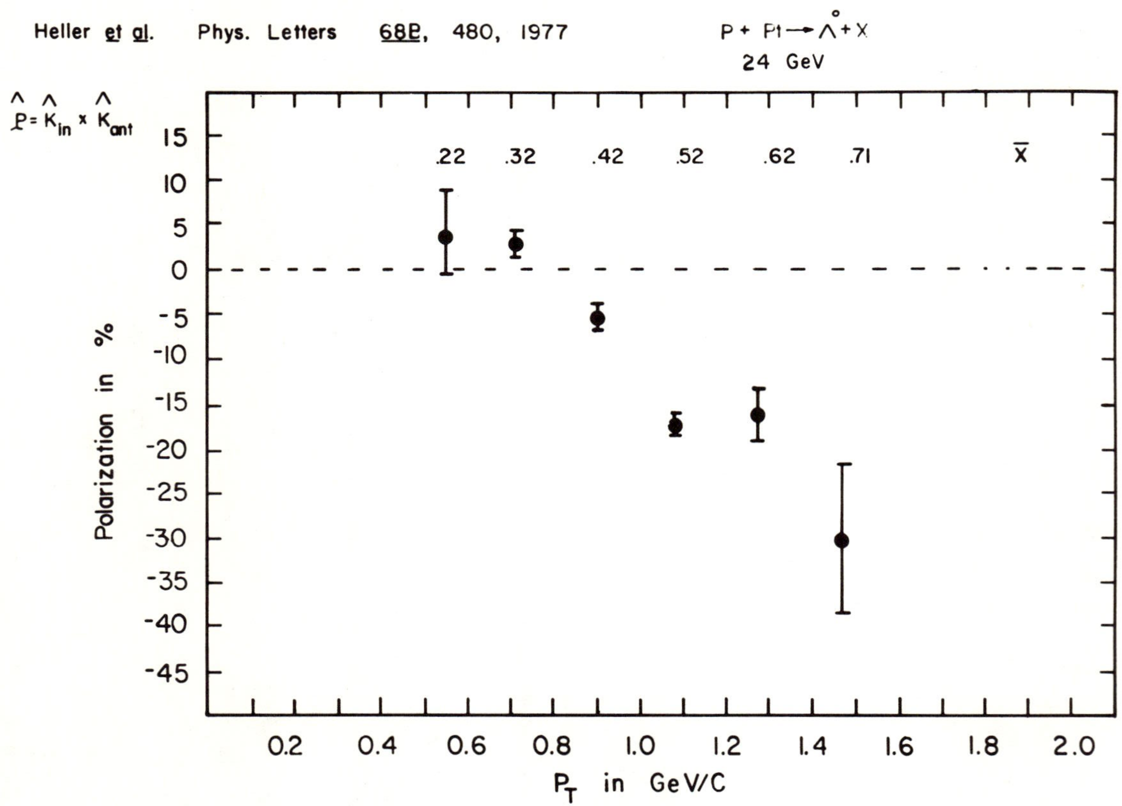

Fig. 22 A Results from inclusive lambda polarization experiment (Ref. 20).

polarization compare with corresponding proton data? In a later section of the paper, I will have some additional comments to make on these questions as we compare the lambda results with the results obtained in a recent Indiana University experiment at Fermilab.

Polarization in Inclusive Proton Production

The Indiana University group is presently engaged in a program to study the inclusive polarization in pp → pX at Fermilab energies. The apparatus being used for these measurements is identical to that described above for the p-p elastic polarization studies in the Internal Target Area at Fermilab. The initial measurements were made using a rotating carbon filament target at incident proton energies of 100, 200, 300, and 400 GeV/c at a transverse momentum of 1.1 GeV/c and x = .88.

The study of proton inclusive processes is a natural extension of the elastic polarization studies. Indeed, as shown in Fig. 23, inclusive final states can be viewed as a massive particle recoiling against the observed proton. For x values near 1, the inclusive cross sections can be calculated as the square of a triple Regge amplitude. Spin flip and non-flip terms can interfere in the triple Regge picture to produce a polarization of the observed proton.

The inclusively-produced protons traverse the superconducting spectrometer and are incident on a carbon analyzing target shown in Fig. 6. The left-right scattering asymmetry of these protons from the carbon analyzer determines the polarization. The analyzing power for the carbon was previously determined in a calibration experiment performed in the Argonne polarized beam. Numerous checks were made to insure that all instrumental asymmetries were cancelled to a high degree.

The results of these measurements[21] are shown in Fig. 24 along with the lambda inclusive polarization results of Bunce _et al._[20] One notes that the lambda and proton polarization are of the same sign and are of the same order of magnitude. The curve shown in

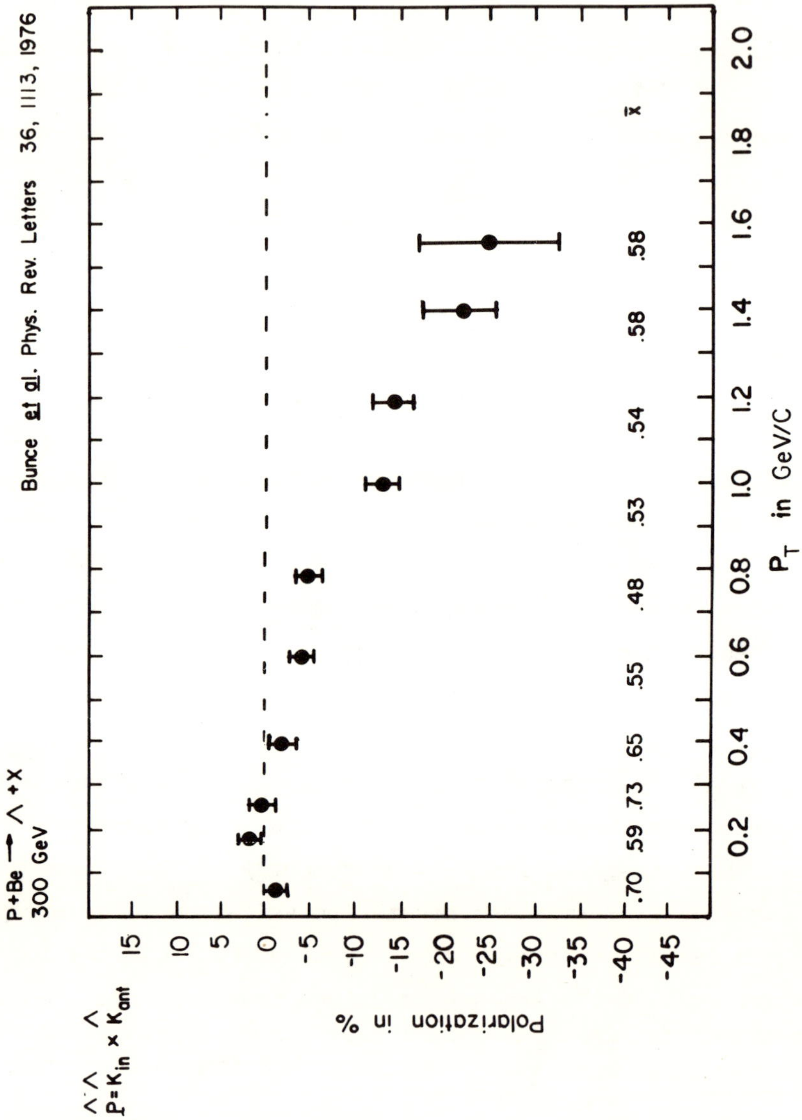

P+Be → Λ° + X
300 GeV
Bunce et al. Phys. Rev. Letters 36, 1113, 1976
P̂ = K̂in × K̂ant
Polarization in %
15
10
5
0
-5
-10
-15
-20
-25
-30
-35
-40
-45
0.2
0.4
0.6
0.8
1.0
1.2
1.4
1.6
1.8
2.0
P_T in GeV/C
.70 .59 .73 .65 .55 .48 .53 .54 .58 .58 x̄

Fig. 22B

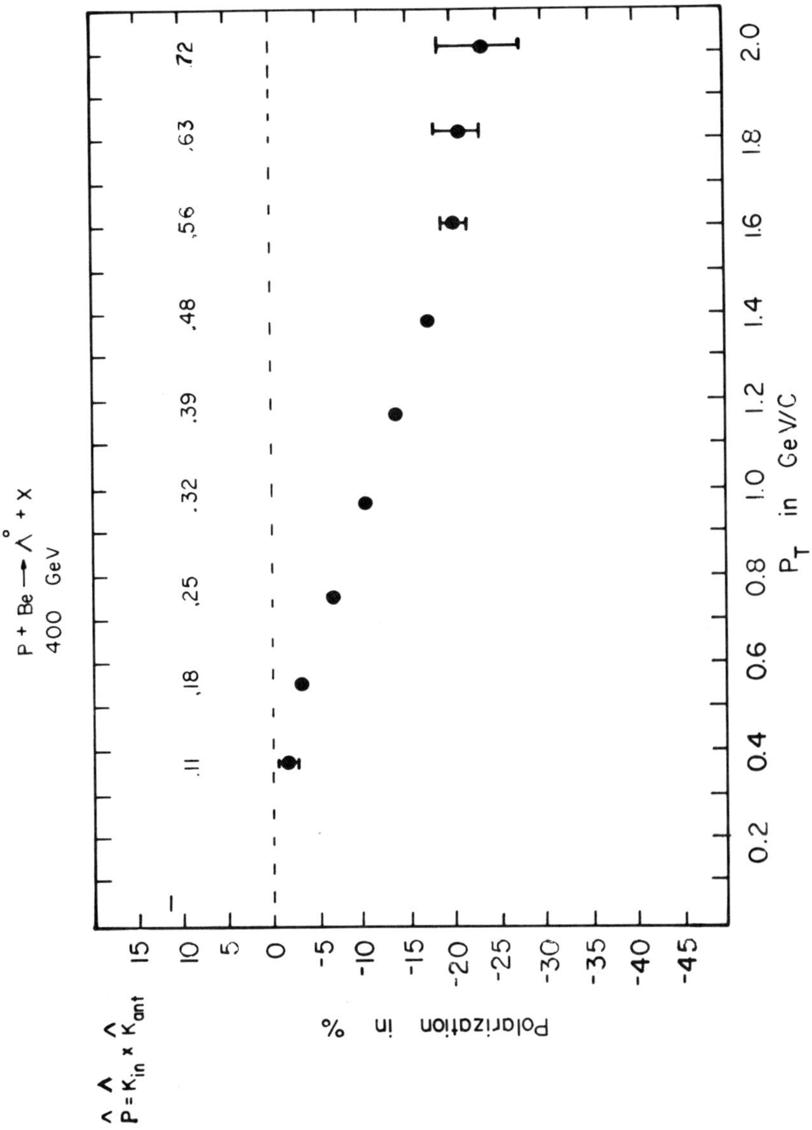
P + Be ⟶ Λ° + X
400 GeV
P̂ = K̂in × K̂ant
Polarization in %
15
10
5
0
-5
-10
-15
-20
-25
-30
-35
-40
-45
.11
.18
.25
.32
.39
.48
.56
.63
.72
0.2
0.4
0.6
0.8
1.0
1.2
1.4
1.6
1.8
2.0
P_T in GeV/C

Fig. 22C

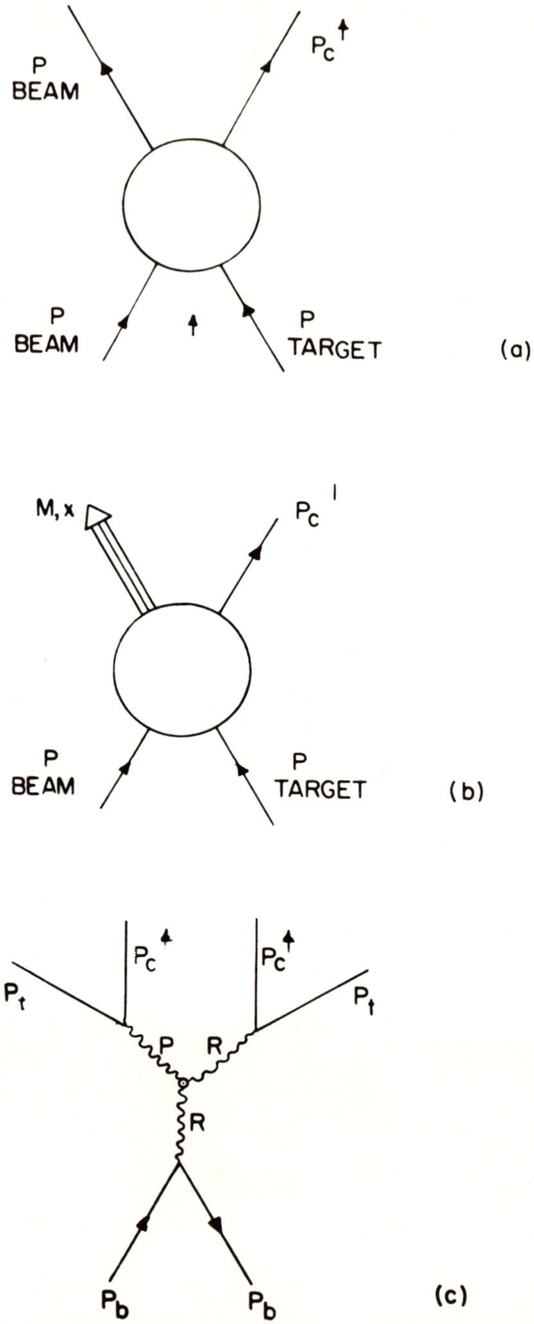

Fig. 23 Illustration of proton inclusive process.

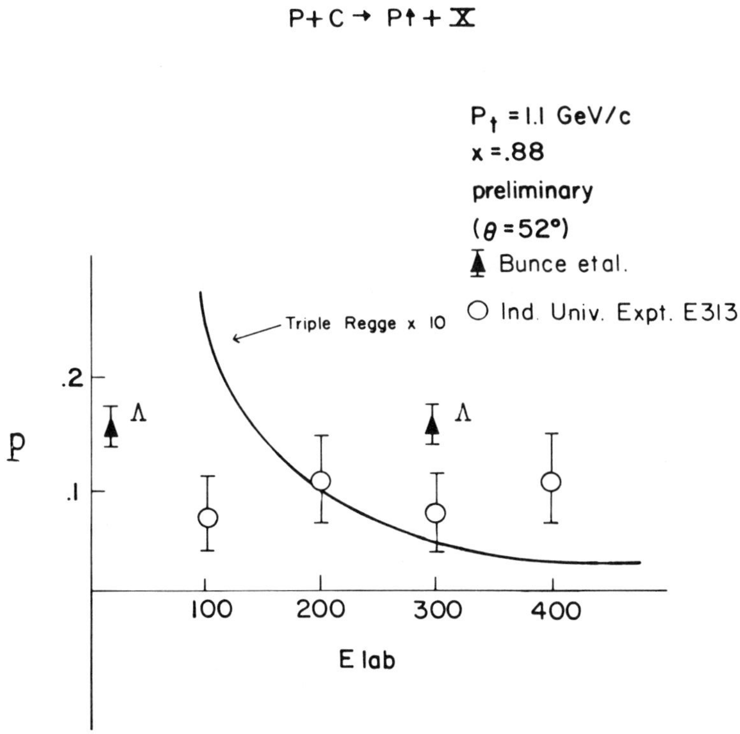

Fig. 24 Inclusive proton polarizations observed in Indiana University experiment E313 at Fermilab (Ref. 21).

this figure is a triple Regge prediction of the type predicted by the calculations of Field and Fox, Salin and Soffen, and Paige and Sidhu.[22] The absolute value of the curve is at least a factor of ten greater than the current prediction of the triple Regge model. The results shown here are not in agreement with the magnitude or energy dependence of any simple triple Regge model. It is interesting to speculate that the proton polarization seen in these measurements may well be related to the large values of the polarization seen in inclusive lambda processes. In the discussion in the next section, we will indeed present a model due to Heller[23] and to Kane and Yao[24] that relates the lambda inclusive polarizations and the proton inclusive polarizations. Before leaving this section, I should comment that we are presently engaged in a major run at Fermilab that should expand the range over which the inclusive proton polarizations are known to higher transverse momenta. It is our hope to have the data from this experiment analyzed by summer.

Theoretical Model of Baryon Inclusive Polarization

In a recent preprint by Ken Heller[23] at the University of Michigan, a mechanism is proposed that allows large polarizations for baryons which are fragments of the projectile or target particle. It is assumed that when one of the quarks in the projectile baryon is scattered by a target nucleon, a vector particle is radiated similar to the process that occurs in bremsstrahlung. Furthermore, the radiated particle is assumed to create a quark-antiquark pair. Then one nember of this pair joins the other two quarks of the projectile to create the outgoing observed baryon. The process is illustrated in Fig. 25. In this example, we are visualizing the inclusive production of a forward lambda hyperon by an incoming proton. The lambda is produced through the interchange of one of the u-quarks in the incoming proton with an s-quark produced by the radiated vector particle.

If it is assumed that the coupling of the quarks to the appropriate vector gluon is independent of color, one may write the wave

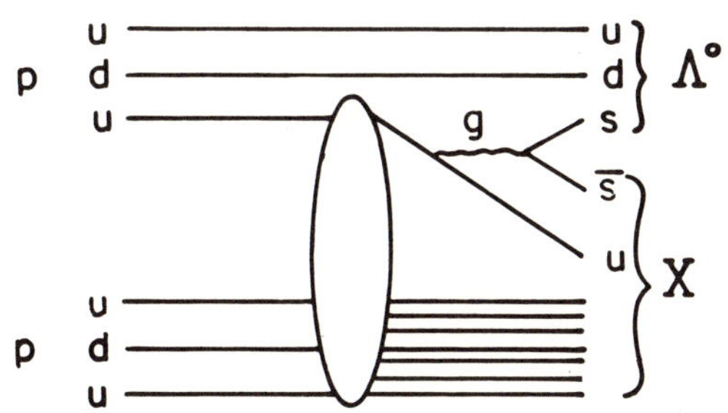

Fig. 25 Illustration of inclusive model of Heller (Ref. 23).

as

$$P_i = \sqrt{2}[u_i u_i d_j - u_i u_j d_i]$$

$$\Lambda_i = \sqrt{3}[u_i d_j s_i - u_j d_i s_i] ,$$

where j is the other spin index.

For an incident proton to produce an outgoing Λ, the unscattered u and d quarks are assumed to be in a singlet state in the incident proton since these quarks are not affected by the proposed transition mechanism. Therefore, the transition matrix for the production of a spin i lambda by a spin i proton is as follows:

$$\langle \Lambda_i|T|p_i\rangle = \sqrt{6}[\langle u_i d_j s_i|T| u_i d_j u_i\rangle - \langle u_j d_i s_i|T| u_i d_j u_i\rangle$$

$$- \langle u_i d_j s_i|T| u_i d_i u_j\rangle + \langle u_j d_i s_i|T| u_j d_i u_i\rangle] .$$

The second and third terms in this expansion are assumed to be negligible since they require spin index changes. Pursuing this calculation leads to the conclusion that, in this model, the polarization of the produced Λ is unity. In this scheme the observation of Λ polarizations smaller than unity would be explained by the dilution created by other competing Λ production processes which decrease in prominence at higher p_t.

This model has other substantial capabilities which we will not investigate in detail here. However, the ratio of polarizations for various baryon production processes is of interest and is reproduced in Fig. 26. One should note that the prediction is that proton inclusive polarization should be approximately one-third smaller than those of the lambda inclusive polarizations. In a more detailed study by Kane and Yao[24] (University of Michigan Report

BARYON POLARIZATION

Reaction	Polarization	Ratio of Polarization to Lambda Polarization
$p+p \rightarrow \Lambda^{\circ} + X$	1	1
$p+p \rightarrow p+X$	$\dfrac{3[(u_i,u_i)+(d_i,d_i)]^2 + [3(u_i,u_j)-(d_i,d_j)]^2}{5[(u_i,u_i)+(d_i,d_i)]^2 + [3(u_i,u_j)-(d_i,d_j)]^2}$	$\dfrac{2}{3}$
$p+p \rightarrow n+X$	$\dfrac{8(d_i,u_i)^2 + 9(d_i,u_j)^2}{10(d_i,u_i)^2 + 9(d_i,u_j)^2}$	$\dfrac{17}{19}$
$p+p \rightarrow \Sigma^{\circ}+X$	$\dfrac{-3(s_i,u_i)^2 + (s_i,u_j)^2}{5(s_i,u_i)^2 + (s_i,u_j)^2}$	$\dfrac{-1}{3}$
$p+p \rightarrow \Sigma^{+}+X$	$\dfrac{(s_i,d_j)^2}{2(s_i,d_i)^2 + (s_i,d_j)^2}$	$\dfrac{1}{3}$
$p+p \rightarrow$ antibaryon $+X$	0	0

Fig. 26 Predicted ratio of various inclusive baryon polarizations in model of Heller (Ref. 23).

HE7-44) it is predicted that the polarization in inclusive nucleon production and lambda production are in the ratio of the non-strange and strange constituent quark masses. From the paper of De Rujula et al. (Physical Review D 12, 147 (1975)) this ratio is predicted to be .62. The SU(6) correction calculated by Heller to account for the extent to which one quark carries the spin of the hadron is two-thirds. These two effects should be combined to obtain the actual ratio of the polarization in pp → pX vs. pp → ΛX. The result of tnis process is that the proton polarizations should be a factor of .42 of the inclusive lambda polarizations. Obviously, there is substantial interest in studying the kinematic region over which relations of this type might be expected to hold. For example, it is important to know whether the inclusive polarization for protons rises with p_t in a manner similar to that for lambda inclusive polarizations. Also, over what range of s values does such a correspondence hold? As can be seen from Fig. 27, the predictions of Heller, Kane, and Yao are remarkably satisfied by the existing data.

Elastic Nucleon-Nucleon Polarization in a Constituent Model

Several years ago in what we believe to be the first attempt to explain the polarization phenomena in a constituent model, Holger Nielsen at the Niels Bohr Institute in Copenhagen and I proposed a scheme to test the dominance of constituent scattering in elastic proton-proton processes.[25] In brief, the model assumed that the various regions of the fixed angle proton-proton differential cross section are due to different numbers of binary parton collisions occurring in the p-p process. If one examines the fixed angle 90° differential cross section as a function of t (Fig. 28), one observes rather distinct breaks near t = - 1, t=-3, and t=-6.5$(GeV/c)^2$. Let us suppose that the first region from 0 < t<-1 is due to the single scattering of one parton in the incident proton from one parton in the target proton, with the partons recombining in the final state to produce the outgoing observed protons as shown in Fig. 29.

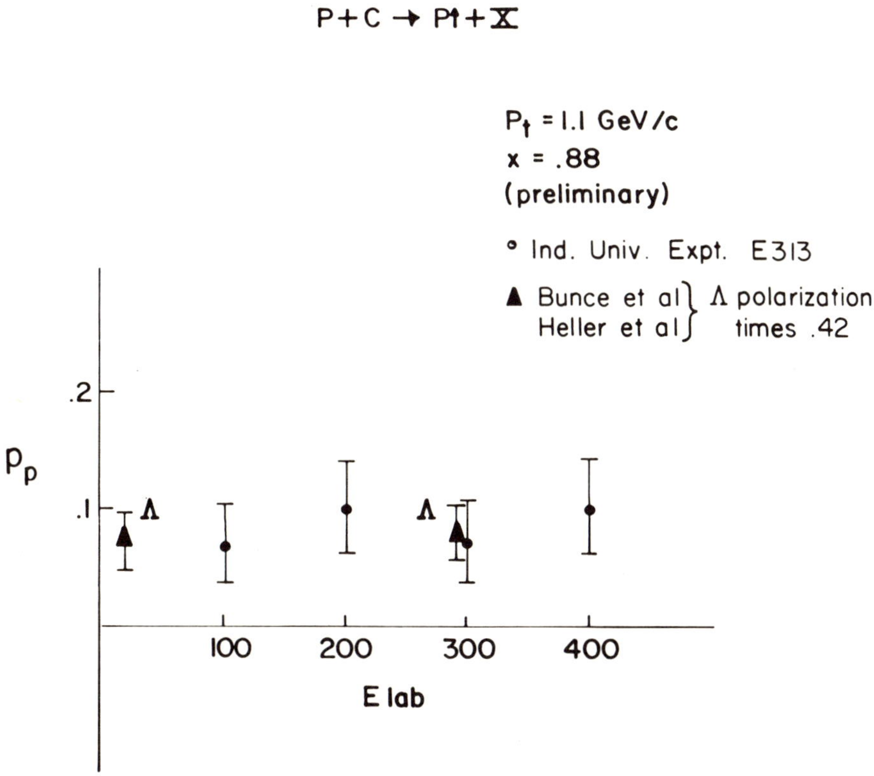

Ref. 27 Comparison of p and Λ inclusive polarizations.

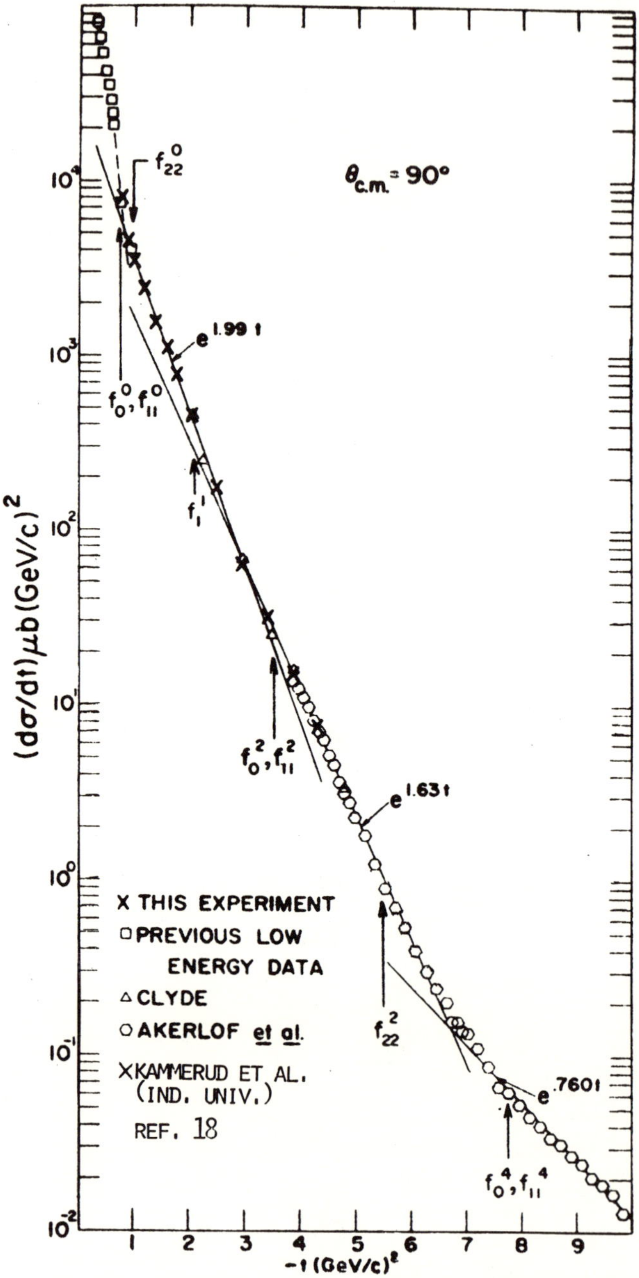

Fig. 28 Fixed angle (90°) differential cross section in p-p elastic scattering.

Furthermore, suppose that the region between t = - 1 and t = - 3 $(GeV/c)^2$ is due to two partons in the incident proton scattering from two partons in the target proton to produce the p-p elastic scattering process at the observed t. At a given value of t in this second region this double scattering process might be preferred over the one described for region 1 since the two partons can scatter through smaller angles to produce the observed t than would be required for a single parton scatter. The description for region 3 and others would be specified in a similar manner. There is one further ingredient in the model. This is the assumption that the polarization of the outgoing protons is the sum of the polarization of the scattered partons. The basic recurrence relation provided by this model is as follows:

$$P(t) = nS(t/n^2) \ ,$$

where n specifies the number of parton collisions taking place, P is the polarization at four momentum transfer squared t, and S is the polarization for parton-parton collisions as determined from proton-proton collisions in region 1.

In order to see if such a relation has any possible validity, one can simply determine S(t) by looking at the polarization in p-p elastic scattering at t values below 1 $(GeV/c)^2$. The value of the predicted polarization in region 2, where two binary parton-parton collisions are assumed to dominate the process, can they be easily generated from the formula above with n = 2. The results of this operation are illustrated in Fig. 30 where one sees the agreement to be unusually good. While this model is obviously extremely crude it does raise the possibility that someday we will be able to understand elastic and inclusive polarizations on the basis of the polarization produced by the constituent particles. One notes here certain similarities between this model and the one utilized by Heller in his description of inclusive polarization. Indeed, one

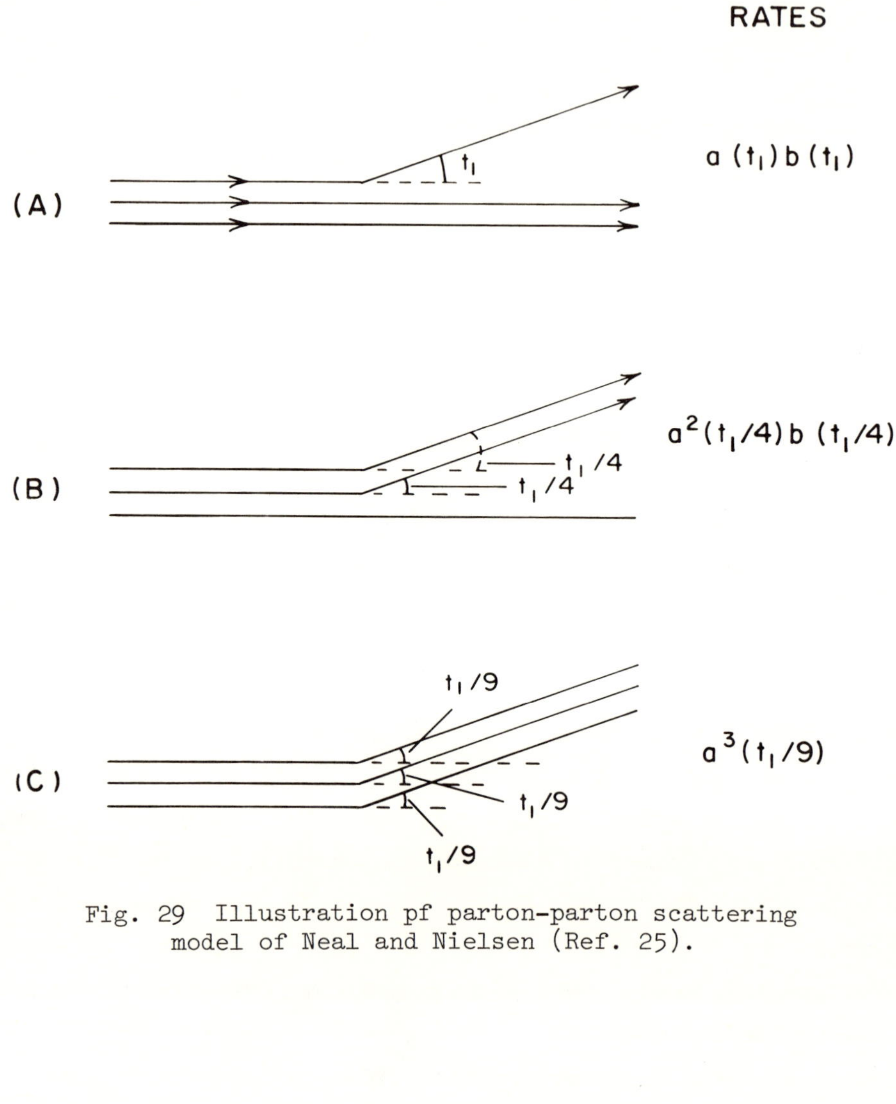

Fig. 29 Illustration pf parton-parton scattering model of Neal and Nielsen (Ref. 25).

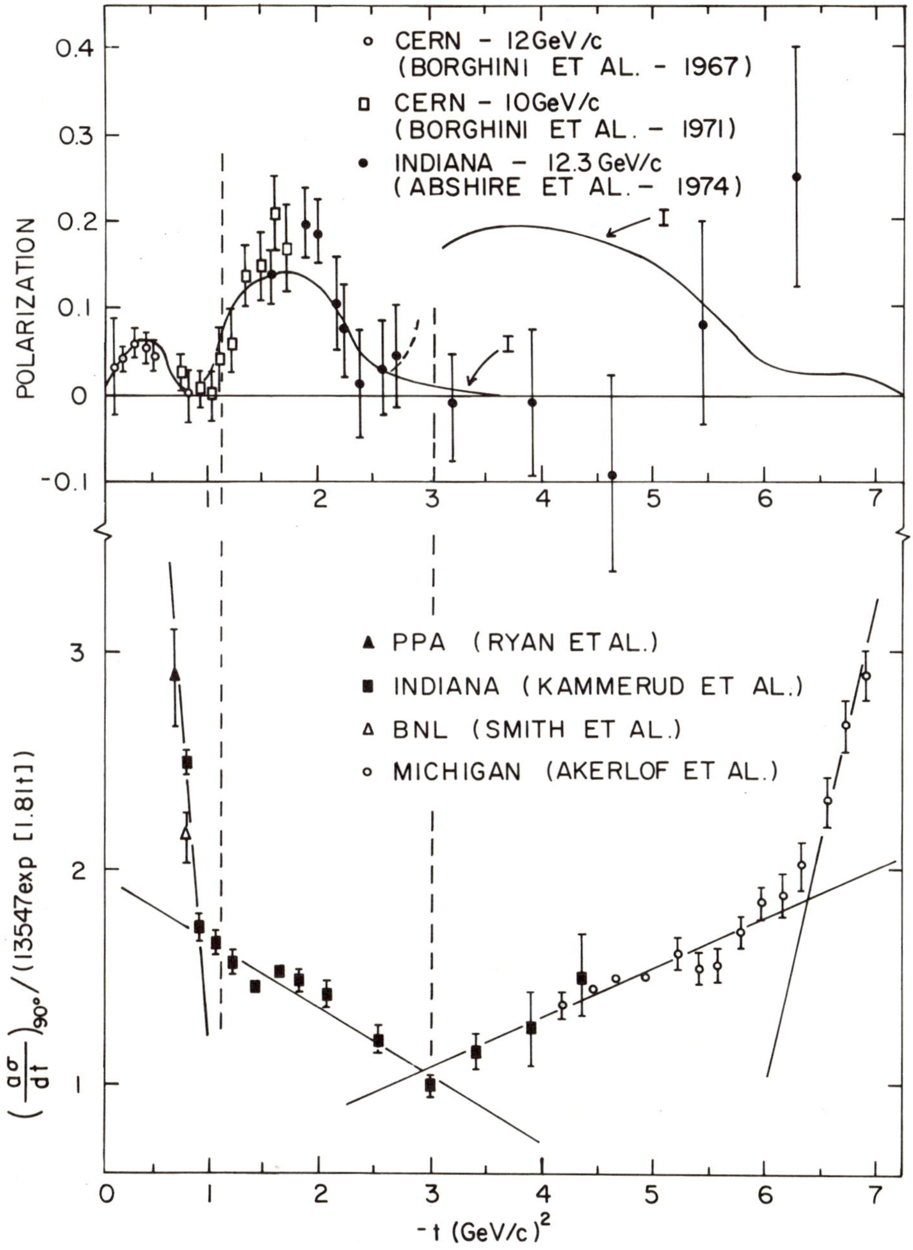

Fig. 30 Fit of model of Ref. 25 to polarization data. The polarization curve in region I generates the curve in region II with no free parameters.

of the interesting features that needs to be explored is the relationship between inclusive and elastic polarization in constituent models.

Amplitude Analysis

From a purely mechanistic view one could say that it is the experimentalist's job to make the measurements required to determine the spin amplitudes and it is the theorist's job to find out what the amplitudes are telling us about nature. We do not yet quite have enough information to fully make an unambiguous amplitude determination--but we are getting close. The p-p elastic amplitude situation at 6 GeV/c and $|t| \approx .3\ (\mathrm{GeV/c})^2$ is shown[12] in Fig. 31. In Fig. 31a we see from the shaded region the range of possible values for the amplitudes N_o, N_1, N_2, U_o, U_2 as a result of constraints imposed by σ, P, D_{nn}, C_{nn}, K_{nn}. In Fig. 31b we see the additional improvement due to knowledge of R. The effects of new information on $I\ (\vec{s},o;o,\vec{s})$ and $I(\vec{s},n;o,\vec{s})$ is illustrated in Fig. 31c. Finally, the effect of adding information on $I(\vec{s},\vec{s};o,o)$ further constrains the amplitudes as shown in Fig. 31d. The goal of having only a zero width dot in each amplitude plot is obviously still not achieved but remarkable progress has been made in the past two years and we can realistically look forward to significant successes in this area in the near future.

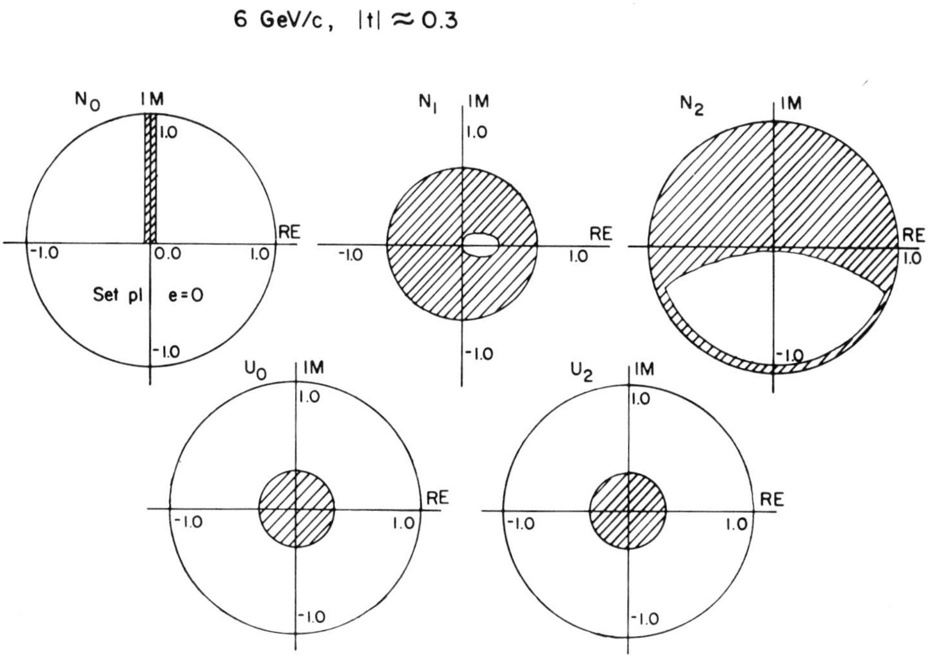

Fig. 31A

Fig. 31 Present status of knowledge of amplitudes.
See text for description.

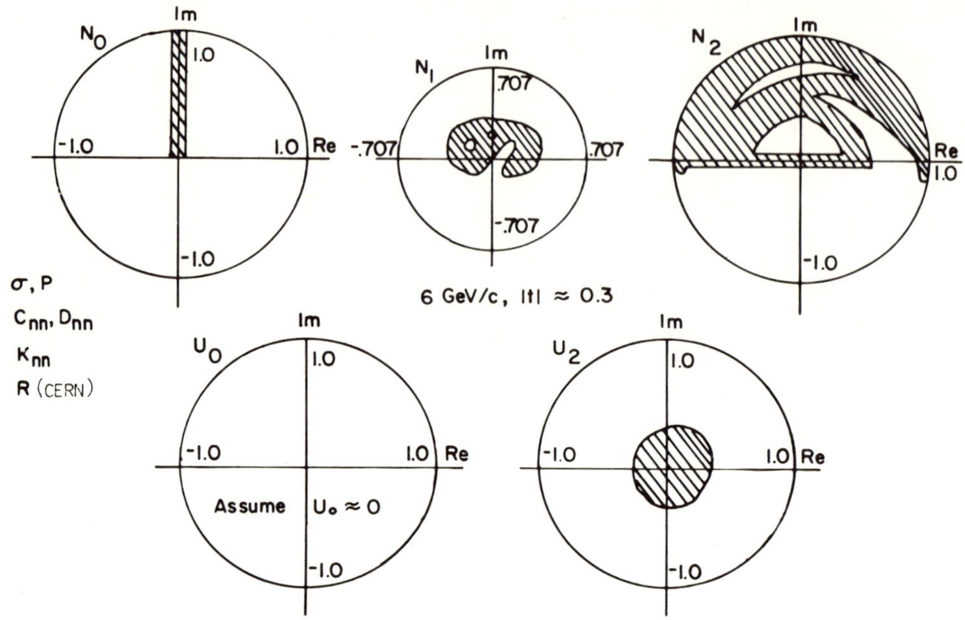

Fig. 31B

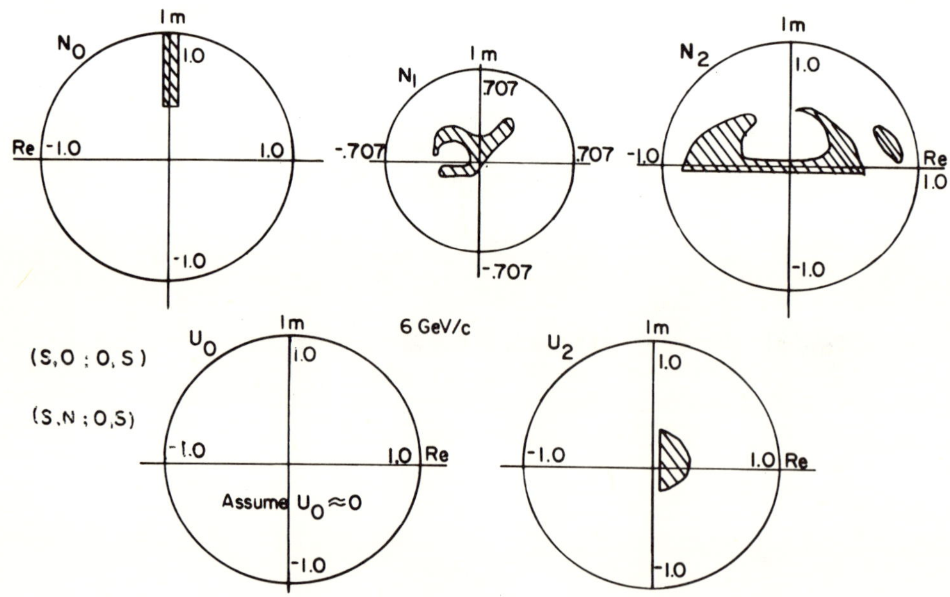

Fig. 31C

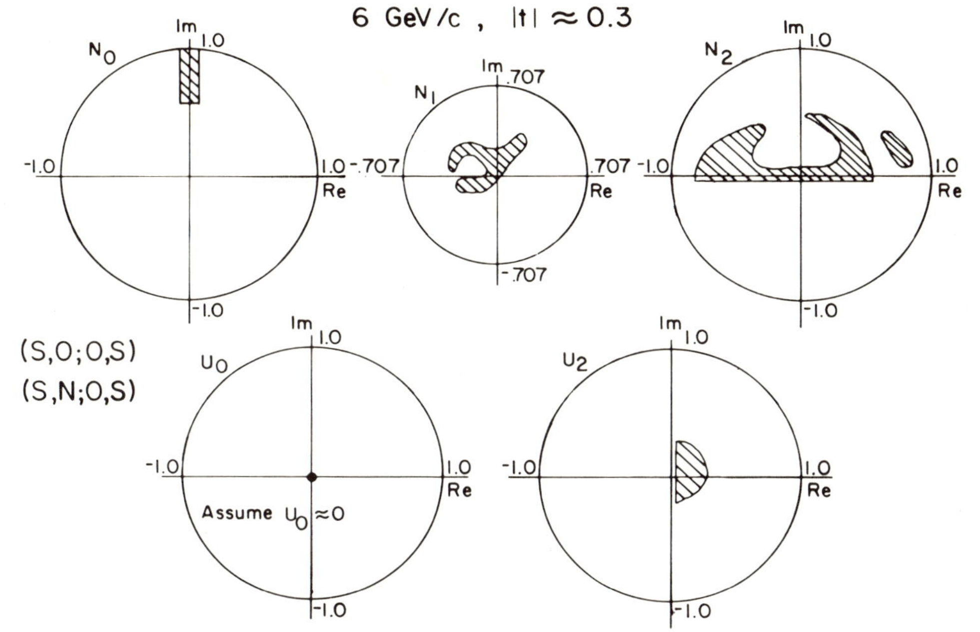

Fig. 31D

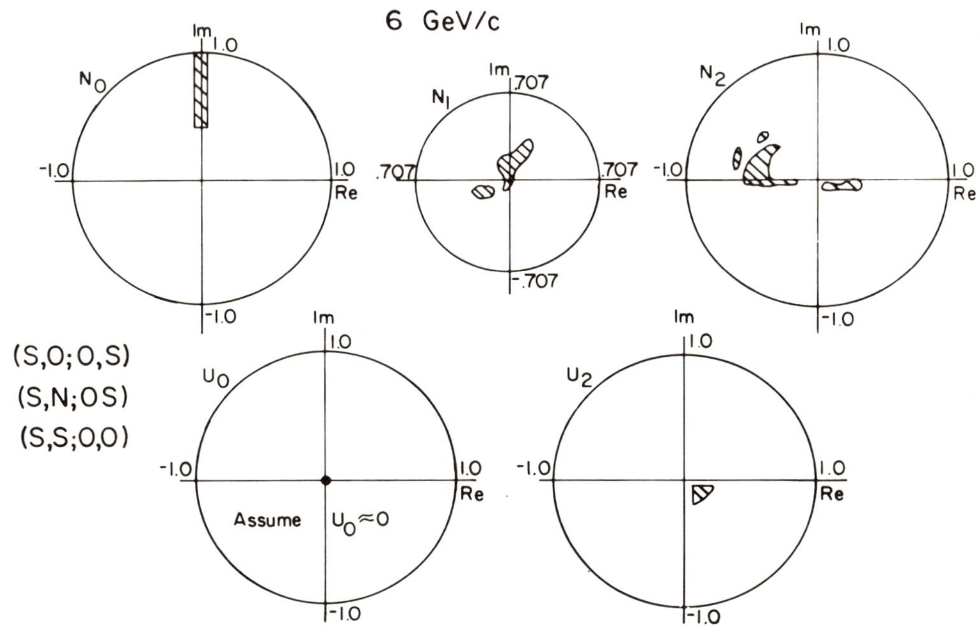

Fig. 31E

REFERENCES

1. M. Goldberger et al., Phys. Rev. 120 (1960).
2. F. Halzen and G. Thomas, Phys. Rev. D10, 344 (1974).
3. H. A. Neal and M. J. Longo, Phys. Rev. 161, 1374 (1967).
4. M. D. Corcoran et al., p-p Polarizations at High Energies, submitted to Phys. Rev. Lett., 1978.
5. J. Pumplin and G. L. Kane, Phys. Rev. D11, 1183 (1975).
6. C. Bourrely and J. Soffer, Nuovo Cim. Lett. 19, 569 (1977).
7. G. Abshire et al., Phys. Rev. D12 3393 (1975).
8. R. C. Fernow et al., Phys. Lett. 52B, 243 (1974).
9. L. G. Ratner et al., Phys. Rev. D15, (1977).
10. M. Marshak, private communications, 1978.
11. A. D. Krisch, Nucleon-Nucleon Elastic Scattering at Large P_t^2 and Spin, Review paper presented at the 1977 APS DPF Meeting, October 1977, Argonne, IL.
12. A. Yokosawa, Proceedings of the Second International Conference on the Nucleon-Nucleon Interaction, Vancouver, Canada (June 1977.
13. J. L. Stone et al.,Phys. Rev. Lett. 38, 1315 (1977).
14. C. E. DeHaven et al., Univ. of Mich. preprint UMHE 77-43.
15. S. L. Kramer et al., Polarization Parameter for Nucleon-Nucleon Elastic Scattering at 11.8 GeV/c, ANL-HEP-PR-77-74, 1977. (Previous meas. referenced therein).
16. U. P. Sukhatme, Some New Aspect of High Energy pp Elastic Scatte: ing, Univ. of Cambridge preprint, 1977.
17. I. P. Auer et al., Phys. Lett. 67B, 113 (1977).
18. R. C. Kammerud et al., Phys. Rev. D4, 1309 (1971).
19. L. Libby and E. Predazzi, Nuovo Cim. Lett. 2, No. 18, 881 (1969)
20. G. Bunce et al.,Phys. Rev. Lett. 36, 1113 (1976); O. Overseth, private communication, 1978.
21. M. D. Corcoran et al., Preliminary Measurements of Proton Polari zation in High Energy Inclusive Processes, Indiana University Internal Report IUHEE #10, 1977.
22. F. E. Paige and D. P. Sidhu, Phys. Rev. D14, 2307, (1976); P.

Salin and J. Soffer, CERN Preprint TH 1769-CERN; R. D. Field and G. D. Fox, Nuc. Phys. B80, 367 (1974).

23. K. Heller, A. Quark-Gluon Mechanism for Inclusive Polarization, Univ. of Michigan preprint, 1977.

24. G. L. Kane and Y. P. Yao, Studying Gluon Properties Experimentally, Univ. of Michigan preprint UM HE 77-44, 1977.

25. H. A. Neal and H. B. Nielsen, Phys. Lett. 51B, 79 (1974).

WHAT CAN ONE LEARN FROM VERY ACCURATE MEASUREMENTS OF THE LEPTON MAGNETIC MOMENTS?*

Toichiro Kinoshita

Cornell University

Ithaca, New York

As you will hear this afternoon, there has been remarkable progress in the measurement of lepton magnetic moments in the last few years. I should like to discuss what can be learned about the fundamental forces of nature from these and forthcoming developments.

Let me begin by summarizing the present experimental situation. The latest published values of the anomalous magnetic moments of the electron[1], positron[2], and negative and positive muon[3] are

$$a_{e^-} = 1\ 159\ 652\ 410\ (200) \times 10^{-12}, \quad (1)$$

$$a_{e^+} = 1\ 160\ \ 2\ (11) \times 10^{-7}, \quad (2)$$

$$a_{\mu^-} = 1\ 165\ 936\ (12) \times 10^{-9}, \quad (3)$$

$$a_{\mu^+} = 1\ 165\ 910\ (12) \times 10^{-9}. \quad (4)$$

All results except (2) have been reported within the last year. Their accuracy represents an improvement of more than an order of

* Work supported in part by the National Science Foundation.

magnitude over that of the previous results. While further refinement will be very difficult for a_μ, no serious obstacle seems to exist for more precise determination of a_e. At present efforts are being made by the Seattle group[4] and the Michigan group[5] to go beyond the result (1). Already the first group has obtained a more accurate (unpublished) value. I should like to ask the moderator of this session, Professor Dehmelt, for permission to quote it here:

$$a_{e^-} = 1\ 159\ 652\ 390\ (50) \times 10^{-12}. \qquad (5)$$

The primary[6] purpose of these measurements is to test the validity of the conventional QED, in particular to verify the presence of higher order terms which necessarily involves renormalization of mass and charge. In addition, inclusion of hadronic effects in a_μ becomes necessary at this level of accuracy. Only after these contributions are fully understood, will one be able to examine more subtle effects such as the one due to weak interactions.

Let me first recall the prediction of the conventional theory.

1. <u>The electron moment.</u> The present QED estimate of a_e is

$$a_e^{QED} = \frac{1}{2}\frac{\alpha}{\pi} - 0.328\ 478\ 445\ (\frac{\alpha}{\pi})^2 + 1.184(7)\ (\frac{\alpha}{\pi})^3\ , \qquad (6)$$

where the last term is the contribution of all 6th order Feynman diagrams, 72 altogether[7-10], of which 51 have thus far been evaluated analytically.[10] The uncertainty in (6) is due to the limited accuracy of the numerical integration of 21 vertex diagrams, shown in Fig. 1, which have not yet been evaluated analytically.[9] If one substitutes in (6) the value of the fine structure constant[11]

$$\alpha^{-1}_{\text{Josephson}} = 137.035\ 987(29) \quad (0.21\text{ppm}) \qquad (7)$$

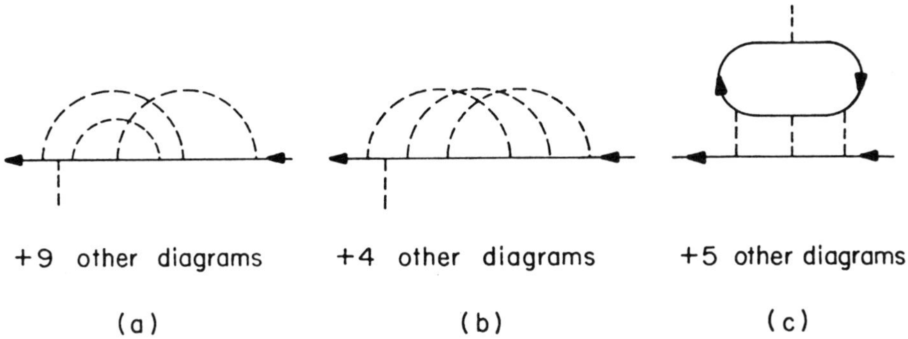

Fig. 1 (a), (b). The diagram not shown explicitly are obtained by moving the external vertex to other electron lines (solid lines), and also changing the direction of electron lines in (a).

(c) The diagram not shown explicitly are obtained by permuting the ends of the photon lines (dotted lines) on the electron loop.

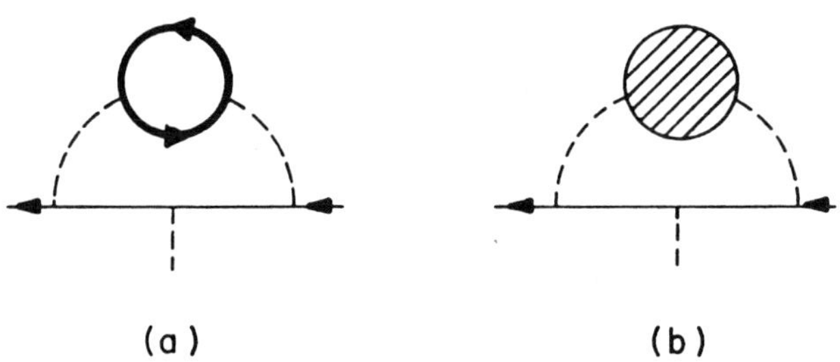

Fig. 2 (a) The heavy solid line represents the muon propagation.
(b) The shaded circle represents the hadronic loop.

determined from the measurements of the ac Josephson effect and the gyromagnetic ratio of the proton, one finds

$$a_e^{QED} = 1\ 159\ 652\ 375(261) \times 10^{-12} \tag{8}$$

in good agreement with (1) or (5). The uncertainty in (8) arises from two sources, one from α (246×10^{-12}) and the other from theory (88×10^{-12}).

Other known contributions to a_e are very small. One is the muon loop contribution of Fig. 2(a)[12]

$$a_e^{\text{muon loop}} = \left(\frac{\alpha}{\pi}\right)^2 \frac{1}{45}\left(\frac{m_e}{m_\mu}\right)^2 = 2.8 \times 10^{-12}. \tag{9}$$

The other is the hadronic contribution of Fig. 2(b)

$$a_e^{\text{hadron}} \simeq 1.6(2) \times 10^{-12} , \tag{10}$$

which can be calculated[13] from the measured hadronic cross section in e^+e^- colliding beam experiments according to the formula

$$a_e^{\text{hadron}} = \frac{1}{4\pi^3} \int_{4m_\pi^2}^{\infty} ds\ G(s) \sigma_{e^+e^- \to \text{hadrons}}(s)$$

with

$$G(s) = \int_0^1 \frac{x^2\ (1-x) dx}{x^2 + (1-x)\ (s/m_e^2)} . \tag{11}$$

The uncertainty is (10) reflects mainly that of the measurement of $\sigma_{e^+e^- \to \text{hadrons}}$. From (8), (9) and (10) we obtain

$$a_e^{\text{theory}} = 1\ 159\ 652\ 379(261) \times 10^{-12} \quad . \tag{12}$$

An alternative, and perhaps more sensible, way to test the theory is to calculate α from theory and experiment of a_e and compare it with $\alpha_{\text{Josephson}}$. To emphasize the advantage of this approach, let us take the preliminary data (5) and assume that the terms of order α^4 and higher are negligible compared with the uncertainty in the α^3 term. Then we find

$$\alpha^{-1}_{g_e-2} = 137.035\ 985(12) \quad (0.09 \text{ ppm}) \ , \tag{13}$$

which is more accurate than $\alpha^{-1}_{\text{Josephson}}$ given in (7).

The effect of weak interaction on a_e will be discussed later.

2. <u>The muon moment</u>. The best QED estimate of a_μ at present is[14,15]

$$a_\mu^{\text{QED}} = \frac{1}{2}\frac{\alpha}{\pi} + 0.765\ 782\ \left(\frac{\alpha}{\pi}\right)^2 + 24.45(6)\ \left(\frac{\alpha}{\pi}\right)^3 + 135(64)\ \left(\frac{\alpha}{\pi}\right)^4$$

$$= 1\ 165\ 852.0(2.4) \times 10^{-9} \quad , \tag{14}$$

where the numerical value is obtained using (7). Note the very large coefficients of the α^3 and α^4 terms. This reflects the fact that the main contribution arises from the diagrams with electron loop insertions in which the expansion parameter is $\alpha\ \ell n(m_\mu/m_e)$ rather than α itself. In fact, this $\ell n(m_\mu/\ m_e)$ dependence enables us to calculate the coefficients of $(\ell n(m_\mu/m_e))^n$, n=1,2,3,..., terms of many, but not all, of these vertex diagrams using a renormalization-group-like argument.[16] The coefficient of the α^4 term in (14) has been estimated in this manner.[14] In contrast the α^3 term in

(14) is obtained by an explicit evaluation of 96 diagrams including 24 diagrams which contain closed electron loops.

For the muon moment the hadronic contribution is not at all negligible. The best estimate of this contribution, obtained from (11) by replacing m_e by m_μ, is[14]

$$a_\mu^{hadron} = 66.7(9.4) \times 10^{-9} . \qquad (15)$$

From (14) and (15) we obtain the theoretical prediction

$$a_\mu^{theory} = 1\ 165\ 919(10) \times 10^{-9} , \qquad (16)$$

which is in excellent agreement with the average of the observed values (3) and (4):[3]

$$a_\mu^{exp} = 1\ 165\ 922(9) \times 10^{-9} . \qquad (17)$$

The remarkable agreement of experiment and theory gives a strong confidence not only in the health of QED but also in the soundness of our treatment of hadronic contributions. We must not forget, however, that one of the rationales for pushing theory and experiment to higher and higher precision is to look for the presence of new and unsuspected phenomena. Strickly speaking, weak interaction does not belong to this category since it undoubtedly has some detectable effect on the lepton magnetic moment. On the other hand, to the extent that we do not yet know the correct form of the weak interaction Hamiltonian the above results for a_e and a_μ can be used, as has been done by many authors, for restricting the range of possible models. For this purpose let us first quote the predictions of several early models of renormalizable spontaneously broken gauge theories. For the moment let us restrict our attention to a_μ.

a) Weinberg-Salam model. In this model three Feynman diagrams of Fig. 3 contribute to a_μ:[17]

$$a_\mu(a) = \frac{Gm_\mu^2}{8\pi^2\sqrt{2}} \frac{10}{3}, \tag{18}$$

$$a_\mu(b) = -\frac{\alpha}{\pi}\frac{1}{3}\left(\frac{m_\mu}{m_Z}\right)^2 + \frac{Gm_\mu^2}{8\pi^2\sqrt{2}}\frac{4}{3}\left(1-\frac{2}{R}\right), \tag{19}$$

$$a_\mu(c) = \frac{Gm_\mu^2}{8\pi^2\sqrt{2}}\, 2\int_0^1 \frac{x^2(2-x)dx}{x^2+s(1-x)}, \tag{20}$$

where

$$R = \left(\frac{M_Z}{M_W}\right)^2, \qquad s = \left(\frac{m_\phi}{m_\mu}\right)^2. \tag{21}$$

If R and s are unrestricted, $a_\mu^W \equiv a_\mu(a) + a_\mu(b) + a_\mu(c)$ has the following upper and lower bounds[14]

$$1.9\times10^{-9} \leq a_\mu^W \leq 8.9\times10^{-9}. \tag{22}$$

If we accept Weinberg's argument[18] on the mass of the Higgs boson, $a_\mu(c)$ is completely negligible. If we further take account of the experimental limits on $\sin^2\theta_W$, we obtain[14]

$$1.9\times10^{-9} \leq a_\mu^W \leq 2.3\times10^{-9}. \tag{23}$$

Clearly, if the Weinberg-Salam model is the correct theory of weak interaction, the theoretical and experimental errors of a_μ must be reduced by an order of magnitude in order to obtain meaningful constraints on the weak interaction from the $g_\mu - 2$ experiment.

b) Lee-Prentki-Zumino model. In this model the contribution

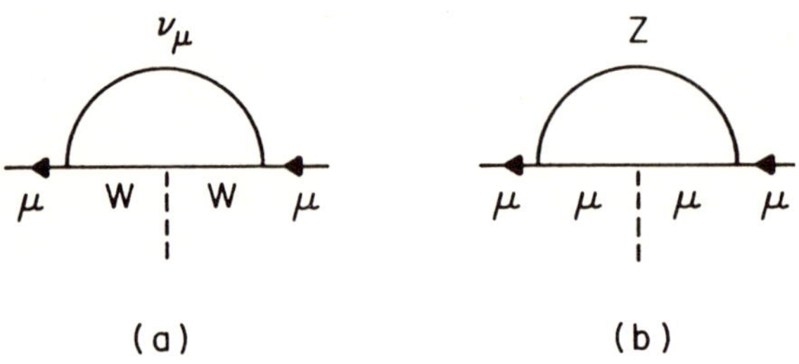

Fig. 3 Diagrams contributing to a_μ in the Weinberg-Salam model. (a) Neutrino exchange diagram. (b) Z-boson exchange diagram.
(c) Higgs scalar exchange diagram.

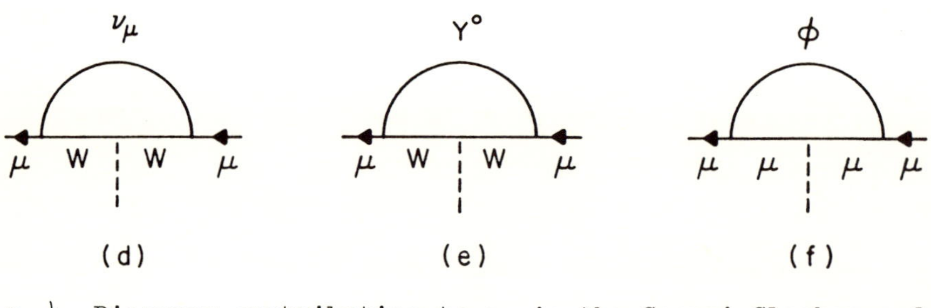

Fig. 4 Diagrams contributing to a_μ in the Georgi-Glashow model. (d) Neutrino exchange diagram. (e) Neutral heavy lepton exchange diagram. (f) Higgs scalar exchange diagram.

to a_μ can be obtained from that of the Weinberg-Salam model by redefining the coupling constant.[19] We get

$$a_\mu^{LPZ} = \frac{Gm_\mu^2}{8\pi^2\sqrt{2}} \left[\frac{10}{3} - \frac{4}{3} \frac{u^2+v^2}{u^2} (1 + \sin^2\theta\cos^2\theta)\right] + \text{Higgs term} \tag{24}$$

where

$$\frac{G}{\sqrt{2}} = \frac{1}{4(u^2+v^2)} . \tag{25}$$

From (24) we obtain (ignoring the Higgs term)

$$a_\mu^{LPZ} \leq 2.3\text{x}10^{-9} . \tag{26}$$

There is no theoretical lower bound. If we regard $\pm 9\text{x}10^{-9}$ as a bound for the contribution of weak interaction, we obtain some restriction on the parameters of theory.

c) Georgi-Glashow model. Three diagrams of Fig. 4 contribute in this model:[20]

$$a_\mu\,(d) = \frac{\alpha \sin^2\beta}{8\pi} \left(\frac{m_\mu}{M_W}\right)^2 \frac{10}{3} , \tag{27}$$

$$a_\mu\,(e) = -\frac{\alpha}{8\pi} \frac{m_\mu(m_{\gamma+} + m_\mu)}{M_W^2} \left(\frac{3}{(1-\rho)^2} (1 - 3\rho - \frac{2\rho^2}{1-\rho} \ln \rho) + 1\right), \tag{28}$$

$$a_\mu(f) = \frac{\alpha}{8\pi}\frac{(m_{\gamma^+} - m_\mu)^2}{M_W^2}\int_0^1 \frac{x^2(2-x)dx}{x^2 + r(1-x)}, \qquad (29)$$

where

$$\rho = \left(\frac{m_{\gamma^o}}{M_W}\right)^2, \quad r = \left(\frac{m_\phi}{m_\mu}\right)^2 \qquad (30)$$

and

$$\frac{\pi\,\alpha\sin^2\beta}{M_W^2} = \frac{G}{\sqrt{2}}, \quad 2m_{\gamma^o}\cos\beta = m_{\gamma^+} + m_\mu . \qquad (31)$$

Nothing definite can be said about the magnitude of $a_\mu^{GG} = a_\mu(d) + a_\mu(e) + a_\mu(f)$ unless $\cos\beta$, m_{γ^+}, and m_ϕ are constrained. It is important to note, however, that $a_\mu(e)$ is very large and negative if $m_{\gamma^+} >> m_\mu$. For instance

$$|a_\mu(e)| \gtrsim 65\times10^{-9} \qquad (32)$$

if $m_{\gamma^+} = 1.8$ GeV/c^2, which is much larger than the experimental uncertainty $\pm9\times10^{-9}$ for a_μ. Since $a_\mu(d)$ is very small, a_μ^{GG} can be made consistent with observation if and only if strong cancellation occurs between $a_\mu(e)$ and $a_\mu(f)$. This imposes a severe restriction on the Higgs meson mass. For instance we have

$$1.2 \text{ GeV}/c^2 < m_\phi < 2.0 \text{ GeV}/c^2 \qquad (33)$$

for $m_{\gamma^+} = 1.8$ GeV/c^2. The size of the range (33) is roughly proportional to $\sin^2\beta$.

Of course, this model by itself cannot be taken seriously since it does not account for the observed weak neutral current.

The effect on a_μ has also been calculated for a number of more recent models of weak interaction.[21] Although most of the physical motivations for constructing these models, such as the neutrino-induced trimuon events, the high-y anomaly, the μ-e conversion, and the apparent absence of parity violation effect in atomic physics, have since evaporated or are not as compelling as they once seemed[22], some features coming out of these calculations are worth mentioning: (i) Our present (lack of) knowledge of weak interactions is such that we cannot exclude the possible existence of weak currents which mix left-handed and right-handed fermions.[23] Such a current will give to a_μ contributions analogous to (28).

(ii) The weak corrections to a_μ (except in models in which it is undetectably small) depend on so many undetermined parameters that these calculations are mainly useful as a framework for narrowing down the choice of possible models rather than as predictions of the magnitude of the correction.

Let us now turn to the effect of the weak interaction on the electron anomaly a_e. This effect is found to be very small in most models. For instance, the Weinberg-Salam model predicts

$$a_e^W \simeq 0.05\times 10^{-12} . \tag{34}$$

The smallness of this contribution can be attributed to the fact that the a_ℓ^W (ℓ=e or μ) is proportional to m_ℓ^2 in this theory. However, this is not necessarily the case in other models. For instance the contribution $a_\ell(e)$ due to neutral heavy lepton exchange of the Georgi-Glashow model depends on m_ℓ linearly:[24]

$$a_\ell(e) \propto m_\ell . \tag{35}$$

Thus, if m_{γ^+} associated with electron and muon are of similar magnitude, we may find

$$a_e^{GG} \sim -300\times10^{-12} \qquad (36)$$

even if $|a_\mu^{GG}| < 10^{-8}$. This is already outside the present experimental uncertainty in a_e.

The behavior (35) is a general feature of weak currents which mix left-handed and right-handed fermions.[23] Because of this feature and because of the extraordinary precision in both experiment and theory, the electron anomaly a_e is a much more sensitive device for detecting the possible presence of such a current than the muon anomaly a_μ. Further improvement of theory and experiment for a_e will thus give meaningful constraints on the general structure of the weak current.

It is clear from these observations that the present theory and experiment for a_e and a_μ are not yet accurate enough to provide useful information on weak interactions or other unknown phenomena. One might thus ask how good is the prospect for further improvement of theory and experiment in the next few years.

For the muon anomaly, the effect of weak interaction (and possibly the effect of the axion[25] if such a particle exists at all) is within a factor 10 of the present experimental accuracy. As far as theory is concerned, a_μ^{QED} is already known with sufficient accuracy, and can be improved further if necessary. The main obstacle in improving a_μ^{theory} is the hadronic contribution (15) whose error is very difficult to reduce. The only way to achieve this is to measure the cross section for the reaction

$$e^+e^- \rightarrow \text{hadrons}$$

very accurately in the energy range between $\pi\pi$ threshold and ρ, ω, ϕ resonances. Of course this effort must go hand in hand with a better determination of a_μ itself. Unfortunately, such a project is not likely to be pushed for some years to come.[26]

For the electron anomaly a_e, the effect of the weak interaction is likely to be very small (see (34)). In fact we do not expect to run into any effect other than those of QED and calculable hadronic effect

(10) until the experimental error is improved by three orders of magnitudes compared with (1). In this sense a_e will remain an ideal ground for checking the higher order calculation of QED.

In view of the anticipated advance in the measurement of a_e within the next few years[4,5], it is clearly useful to sharpen the QED prediction of a_e beyond the present result (6). To see exactly where we stand now, let me write down explicitly where the error of the α^3 term in (6) comes from. It is a composite of errors from the three types of diagrams of Fig. 1(a),(b),(c):[9,27]

$$\Delta a_e(2a) = 2.3469(56)\left(\frac{\alpha}{\pi}\right)^3 , \qquad (37)$$

$$\Delta a_e(2b) = 1.7243(12)\left(\frac{\alpha}{\pi}\right)^3 , \qquad (38)$$

$$\Delta a_e(2c) = 0.3717(27)\left(\frac{\alpha}{\pi}\right)^3 , \qquad (39)$$

Analytic or semianalytic evaluation of $\Delta a_e(2a)$ and $\Delta a_e(2c)$ is in progress.[28] Thus (37) and (39) will be improved substantially before long. On the other hand, no one has found a way to evaluate $\Delta a_e(2b)$ analytically. At present, the only way to improve it is by numerical means, and this is what we are pursuing.[9] Eventually we will be able to reduce the uncertainty of the entire α^3 term to

$$0.0003\left(\frac{\alpha}{\pi}\right)^3 \sim 4\text{x}10^{-12} . \qquad (40)$$

Once this goal is achieved the main source of theoretical uncertainty is the α^4 term since

$$\left(\frac{\alpha}{\pi}\right)^4 \simeq 29.1\text{x}10^{-12} . \qquad (41)$$

Clearly the α^4 term must be evaluated in order to take full advantage of the forthcoming measurements of a_e. We have recently decided to compute the complete α^4 terms. Let me briefly outline this project.

There are altogether 891 Feynman diagrams that contribute to a_e in the order α^4. Of these 373 are vertex diagrams of (mostly) lower orders in which closed electron loops are inserted. Most of them are not much more complicated than those of α^3 diagrams and are relatively easy to handle. The remaining 518 diagrams have no electron loop insertions and are the hard core of the α^4 contribution. They can be classified into 74 groups, each group consisting of 7 vertex diagrams generated by inserting an external vertex in an α^4 electron self-energy diagram in all possible ways. Of these only 47 require independent evaluation.

We have combined all diagrams in each group into one integral by the method given in Sec. 6, ref. 8. We have written a SCHOONSCHIP program for projecting out the magnetic moment contribution of the combined expression. It generates an integrand of 7 to 10 dimensional Feynman integral in 2 to 10 minutes of computing on CDC-7600. The size of the generated terms after some refinements, is from about 500 to more than 10,000 terms, each term being a product of up to seven factors. This is 10∿20 times larger than a typical α^3 integrand. Pointwise subtraction of ultraviolet and infrared divergences must be carried out before the integral is evaluated on the computer.

It is expected that several hundred hours of numerical integration on CDC-7600 will give the α^4 term with an accuracy better than 10%. The whole project will take 4 to 6 man-years.

Such an extensive calculation would be somewhat futile unless the fine structure constant α is measured very accurately by some other means. Fortunately it appears that $\alpha_{\text{Josephson}}$ can in fact be determined to a much higher accuracy.[29] The measurement of $\alpha_{\text{Josephson}}$ is based on the formula

$$\alpha^{-1}_{\text{Josephson}} = \left(\frac{c}{4R_\infty \gamma'_p} \frac{\Omega_{\text{ABS}}}{\Omega_{\text{NBS}}} \frac{\mu'_p}{\mu_B} \frac{2e}{h} \right)^{1/2} , \qquad (42)$$

where Ω_{ABS} is the absolute ohm and Ω_{NBS} is the standard ohm at the National Bureau of Standards. The present ppm errors in the measurements of various factors are [11]

c	R_∞	$\Omega_{ABS}/\Omega_{NBS}$	μ'_p/μ_B	$2e/h$	γ'_p	
0.004	0.009	0.054	0.014	0.030	0.42	(ppm)

Olsen and Williams[29] are planning to reduce the error in γ'_p and hence in α to 0.02 ppm.

Thus we will be able to push the test of QED by another order of magnitude within the next few years. Finally we note that the comparison of $\alpha_{\mathrm{Josephson}}$ and α_{g_e-2} provides a serious experimental test for the BCS theory of superconductivity. Such a confrontation might force us to raise questions such as "Is the charge carried by a Cooper pair exactly equal to 2e?" and "If not, how large will the correction term be?".

REFERENCES

1. R. S. Van Dyck, Jr., P. B. Schwinberg, and H. G. Dehmelt, Phys. Rev. Lett 38, 310 (1977).
2. J. Gilleland and A. Rich, Phys. Rev. Lett. 23, 1130 (1969).
3. J. Bailey et al., Phys. Lett. 68B, 191 (1977).
4. H. Dehmelt, invited talk presented at this Conference.
5. A. Rich, invited talk presented at this Conference.
6. These experiments can also be used to test other theoretical predictions, e.g. the CPT Theorem. From (3) and (4) we obtain $(g_{\mu-} - g_{\mu^+})/g_\mu = 2.6(1.7)\times10^{-8}$ which provides the most accurate test of the CPT Theorem applied to muons. To achieve similar accuracy for electrons, the result (2) for $a_{e\pm}$ must be improved substantially. Note, however, that the work of S. I. Serednyakov et al. (Phys. Lett. 66B, 102 (1977) has established the equality of a_{e-} and a_{e+} to a very high accuracy: $|a_{e-} - a_{e+}|/a_e \leq 1.0\times10^{-5}$. This is equivalent to $|g_{e-} - g_{e+}|/g_e \leq 1.2\times10^{-8}$.
7. All 72 diagrams have been evaluated by numerical integration, most of them by several groups. The most extensive work has been done by M. J. Levine and J. Wright (Phys. Rev. D8, 3171 (1973)),R. Carroll (Phys. Rev. D12, 2344 (1975), P. Cvitanovic and T. Kinoshita (ref. 8), and T. Kinoshita and B. Lindquist (ref. 9).
8. P. Cvitanovic and T. Kinoshita, Phys. Rev. D10, 4007 (1974).
9. T. Kinoshita and B. Lindquist, Cornell preprint, CLNS-374 (1977).
10. See M. J. Levine and R. Roskies, Phys. Rev. D14, 2191 (1976) and references therein.
11. P. T. Olsen and E. R. Williams, Proc. of the Fifth International Conference on Atomic Masses and Fundamental Constants, J. H. Sanders and A. H. Wapstra, eds. (Plenum Press, New York, 1976).
12. B. E. Lautrup and E. de Rafael, Phys. Rev. 174, 1835 (1968).
13. Actually (10) is obtained from (15) by noting that $a_e^{hadron} \simeq (m_e/m_\mu)^2\, a_\mu^{hadron}$.

14. J. Calmet et al., Rev. Mod. Phys. 49, 21 (1977)

15. C. Chlouber and M. A. Samuel, Phys. Rev. D16, 3596 (1977).

16. T. Kinoshita, Nuovo Cimento 51B, 140 (1967); B. E. Lautrup and E. de Rafael, Nucl. Phys. B70, 317 (1974).

17. R. Jackiw and S. Weinberg, Phys. Rev. D5, 2473 (1972); I. Bars and M. Yoshimura, Phys. Rev. D6, 374 (1972); W. A. Bardeen, R. Gastmans, and B. E. Lautrup, Nucl. Phys. B46, 319 (1972).

18. S. Weinberg, "Mass of the Higgs Boson", Harvard preprint, 1975.

19. K. Fujikawa, B. W. Lee, and A. I. Sanda, Phys. Rev. D6, 2923 (1972).

20. J. R. Primack and H. R. Quinn, Phys. Rev. D6, 3171 (1972).

21. See, for example, J. D. Bjorken, K. Lane, and S. Weinberg, Phys. Rev. D16, 1474 (1977); B. W. Lee and R. E. Shrock, FERMILAB-Pub-77/48-THY, 1977; V. Barger and R.J.N. Phillips, Univ. of Wisconsin preprint COO-881-17,1978.

22. For example see M.A.B. Beg, invited talked presented at this Conference.

23. See, for instance, B. W. Lee and R. E. Shrock, Phys. Rev. D16, 1444 (1977).

24. The contribution $a_\ell(f)$ behaves as m_ℓ^2 for $m_\phi^2 >> m_\ell^2$.

25. F. Wilczek, Phys. Rev. Lett. 40, 279 (1978); S. Weinberg, Phys. Rev. Lett. 40, 223 (1978).

26. R. W. Williams, invited talk presented at this Conference.

27. M. A. Samuel and C. Chlouber (Oklahoma State Univ. preprint, 1977) have obtained $\Delta a_e(2c) = 0.400(4)\ (\frac{\alpha}{\pi})^3$. Further study is needed to resolve the discrepancy between this result and (39).

28. Private communication from R. Roskies.

29. Private communication from B. N. Taylor

LEPTON SPIN MOTION IN WEAK MAGNETIC MIRROR TRAPS

G. W. Ford

University of Michigan

Ann Arbor, Michigan

My topic is a precision calculation of the spin and orbital motion of leptons in weak magnetic mirror traps of the sort used in precision g-2 experiments, done in collaboration with my student, Sara Granger.[1,2] The calculation itself is rather complicated and technical but the methods used, the averaging methods of non-linear mechanics, are of general interest with a broad range of applications. Since these methods are not familiar to most physicists, and since one of the topics of this year's Orbis Scientiae is "Nonlinearities in Natural Sciences", a discussion of these methods should be appropriate.

I begin, therefore, with a brief description of two of these averaging methods: The Method of Averaging and the Method of Rapidly Rotating Phase. Then I sketch our calculations of the spin motion, showing where these methods are used. For details I refer to our papers. Finally, I say something about the application of the results to the last two precision electron g-2 measurements.

A. THE METHOD OF AVERAGING[3,4]

Consider an ordinary differential equation of the form

$$\frac{dx}{dt} = \varepsilon F(x,t), \tag{1}$$

where ε is a small parameter. In general x will be a vector with, say, n components but for simplicity I will consider it to be just a single variable. If ε is small the solution of this equation will consist of a slowly varying part about which there will be small amplitude, rapid oscillations arising from the explicit time dependence in F(x,t). We accordingly seek a solution in the form

$$x = \bar{x} + \varepsilon X^{(1)}(\bar{x},t) + \varepsilon^2 X^{(2)}(\bar{x},t) + \ldots, \tag{2}$$

where $\bar{x}$ is the slowly varying mean and satisfies an equation of the form

$$\frac{d\bar{x}}{dt} = \varepsilon\, \Omega^{(1)}(\bar{x}) + \varepsilon^2\, \Omega^{(2)}(\bar{x}) + \ldots, \tag{3}$$

with no explicit time dependence.

Inserting (2) in (1), using (3) and then equating coefficients of equal powers of ε on either side, we get the following sequence of equations for determining the X's and Ω's.

$$\begin{aligned}
\Omega^{(1)}(\bar{x}) + \frac{\partial X^{(1)}}{\partial t}(\bar{x},t) &= F(\bar{x},t) \\
\Omega^{(2)} + \frac{\partial X^{(2)}}{\partial t} &= \frac{\partial F(\bar{x},t)}{\partial \bar{x}} X^{(1)} - \Omega^{(1)} \frac{\partial X^{(1)}}{\partial \bar{x}}, \\
\Omega^{(3)} + \frac{\partial X^{(3)}}{\partial t} &= \frac{\partial F}{\partial \bar{x}} X^{(2)} + \frac{1}{2}\frac{\partial^2 F}{\partial \bar{x}^2}(X^{(1)})^2 \\
&\quad - \Omega^{(2)}\frac{\partial X^{(1)}}{\partial \bar{x}} - \Omega^{(1)}\frac{\partial X^{(2)}}{\partial \bar{x}}
\end{aligned} \tag{4}$$

Each of these equations is of the same form, with the right hand side known in terms of the solution of the previous equation. The solution of the first is therefore typical. We determine $\Omega^{(1)}(\bar{x})$ so as to cancel the mean behavior of the right hand side,

$$\Omega^{(1)}(\bar{x}) = \langle F(\bar{x},t)\rangle \equiv \lim_{t\to\infty} \frac{1}{T} \int_0^T dt\, F(\bar{x},t) \quad , \tag{5}$$

$$X^{(1)}(\bar{x},t) = \int dt\, [F(\bar{x},t) - \langle F(\bar{x},t)\rangle] \; . \tag{6}$$

We see from (5) why this is called the method of averaging; in each order $\Omega^{(n)}$ will be the time average of the right hand side. In higher orders this will reflect the interference of oscillations in the solution.

Equations (2) and (3) represent a separation of the slowly varying mean notion from the rapid oscillations about the mean. In applications this often has an important physical significance since measurements of the motion will use "low band pass" instruments and the mean motion will therefore be the observed motion.

B. THE METHOD OF RAPIDLY ROTATING PHASE[3,5,6]

In this method we have a set of equations of the form

$$\begin{aligned} \frac{dx}{dt} &= \varepsilon\, G\,(x,\, \psi) \; , \\ \frac{d\psi}{dt} &= \omega\,(x) + \varepsilon H\,(x,\psi) \end{aligned} \tag{7}$$

where ε is again a small parameter and where H and G are, for fixed x, periodic functions in ψ with period 2π. Again, in general x and also ψ can have several components, but for simplicity I consider the case of only one each. The variable ψ is called the phase, and clearly for ε small it undergoes the most rapid motion of the system.

In a way similar to that for the systems to which the Method of Averaging applies, the solution of equations (7) will exhibit a slowly varying mean motion about which there will be small amplitude, rapid oscillations. This time, however, we seek a solution in the form

$$x = \bar{x} + \varepsilon A^{(1)}(\bar{x}, \bar{\psi}) + \varepsilon^2 A^{(2)}(\bar{x}, \bar{\psi}) + \ldots ,$$
$$\psi = \psi + \varepsilon B^{(1)}(\bar{x}, \bar{\psi}) + \varepsilon^2 B^{(2)}(\bar{x}, \bar{\psi}) + \ldots , \tag{8}$$

where $A^{(n)}$ and $B^{(n)}$ are periodic functions of $\bar{\psi}$ with period 2π, and where $\bar{x}$ and $\bar{\psi}$ are the mean motion and satisfy equations of the form:

$$\frac{d\bar{x}}{dt} = \varepsilon C^{(1)}(\bar{x}) + \varepsilon^2 C^{(2)}(\bar{x}) + \ldots ,$$
$$\frac{d\bar{\psi}}{dt} = \omega(\bar{x}) + \varepsilon D^{(1)}(\bar{x}) + \varepsilon^2 D^{(2)}(\bar{x}) + \ldots , \tag{9}$$

with no oscillating terms on the right hand side.

Inserting (8) in (7), using (9), and equating the coefficients of equal powers of ε on either side, we get again a sequence of equations. In first order we get

$$C^{(1)}(\bar{x}) + \omega(\bar{x}) \frac{\partial A^{(1)}(\bar{x}, \bar{\psi})}{\partial \bar{\psi}} = G(\bar{x}, \bar{\psi}),$$
$$D^{(1)}(\bar{x}) + \omega(\bar{x}) \frac{\partial B^{(1)}(\bar{x}, \bar{\psi})}{\partial \bar{\psi}} = H(\bar{x}, \bar{\psi}) \tag{10}$$

and in second order

$$C^{(2)} + \omega \frac{\partial A^{(2)}}{\partial \bar{\psi}} = A^{(1)} \frac{\partial G}{\partial \bar{x}} + B^{(1)} \frac{\partial G}{\partial \bar{\psi}} - C^{(1)} \frac{\partial A^{(1)}}{\partial \bar{x}} - D^{(1)} \frac{\partial A^{(1)}}{\partial \bar{\psi}} , \tag{11}$$

$$D^{(2)} + \omega \frac{\partial B^{(2)}}{\partial \bar{\psi}} = A^{(1)} \frac{\partial H}{\partial \bar{x}} + B^{(1)} \frac{\partial H}{\partial \bar{\psi}} - C^{(1)} \frac{\partial B}{\partial \bar{x}} - D^{(1)} \frac{\partial B}{\partial \bar{\psi}} ,$$

and so on. Again, in each order these equations are of the same form, with the right hand side known in terms of the solution of the lower order equations. The solution of the first order equation (10) is therefore typical. If $A^{(1)}$ and $B^{(1)}$ are to be periodic in $\bar{\psi}$, then clearly $C^{(1)}$ and $D^{(1)}$ must be chosen to cancel the phase average of the right hand side,

$$C^{(1)}(\bar{x}) = \langle G(\bar{x},\bar{\psi})\rangle_{phase} \equiv \frac{1}{2\pi}\int_0^{2\pi} d\bar{\psi}\, G(\bar{x},\bar{\psi}) ,$$
$$D^{(1)}(\bar{x}) = \langle H(\bar{x},\bar{\psi})\rangle_{phase} . \tag{12}$$

Then

$$A^{(1)}(\bar{x},\bar{\psi}) = \frac{1}{\omega(\bar{x})} \int d\bar{\psi}\, [G(\bar{x},\bar{\psi}) - \langle G(\bar{x},\psi)\rangle_{phase}] , \tag{13}$$
$$B^{(1)}(\bar{x},\bar{\psi}) = \frac{1}{\omega(\bar{x})} \int d\bar{\psi}\, [H(\bar{x},\bar{\psi}) - \langle H(\bar{x},\bar{\psi})\rangle_{phase}] .$$

In a manner similar to that of the Method of Averaging, equations (8) and (9) represent a separation of the slowly varying mean motion from the rapid motion of the phase. Here, too, in physical applications the mean motion will be identified with the observed motion.

It may appear that the Method of Rapidly Rotating Phase and the Method of Averaging are equivalent, in the sense that if one divides the first of equations (7) by the second one gets an equation of the form (1), with the time replaced by the phase. Application of the Method of Averaging to the resulting equation results in lowest order in an equation of mean motion which is the same as that obtained by dividing the first of equations (9) by the second. But in higher orders the results will be different.

The point is that in the Method of Rapidly Rotating Phase, the phase itself fluctuates about its mean.

A final remark about physical applications of these Averaging Methods. The equations of motion of the physical systems must be prepared in the appropriate form, such as (1) or (7), for application of the method. In general the key feature which enables one to do this is the existence of well-separated fast and slow motions of the system. Still, the discovery of the appropriate transformation to put the equations in standard form can require considerable ingenuity. This is especially so if we are to identify the mean motion as the observed motion since the quantity x must then be the physically observed quantity.

C. SPIN AND ORBITAL MOTION IN A WEAK TRAP[8]

In the typical g-2 measurements at Michigan and CERN the lepton moves in a nearly uniform magnetic field and one measures the rate of precession of the lepton spin relative to its velocity. This so-called difference frequency is, in the ideal case of motion perpendicular to a uniform magnetic field B_o, given by

$$\omega_D = a \frac{eB_o}{mc} \equiv a\omega_o \quad , \tag{14}$$

where a = g/2-1 is the g-factor anomaly. In the actual experiments the leptons must be held in a weak trap formed by slightly pinching the magnetic field or by applying a weak electric field. In either case there are small corrections to the simple formula (14) and our aim was to accurately and rigorously calculate these corrections up to terms of second order in the trap strength. Here as the quantitative measure of trap strength we choose the ratio of magnetic field variation to the applied magnetic field, $\Delta B/B_o$, for the magnetic trap or the ratio of electric trapping field to applied magnetic field, E/B_o, for the electric trap.

The equations of orbital motion of the lepton are:

$$\frac{d\vec{r}}{dt} = \vec{v} \quad ,$$

$$M \frac{d\gamma\vec{v}}{dt} = - e \left(\vec{E} + \frac{\vec{v}}{c} \times \vec{B}\right) . \qquad (15)$$

From these equations one can show that the lepton velocity rotates with (modified) cyclotron angular velocity

$$\vec{\omega} = \frac{e}{mc} \left(\frac{1}{\gamma} \vec{B} - \frac{1}{(\gamma^2-1)^{\frac{1}{2}}} \hat{v} \times \vec{E}\right). \qquad (16)$$

In a frame rotating with this angular velocity, the lepton spin precesses with angular velocity[9]

$$\vec{\Omega} = a \frac{e}{mc} \left(\vec{B} - \frac{\gamma -1}{\gamma} \hat{v}\cdot\vec{B}\, \hat{v} + \frac{1-a(\gamma^2-1)}{a\gamma(\gamma^2-1)^{\frac{1}{2}}} \hat{v} \times \vec{E}\right) . \qquad (17)$$

Clearly the spin precession will reflect the orbital motion.

For the case of motion nearly **perpendicular** to the applied magnetic field the lepton's orbital motion will consist of a rapid cyclotron rotation about the field direction, a slower longitudinal oscillation along the field direction of the plane of the cyclotron orbit, and a still slower motion of the center of the cyclotron orbit in its plane. We denote the angular frequencies of these three motions by ω_c, $\omega_{||}$ $\omega_{\perp}$, respectively. An inspection of the equations of motion show us that $\omega_{||}/\omega_c$ and $\omega_{\perp}/\omega_{||}$ are both of the order of the square root of the trap strength, so that for weak traps these three frequencies are well separated. We begin by introducing appropriate "guiding center" coordinates in order to put equations (15) into a form appropriate for the Method of Rapidly Rotating Phase, with the phase being the angle defining the position of the lepton in the plane of the cyclotron orbit. Application of the method allows us to separate the slower longitudinal and transverse motions of the cyclotron orbit from the rapid rotation of the phase at the cyclotron frequency. Our first important result is a theorem which states that the equations of mean motion obtained in lowest order from the Method of Rapidly Rotating Phase, that is the equations of motion obtained by simply phase-averaging the guiding center equations of motion, are correct up to second order in the trap strength. This result is not so obvious as it sounds since the original equations contain terms of the order of the square root of the trap strength. Thus, one should expect it would be necessary to carry the method through third order to get all the terms up to second order in the trap strength.

Having removed the rapid cyclotron motion, we are left with equations of mean guiding center motion which exhibit only the slow longitudinal oscillation and the still slower transverse motion. Again, for weak traps the frequencies, $\omega_{||}$ and $\omega_{\perp}$, of these motions are well separated so that averaging methods can be used, this time the Method of Averaging. The result is the demonstration that, to second order in the trap strength, the mean longitudinal motion is simply periodic with angular frequency $\omega_{||}$ and the mean transverse motion is a pure rotation about the applied field direction with angular frequency $\omega_{\perp}$.

When the spin equation of motion,

$$\frac{d\vec{S}}{dt} = \vec{\Omega} \times \vec{S} , \tag{18}$$

is appended to the equations of orbital motion, the above arguments can be repeated without change. The result is a demonstration that, still to second order in the trap strength, the spin motion is a precession with angular velocity $\langle\vec{\Omega}\rangle_{cyc}$, where the phase average is with respect to the cyclotron angle. Here there is an important technical point that this phase average must be performed in a coordinate frame rotating with <u>instantaneous</u> angular velocity

$\vec{\omega}$, given by (16). This is because one measures the spin direction relative to the velocity. This leads to some awkwardness, but does not change the essential result, which is that $\langle\bar{\Omega}\rangle_{cyc}$ is simply periodic with angular frequency $\omega_{||}$.

The spin motion therefore exhibits motion at two frequencies; a precession at the difference frequency $\omega_D \approx a\omega_o$, and a nutation driven at angular frequency $\omega_{||}$. This nutation arises from the small terms in $\langle\vec{\Omega}\rangle_{cyc}$ which oscillate with angular frequency $\omega_{||}$. Our analysis shows that these terms are smaller than $a\omega_o$, the nominal magnitude of $\langle\vec{\Omega}\rangle_{cyc}$, by a factor of the order of the square root of the trap strength. Therefore, when we transform to a coordinate frame rotating with angular frequency $a\omega_o$, the resulting equation is of the form (1) suitable for the Method of Averaging. The explicit time dependence of the right hand side is doubly periodic with angular frequencies $a\omega$ and the nominal magnitude of the right hand side is of the order of the square root of the trap strength. This means that, to get a result for the mean motion corrected up to second order in the trap strength, we must carry the Method of Averaging through third order. When we do this we get the following expression for the difference frequency, valid for the case of magnetic trapping.

$$\omega_D = a\frac{e}{mc}\left\{\left\langle\left\langle\hat{Z}\cdot\vec{B} + \frac{1-a(\gamma^2-1)}{a(\gamma(\gamma^2-1))^{\frac{1}{2}}}\hat{\rho}\cdot\vec{E}\right\rangle_{cyc}\right\rangle_{||} - \frac{1}{2}\sum_{n=-\infty}^{\infty}\frac{(1+2a)n^2\omega_{||}^2 - a^2\omega_o^2(\gamma^2-1)/\gamma^2}{n^2\omega_{||}^2 - a^2\omega_o^2}|\bar{\sigma}_n|^2\right\}. \tag{19}$$

Here $\hat{Z}$ points along the axis of the instrument, i.e., along the applied uniform magnetic field, and $\hat{\rho}$ points radially from the center of the cyclotron orbit and rotates with the cyclotron phase. In this expression $\langle\rangle_c$ is the cyclotron phase average, and $\langle\rangle_{||}$ is the time average over one period of the longitudinal motion. Finally,

$$\bar{\sigma}(t) = \sum_{n=-\infty}^{\infty} \bar{\sigma}_n \, e^{-in\omega_{||}t} \tag{20}$$

is the sine of what is called the pitch angle, the angle between the lepton's velocity and the plane of the cyclotron orbit. There is a similar formula for the case of electric trapping.

The formula (19) clearly shows the reasonances which occur when $n\omega_{||} = a\omega_o$. In the neighborhood of such a resonance the spin motion is more complicated than the simple difference frequency motion described by this formula, and we discuss it in our paper. But resonances are not important in g-2 machines, except that they must be avoided, so I will say nothing more of them here. However, the formula (19) gives rather different answers when the longitudinal frequency is far above or far below the difference frequency. Thus

$$\omega_D = a\frac{e}{mc} \Big\langle\Big\langle \hat{Z}\cdot\vec{B} + \frac{1-a(\gamma^2-1)}{a\delta(\gamma^2-1)^{\frac{1}{2}}}\, \hat{\rho}\cdot\vec{E}\Big\rangle_{cyc} - \frac{1}{2}\bar{\sigma}^2\Big\rangle_{||}, \qquad \omega_{||} \gg \omega_o, \tag{21}$$

and

$$\omega_D = a\frac{a}{mc} \Big\langle\Big\langle \hat{Z}\cdot\vec{B} + \frac{1-a(\gamma^2-1)}{a\gamma(\gamma^2-1)^{\frac{1}{2}}}\, \hat{\rho}\cdot\vec{E}\Big\rangle_{cyc} - \frac{\gamma^2-1}{2\gamma^2}\,\bar{\sigma}^2\Big\rangle_{||}, \qquad \omega_{||} \ll \omega_o . \tag{22}$$

This last expression is equivalent to that obtained by averaging the motion in uniform fields. The general expression (19) can be regarded as the correct interpolation between these limits.

The most important success of our calculation was the resolution of the outstanding discrepancy between the earlier electrons g-factor experiment of Wilkinson and Crane[11] and the later, more precise experiment of Wesley and Rich[12]. These experiments had both been analyzed using the expression (22) for the difference frequency, and the discrepancy was at the level of three standard deviations. But the Wilkinson-Crane instrument was such that the expression(21) applied. When this expression was used in the analysis the discrepancy disappeared.

In the analysis of the Wesley-Rich experiment, the general expression (19) was used. This resulted in only a very slight change in the measured g-factor anomaly. The analysis involves a solution of the equations of longitudinal motion in the effective trap potential, obtained from measurements of the azimuthally averaged magnetic field. We estimate the uncertainty in this solution (including magnetic field uncertainty) at less than 10%. Since the magnitude of the terms resulting from this solution was less than 10 parts per million (ppm) of the experimental number, we estimated the error from this source at less than 1 ppm. This was small compared with the overall experimental uncertainty of 3.5 ppm.

What about the next generation of experiments, which aim for an order of magnitude increase in precision? The trapping fields will be more precisely determined, which should allow an improvement in the accuracy of the solution of the equations of longitudinal motion. But an improvement by an order of magnitude to yield an uncertainty of less than 1% in the solution is I think too optimistic. Instead the traps will have to be weaker. In the proposed experiment at Michigan the trap strength (defined as ratio of trapping field to uniform field) will be reduced by a factor of 15 over that in the Wesley-Rich experiment,[13] allowing us to calculate the trapping corrections with confidence.

REFERENCES

1. Sara Granger and G. W. Ford, Phys. Rev. Lett. 28, 1479 (1972).

2. Sara Granger and G. W. Ford, Phys. Rev. 13D, 1897 (1976).

3. N.N. Bogoliubov and Y. A. Mitropolsky, "Asymptotic Methods the Theory of Non-linear Oscillations" (Gordon and Breach, New York, 1951).

4. K. M. Case, Prog. Theor. Phys. Suppl. 37, 1 (1966).

5. N. N. Bogoliubov and D. N. Zabarev, Ukrain. Mat. Zh. VII, 5, (1955).

6. N. Minorsky, "Nonlinear Oscillations" (D. van Nostrand Co., Princeton, N. J., 1962).

7. T. P. Coffey and G. W. Ford, J. Math. Phys. 10, 998 (1969).

8. For a description of the experiments see the accompanying talk by A. Rich.

9. There are two general ways to obtain this formula. The first is to treat the spin as a classical dynamical variable, and then to write down the most general equation of motion consistent with relativistic invariance. This was the method of L. H. Thomas, Phil. Mag. 3, 1 (1927), and J. Frenkel, Zeit. F. Physik 37, 243 (1926). The second method is to use the Foldy-Wouthuysen transformation of the Dirac equation. This was first done by H. Mendlowitz and K. M. Case, Phys. Rev. 97, 33 (1955). The formula is sometimes called the Bargmann-Michel-Telegdi formula, since the publication (Phys. Rev. Lett. 2, 435 (1959))of an equivalent result obtained by those authors using the first method above. It would be better to call it the Thomas formula since it appears explicitly and in full generality in the paper of Thomas.

10. G. R. Henry and J. E. Silver, Phys. Rev. 180, 1262 (1969); M. Fierz and V. L. Telegdi in Quanta, ed. P. G. O. Fruend (Chicago Univ. Press, Chicago 1970).

11. D. T. Wilkinson and H. R. Crane, Phys. Rev. 130 , 852 (1962).

12. J. C. Wesley and A. Rich, Phys. Rev. A4, 1341 (1971)

13. D. Newman, Private communication.

ELECTRON MAGNETIC MOMENT FROM GEONIUM SPECTRA*

Robert S. Van Dyck, Jr., Paul B. Schwinberg, and Hans G. Dehmelt (Presented by Hans Dehmelt)

University of Washington

Seattle, Washington

INTRODUCTION

Important preparatory contributions to the experiments to be described here were made by Philip A. Ekstrom, David J. Wineland, Fred L. Walls, and Talbert S. Stein. Shorter accounts have been published (Van Dyck et al., 1976B and 1977).

The magnetic resonance art is highly developed today. A good example is the resolution of 10^{10} now obtainable for proton NMR lines in the 10^{8} Hz region with commercial equipment. For the electron in similar fields spin and cyclotron frequencies ν_s, ν_c are $\simeq 10^{11}$ Hz with $\nu_a \equiv \nu_s - \nu_c \simeq 10^{8}$ Hz. All three are directly accessible. Forming $a \equiv (\nu_s - \nu_c)/\nu_c$ one obtains an important empirical dimensionless structural parameter for this truly elementary particle. It is amusing to compare a - values for e^-, e^+, μ^-, μ^+. The most accurate theoretical a - values have been obtained on the basis of Josephson effect input data. We quote here that given by Kinoshita and coworkers (1974) using analytical calculations by Levine and Roskies (1976),

$$a_{KJ} = 1\ 159\ 652\ 340\ (310) \times 10^{-12}.$$

* Work supported by the National Science Foundation.

The previously most accurate experimental value was that obtained by Wesley and Rich (1971),

$$a_{WR} = 1\ 159\ 656\ 700\ (3500) \times 10^{-12}.$$

We note that the error limit is $\simeq 3 \times 10^4$ larger than the cited NMR line width, suggesting ample room for possible improvement. Some of this has already been realized in our recent experiments at the University of Washington and our ratio

$$(\nu_s/\nu_c)_{UW} = 1.\ 001\ 159\ 652\ 410\ (200)$$

which is identical with the value of the electron magnetic moment in Bohr magnetons is now the most exactly measured parameter of an elementary particle.

PRINCIPLE OF EXPERIMENT

It would have been nice to do the experiment on a single electron at rest in free space. Actually, we settled for a single electron confined in a Penning trap at liquid helium temperature, Fig. 1. Such a trap provides an axial electric parabolic potential well, of course accompanied by a radial hill, superimposed on a strong (18.3 kG) magnetic field. It was our intent to study in maximum detail the electron orbit in the trap as to frequencies and amplitudes. The axial resonance at ν_z was directly observed continuously. The other resonances at ν_c' and ν_a' -- both slightly shifted by the electric field-- and at ν_m, the frequency of the magnetron or drift motion caused by the electric field, were observed indirectly by a kind of Stern-Gerlach effect. The latter was realized by a weak magnetic bottle, which caused ν_z to be slightly dependent on the spin state and the diameters of cyclotron and magnetron orbits. The ν_m -, ν_c' - , ν_a' - resonances would change one of the latter and therefore be signaled by a detectable shift in ν_z.

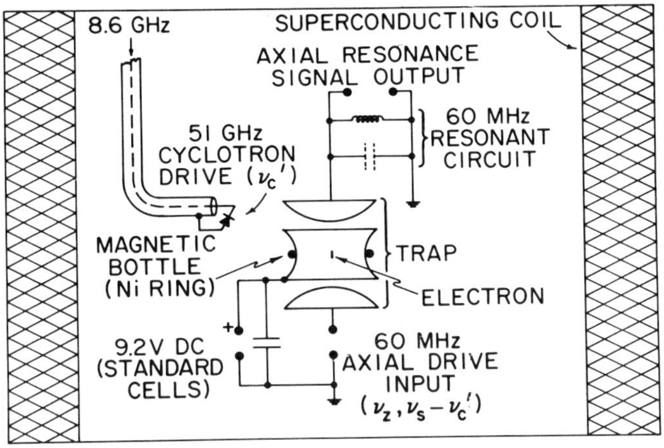

Fig. 1 Geonium spectroscopy experiment (schematic). This apparatus allows the measurement of the cyclotron frequency, ν_c' , and the spin-cyclotron-beat (or anomaly) frequency, $\nu_a' = \nu_s - \nu_c$, on a single electron stored in a Penning trap at $\simeq 4^oK$ ambient. Detection is via Rabi-Landau level-dependent shifts in the continuously monitored axial resonance frequency, ν_z , induced by a weak magnetic bottle.

THEORETICAL BACKGROUND

The damping times for the axial motion at ν_z coupled to the external LC circuit, τ_z, and the cyclotron motion at ν_c' coupled to the (free space) radiation field, τ_c, are about .02 sec and 1 sec respectively. Therefore these motions quickly reach thermal equilibrium and the axial and cyclotron orbit sizes in our experiment should be $\sim$ .1 mm and 600 Å, Fig. 2. The spin relaxation time should be $\simeq (r_o/\lambda\!\!\!^-_C)^2\tau_c \simeq 10^{10}$ sec; r_o and $\lambda\!\!\!^-_C$ are defined later. The characteristic decay time due to coupling with the (free space) radiation field for the magnetron motion should be $(\nu_c/\nu_m)^3\tau_c \simeq 10^{18}$ sec. The latter two motions should therefore be practically totally decoupled from the thermal bath. Nevertheless, by a special technique a magnetron radius $r_m \simeq .01$ mm has been realized so far.

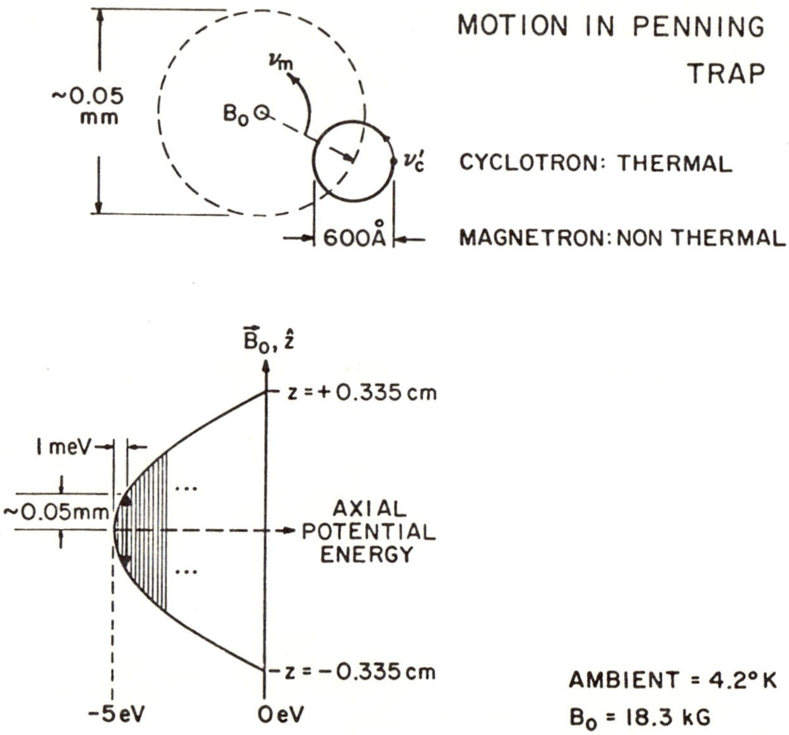

Fig. 2 Geonium orbits under thermal excitation at ≃ 4°K.

The equations of motion for the electron in the trap may be written:

$$\ddot{x} - \omega_z^2 x/2 + \omega_c \dot{y} = 0$$

$$\ddot{y} - \omega_z^2 y/2 - \omega_c \dot{x} = 0$$

$$\ddot{z} + \omega_z^2 = 0$$

with $\omega_z \equiv 2\pi\nu_z$, $\omega_c \equiv 2\pi\nu_c$. They yield the orbits, Fig. 2,

$$x + iy = r_c e^{i\omega_c' t} + r_m e^{i\omega_m t}$$

with $\nu_c' \equiv \omega_c'/2\pi \equiv \nu_c - \delta_e$, $2\delta_e \nu_c' = \nu_z^2$. For a perfect trap holds $\nu_m \equiv \omega_m/2\pi = \delta_e$. Misalignment of electric and magnetic axes leads to $\nu_m - \delta_e > 0$. Nevertheless the relation $2\delta_e \nu_c' = \nu_z^2$ is practically invariant to small misalignments and calculations (Brown, 1976) confirm $\{\text{correction to } \delta_e\} \ll \nu_m - \delta_e$. The quantum mechanics of the system (Sokolov and Pavlenko, 1967; Landau and Lifshitz, 1965; Rabi, 1928) holds no particular surprises. It yields the equidistant energy levels, see Fig. 3,

$$h^{-1}E_{mnkq} = m\nu_s + (n + 1/2)\,\nu_c' + (k + 1/2)\,\nu_z - (q + 1/2)\,\nu_m$$

$$m = \pm 1/2\ ;\ n, k, q = 0, 1, 2, 3 \ldots\ .$$

Also one obtains for the cyclotron and magnetron orbit radii:

$$r_c \simeq (2n + 1)^{1/2} r_o\ ,\ r_m \simeq (2q + 1)^{1/2} r_o\ ,\ r_o \simeq (\lambda\!\!\!^{-}_c \lambda\!\!\!^{-}_C)^{1/2} \simeq 200\ \text{Å},$$

where $\lambda\!\!\!^{-}_c$ and $\lambda\!\!\!^{-}_C$ are Cyclotron and Compton wavelengths respectively. We look upon the single electron bound via magnet and trap structure to the earth as a metastable pseudo-atom which has been synthesized optimally for the measurement of the g - factor. We call this pseudo-atom "geonium".

When a shallow magnetic bottle with field components

$$b_z = \beta\,(z^2 - r^2/2),\ b_y = -\beta zy,\ b_x = -\beta zx\ ,\ \beta \simeq 120\ \text{G/cm}^2$$

is superimposed on the large homogeneous field $B_z = B_o$, $B_x = B_y = 0$, the interaction of the spin magnetic moment, the cyclotron and

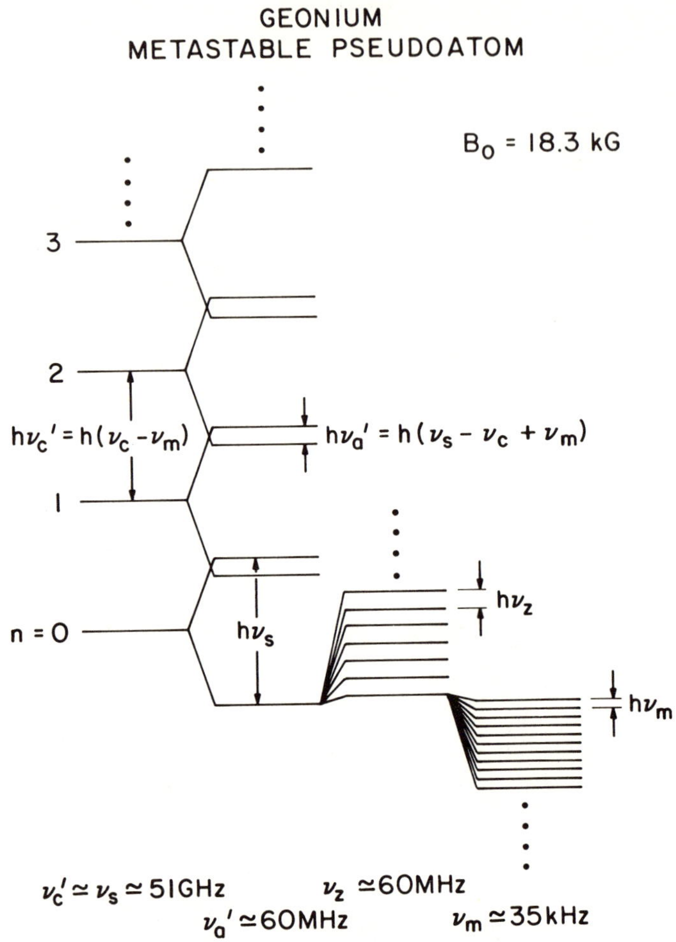

Fig. 3. Geonium levels for ≃ 5 eV axial well and 18 kg magnetic field.

magnetron orbits with the inhomogenous bottle field leads to slight additional axial restoring forces which shift the axial frequency by

$$\delta\nu_z \simeq [m + n + 1/2 + (\nu_m/\nu_c)q]\,\delta$$

$$\delta \simeq \mu_B\beta/2\pi^2 m\nu_z \simeq 1\text{Hz}.$$

For large n, q this becomes

$$\delta\nu_z \simeq [r_c/r_o)^2 + (\nu_m/\ \nu_c)(r_m/r_o)^2]\delta/2 \ .$$

THE AXIAL RESONANCE

Inspection of the differential equation describing the interaction of the elastically bound electron with a rf generator connected to the cap electrodes shows that it behaves effectively like a LC circuit series-resonant at ν_z (Wineland and Dehmelt, 1975). This is exploited in the detection circuit shown in Fig. 4. The output voltage from a very stable frequency synthesizer is applied through an attenuator to one of the caps of the Penning trap. On resonance current flows through this LC circuit to the other cap and the signal voltage is developed across the resistive shunt impedance of the external physical LC circuit. The signal is amplified by a low noise Ga As field-effect transistor submerged in liquid He followed by more conventional amplifiers and fed to one input of a phase-sensitive detector. An output of the frequency synthesizer is applied through a phase shifter to the other input. The phase-sensitive detector may be thought of as a synchronous commutator whose switching action rectifies the signal which may be applied to a pen-recorder. By sweeping the synthesizer frequency the resonance curves shown in Fig. 5 have been obtained. In the normal operation of the circuit the phase shifter is set for the dispersion mode and the DC signal is integrated and added to the trap voltage closing the feed back loop. In this way the frequency ν_z of the electron is locked to that of the synthesizer and also a DC output proportional to any shift $\delta\nu_z$ in the natural frequency is made available. This shift $\delta\nu_z$ is a most important observable in the following. Focussing on the floor of the noise pattern, Fig. 6 demonstrates the capacity of the apparatus to detect $\delta\nu_z$ shifts <<1 Hz out of 60 MHz.

An important check of the electric potential actually realized

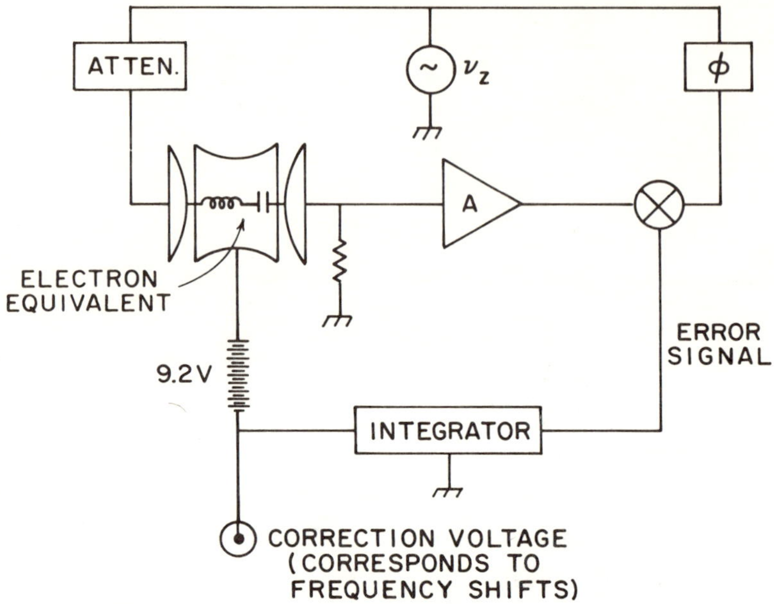

Fig. 4 Detection circuit for the axial resonance at ν_z. The electron acts effectively like an LC series resonant circuit connecting the cap electrodes. The circuit also locks the electron frequency to that of the very stable generator.

in the trap in the presence of contact potentials not easily controlled may be executed by varing the trapping potential V_o and fitting the data to $\nu_z^2 = k(V_o + V_c)$. A typical result was $V_c = 62 \pm 1$ mV vs $V_o \simeq 10$ V.

THE PENNING TRAP TUBE

Things were not quite as easy as it might appear from the preceding sections. In a standard cap-ring-cap geometry, the potential will be given by

$$\phi = A\,(r^2 - 2z^2) + B\,(r^4 - 24r^2z^2 + 8z^4) + \ldots$$

Because of the truncation of the electrodes, it was found empirically

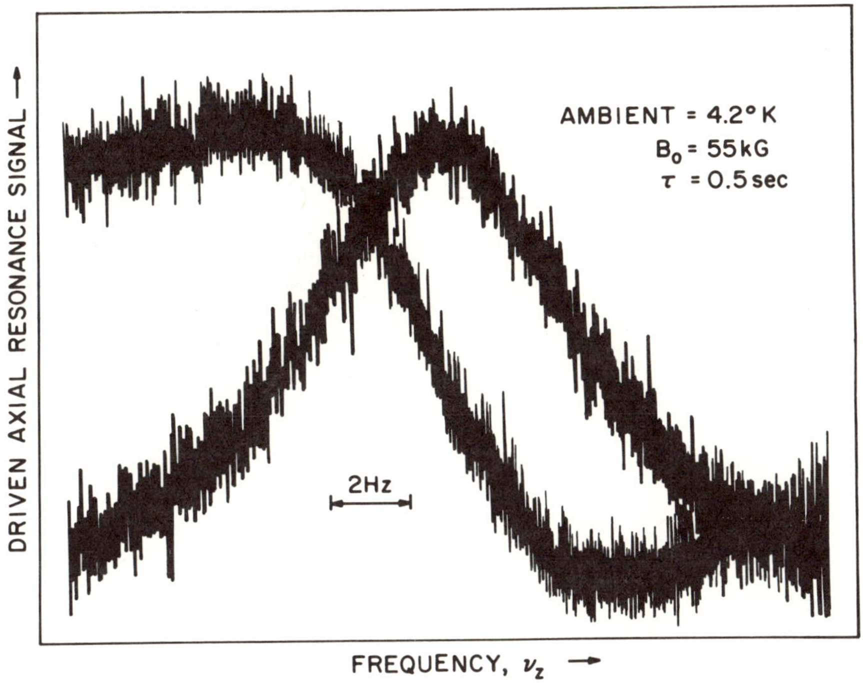

Fig. 5 Axial resonance signals at ≃ 60 MHz, note 8 Hz width. Absorption and dispersion modes are shown.

that the z^4 term led to excessive broadening of the axial resonance and a new anharmonicity-compensated 5-electrode tube (Van Dyck et al., 1976A), Fig. 7, was developed, reducing the line-width 100-fold. Besides the guard rings serving to compensate the B-term, the tube incorporates many other innovations. The field emission cathode, when pulsed on, emits a beam of 1 KeV electrons whose impact liberates adsorbed gas and then knocks off slow electrons from the gas inside the trap, filling it. A photograph of the tube built by Van Dyck and Schwinberg, which justifiedly may be called a masterpiece of the experimental art, is shown in Fig. 8. A single electron is isolated in a fashion similar to that used by Millikan, see Fig. 9.

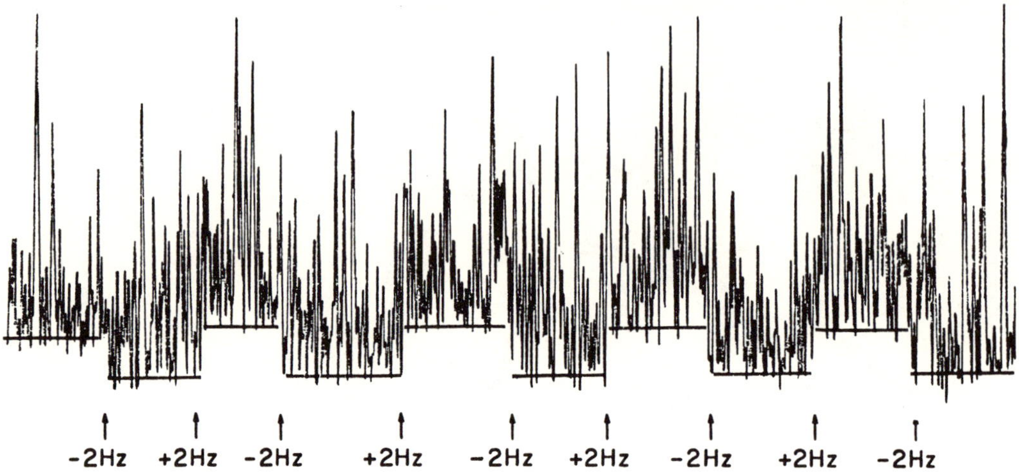

Fig. 6 Control and calibration run obtained with earlier 2.5 Hz magnetic bottle. The asymmetric noise spikes are due to thermal excitation ($\simeq 4^{o}K$) of the cyclotron motion. The 2 Hz changes in the floor were obtained by changing the generator frequency by 2 Hz.

THE MAGNETRON RESONANCE

The magnetron resonance at ν_m was observed making use of the previously discussed slight dependence of ν_z on the magnetron orbit radius,

$$\delta\nu_z \simeq (\nu_m/\nu_c)\ (r_m/r_o)^2\delta/2.$$

When the (undamped) resonance is driven near ν_m, both driven and free motions are excited. Their beat modulates r_m which in turn

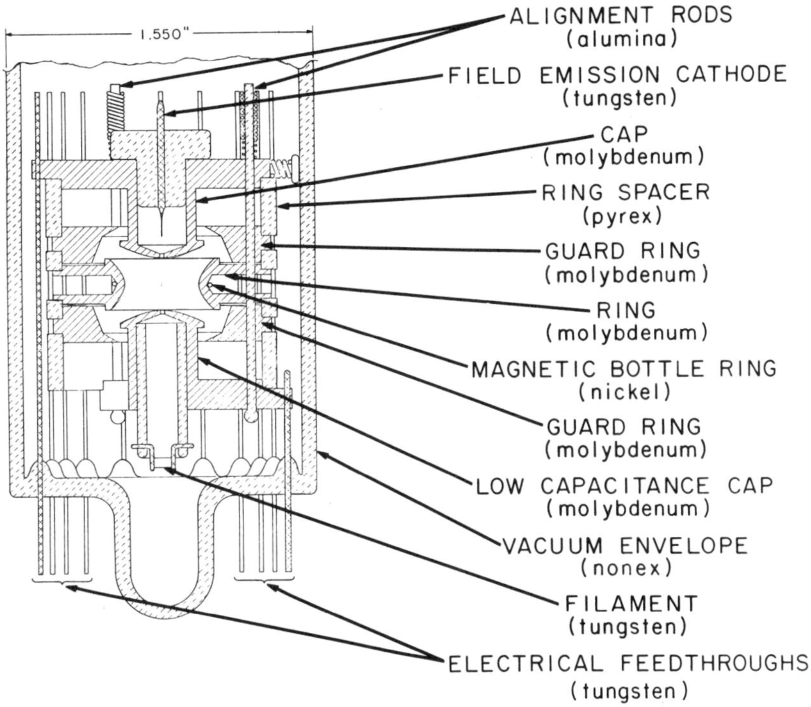

Fig. 7 5-electrode 1975 tube. The standard cap-ring-cap Penning trap configuration is augmented by two guard rings. By critically adjusting the guard potential the ν_z - line width is reduced 100-fold.

Fig. 8 Photograph of 5-electrode 1977 tube. The left half houses the trap, the right ion-getter and cryosorption pumps capable of < 10^{-14} torr vacua.

modulates $\delta\nu_z$, see Fig. 10. The necessary rf field near ν_m is generated by an antenna slightly protruding between one cap and ring into the trap region.

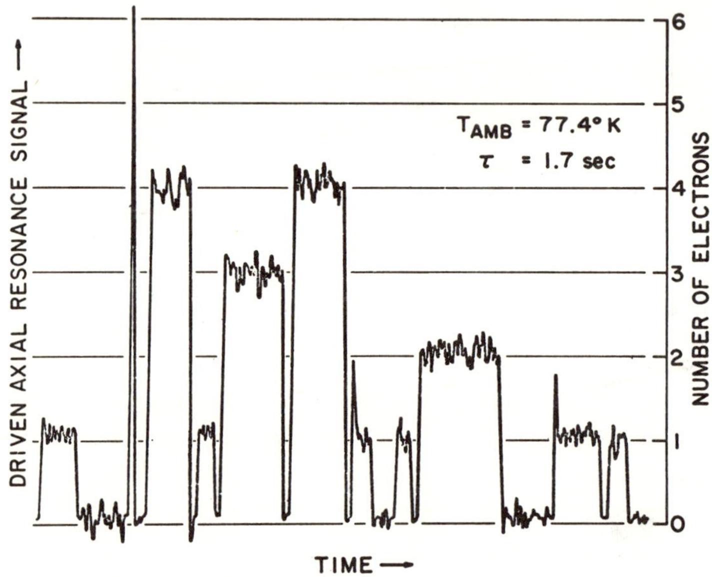

Fig. 9 Identification of a single electron in the trap. The detection circuit is adjusted to work on the wing of the ν_z-resonance, yielding a signal proportional to the total trapped electron number. Electrons are alternately liberated in and ejected from the trap, creating the quantized pattern.

So that the electron may see reproducible fields it would be highly desirable to center it at the bottom of the electric and magnetic wells. This may indeed be achieved by a new technique developed by us. When the damped axial motion is driven by an auxiliary inhomogeneous rf field $\nu_z + \nu_m$, the electron will see a sideband at

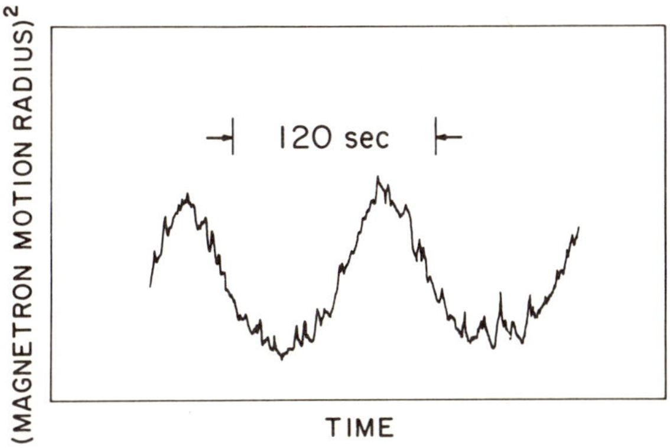

Fig. 10 Beat between free and driven magnetron motions at 35,052.628 and 35,052.620 Hz. Driving the completely undamped magnetron motion only ≃ 10mHz away from the resonance creates a slow beat from which the resonance frequency may be determined to ≃ 1 mHz.

ν_z because of the ν_m motion through the rf field. As it is absorbing $h(\nu_z + \nu_m)$ quanta, while the damping turns $h\nu_z$ quanta into heat, the balance $h\nu_m$ can only go into the magnetron motion, shrinking r_m. Experiments, see Fig. 11, confirm this strikingly, and also the expanding of r_m when the drive occurs at $\nu_z - \nu_m$. The observed (expected) exponential growth of r_m may be analyzed and thereby the minimum value r_m (min) ≃ .014 (7) mm, realized by a previous shrinkage treatment, may be obtained.

THE CYCLOTRON RESONANCE

As in our experiment we have $h\nu_c/kT \simeq .61$ a few low cyclotron levels are randomly excited thermally, leading to corresponding asymmetric fluctuations in $\delta\nu_z$ because of

$$\delta\nu_z \simeq (n + 1/2)\,\delta \ .$$

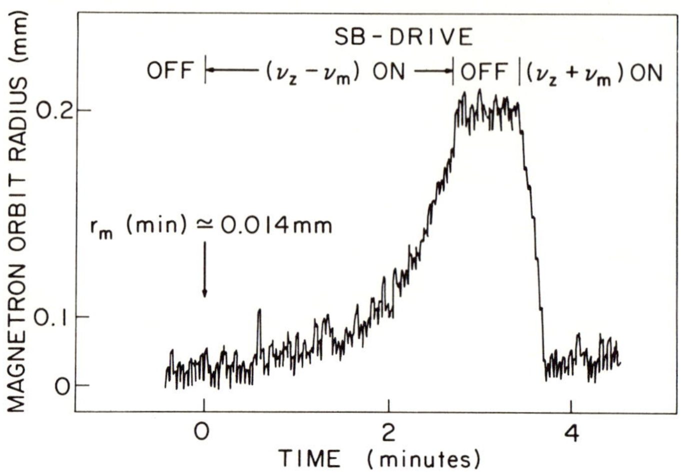

Fig. 11 Expanding and shrinking of magnetron orbit radius. By driving the axial motion on the side bands $\nu_z \pm \nu_m$, it is possible to force the magnetron motion at ν_m to absorb/provide the energy balance $h\nu_m$ and thereby shrink/expand the magnetron orbit.

Corresponding up-spikes are seen in Figs. 6, 14a and c. Much higher n-levels may be excited when an external rf drive near ν_c' is employed. This 51 GHz drive was provided by feeding a small Schottky diode mounted near the cap-ring gap, which acted simultaneously as a frequency multiplier and antenna, 8.6 GHz power through a small coaxial cable, Fig. 1. In order to obtain the smallest line width possible in the presence of the magnetic bottle, the ν_z drive was reduced to the lowest value at which the frequency locking circuit, Fig. 4, would operate. Tests showed that doubling the ν_z - drive power would cause no noticeable shift of the resonance pattern, Fig. 14c. This implied that the driven excitation was appreciably smaller than the thermal one. It is instructive to estimate the magnetic bottle related shifts in the ν_c' - value obtained in this fashion. The finite $r_m \simeq .014$ mm gives with $b_z \simeq -\beta r^2/2$, β=120 G/cm^2, B $\simeq$ 18 kG,

$$(\delta\nu_c')_m/\nu_c' \simeq - 7 \times 10^{-9}.$$

Thermal excitation, $kT_z \simeq$ 1meV , of the axial motion in the eD=5eV deep, $2z_o$ =.67 cm wide well gives

$$(\delta\nu_c')_z/\nu_c' \simeq (kT_z/eD)\ (z_o^2\ \beta/B_o) \simeq 1.5 \times 10^{-7}$$

for the shift of the center of gravity of the line. The relativistic shift, assuming an average value $\bar{n} \simeq 15$ under excitation, Fig. 14c, is

$$(\delta\nu_c')_r/\nu_c' \simeq - mv^2/2mc^2 \simeq 15h\nu_c/mc^2 \simeq - 6 \times 10^{-9}.$$

Comparing the measured ν_m - value with δ_e from $2\delta_e\ \nu_c' = \nu_z^2$ yields a valuable check of the trap symmetry. A typical value found was

$$\nu_m - \delta_e \simeq 3.7 \text{ Hz}$$

implying

$$(\text{correction to } \delta_e) \ll 3.7 \text{ Hz}.$$

THE ANOMALY RESONANCE

For the purpose of flipping the spin and measuring the difference $\nu_s - \nu_c' = \nu_a'$ the magnetic bottle previously introduced is put to a completely different use, see Fig. 12. Assuming $r_m = 0$ the electron moving in its cyclotron orbit through the magnetic bottle will see the radial field components as a field rotating at ν_c'. In addition the sign of the radial field components will reverse each time the electron crosses the z = 0 plane. A forced auxiliary axial oscillation at ν_a' will therefore split the rotating field into two components at $\nu_c' \pm \nu_a'$. As by definition $\nu_a' \equiv \nu_s - \nu_c'$ it follows that $\nu_c' \pm \nu_a' = \nu_s$ and spin flips will now occur. Since choosing $\nu_a' = \nu_z$ and relying on thermal excitation of the axial motion led to an excessive rate of spin flips, it follows that for $\nu_a' \neq \nu_z$ the auxiliary axial motion induced at ν_a' need only have an amplitude much smaller than the thermal one. As $\delta\nu_z$ (spin-flip) amounts to only δ = 1 Hz it was desirable to operate the locking circuit with a maximum signal/noise ratio which implied a relatively large axial drive, causing undesirable bottle shifts and broadening of the ν_a' resonance. The following scheme was therefore adopted: 1/2-minute periods in which the drive near $\nu_a' \simeq \nu_z$ - 200 KHz is on and the axial motion is practically thermal alternate with periods in which the ν_a' drive is off and the strong ν_z - drive desirable for optimal resolution in $\delta\nu_z$ - shifts is pulsed on. The ν_a' resonance is associated with a simultaneous $\Delta n = -1$, $\Delta m = +1$ transition which does not cause an immediate change in $\delta\nu_z$, see Fig. 13. The effect which we observe, see Fig. 14a, is that the floor of the noise pattern created by thermal cyclotron excitation changes by 1 Hz when the spin state changes from m =-1/2 to m = +1/2. By varing the ν_a' drive frequently in small steps, and measuring the rate of spin flips, the resonance curve shown in Fig. 14b has been obtained. The shift in ν_a' due to the thermal axial motion in the magnetic bottle should be the same as that estimated for ν_c', likewise that for finite r_m. It is, however,

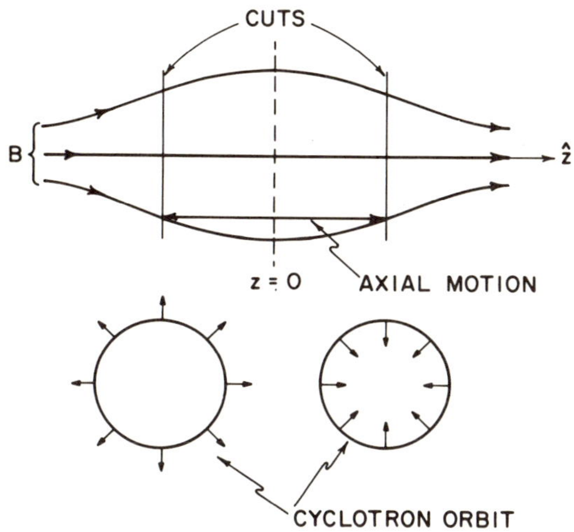

Fig. 12 Mechanism for inducing spin flips by the electron's axial motion in the magnetic bottle. From the electron's frame of reference, a magnetic field is seen rotating at ν_c', but modulated by the axial motion at ν_a', yielding sidebands at $\nu_c' \pm \nu_a'$ with $\nu_s = \nu_c' + \nu_a'$.

necessary to discuss the relativistic shift in ν_a' even at our low energies. Mendlowitz and Case (1955) have shown that only the axial kinetic energy $mv_z^2/2 = kT_z/2$ causes a relativistic shift.

$$(\delta\nu_a'/\nu_a')_r \simeq - kT_z/2mc^2 \simeq - 10^{-9} \quad .$$

RESULTS AND DISCUSSION

One or more runs are carried out on the same electron. Important experimental results are summarized in Table 1. Sample data for the frequencies ν_z, ν_m, ν_c', ν_a' measured in one run are given first.

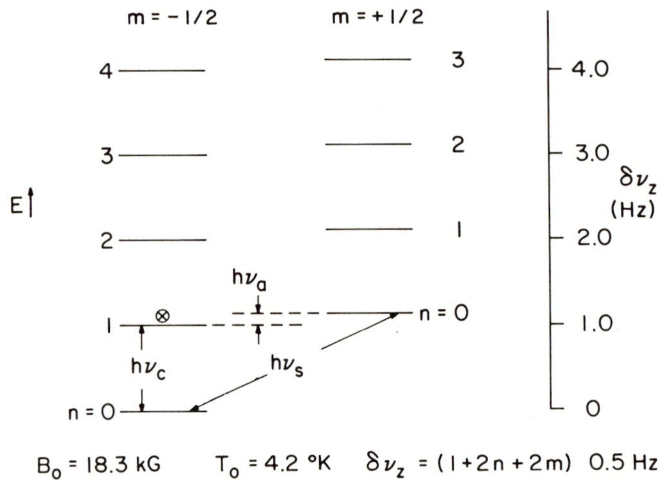

Fig. 13. Electron Rabi-Landau levels and associated magnetic bottle shifts $\delta\nu_z$.

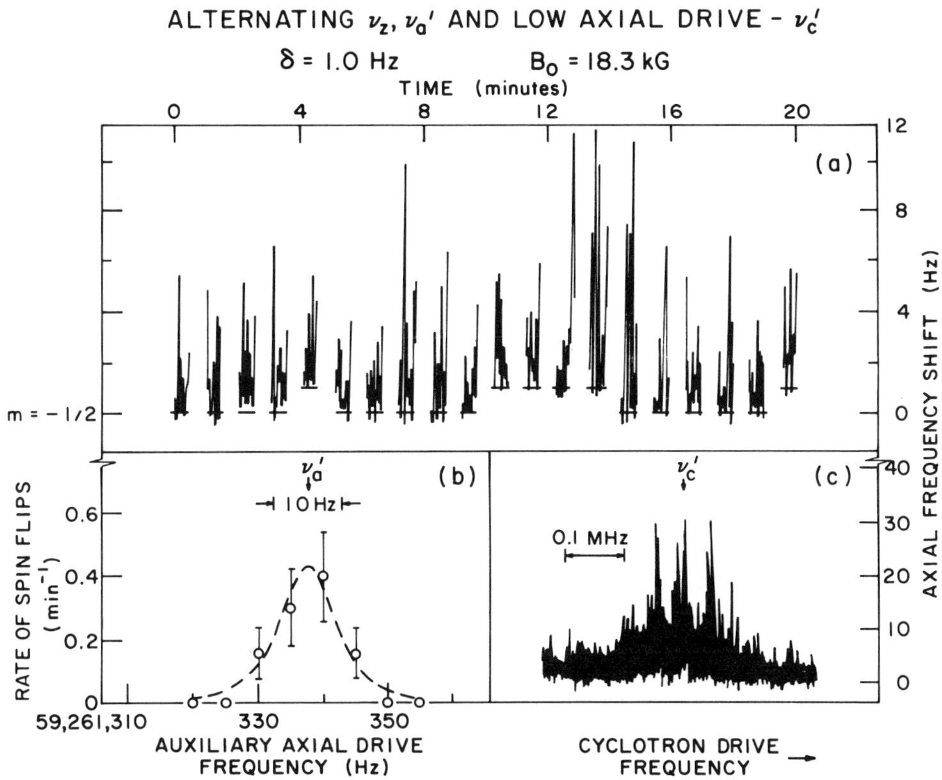

Fig. 14 Geonium anomaly and cyclotron resonance data. In (a) excitation of the transition at $\nu_a' \equiv \nu_s - \nu_c'$ is alternated with recording the axial frequency shift $\delta\nu_z$. Random 1Hz shifts of the noise pattern floor appear, corresponding to a spin flip. In (b) a plot of the rate of spin flips vs. excitation frequency is shown. The sharp cyclotron resonances in (c) at ν_c' and $\nu_c' \pm \nu_m$ were obtained by minimizing the axial motion through the inhomogenous magnetic bottle field through the use of a very weak axial locking drive.

Table 1

SAMPLE DATA & RESULTS

MEASURE (1 RUN)

$\nu_z = 59\ 336\ 170.14 \pm .10$ Hz

$\nu_m = 34\ 471.9 \pm .1$ Hz

$\nu_c' = 51\ 072\ 915 \pm 10$ Hz

$\nu_a' = 59\ 261\ 337.5 \pm 4.5$ Hz

CALCULATE

FROM $2\delta_e \nu_c' = \nu_z^2$

$\delta_e = 34\ 468.18 \pm .1$ Hz

MEASURE TRAP INPERFECTIONS

$\nu_m - \delta_e = 3.7$ Hz , $(\nu_m - \delta_e)/\delta_e \simeq 10^{-4}$

(CORRECTION TO δ_e) << 3.7 Hz

OBTAIN (8 RUNS)

WITH $\nu_c = \nu_c' + \delta_e$, $\nu_s - \nu_c = \nu_a' - \delta_e$,

$a \equiv (\nu_s - \nu_c)/\nu_c$

$a = 1\ 159\ 652\ 410\ (200) \times 10^{-12}$

200×10^{-12} IN a → 10 Hz IN ν_a'

The electric field correction δ_e is calculated from ν_c' and ν_z. Comparison of δ_e with the measured ν_m shows that corrections due to trap imperfections to δ_e should be a fraction of 1 Hz. Zero E-field values ν_c and $\nu_s - \nu_c \equiv \nu_a$ are then calculated from the raw ν_c', ν_a', δ_e data which vary slightly from run to run. For each run the a-value is determined. The average a-value obtained for eight 18.3 kG-runs is displayed as our final result. No corrections corresponding to the previously discussed small shifts in ν_c' and ν_a' need be applied at this stage of accuracy claimed, .2 ppm, because the shifts are either too small or affect ν_c' and ν_a' in the same way. Efforts to analyze line shapes are under way, but have so far not been relied on, .2 ppm being equivalent to the full 10 Hz line width realized for ν_a'. A plot comparing our data with theoretical values obtained by three different groups is shown in Fig. 15, demonstrating generally good agreement. The preliminary

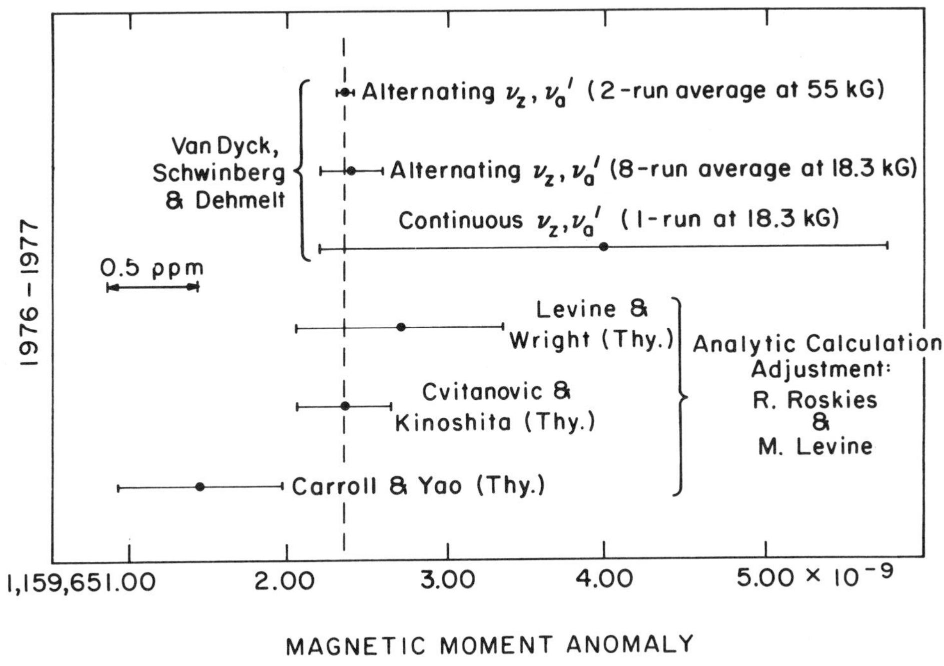

Fig. 15 Plot of recent theoretical and experimental anomaly data.

data obtained from two 55 kG-runs have not been incorporated in the a-value quoted by us. Further efforts are under way to approach closer the shining example of 10^{10} resolution realized in NMR work. Work has also begun on a single positron. The ramifications of the higher accuracies now available in the experimental a-values of leptons is discussed in a paper by Konoshita, also published in these proceedings.

REFERENCES

1. Brown, L., (1976). Private communication
2. Carroll, R., (1975). Phys. Rev. D12, 2344.
3. Cvitanovic, P., and Kinoshita, T. (1974). Phys. Rev. D10, 4007.
4. Landau, L, and Lifschitz, E., (1965). Quantum Mechanics, Pergamon Press, § 111.
5. Rabi, I., (1928). Z. Physik 49, 507.
6. Levine, M., and Roskies, R. (1976). Phys. Rev. D14, 2191.
7. Mendlowitz, H., and Case, K., (1955). Phys. Rev. 97, 33.
8. Van Dyck Jr., R., Wineland, D., Ekstrom, P., and Dehmelt, H. (1976A). Applied Physics Letters 28, 446.
9. Van Dyck Jr., R., Ekstrom, P., and Dehmelt, H. (1976B). Nature 262, 776.
10. Van Dyck Jr., R., Schwinberg, P., and Dehmelt, H. (1977). Phys. Rev. Letters 38, 310.
11. Wesley, J., and Rich, A., (1971). Phys. Rev. A4, 1341.
12. Sokolov, A., and Pavlenko, Y., (1967). Optics and Spectroscopy 22, 1.
13. Wineland, D. and Dehmelt, H., (1975). J. Applied Physics 46, 919.

MUON ANOMALOUS MAGNETIC MOMENT - THE LAST WORD

Robert W. Williams

University of Washington

Seattle, Washington

The great series of muon g-2 experiments at CERN has come to an end with final publication[1] of the results of the third experiment (Table 1). The accuracy of 8 parts per million (ppm) in the anomaly, or 9 parts per billion in the g factor, is better than was forecast in the proposal and, in fact, challenges the accuracy of the purely QED part of the theoretical calculation and equals the accuracy of the calculated hadronic correction (Table 2).[2] The g-2 experiments have not yielded clues to new physics but instead have strongly reinforced confidence in the possibility of carrying particle physics of our presently understood type to great heights of subtlety and sophistication. Unconventional effects tend to change the anomaly by an amount proportional to m^2, which led to the hope that the muon might reveal something remarkable if pressed hard enough. It now appears that further refinement in the measurement is unlikely for a long time and that the remarkable agreement of theory and experiment will remain at the 10 ppm level. This agreement can be variously interpreted: The sixth-order QED corrections are verified to a few percent; a cutoff, Λ, in the photon propagator would have to be more than 18 GeV; the existence of hadronic polarization of the vacuum is established, and its magnitude verified to ~ 15%; various off-beat theoretical schemes (e.g., Georgi-Glashow neutral lepton gauge theory) can be ruled out or severely restricted; a large (~10^{-18} e·cm) electric dipole moment of the muon is ruled out; CPT for muons is verified by equality of g_{μ^-} and g_{μ^+} to 0.02 ppm.

I will explain the principles of this recent measurement and show why, in my opinion, it is not likely to be soon repeated with much greater precision.

Experimentally, the muon problem is dominated by four facts: Muons are rare; they are short-lived; they are born with longitudinal polarization; and they yield statistical information on their polarization direction when they decay. Experiments on them are typically high energy, so that the classical description of motion of the polarization vector is appropriate. The difference frequency

$$\omega_a = \left(\frac{g-2}{2}\right)\frac{eB}{mc} = a\,\frac{eB}{mc}$$

is directly observable, and is independent of momentum. By comparing it with the spin resonance frequency, $\omega_s = (1+a)\,\frac{eB}{mc}$, of muons at rest in the magnetic field, we find the anomaly as the ratio of the two frequencies,

$$a = \frac{\omega_a/\omega_s}{1-(\omega_a/\omega_s)}\,.$$

The principle of observing ω_a is as follows: Collect a sample of N longitudinally polarized muons trapped in a magnetic field. At t = 0 the polarization vector is parallel to $\vec{v}$, say, and the sample begins its history of revolving in a cyclotron orbit and slowly decaying (Fig. 1). The polarization vector turns a little faster than the velocity so that the picture as time advances appears like the subsequent illustrations in Figure 1. When one of the muons in the sample decays the momentum of the decay electron is correlated with the spin direction of that muon in the typical $1+A\cos\phi$ manner. Collecting this information as the sample decays, we have a way of detecting the growth of the angle between $\vec{P}$ and $\vec{v}$, an angle which is given by $\theta = \omega_a t$. The problem is to infer something about the electron momentum $\vec{P}_e$ in the muon rest frame from the observed electron momentum in the laboratory. This appears unpromising since the very large Lorentz boost (in our experiment $\gamma = 29$) projects all momenta into the forward direction. However, Francis Farley noted that the energy of the electron in the lab is an excellent indicator of the direction of its momentum in the decay frame, and if one requires that the lab energy be near its maximum value, one is surely restricting the sample to those decay electrons which were projected forward in the muon's rest frame. Thus, instead of observing the growth of the angle theta directly (Figure 2), one would observe the modulation of the high energy component of decay electrons, which is caused by the rotation of the polarization vector with respect to the forward direction.

Measuring time from the injection of the sample, the counting rate of high energy decays will go as

$$N_E(t) = N_o \exp(-t/\tau)[1-A_E\cos(\omega_a t + \phi_o)] \quad . \quad (1)$$

Obviously if one counts all decay electrons with no energy discrimination there is no modulation. The optimum stastical information is obtained in this method if data from a large number of energy bands is collected independently. In fact, this is the method used in CERN experiment number 3. Some statistical power is lost because $\vec{P}_{e,cm}$ is not observed directly, but it is not a large effect and in any case is probably not recoverable in a practical experiment.

From a modulated exponential of the type shown in Figure 2, the relative accuracy of the frequency will depend on the number of cycles per mean life, $\omega\tau$. It is not hard to show that the approximate accuracy to be expected from extracting ω from the expression shown in Eq. (1) is

$$\frac{\sigma_\omega}{\omega} \simeq \frac{\sqrt{2}}{\omega\tau A\sqrt{N}} \quad . \tag{2}$$

This expression includes the effect of finding the phase ϕ_o, which cannot be determined *a priori* with sufficient accuracy, but must be obtained from the data (fixing ϕ_o, if possible, would reduce $\sqrt{2}$ to 1).

Of the factors in Eq. (2), τ is fixed because, as we shall see, γ is fixed. A large value of A requires primarily that the polarization of the sample be near unity and the energy resolution be good. Large ω requires large B. The total number of events observed in the experiment, N, is restricted by the rate at which events can be collected, multiplied by the longevity of the experimenters. In the recent CERN experiment, the process of trapping a sample of muons, setting the clock to zero, and observing the decay times of the energetic electrons was repeated several million times, so that the total then was approximately 10^8 events. A corresponding average value of A would be 0.2. With B = 14 KG, and γ = 29, $\omega\tau$ is ~100, and $\frac{\sigma_\omega}{\omega}$ is about 1.4 x 5 = 7 ppm. Other sources of uncertainty raise this to the final error of 8 ppm.

The realization of the experiment is illustrated in Figure 3. An annular vacuum chamber 14 meters in diameter was contained in a ring of 40 magnets which were shimmed to an extraordinary degree of uniformity. Pions from the CERN proton-synchrotron were injected into the vacuum chamber and a small fraction of them decayed into forward-going muons which remained trapped in the chamber

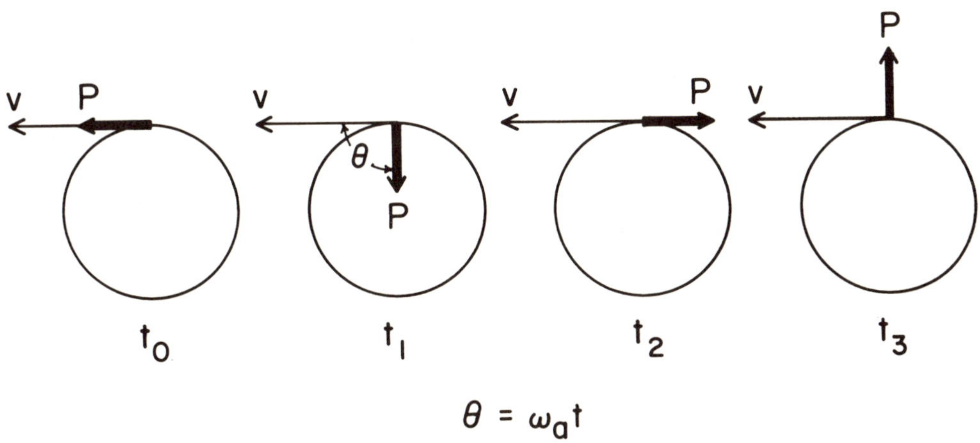

Figure 1. Successive snapshots of the polarization vector as a muon sample circles in the magnetic field. There are about seven rotations between snapshots.

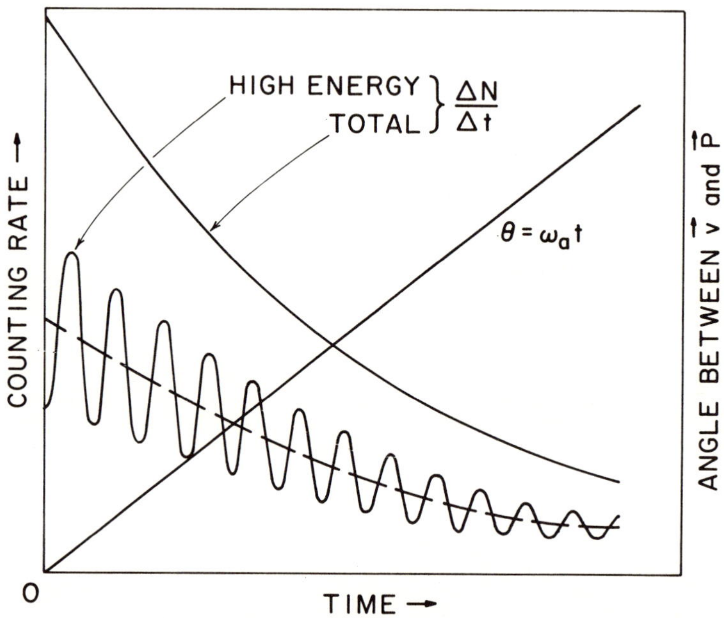

Figure 2. Time variation (schematic) of the total decay-electron counting rate, the counting rate of high-energy electrons, and the angle between < $\vec{P}$ > and <$\vec{v}$>.

with $\vec{P}$ parallel to $\vec{v}$. The muons were prevented from escaping vertically by a vertical focusing electric quadrupole field. This meant that transverse electric fields were present and it is evident that in general such fields could upset the relationship between $\vec{v}$ and $\vec{P}$. However, the equations of motion show that for a positive anomaly (that is, the case where the polarization turns faster than the velocity), the effect of transverse electric fields changes sign at a value of γ given by

$$\gamma_{magic} = \sqrt{\frac{1}{a} + 1} = 29.3 \quad .$$

Remarkably, this value of gamma corresponds for muons to a momentum of 3.1 GeV/c and therefore a very practical working energy. Central momentum for the trapped sample was therefore fixed at this "magic" value of gamma and electric field effects on those muons which were somewhat off momentum were only second order and below the 1 ppm level. Figure 3 shows a cross section of the vacuum chamber with its specially shaped electrodes. Figure 4 shows some of the energy-sensitive electron detectors mounted next to the inner wall of the vacuum chamber.

Possibly the most heroic effort in the 5-year life of this massive experiment was the magnetic field, its shimming, regulation, and measurement. Regulation and measurement involved over 80 NMR probes, and a full field map placed a quarter of a million measurements on tape via computer control. The azimuthally-averaged field was uniform to a few parts per million over the significant regions of the chamber. The contributions of various field uncertainties can be understood from Table 3. Proton resonance was used as the transfer frequency between muon spin resonance, which had been measured in an entirely different precision experiment, and the field in which the g-2 observations were made.

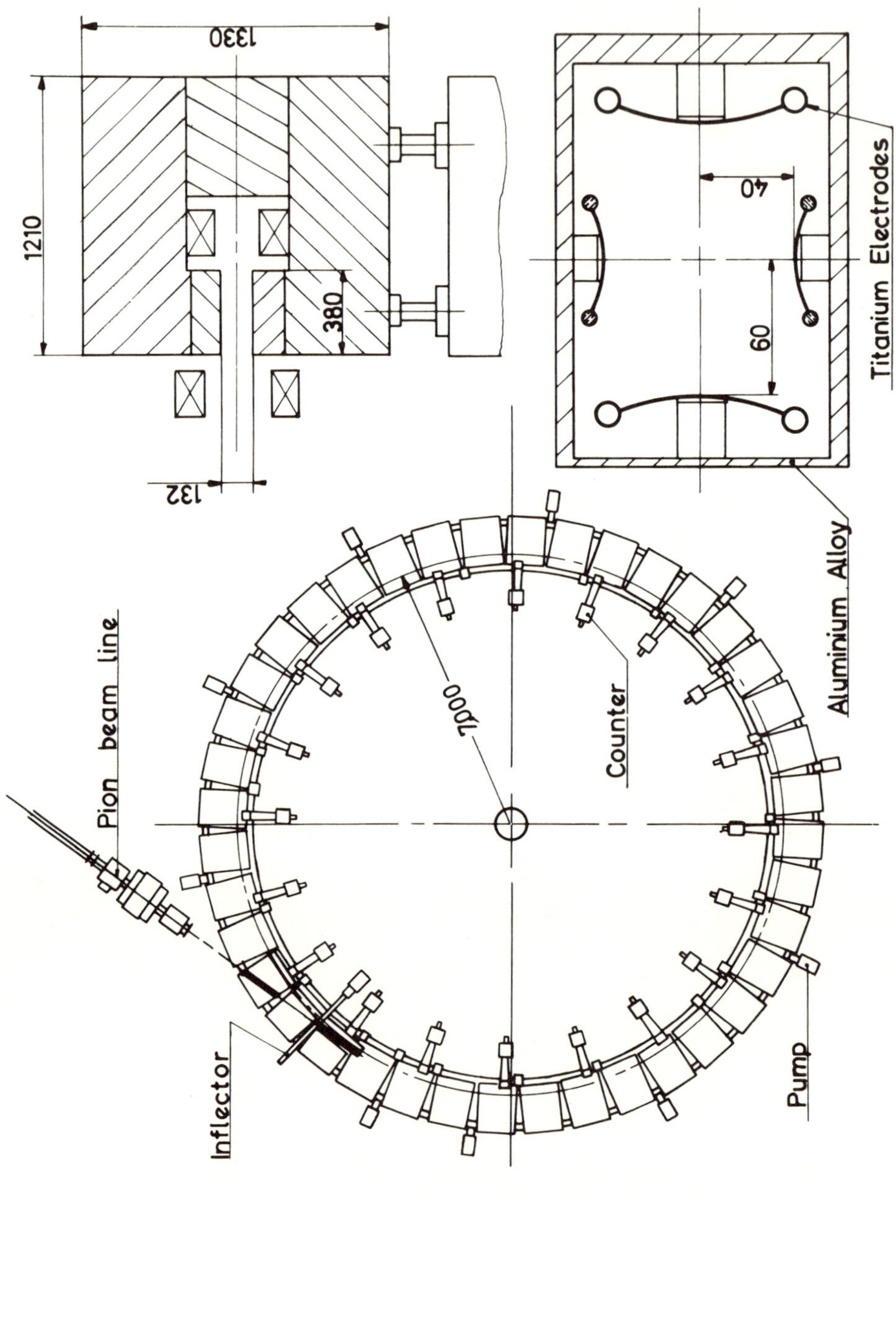

Figure 3. Plan view of the 14-meter storage ring. Vertical sections through the ring and through the vacuum

Figure 4. The storage ring from inside, showing some of the electron detectors.

As noted, the final error was due primarily to the counting statistics of the experiment, rather than the systematic uncertainties, and the final result is very close to the best theoretical calculation (Tables 1 and 2). Theories may change or improve, of course, and I would like to raise the question whether this result could be sufficiently improved to warrant the great effort which would presumably be involved. We must look to Eq. (2), since it appears that no principle is available which would give a better measurement of ω_a. First, going to high field, for example with a superconducting ring, would be directly beneficial and one might guess that our 14 kG could be raised by perhaps a factor of 3. To go to higher energy, thereby raising τ, means to leave the magic value of γ and therefore to pay a penalty in systematic uncertainties which rapidly becomes unpleasant. Experimental problems of precision determination of the product of electric fields and average muon-sample densities in such a system seem particularly intractable; returning to weak magnetic focusing as in the case of CERN Exp. #2, is quite out of the question for 1 ppm precision; in fact, the uncertainty introduced in CERN Experiment 2 by the nonuniform magnetic field was 170 ppm. A might be raised by very careful design, by at most a factor of 2 (one has in mind, of course, that what is really at stake is a number of values of A, for each of which the corresponding value $\sqrt{N_i}$ obtains). The instantaneous rate, and therefore ultimately N, is also limited by considerations of systematics. Once one has trapped a sample, it is essential that the measurement conditions remain as nearly unchanged as possible so that there is no time distortion during the observation of the sample decay, in our case a time of 650 micro-seconds. Any counting-rate loss at early time will violate this principle, since a second order time shift accompanies the counting-rate loss, and since the loss will diminish as the sample decays. This limits initial rates to around 10^6 per second, which with a lifetime of 64 microseconds means possibly 100 counts in 1 proton-synchrotron

TABLE 1

Muon's Progress: g-2 at CERN.

Experiment 1: Gradient magnet (1959-62)

Jan., '61	$\frac{g-2}{2} \equiv a_\mu$ = .001145(22)	2%
Mar., '62	= .001162(5)	0.4%

Experiment 2: Storage ring, magnetic weak focusing (1965-68)

Oct., '68	a_μ = .00116616(31)	(270 ppm)

Experiment 3: Storage ring, electrostatic focusing, "magic" γ. (1972-77)

Feb., '75	$a_\mu = 1165895(27)\times10^{-9}$	(23 ppm)
Feb., '77	$a_\mu = 1165922(9)\times10^{-9}$	(8 ppm)

TABLE 2

Contributions to the theoretical value of a_μ.

a_μ (QED)	$1165851.8(2.4) \times 10^{-9}$
a_μ (hadronic)	$66.7(9.4) \times 10^{-9}$
a_μ (weak)	$2(2) \times 10^{-9}$
a_μ (theory)	$1165921(10) \times 10^{-9}$

TABLE 3

Corrections and errors to $\bar{f}_p$, the mean field expressed in proton resonance frequency, for a particular run.

	Frequency (MHz)	Relative error (ppm)
Weighted mean of map	62.78167(4)	0.7
Calibration and diamagnetic shielding	0.00148(4)	0.6
Time variation of field and clock drift	-0.00004(6)	1.0
Vacuum tank, electrodes and inflector	0.00000(4)	0.6
Non-ideal phase space muon distribution	0.00001(1)	0.1
Electric field effect	-0.00010(2)	0.2
Pitch correction	-0.00005(1)	0.1
$\bar{f}_p$	62.78297(10)	1.5

pulse. More than a few times 10^6 machine pulses in the lifetime of an experiment becomes unreasonable for the machines that are presently available. We might expect a factor of 3 in N or $\sqrt{3}$ in the error. Taking into account the factor of 2 in A and perhaps 3 in field, it appears that the statistical error could be driven under 1 ppm. The error in the weighted-average magnetic field is unknown for such a large and unusual supermagnet, but it is unlikely to be better than that of the present experiment, 1½ ppm.

I conclude that a super effort might lead to a determination of a_μ to between 1 and 2 parts per million. Looking at Table 2, we see that this is not good enough to measure the anticipated effects of weak interactions, even if one could somehow, by new calculations and new measurements, master the uncertainties in the hadronic part of the theoretical calculation. The hope for pioneers setting out on this long and difficult road would only be that something remarkable would show up between 8 ppm and 1 ppm, and I doubt that they would make the trip. It is the end of an era; the muon g-2 worker is no longer a Kit Carson blazing a trail in the wilderness, but more like a J. Edgar Hoover keeping a watchful eye on the imaginative excesses of theorists.

ACKNOWLEDGEMENTS

The experiment described here was carried out by the authors of Reference 1. It has been a pleasure to work with them.

REFERENCES

1. J. Bailey, K. Borer, F. Combley, H. Drumm, F. J. M. Farley, J. H. Field, W. Flegel, P. M. Hattersley, F. Krienen, F. Lange, E. Picasso and W. von Ruden, Phys. Lett. 67B, 225 (1977). A complete report is in preparation.
2. J. Calmet, S. Harrison, M. Perottet, and E. de Rafael, Rev. Mod. Phys. 49, 21 (1977).

PROPOSED MEASUREMENTS OF THE ANOMALOUS MAGNETIC MOMENT OF THE ELECTRON AND POSITRON

D. Newman, A. Rich, E. Sweetman

(presented by A. Rich)

University of Michigan, Ann Arbor, Michigan

ABSTRACT

We will describe two new experiments at the University of Michigan to measure the g-factor anomalies of the electron and positron to accuracies of 25 parts per billion (ppb) and 3 parts per million (ppm) respectively. The features of these experiments that differ from previous g-factor measurements and allow higher statistical and systematic accuracy will be emphasized. We will also point out several theories of current interest which will be tested by these results. In particular, we describe in the Appendix how these results will constitute the most precise test of special relativity to date. First, however, we outline recent theoretical and experimental progress in lepton g-factor research.

I. CURRENT STATUS OF g-FACTOR RESEARCH

Measurements of the electron and positron g-factors are the most precise experimental tests of quantum electrodynamics (QED) and of the CPT theorem applied to leptons. The magnetic moment $\vec{\mu}$ is related to the g-factor by:

$$\vec{\mu} = \frac{g}{2}\frac{e}{mc}\vec{s} \ , \ |\vec{s}| = h/2 \quad . \tag{1}$$

In the Dirac theory, g is exactly 2. When the effects of the radiative corrections of QED are included, a deviation of order 10^{-3} is expected. This deviation, called the anomaly a, is defined as:

$$\underline{a} \equiv (g-2)/2 \quad . \tag{2}$$

QED predicts the value of a in terms of a power series in α/π, α being the fine structure constant e^2/hc:

$$\underline{a} = A(\frac{\alpha}{\pi})+B(\frac{\alpha}{\pi})^2+C(\frac{\alpha}{\pi})^3+D(\frac{\alpha}{\pi})^4+ \ldots$$

$$+ \text{ weak interaction corrections} \tag{3}$$

$$+ \text{ muonic corrections} + \text{hadronic corrections}$$

where

A	$= 0.50$	
B	$= 0.32848$	(1)
C	$= 1.179 \pm 0.010$	(2,3)
D	$=$ unknown	(3)
α^{-1}	$= 137.035987(28)$	(4) .

The coefficient C given above is the weighted average of the published theoretical predictions. Assuming that sufficiently accurate measurements of α will become available, a determination of a to 25 ppb will give the C coefficient in Eq. (3) to about 0.2%. We note that $(\alpha/\pi)^4$ is 30 ppb of a, so the proposed measurement would set limits on this term also.

Numerical predictions of a from QED are at present limited by an uncertainty of 210 ppb in α. Thus a new measurement of a to 25 ppb would yield a value of α to the same accuracy, assuming QED. The current experimental program at the National Bureau of Standards to measure α is expected to yield an accuracy of 100 ppb in the near future and 10 ppb by the time this g-factor experiment is complete (5).

The hadronic correction to Eq. (3) is 1.4 ± 0.2 ppb while muonic vacuum polarization contributes 2.5 ppb (3). Thus these effects would not limit the precision with which QED will be tested in this experiment. Although most weak interaction theories predict effects of order 1 ppb or less, certain new weak interaction theories predict very large effects which may be tested by the proposed experiment. As a result of the failure of recent experiments to detect parity violation in heavy atoms, and in order to explain high energy neutrino results such as the unexpectedly large number of trimuon events observed (6), new weak interaction theories based on SU(3)×U(1) have been presented (7). One of these predicts an effect of order 70 ppb in the anomaly.

This would be detectable in the proposed experiment. It is thus expected that the new g-factor results could shed light on the fundamental structure of the weak interactions, whether the measurement is in agreement with QED or not!

We note further that g-factor measurements over a range of energies test relativistic kinematics. As shown in the Appendix, any error in the special relativity prediction of Thomas precession of order 4×10^{-11} would be seen in the proposed experiment. This would constitute the most precise test to date of relativistic kinematics (8). It is independent of QED or weak interaction assumptions.

The g-2 technique is also sensitive to the possible electric dipole moment (EDM) of the electron (9,10). Measurements of ω_D with the anticipated accuracy for different magnetic fields can be analyzed to set a limit on the EDM of the electron of order 10^{-18}e-cm. Although this is much larger than the limit (2×10^{-24}e-cm) set by atomic beam techniques (11), it would be the best measurement of the EDM of a free electron.

The CPT theorem predicts that all properties of particles and anti-particles are identical in magnitude. We test this prediction by measuring the magnetic moment of both the electron and positron. Note that a measurement of the anomalies $\underline{a}^+$ and $\underline{a}^-$ to 3 ppm accuracy constitutes a test of CPT to 3 ppb.

II. HISTORY

The electron g-factor has been measured to high accuracy by use of two techniques in recent years. Wesley and Rich (12) at Michigan trapped electrons in a magnetic well at 1.2 kG. They used Mott scattering to polarize and analyze the electron spin, and observed the difference frequency ω_D between the spin and cyclotron motions, obtaining an uncertainty of 3.5 ppm in 1971. Walls and Stein (13) at the University of Washington applied a bolometric technique to electrons in a Penning trap to obtain an uncertainty of 24 ppm in 1973. In an extremely sensitive experiment, Van Dyck, Schwinberg and Dehmelt (14) at Washington have recently succeeded in driving the g-2, spin, cyclotron, and other resonances with single electrons in a Penning trap. By precisely controlling the anharmonicities of the trap, the electric field, and other systematic parameters, they have obtained an uncertainty of 200 ppb in $\underline{a}^-$!

Progress in the theoretical prediction of $\underline{a}$ has been due to improvements in the numerical techniques by which QED divergences are treated, and by the increasing number of sixth-order Feynman diagrams that can be calculated analytically. Uncertainty in $\underline{a}$ due to the sixth-order diagrams is now 114 ppb. These calculations are being vigorously pursued using both analytic and numerical techniques. It is hoped that in the near future all but five of the Feynman diagrams in the C term will have been determined analytically, and that a numerical evaluation of the remaining five will reach an accuracy of order 5 ppb (3).

The D term in Eq. (3) will be much more challenging because 891 diagrams contribute to it. If D is of order 1, the effect of eighth order terms on $\underline{a}$ is 30 ppb, comparable to the resolution of the next generation of g-factor experiments. It is hoped that an evaluation of the D term to an accuracy of order 10% (3 ppb) will be possible in the next several years (3).

The g-factor of the positron is predicted by CPT to be equal to that of the electron. The positron anomaly $\underline{a}^+$ was measured by Gilleland and Rich (15) at the University of Michigan in 1969. They trapped positrons which were polarized from nuclear beta decay and analyzed the polarization with an early version of our positron polarimeter. The accuracy of 1000 ppm in the anomaly $\underline{a}^+$ (1000 ppb for g^+) was the most precise test of CPT for leptons at the time. More recently, Serednyakov, et al. (16) at Novosibirsk compared $\underline{a}^+$ and $\underline{a}^-$ directly by trapping polarized positrons and electrons in the same storage ring. They observed depolarization of both beams at the same frequency, indicating that the anomalies are equal within a precision of 10 ppm, thus testing CPT to 10 ppb.

The muon g-factor anomalies $\underline{a}_\mu^+$ and $\underline{a}_\mu^-$ have recently been measured to 10 ppm by Bailey, et al. (17) by precise analysis of the g-2 frequency of muons in the CERN storage ring. The theoretical prediction of the muon anomaly is extremely challenging because hadronic and weak interaction effects are expected at about 67 ppm and 2 ppm, respectively (18). Current calculations of $\underline{a}_\mu$ are in agreement with the experiments, and have 10 ppm uncertainty in the hadronic corrections. These results represent a test of QED to a precision of 10 ppm and a test of CPT to 10 ppb.

III. THE POSITRON g-FACTOR EXPERIMENT

The method that we will use to measure the positron g-factor is called resonant spin rotation, a novel technique that was demonstrated using a modification of the last Michigan electron g-factor experiment (19). The experiment is shown schematically in Fig. 1.

The positrons are produced in ^{68}Ga decay with polarization $\vec{P} = \vec{v}/c = .95\ \hat{v}$ in the energy range (1.0 ± 0.1 MeV). They pass

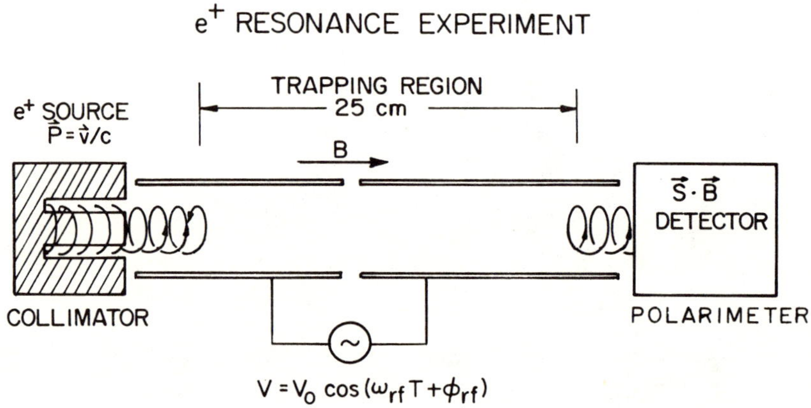

Figure 1. Apparatus of the positron g-factor experiment. Positrons from the source at left pass through the annular collimator and are trapped in the central region. Their spins are rotated into the z direction by an electric field applied between the electrodes as shown. The positrons are then ejected into the polarimeter.

through an annular collimator and enter a shallow magnetic well (10-100 ppm) produced by the 10 kG solenoid, where they are trapped by a retarding electric pulse.

Positrons remain in the trap for 0.1 to 10 msec. They execute cyclotron motion at frequency ω_c and their spins precess at ω_s:

$$\omega_c = \frac{eB}{\gamma mc} \tag{4}$$

$$\omega_s = \frac{g}{2}\,\omega_o + \omega_o\left(\frac{1-\gamma}{\gamma}\right) \quad . \tag{5}$$

The first term in Eq. (5) is equal to the precession frequency of an electron at rest in the laboratory magnetic field. The second term is Thomas precession, an effect of relativistic kinematics due to the electron's instantaneous acceleration. The difference frequency ω_D is the requency at which the spin precesses relative to the momentum:

$$\omega_D = \omega_s - \omega_c = \frac{eBa}{mc} \quad . \tag{6}$$

In the technique of resonant spin rotation, an externally applied rf electric field at a frequency $\omega_{rf} = \omega_D$ rotates the spin of the positrons out of the initial orientation in the orbital plane and into the $\hat{z}$ direction. The electric field is applied in the $\hat{z}$ direction between a pair of cylindrical electrodes which surround the trapping region. This appears as a radial magnetic field in the frame of the positrons and exerts a torque $\vec{\tau} = \vec{\mu} \times \vec{B}'$, where $\vec{B}' = \hat{r}\gamma E_z(t)\beta$ and $E_z(t) = E_o\cos(\omega_{rf} + \phi_{rf})$ remains in the same direction as the spins precess, and the spins are thus forced to rotate into the $\hat{z}$ direction. The positrons are then ejected from the magnetic well by an electric pulse and enter the positron polarimeter. The polarimeter is based on the formation of positronium in MgO powder in the solenoid magnetic field and on the analysis of its field perturbed decay components (20).

The anomaly is determined by observing the $\hat{z}$ component of the positron polarization as the driving frequency ω_{rf} is varied. Note that the positron spins can be easily rotated into the $-\hat{z}$ direction by reversing the phase ϕ_{rf} of the driving signal. The full width of the resonance peak is expected to be 100 ppm or less.

The statistical error in extracting ω_D from the peak in ω_{rf} is given by:

$$\frac{\sigma}{\omega_D} = \frac{1.7}{2\pi N_D \delta} \sqrt{\frac{2}{N_T}} \tag{7}$$

where N_D = number of ω_D periods in trap (5000)

δ = asymmetry when spin is reversed (0.10)

N_T = number of detected events in the run.

For a statistical accuracy of 2×10^{-6}, we will need $N_T = 1.5\times10^5$ events. The number of events expected in time T_T is determined by the source strength and the trapping and detection efficiencies. We estimate that runs of order 8 days will be required to accumulate the required statistical accuracy.

Other sources of error come from (1) NMR calibration from which B is derived, (2) possible shifts in ω_D due to stray electric fields in the trap (see Eq. (10)), and (3) a correction for the axial motion, which depends on well depth.

Electric field effects can be checked by taking runs at various magnetic fields. The extrapolation in $\underline{a}^+$ is linear for fixed electric fields when ω_D is plotted against the quantity $(\beta\gamma^2 B)^{-1}$. Correction for the axial velocity can be done as in past experiments (12) by measuring the number of particles at various levels of the trap and explicitly calculating the time-average velocity.

These errors are summarized as follows:

σ_{stat}	2 ppm	
σ_{NMR}	< 1	
σ_{Elec}	< 1	
σ_{Axial}	2	Net σ: 3 ppm.

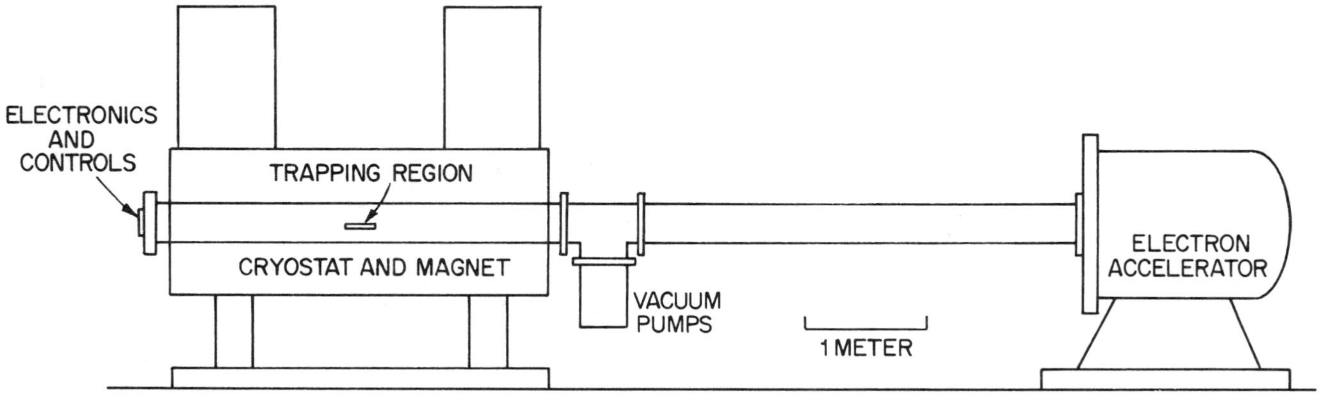

Figure 2. Schematic diagram of experimental apparatus. Electrons from the accelerator at right enter the solenoid parallel to the field axis and are Mott scattered into the trapping region.

IV. THE ELECTRON g-FACTOR EXPERIMENT

The proposed electron g-factor experiment (which we call g-V) is an extension of the basic Michigan g-factor precession technique used in the previous measurement (g-IV) (12). As shown schematically in Fig. 2, electrons of about 200 keV energy will be injected axially into a 10 kGauss superconducting solenoid which is about 2 m long by 0.6 m in diameter. Near the center of the solenoid the electrons are Mott scattered by a thin gold foil which both deflects them into cyclotron orbits and polarizes them in the transverse (orbital) plane. The trapping and detection apparatus is shown in Figure 3. A 10 ns pulse of unpolarized electrons is incident parallel to the magnetic field from the right on the first Mott foil, supported perpendicular to the field. Most of the electrons will pass through the foil with rms multiple scattering angle of 5° (21) and an energy loss of 0.5 keV (22). Electrons that Mott scatter in the foil will enter cyclotron orbits with a diameter of 3 mm and will drift through a beam defining hole and into the trapping region, where a retarding electric pulse will drop them into the well. While the electrons are in the well their spins precess as described in equation 6.

After remaining in the trapping region for a predetermined time interval, the electrons are ejected and strike the second Mott foil, which is supported nearly perpendicular to the field, but rotated by 5° around a vertical axis. Multiple scattered electrons leave the second Mott foil with an rms pitch of 9°, while Mott scattered electrons are travelling nearly parallel to $\hat{z}$. Thus in the $\vec{E}\times\vec{B}$ drift region (Fig. 3) the multiple and Mott scattered electrons will be separated transversely by about 3 cm at the detector.

As a result of the spin dependence of Mott scattering, the number of Mott electrons arriving at the detector varies as $\hat{S}.\hat{v}$ where $\hat{v}$ is the direction of velocity just before scattering. The counting rate is given by:

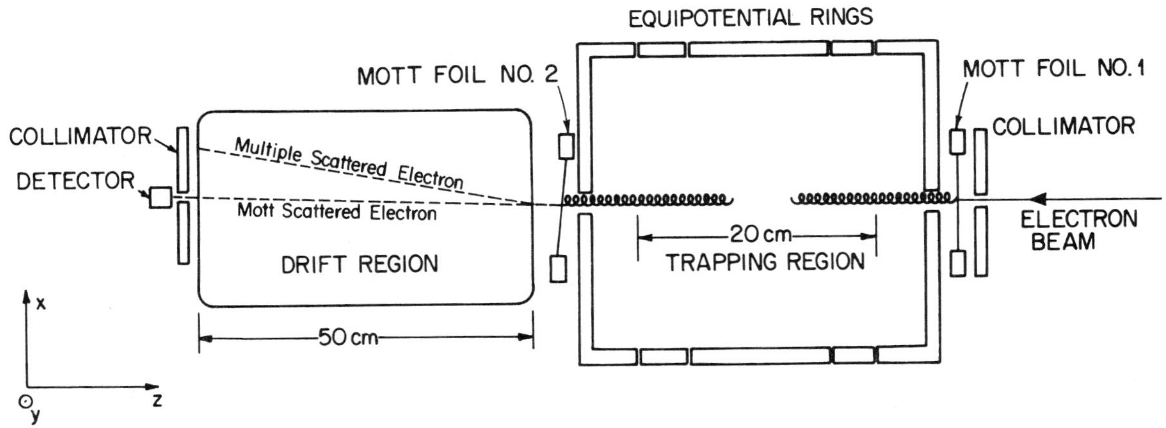

Figure 3. The trapping and detection apparatus (not to scale). The magnetic field is in the z direction. Electrons are polarized and scattered into cyclotron orbits at the first foil. The equipotential rings apply the electrostatic trapping field. After scattering in the second foil, multiple scattered electrons are deflected as shown in the E×B drift region (electric field in y direction). Mott scattered electrons pass through the drift regi(with less deflection and reach the detector.

$$R(T) = R_o[1 + \delta_M \cos(\omega_D T + \phi)] \quad , \tag{8}$$

where R_o is the rate for unpolarized electrons, ϕ is an arbitrary phase angle, and δ_M is the Mott scattering asymmetry which is expected to be about 0.035 (see Fig. 4). The difference frequency is measured by determining R(T) over short intervals at two times T_1 and T_2 and then determining the number of cycles N between them. Thus we have:

$$\omega_D = \frac{2\pi N}{T_2 - T_1} \quad . \tag{9}$$

Unfortunately Eqs. (6) and (8) are only exact for planar cyclotron motion in a uniform magnetic field. The difference frequency is perturbed by the axial component of velocity (β_z) and by stray electric fields, as well as by the trapping magnetic field $\Delta B(z) = B(z) - B(o)$. When these effects are included, we find (23,24):

$$[\omega_D] = a\omega_o \left(1 + \frac{[\Delta B]}{aB_o} - \frac{[\beta_z^2]}{2} + \frac{1}{a\beta\gamma^2} \frac{[E_r]}{[B_z]} + \frac{1}{2a^2} \left(\frac{1}{\beta\gamma^2}\right)^2 \frac{[E_z^2]}{[B_z^2]}\right) \tag{10}$$

where [...] indicates the time average over the axial oscillation in the well, E_r and E_z are the radial and axial components of the electric field, and $\omega_o = eB(0)/mc$.

In g-IV the time averaged quantities in Eq. (10) were determined by observing the number of electrons that came out of the well when the amplitude of the ejection pulse was varied. From this and a map of the magnetic field, a correction Δa to the observed anomaly was calculated. This technique is sensitive to errors in determination of the population distribution as well as the beam energy and stray electric fields in the trapping region.

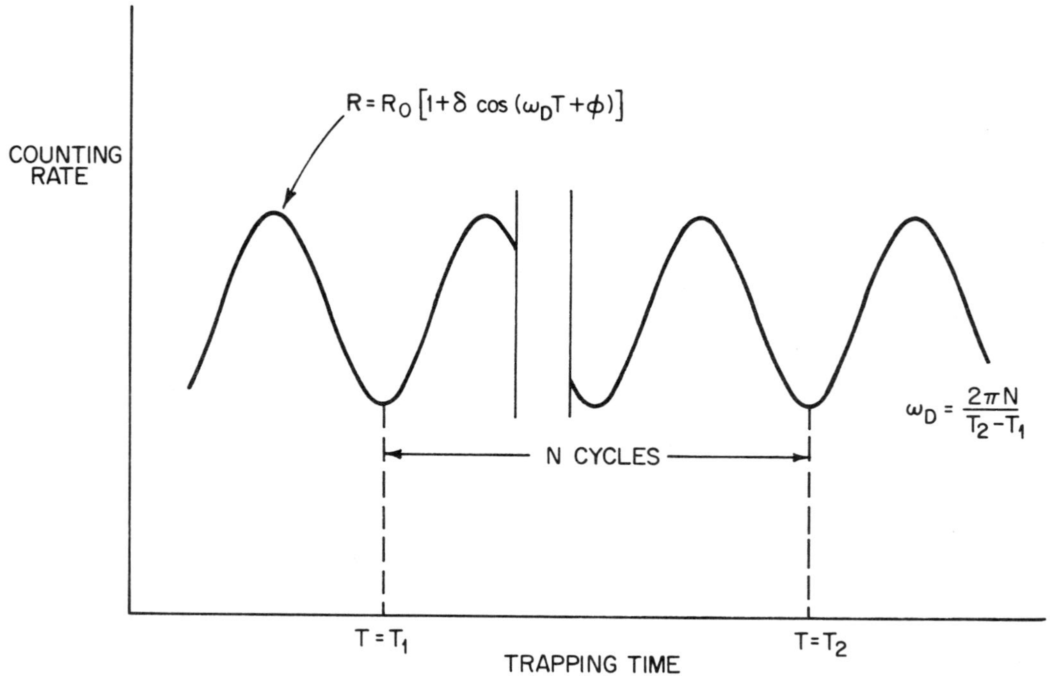

Figure 4. The data consists of a sinusoidally varying counting rate in the detector as the spins of trapped electrons precess relative to their velocities. The difference frequency is precisely determined by matching the data around T_1, a short trapping time, with the data around T_2= 30 ms, corresponding to $N=10^6$ difference cycles.

We now propose to use a new configuration of trapping fields which greatly reduces systematic errors in a due to these parameters. The new trap, which we call a hybrid well, consists of both electric and magnetic trapping fields that are shaped to cause many of these errors to cancel out. The trap will be axially symmetric, 1 volt deep, and 20 cm long.

As the electron oscillates in the z direction with period 2 μs, it samples different values of ΔB and E_r at different axial positions, just at β_z^2 varies from a maximum at the center of the trap, to zero at its turning points z_{min} and z_{max} (12). By appropriate shaping of magnetic and electric fields, the systematic corrections due to ΔB, E_r and β_z^2 can be arranged to cancel each other out.

The analysis is greatly simplified by separating the motion into independent orbital (x,y) motion and axial (z) oscillation. The energy in the axial motion is given by:

$$\frac{1}{2}\gamma m v_z^2(z) = \frac{1}{2}\gamma m v_z^2(0) - \frac{1}{2}\gamma m v^2(z)\frac{\Delta B(z)}{B(z)} - eV(z) \qquad (11)$$

where V(z) is the electric potential (12). To satisfactory approximation we set $v^2(z) \simeq v^2(0) \simeq$ const and $B(z) \simeq B(0)$.

The effect of each term in Eq. (10) was calculated for each point in the well using a numerical routine which integrated the particle and spin motion for a given well shape. The net effect of all these perturbations is a shift Δ in the instantaneous difference frequency. The contributions to Δ from each step were added up to give the time averaged effect [Δ]. The procedure was then repeated for each level of the well to obtain the ensemble averaged correction <[Δ]>. This calculation was repeated for various values of the total electron energy to obtain the sensitivity of $<[\omega_D]>$ to γ. Stray electric fields were simulated by varying V(z).

Many different well shapes for electric, magnetic and hybrid confinement were investigated. An overall quality factor was found for each well, based on the variation of $[\omega_D]$ with β_z, E_r, and γ. The optimization process is still going on, and it is to be expected that better wells will be found before the experiment is executed. The best well to date is a hybrid well composed primarily of a standard parabolic electrostatic confining potential. In this well, a maximum electric field in the z direction of 0.2 V/cm, a maximum radial electric field of 0.002 V/cm at the orbit radius, and a maximum magnetic field gradient of 0.003 G/cm exist in the trapping region. This well has an average correction of $<[\Delta]>$ = 560 ppb relative to the unperturbed difference frequency based on the magnetic field at the center of the well:

$$<[\omega_D]> = \frac{eB_o a}{mc} + <[\Delta]> \quad . \tag{12}$$

The uncertainty in this correction is calculated by letting β_z, E_r, and γ vary within anticipated limits. The expected error in ω_D due to the combined uncertainties above is 15 ppb. This is more than 100 times smaller than analogous errors in g-IV and 10 times smaller than would be expected in g-V without hybrid traps.

The run time required for a statistical accuracy of 10 ppb can be calculated from the known parameters of the experiment. If we fit the asymmetry function (Fig. 4) with data points spanning a few difference frequency cycles at two widely spaced times T_1 and T_2, then it can be shown that the uncertainty in the frequency is given by (12):

$$\frac{\sigma_\omega}{\omega} = \frac{1}{2\pi} \frac{1}{N\delta_M \sqrt{N_T}} \tag{13}$$

where N is the number of difference frequency cycles in the time T_2-T_1 and N_T is the number of counts in the entire data set. The difference frequency ($\omega_D/2\pi$) at 10 kG is 30 MH_z, so that in about 30 msec we will get $N=10^6$ difference periods.

The Mott double scattering asymmetry δ_M is predicted (25) to be 7% for 200 keV electrons and gold foils. In g-IV the observed asymmetry was only $\frac{1}{2}$ of the predicted value, probably due to multiple scattering and background. We will assume that the same factor of $\frac{1}{2}$ will apply in g-V and use 3.5% as the asymmetry henceforth.

The total number of counts N_T collected in time T_T is determined by the ejection and detection efficiency, the repetition rate, and the number of electrons in the trap. The number n of electrons in the trap will be limited by the fact that the observed difference frequency ω_D is shifted by radial electric fields. We conservatively account for this effect by limiting n to 3000, which limits the maximum calculated space charge shift in ω_D to the same order as the statistical precision of the experiment. The necessary number of detected counts, assuming $N=10^6$, is $N_T=2\times10^5$. With a calculated data rate of 7.5/sec, a very conservative estimate of run time is 8 hours to achieve a statistical accuracy of 10 ppb.

The expected sources of systematic error in this experiment are due to uncertainties in (1) β_z, the axial component of velocity while in the trap, (2) E_r, the radial electric field in the trap, (3) B_o, the magnetic field at the center, (4) $\Delta B(z)$, the shape of the trapping field, (5) γ, the total energy of the electrons and (6) T_{12}, the time interval between short and long trapping times. Using the newly developed hybrid wells we will minimize the effect of most of these parameters on ω_D. Based on numerical simulations of particle motion in the well, we conservatively estimate that errors of 10 ppb from the $\beta_z^{\,2}$ term can be achieved.

Uncertainty in the radial electric field E_r could come from (1) stray charge accumulation on insulating surfaces near the trap, (2) contact potentials generated by different material in the trapping region, (3) misalignment in the system of equipotential rings that will be used to form the electric trap, and (4) the space charge of the trapped beam itself. Stray charge accumulation

and contact potentials will be minimized by spraying all surfaces in the vicinity of the trapping region with conducting graphite, as was done in g-IV. The size of these effects at the orbit radius will be further reduced by placing the rings at as large a radius as possible and by forming the fields with as large a potential difference between equipotential rings as possible.
We estimate that the maximum net effect of items (1), (2), and (3) will be equivalent to changing the potential of the middle of the trap by up to ±100 mV with respect to the ends. We calculate that this will have an effect of only ±4 ppb on ω_D with our hybrid well. A conservative error of ±10 ppb will be assumed for the space charge shift. The calibration of the absolute magnetic field is expected to be limited by the ratio of the magnetic moment of protons in water to the nuclear magneton (26), known to 11 ppb.

Electrons with different γ will experience different E_r, as shown by Eq. (10). The uncertainty in energy due to the resolution of the generating voltmeter that measures the accelerator voltage and to energy loss while passing through the first Mott foil is conservatively estimated to be ±1%. Using hybrid wells, this will cause 12 ppb uncertainty in ω_D.

The time base of the trapping interval T_{12} and of the NMR system will be a precision 1 MHz crystal oscillator, calibrated against WWV as the absolute time standard. Since the same standard will be used in both the NMR and T_{12} measurement, any error in T_{12} will be exactly compensated by an error in B_o. Even without this effect, however, errors of only 1-2 ppb are anticipated.

These errors are summarized as follows:

Source	Contribution (ppb)
Statistics	10
β_z	10
E_r	4
Space Charge	10
μ_p'/μ_B	11

Energy	12
Timing	2

Since there are many independent sources of possible systematic errors at about the same level, we take the rms sum of these contributions and estimate an anticipated accuracy of 25 ppb in the anomaly.

V. CONCLUSION

The results of this program will allow us to test four fundamental theories of current interest to higher precision than has been previously achieved. The positron g-factor can be compared with the electron g-factor to test the CPT theorem to an accuracy of 3×10^{-9}. This will be the most precise leptonic test of CPT to date. The electron g-factor experiment will determine the anomaly $\underline{a}$ to an accuracy of 25 ppb. Assuming that theoretical calculations of the sixth order contributions to $\underline{a}$ will reach this level and that measurements of α will become equally precise, the g-factor result will then reveal the order of magnitude of the eighth order term, which is well below the resolution of current experiments. The result is also expected to test the Lee-Shrock gauge theory of weak interactions which predicts an effect of 2-3 times the expected uncertainty in $\underline{a}^-$. In addition, a comparison of the results of this experiment with g-2 measurements at very low energies will test Thomas precession to higher precision than has been achieved previously. As a further product of the experiment, precise measurement of the difference frequency over a range of magnetic fields can be analyzed to yield the electric dipole moment of the free electron to an anticipated accuracy of order 10^{-18} e-cm.

APPENDIX

RELATIVITY TESTS

There are few precise experimental tests of the predictions of special relativity (8). We will show that the agreement between g-IV results and the recent low energy g-factor experiment of Van Dyck, Schwinberg, and Dehmelt (14) at the University of Washington verifies special relativistic kinematics with an accuracy of 5×10^{-9}. Alternatively, if relativity is assumed, the comparison can be viewed as verification of the relativistic electromagnetic theory of particle motion and spin precession in magnetic fields to the same precision.

The electrons in g-IV had an energy of 110 keV ($\gamma = 1.2, \beta = 0.57$). The measured anomaly was $\underline{a} = 0.00115965770(350)$. This is to be compared with the Washington results of $\underline{a} = 0.00115965241(20)$, obtained with single electrons in a Penning trap. Their electrons had an energy of about 5×10^{-4} eV ($\beta = 5\times10^{-5}$, $\gamma = 1 + 10^{-9}$). The observed difference is $\Delta a = (5.3 \pm 3.5) \times 10^{-9}$.

To analyze the implications of these measurements, we note that the inertial rest mass of the electron is given by:

$$\frac{1}{m} = \lim_{p\to 0} \frac{1}{p}\frac{dE}{dp} \quad . \tag{14}$$

The relation between E and p could be different from the usual relativistic form if, for example, there were a periodic structure to space. Thus we write the parameter $\tilde{\gamma}$ (equal to γ in special relativity) as:

$$\tilde{\gamma} = \frac{p}{m}\frac{dp}{dE} \quad . \tag{15}$$

The cyclotron frequency is then

$$\omega_c = \frac{eB}{\tilde{\gamma}mc} \quad . \tag{16}$$

The quantity $\tilde{\gamma}m$ is analogous to the effective mass of an electron in a crystal.

The spin precession frequency is

$$\omega_s = \frac{geB}{2mc} + (1-\gamma)\omega_c \tag{17}$$

where $\gamma = (1-\beta^2)^{-\frac{1}{2}}$ and $\beta = \frac{1}{c}\frac{dE}{dp}$. Note that the γ appearing in Eq. (17) is a direct result of relativistic kinematics (Thomas precession (27)) and thus is equal to the usual $1/\sqrt{1-\beta^2}$ which appears in the Lorentz transformation. The $\tilde{\gamma}$ in Eq. (15), however, arises from electron dynamics and need not be the same as γ.

The anomaly a is found from the difference frequency ω_D between the spin and cyclotron frequencies:

$$\omega_D = \omega_s - \omega_c = \left(\frac{g}{2} - \frac{\gamma}{\tilde{\gamma}}\right)\frac{eB}{mc} \quad . \tag{18}$$

Thus the experiments place the limit:

$$(\tilde{\gamma}-\gamma)/\gamma \leq 5\times10^{-9} \quad . \tag{19}$$

This is comparable in accuracy with recent work by Brecher (28) and is the most precise verification of a relativistic prediction at high velocity yet reported.

It is clear that further refinement in this relativity test requires a higher precision experimental determination of a at relativistic velocities. We plan to run g-V at a variety of energies to check for systematic errors, and the values of a obtained may then be compared to test relativity predictions to 4×10^{-11}, an improvement of two orders of magnitude over the results presented to date, and would have the advantage that all of the data would come from the same experimental technique. Comparison with any improved Washington data available would also, of course, be possible.

REFERENCES

1. C.M. Sommerfield, Phys. Rev. 107, 328 (1957); A. Peterman, Helv. Phys. Acta 30, 407 (1957).
2. M. Levine and R. Roskies, Phys. Rev. D14, 2191 (1976).
3. T. Kinoshita and W. Lindquist, CLNS-374 (Cornell), and private communications (1977).
4. P.T. Olsen and E.R. Williams, Proceedings of the Fifth Int. Conf. on Atomic Masses and Fundamental Constants, Paris, 2-6 June 1975.
5. E.R. Williams, private communication (1977).
6. A. Benvenuti, et al., Phys. Rev. Letters 38, 1110, 1183 (1977).
7. B. Lee and R. Shrock, Fermilab-Pub-77/48-THY (June 1977).
8. D. Newman, G.W. Ford, A. Rich, and E. Sweetman, Univ. of Mich. preprint, submitted for publication January 1978.
9. G.W. Ford and C.W. Hirt, Contract NONR 1224 (15) unpublished.
10. D.F. Nelson, A.A. Schupp, R.W. Pidd, and H.R. Crane, Phys. Rev. Letters 2, 492 (1959).
11. M.A. Player and P.G.H. Sandars, J. Phys. B 3, 1620 (1970).
12. J. Wesley and A. Rich, Phys. Rev. A4, 1341 (1971).
13. F. Walls and T. Stein, Phys. Rev. Letters 31, 975 (1973).
14. R. Van Dyck, P. Schwinberg, and H. Dehmelt, Phys. Rev. Letters 38, 310 (1977).
15. J. Gilleland and A. Rich, Phys. Rev. Letters 23, 1130 (1969).
16. S. Serednyakov, et al., Phys. Letters 66B, 102 (1977).
17. J. Bailey, et al., Phys. Letters 68B, 191 (1977).
18. J. Calmet, S. Narison, M. Perrottet, E. deRafael, to be published in Rev. Mod. Phys.
19. G.W. Ford, J. Luxon, A. Rich, J. Wesley, and V. Telegdi, Phys. Rev. Letters 29, 1691 (1972).
20. G. Gerber, D. Newman, A. Rich, and E. Sweetman, Phys. Rev. D15, 1189 (1977).
21. E. Segre, Ed., "Experimental Nuclear Physics", (1953).

22. M.J. Berger and S.M. Seltzer, "Tables of Energy Losses and Ranges of Electrons and Positrons", NASA SP-3012 (1964).

23. A. Rich and J. Wesley, Rev. Mod. Phys. 44, 250 (1972).

24. S. Granger and G.W. Ford, Phys. Rev. D13, 1897 (1975).

25. N.F. Mott, Proc. Roy. Soc. 124, 425 (1929); 135, 429 (1932).

26. W.D. Phillips, W.E. Cooke, and W. Kleppner, Phys. Rev. Letters 35, 1619 (1975).

27. C. Møller, "The Theory of Relativity", Oxford University Press, London, 1952.

28. K. Brecher, Phys. Rev. Letters 39, 1051 (1977).

INTRODUCTORY REMARKS - QUARK FLAVORS*

Sydney Meshkov

California Institute of Technology, Pasadena, CA 91125

and National Bureau of Standards, Washington, D.C. 20234

ABSTRACT

The question of where the next quark flavor n should occur is discussed. It is guessed that it will have a mass of 14-15 GeV by extrapolating from lower heavy quark mass ratios. The corresponding $\bar{n}n$ vector meson would have a mass of 30 GeV. The excitation spectrum of quarks is contrasted with that of the hydrogen atom. The possibility of regarding flavor as an excitation of some new degree of freedom in the system is also discussed.

I want to open this session by making a few observations and raising a few questions about flavors, especially about quark flavors. By now, we know of at least five and possibly six flavors of quarks. They are the u,d,s,c and b (or $\Upsilon/2$) quarks, displayed in Fig. 1. They have constituent quark masses of $\sim \omega/2, \omega/2, \phi/2, \psi/2$ and $\Upsilon/2$, if we use the vector mesons to determine the quark masses. There is still some ambiguity about how the Υ' at $\sim$ 10 GeV should be described, i.e., whether it is a radial excitation

*Work supported in part by the U.S. Department of Energy under Contract No. EY76-C-03-0068.

of the Υ, or whether it corresponds to a different quark - say the t - or even whether it is an I = 1 state. In any case, with the array of quark flavors shown in Fig. 1, we might ask, "where does the next flavor of quark occur?" In fact, with the apparently growing number of flavors, can we regard flavor as an excitation of some new degree of freedom in the system?

From the quark mass levels, displayed in Fig. 1, notice that the ratios, c/s and b/c of the constituent masses are ≈ 3, while the s/d and s/u ratios are ≈ 1.3. If I neglect the s/d,u ratio on the grounds that the u,d system is more relativistic than the high quark mass systems, then I can guess that the next quark mass n will be ~ 3 times that of the b quark, i.e., 14-15 GeV, and that the corresponding vector meson $\bar{n}n$, will be at ~ 30 GeV.

As each new quark comes in, it appears that we have a new broken internal symmetry that we can use to describe the dynamics of the system, i.e., SU(2) for the u,d quarks, SU(3) when the s quark is added, SU(4) when the c quark is reached, and so on (Fig. 2).

The excitation spectrum of quarks is a very peculiar one. It looks like the inverse of the hydrogen atom. The hydrogen atom, with a $1/n^2$ spacing starts with large level spacings; then the excited levels crowd to a continuum limit indicating the possibility of ionization. In the quark system, the spacings seem to start out small and then to grow geometrically (Fig. 3), consistent with the idea of confinement (or no ionization).

I don't know what kind of potential or, in fact, whether any potential can be invoked to give this apparent geometric increase. The intriguing questions of

1) where the next flavors will come in, and
2) whether they are due to further substructure of the quark, or perhaps come from some other kind of excitation seem to be outside the realm of QCD for the moment.

I don't know the answers to these questions, but I think that they are worth pursuing.

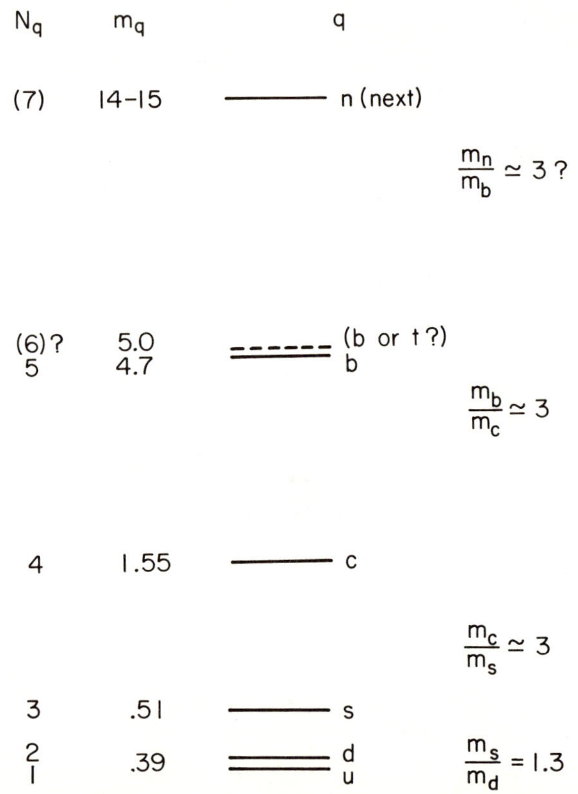

Fig. 1: Quark Flavor Spectrum. The presently known quarks are displayed, along with their constituent quark masses. A new quark n is guessed to occur at 14-15 GeV, if the ratio m_n/m_b is $\simeq 3$ as are the ratios m_b/m_c, and m_c/m_s.

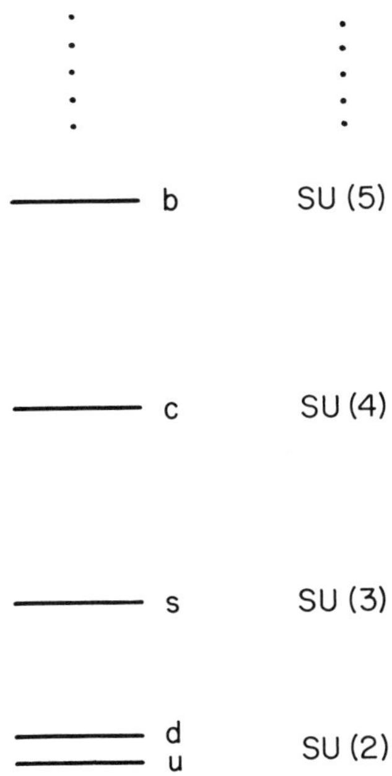

Fig. 2: Internal Symmetries. As the excitation energy for each quark is reached a new broken internal symmetry describes the dynamics of the system.

Fig. 3: Quark flavor spectrum compared to the hydrogen atom spectrum. The quark flavor spectrum is the inverse of the hydrogen atom spectrum. The quark spectrum spacings grow geometrically whereas the hydrogen atom spectrum ($1/n^2$) crowds to a continuum limit.

NEUTRINO-INDUCED MULTIMUON EVENTS FROM THE CDHS EXPERIMENT

K. Kleinknecht

Institut für Physik der Universität Dortmund

Dortmund, Germany

ABSTRACT

Experimental results on multimuon production by neutrinos and antineutrinos from the CERN - Dortmund - Heidelberg - Saclay neutrino are presented:

(a) 257 (58) fully reconstructed opposite-sign dimuon events (antineutrinos) are interpreted as due to charm production and decay;

(b) 28 trimuon ($\mu^-\mu^-\mu^+$) events are compared to different models assuming an electromagnetic, hadronic or purely leptonic origin of the events. The data point towards hadronic μ-pair production as the main source of trimuons;

(c) a first tetramuon event $\nu + N \rightarrow \mu^-\mu^+\mu^-\mu^+ X$ has been observed.

I. INTRODUCTION

The neutrino detector[1] of the CDHS Collaboration[x] has now

[x]The members of this group are: M. Holder, J. Knobloch, J. May, H. P. Paar, P. Palazzi, D. Schlatter, J. Steinberger, H. Suter, H. Wahl, E.G.H. Williams (CERN), F. Eisele, C. Geweniger, K. Kleinknecht, G. Spahn, H.-J. Willutzki (Universität Dortmund), W. Dorth, F. Dydak,

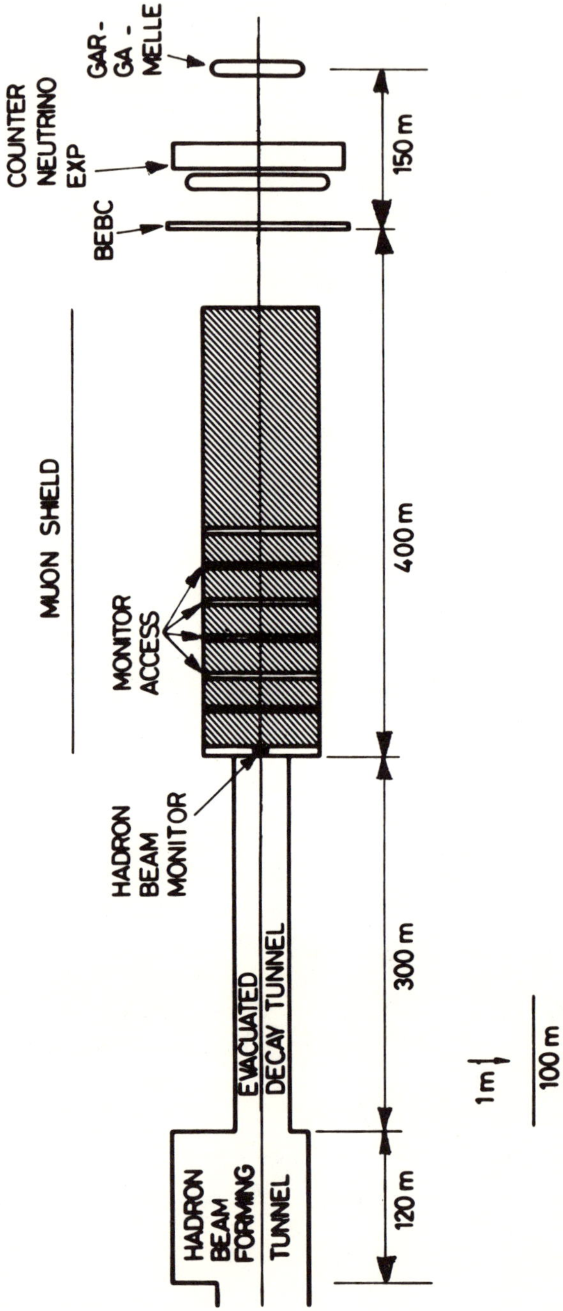

Fig. 1 Layout of neutrino beam area at the CERN SPS.
The CDHS detector is labeled "counter neutrino experiment"

been in operation for one year at CERN, and results on neutral currents[2,3], charged currents[4] opposite-sign dimuons[5], like-sign dimuons[6], trimuons[7], and a tetramuon event[8] have been published. This report will be devoted to the multimuon events, i.e., events with two, three and four muons in the final state.

2. BEAMS AND DETECTOR

There are two neutrino beams at CERN (fig. 1). In the narrow-band beam, π and K mesons produced by 400 GeV protons impinging on a Be target are entering a beam line of 120 m length which selects particles of given electric charge in a narrow momentum band around $p = (200 \pm 10)$ GeV/c. These mesons are then allowed to decay in an evacuated beam pipe of 300 m length and 1.5 m diameter. Behind this tube hadrons and nearly all muons are absorbed in 350 m of shielding material consisting mainly of iron. The neutrino energy spectrum in this narrow-band beam contains two components: the neutrinos from π decay extending from 5 - 86 GeV and the ones from K decay between the maximum energy of 191 GeV (for 200 GeV parent momentum) and about 100 GeV where the neutrinos start missing the detector sideways. The resulting energy spectrum of charged-current events is shown in fig. 2 for incident neutrino and antineutrino beams.

On the other hand, in the wide-band-beam (WBB) there is no momentum selection of secondary mesons behind the proton target. Instead, a magnetic horn and a reflector are used for focussing of positive (neutrino beam) or negative (antineutrino beam) mesons. Decay pipe and shielding remain the same as in the narrow band beam.

A spectrum of total visible energy for charged current events for the WBB focussing positive secondaries is shown in fig. 3.

The neutrino beams impinge on several detectors, as indicated in fig. 1. The one labeled "counter neutrino experiment" is the

---

V. Hepp, K. Tittel, J. Wotschak (Universität Heidelberg), P. Bloch, B. Devaux, M. Grimm, J. Maillard, B. Peyaud, J. Rander, A. Savoy-Navarro, R. Turley (CEN-Saclay), F.L. Navarria (Universita Bologna)

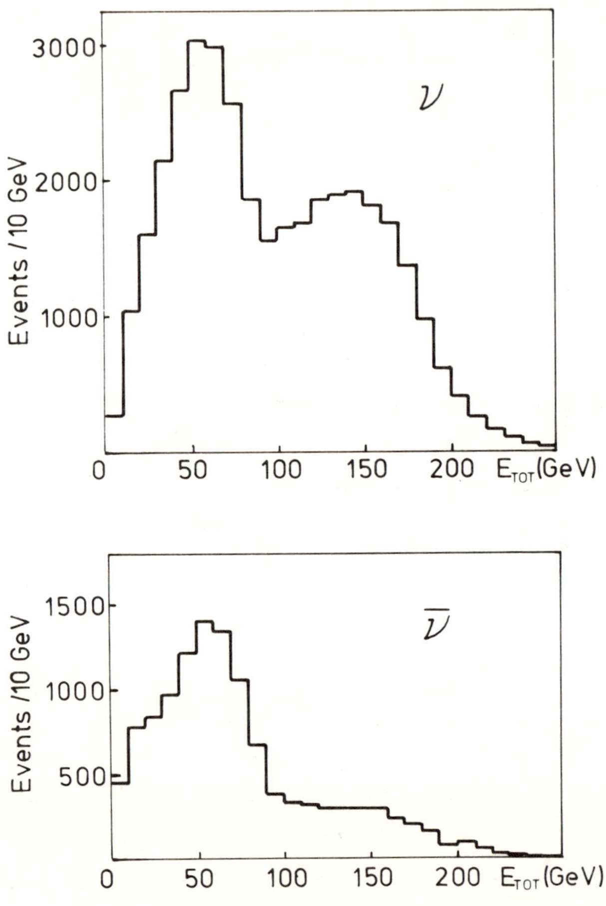

Fig. 2 Charged-current event rate as function of neutrino energy for the neutrino and antineutrino narrow-band beam.

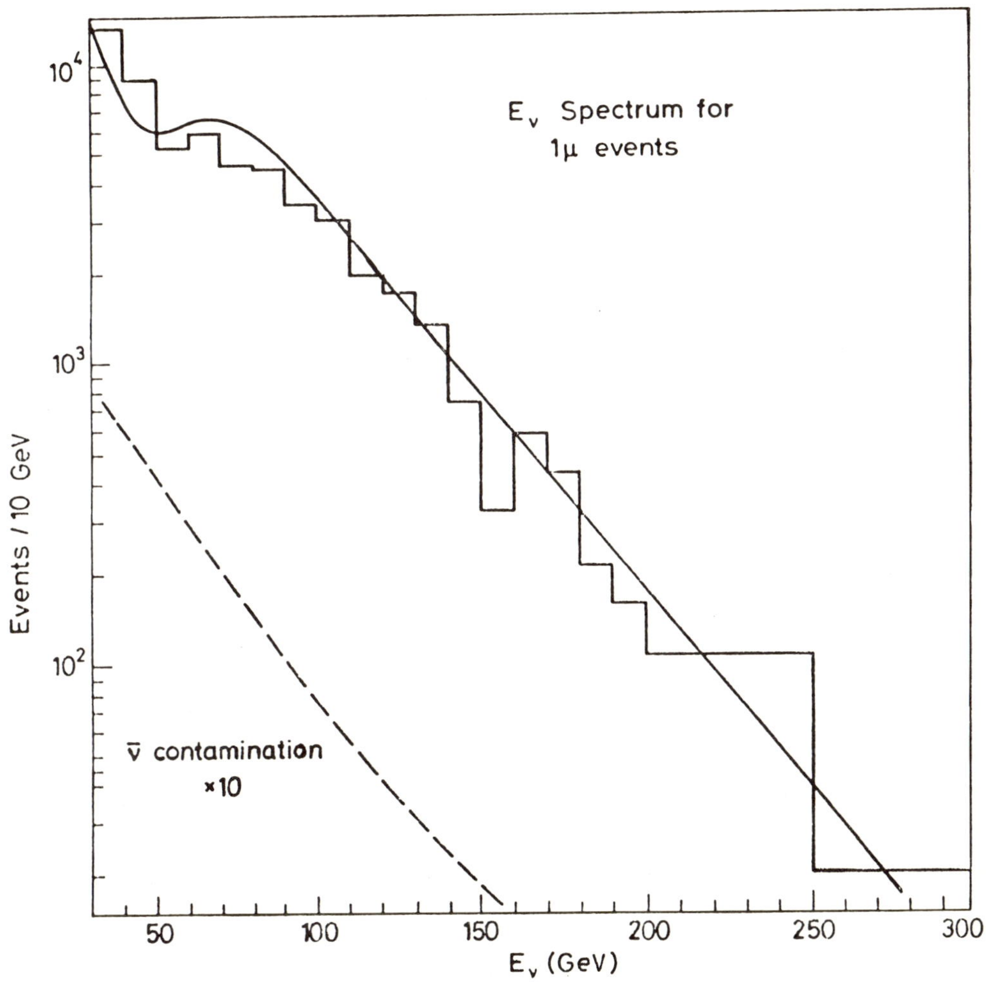

Fig. 3 Charged-current event rate as function of neutrino energy for the neutrino wide-band beam.

detector of the CDHS collaboration.

In this detector (fig. 4,5) the functions of neutrino target, hadron calorimeter, muon identifier and muon magnetic spectrometer are integrated. It consists of 19 toroidal modules of magnetized iron plates interspaced with 19 triple plane drift chambers. The diameter of the toroids, 3.75 m is matched to the 90° c.m. decay angle of neutrinos from $\pi \to \mu\nu$ decay such that nearly all π decay neutrinos hit the detector, while about half of the neutrinos from K decay miss the apparatus. The thickness of these magnets is 75 cm, composed of 15 plates of 5 cm thickness for the first seven modules and of 15cm plates for the other twelve . In each gap between two plates is inserted a plane of eight 6 mm plastic scintillators viewed by two phototubes at each end. The sum of right and left pulse heights is used for calorimetry, while their ratio serves for determining the shower position along the counter. Pulse height calibration and measurement of light attenuation in the counters is done using cosmic muons continuously between machine bursts.

The iron weight of the detector is 1240 t, of which 800 t are used as a fiducial target. The average magnetic field in the toroids amounts tc 16.5 kG. An average density of 5.3 g/cm^3 is obtained, which is important for the suppression of $\pi \to \mu$ and $K \to \mu$ decays of π and K mesons from the hadron shower.

The drift chambers are hexagonal and consist of three independent gaps with wires in the vertical direction and at $\pm$ 60° relative to the vertical. The wire spacing is 6 cm, the measurement accuracy 1 mm and the efficiency typically 99.5% per gap. We study the inclusive processes $\nu + Fe \to n(\mu^\pm) + X$ and $\bar{\nu} + Fe \to n(\mu^\pm) + X$ (n = 2,3,4), where X is any hadronic final state. The hadron energy E_h of X is measured in the Fe-scintillator calorimeter. The response of this calorimeter to incident hadrons (and electrons) has been determined in a test calorimeter of smaller dimensions (150 cm long, 60 x 80 cm transverse size) exposed to hadron beams

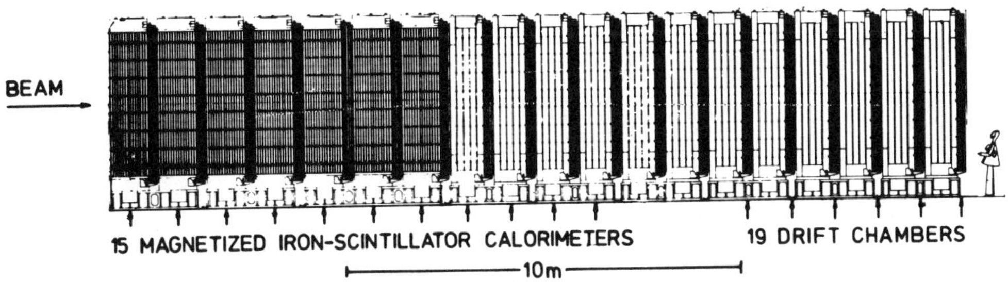

Fig. 4 Side view of the CDHS detector

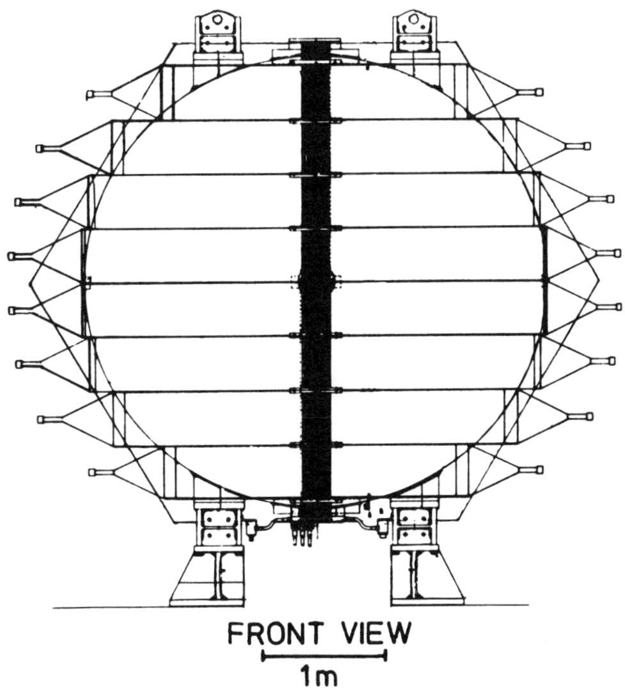

Fig. 5 Front view of module of the CDHS detector

of energy between 15 and 140 GeV. The resolution is $\Delta E_h/E_h = 0.9/\sqrt{E_h}$ (GeV) for 5 cm sampling and $\sqrt{3}$ times larger for 15 cm sampling.

The resolution in muon momentum as measured in the magnetic spectrometer is determined mainly by the multiple scattering in the iron and is depending on the length L of the muon track. Using the measured value of the magnetic field, one obtains $\Delta p/p \simeq 0.2/\sqrt{L(m)}$.

The angular acceptance of the spectrometer is large due to the integration of the target and the spectrometer. Since we require an event to be detected in at least 5 consecutive drift chambers, this corresponds to an effective cut in the muon laboratory angle around 400 mrad and to a cut in muon momentum at about 4.5 GeV/c.

3. OPPOSITE SIGN DIMUON EVENTS

This type of event has been discovered by the HPWF group[9], and further experiments[10,11] gave support to the explanation of a charm origin of these events. Our data were taken in the narrow-band-beam in a neutrino exposure corresponding to 3×10^{17} protons and an antineutrino exposure of 6×10^{17} protons, yielding 53,000 charged current events for ν and 15,000 for $\bar{\nu}$. Fig. 6 shows the spatial reconstruction of a dimuon event. From the reconstructed vector momenta of the two mesons, the neutrino direction and the measured hadron energy E_h we obtain the following kinematical variables:

- $E_{\mu 1}$: the energy of the leading muon, i.e. the one carrying the same lepton number as the incident (anti) neutrino
- $E_{\mu 2}$: the energy of the non-leading muon
- $x = (p_\nu - p_{\mu 1})^2/[2M_p(E_h + E_{\mu 2})]$
- $\vec{W} = \vec{\nu} - \vec{\mu}_1$ the direction of the hadron shower
- ϕ : the azimuthal angle between the projections of the momenta of μ_1 and μ_2 onto the plane perpendicular to the neutrino direction

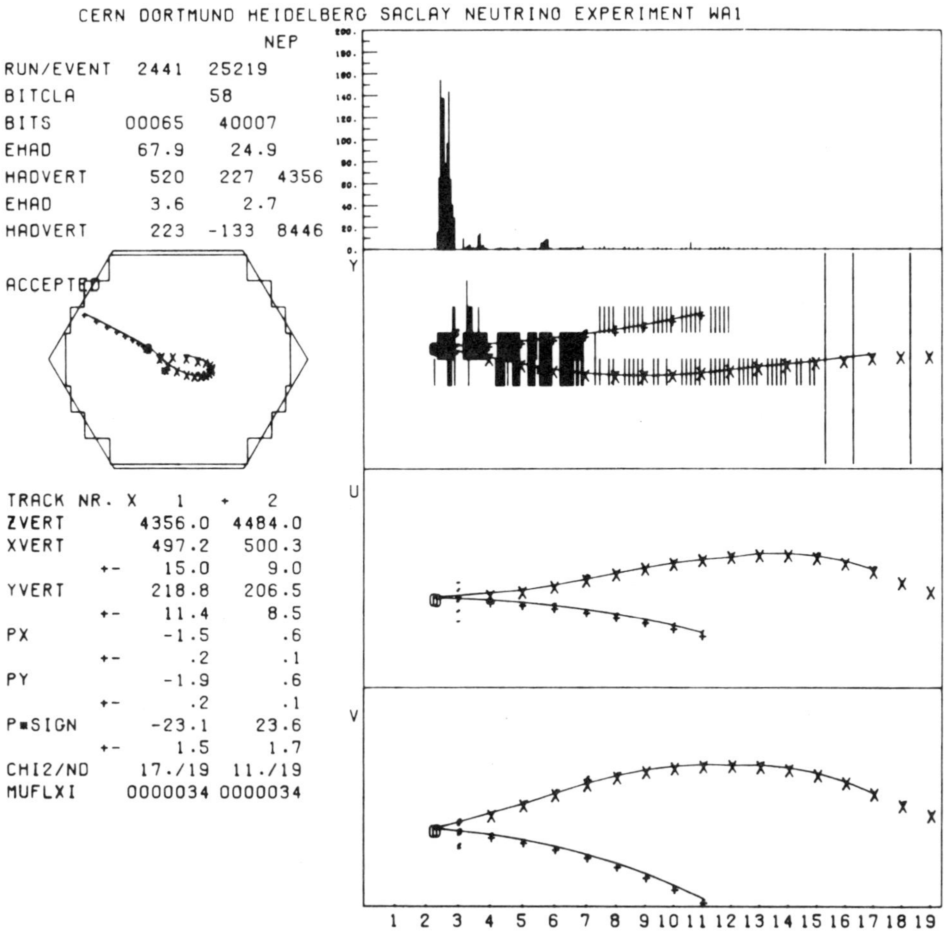

Fig. 6 Dimuon event display. The two lowest sections show $\pm 60^{\circ}$ side views of the drift chamber wire hits and the reconstructed muon tracks, the top one gives summed pulse heights in each scintillator plane, and the one between combines spatial information from scintillators and horizontal drift chamber wires.

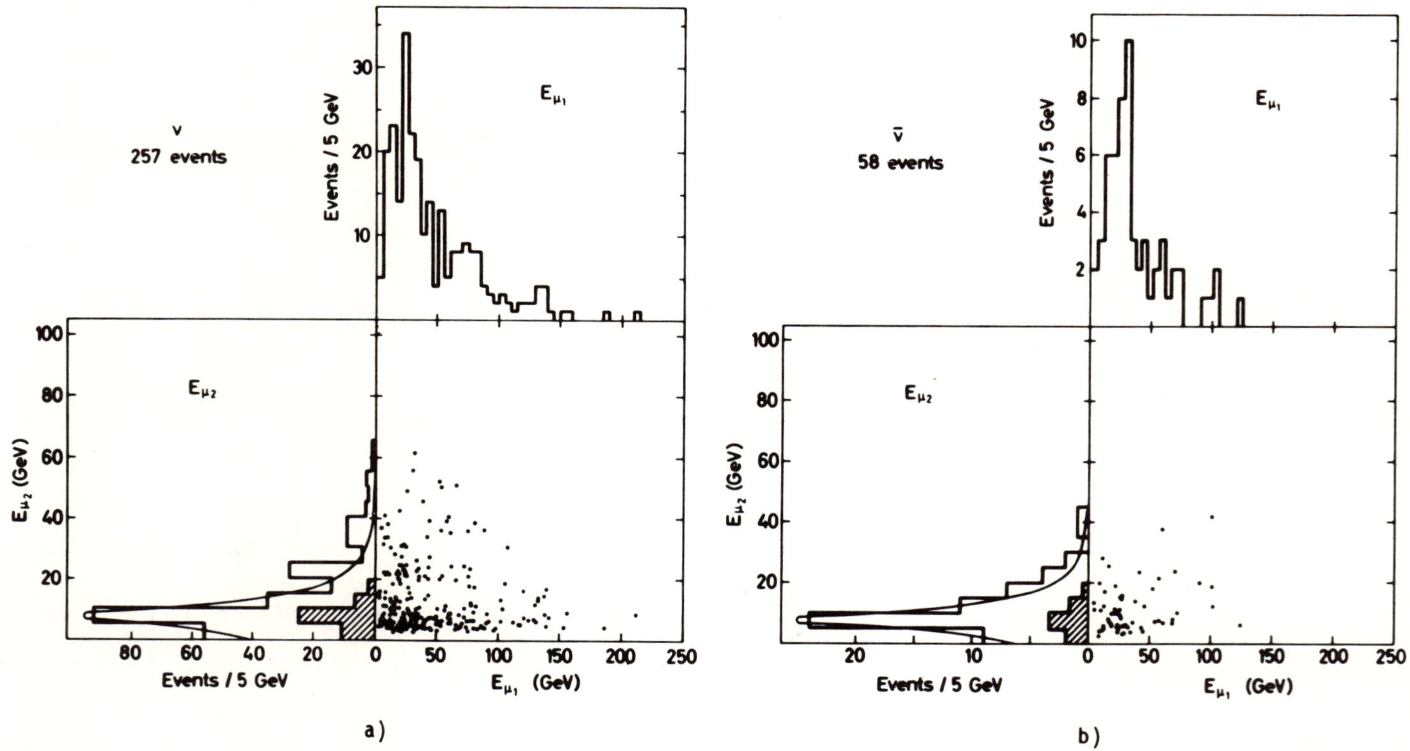

Fig. 7 Scatter plots in $E_{\mu1} - E_{\mu2}$ plane for ν and $\bar{\nu}$ opposite sign dimuons. Background from π and K decay is shown. The line is the prediction of a charm model.

P_{TSH}: the transverse momentum of μ_2 relative to the shower axis $\vec{W}$.

Fig. 7 is a scatter plot of $E_{\mu 1}$ vs. $E_{\mu 2}$ for neutrino and antineutrino events showing a marked asymmetry between the two muons; the averages (for ν) $\langle E_{\mu 2} \rangle = (13.7 \pm 1)$ GeV and $\langle E_{\mu 1} \rangle = (45 \pm 2)$ GeV differ considerably, in line with a reaction mechanism where the leading μ_1 is produced at the lepton vertex, and the other μ_2 at the hadron vertex. The data are contaminated with a background where μ_2 comes from the decay of π and K mesons in the hadron shower of a charged-current event. This is calculated to be (13 ± 4) % of the events.

In order to investigate the nature of the dimuon signal, we then plot the x distributions for neutrino-and antineutrino induced events (fig. 8). The distribution for antineutrinos is shifted towards low x, $\langle x \rangle_{\bar{\nu}} = 0.15 \pm 0.02$ compared to the neutrino distribution $\langle x \rangle_{\nu} = 0.24 + 0.01$, indicating that for $\bar{\nu}$, the dimuons are produced predominantly off the sea quarks. The x distribution for $\bar{\nu}$ is compatible with a form $(1-x)^7$ which agrees with the x distribution of the sea as obtained from charged current events.[12]

Further clarification of the origin of the second muon comes from the distributions shown in Fig. 9, where a) is a scatter plot of $E_{\mu 2}$ vs. the azimuthal angle ϕ between the two muons, and b) its projection onto the ϕ axis. If the origin of the two muons would be[13] the production and decay of a new heavy lepton L^o via $\nu + N \rightarrow L^o + X$ and $L^o \rightarrow \mu^+ \mu^- \nu$, then one would expect in general a rather flat distribution in ϕ, independent of $E_{\mu 2}$. If on the other hand, the second muon is the decay product of a particle produced at the hadron vertex, then one expects an anticorrelation in ϕ between leading μ_1 and the μ_2 following the hadron shower direction. The most probable ϕ-angle is then 180°, i.e. back-to-back emission of the two muons in the transverse direction. The correlation becomes stronger when $E_{\mu 2}$ increases because then μ_2 and its parent carry a larger fraction of the total hadron momentum.

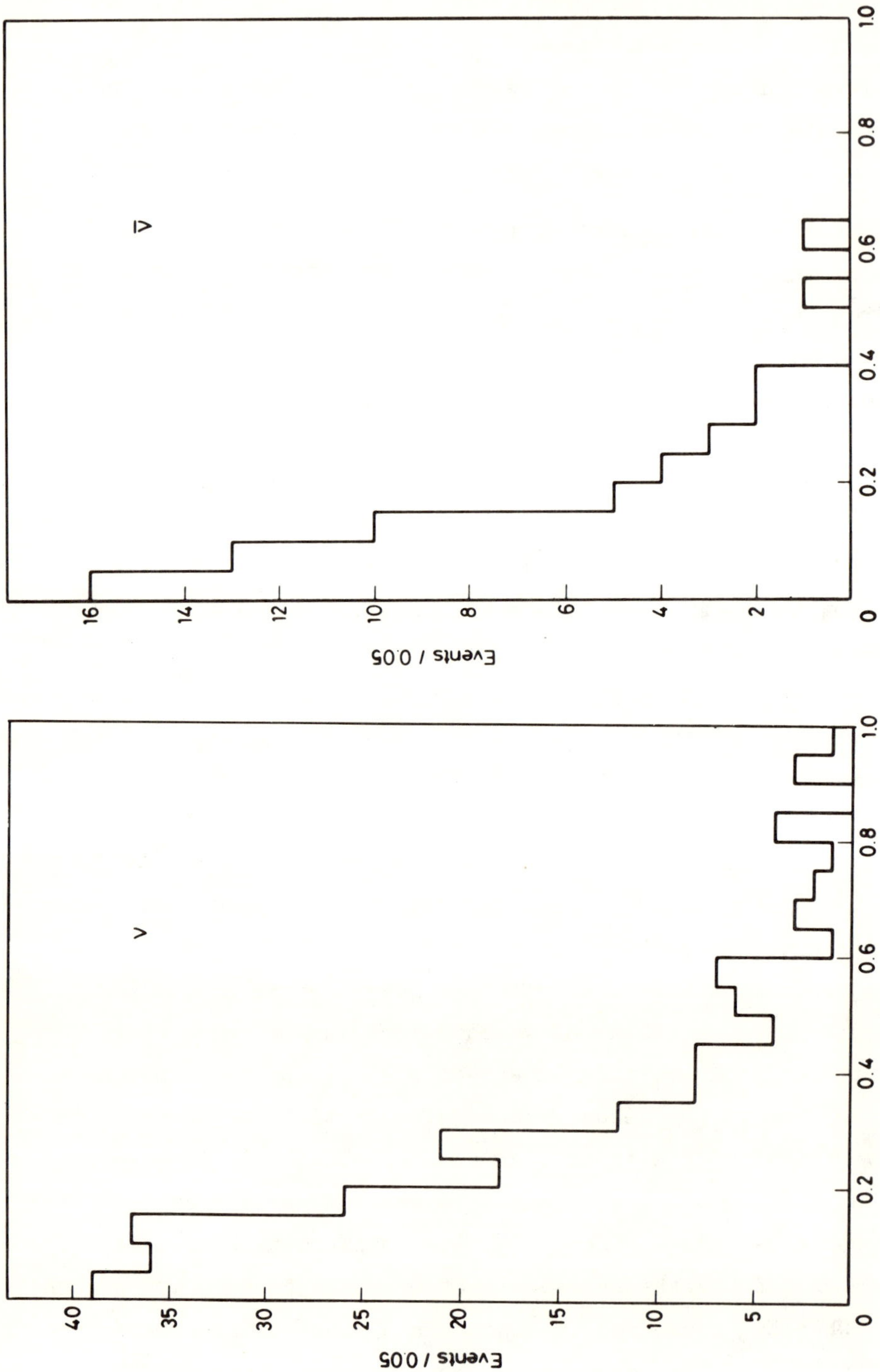

Fig. 8. x distribution for neutrino (a) and antineutrino (b) dimuons

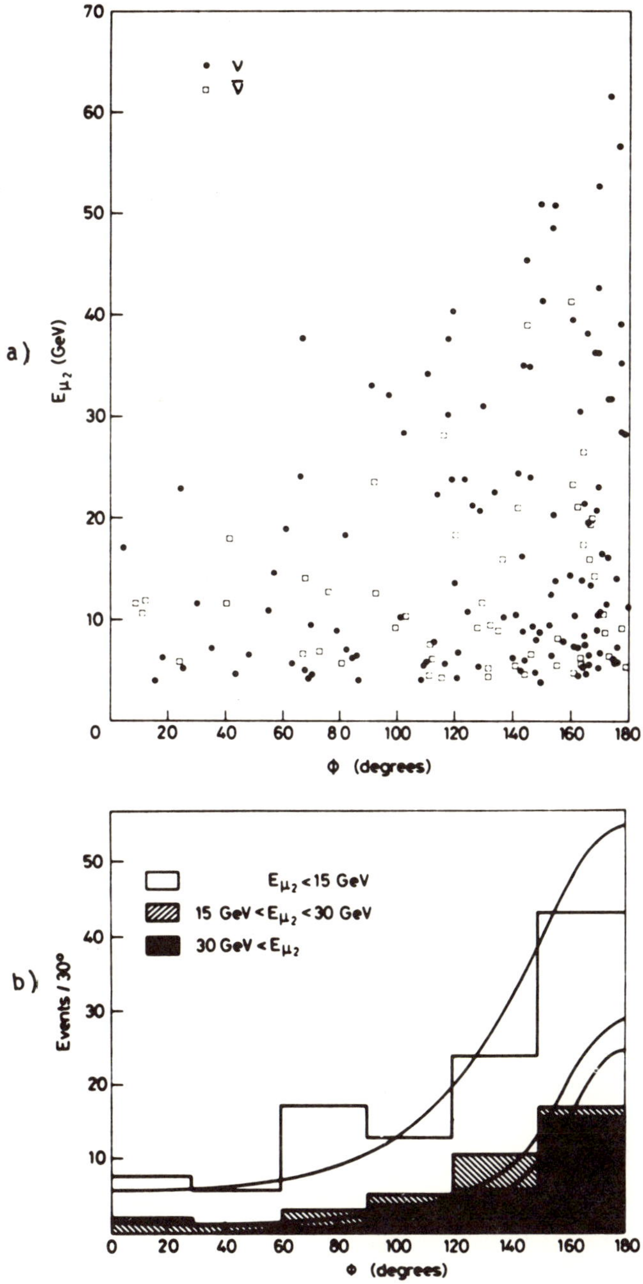

Fig. 9 (a) Scatter plot in azimuthal angle ϕ versus $E_{\mu 2}$ for ν and $\bar{\nu}$ dimuons (b) Projection on ϕ axis with predictions of charm model (lines).

The data of fig. 9 favour clearly models with hadronic origin of the second muon, and the curves shown are calculated according to such a model of charm production[14] with a fragmentation function $D(z) \propto (1-z)^2$.

The same conclusion can be drawn from fig. 10 which shows that the transverse momentum of the second muon relative to the shower axis is small even if $E_{\mu 2}$ becomes large. This is not so if one plots the transverse momentum of μ_2 relative to the μ_1 direction.

The rates of dimuon production relative to charged-current reactions are shown in fig. 11 together with the experimental detection efficiency calculated according to the charm production model mentioned above. The uncertainty in this model lies in the choice of the quark fragmentation function D(z). Since D(z) is related to the spectrum of $E_{\mu 2}$, we can limit the ranges of possible functions D(z). For those fragmentation functions compatible with the $E_{\mu 2}$ spectrum we find that the corrected cross-section ratios $R_\nu = \sigma(2\mu)/\sigma(1\mu)$ are flat within our neutrino energy range for ν and $\bar{\nu}$.

All of these observations are in agreement with the GIM model, where neutrinos can produce charm via $\nu + d \to \mu^- + c$ and $\nu + s \to \mu^- + c$ with relative rates of $\sin^2\theta_c$ and $\xi_s \cos^2\theta_c$ (where $\xi_s = \int x\, d(x)\, dx$ is the amount of strange sea in the nucleon), while antineutrinos can only produce anticharm from the sea $\bar{\nu} + \bar{s} \to \mu^+ + \bar{c}$ with a rate of $\xi_s \cos^2\theta_c$. From the measured dimuon rates R_ν and $R_{\bar{\nu}}$ we therefore deduce the amount of sea, $\xi_s = 0.035 \pm 0.01$.

The dimuon production from antineutrinos would increase drastically if a right-handed b-quark would exist, which was suggested[16] as an explanation of anomalies in the antineutrino scattering[17]. In fig. 12 the predicted increase in the double ratio $D = R_{\bar{\nu}}/R_\nu$ for b-quark masses of 5 GeV and 7 GeV are compared to our data. These results do not support the esistence of an $(ub)_R$ current, or if this b quark would exist with $m_b = 5$ GeV, then the branching ratio for

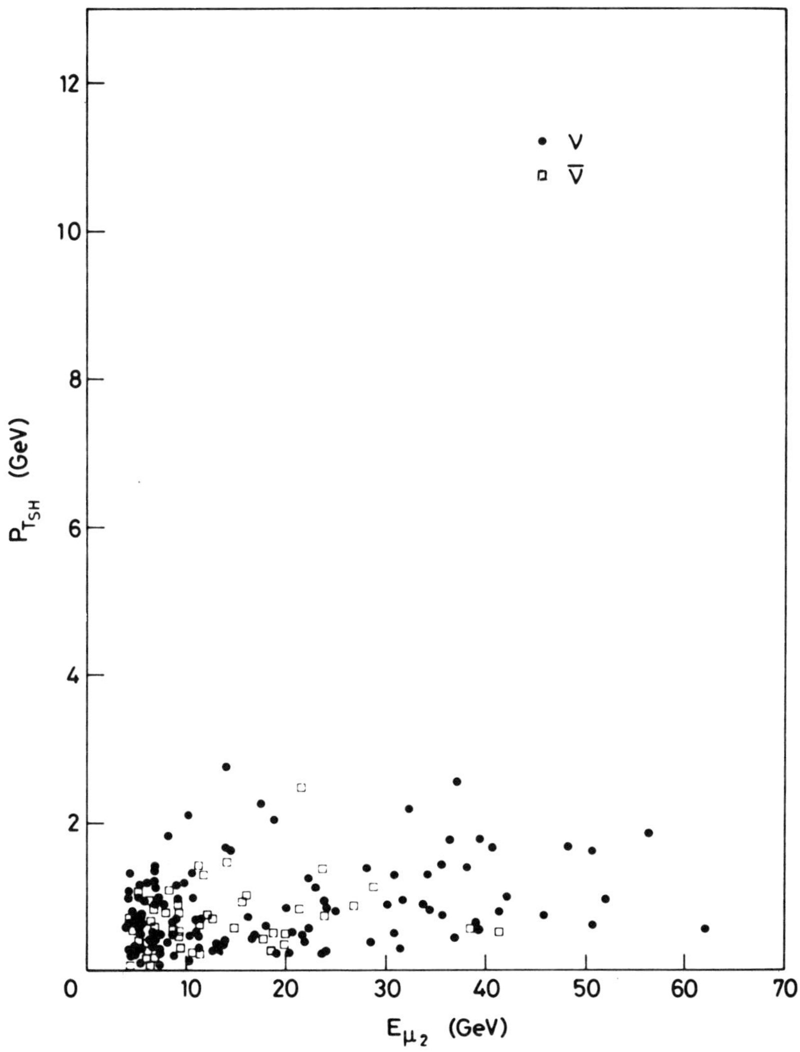

Fig. 10 Scatter plot in the transverse momentum of the second muon with respect to the hadron shower (P_{TSH}) versus the energy of the second muon($E_{\mu 2}$).

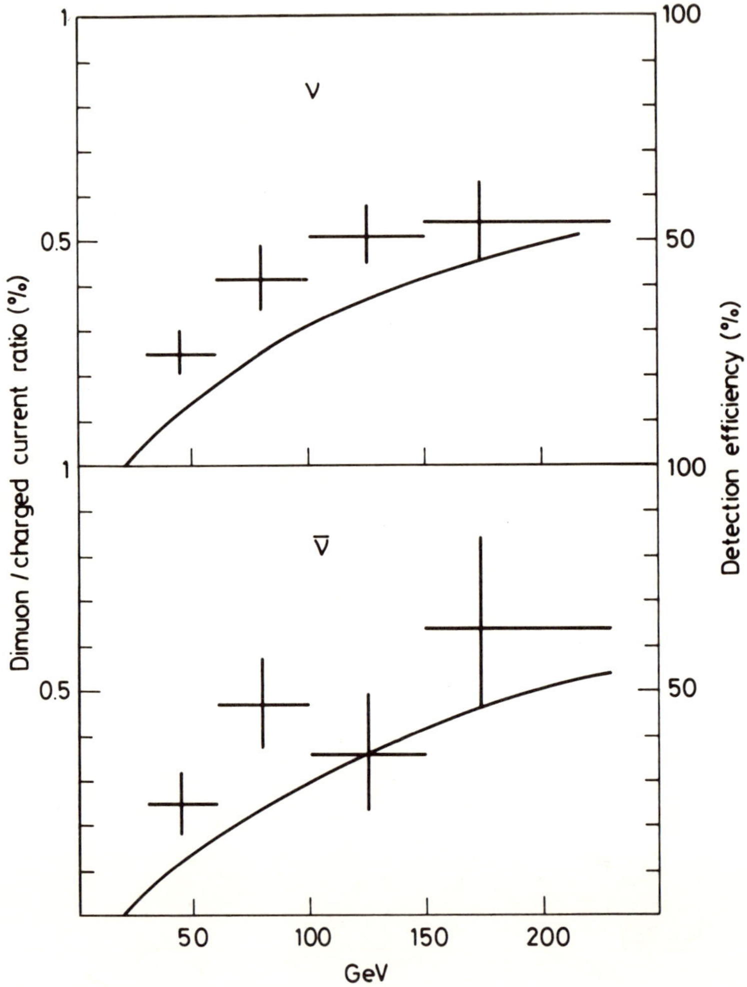

Fig. 11 Ratio of observed dimuon rates to charged current rates as function of energy. The curve shows the calculated detection efficiency for dimuons in a charm model.

muonic b-decay must be less than 8% of the one for semileptonic charm decay.

Similar results have been reported recently by the CITFR group[18].

4. TRIMUONS

Trimuons were first seen in the CITF[19] and HPWF[20] experiments. The CDHS collaboration has obtained 3 events in the narrow-band beam[7] and 28 events in the wide-band beam (a more complete report with 76 events is being prepared[21]). Apart from the trivial case of dimuons with an additional muon from a pion or kaon decaying in the shower, trimuons may originate in several kinds of processes.

a) a charged-current interaction with electromagnetic production of a muon pair, either from the outgoing muon or from the incoming or outgoing quark[22, 23]

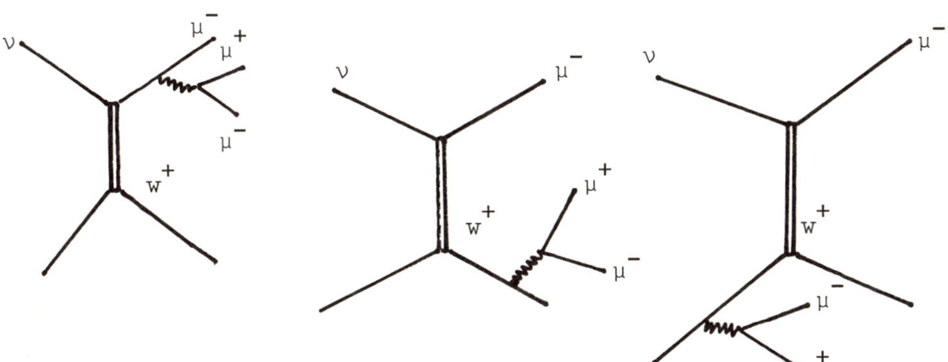

b) a charged-current interaction with hadronic production of a muon pair, e.g. by the decay of a vector meson

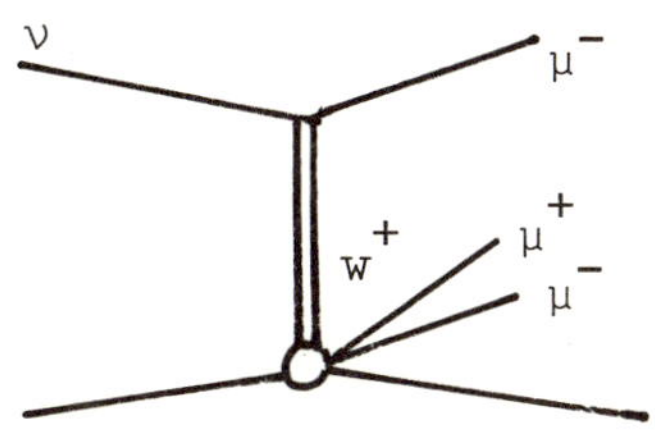

c) a charged-current interaction with charm-anticharm production at the hadron vertex and subsequent decay of both c and $\bar{c}$ to muons[24]

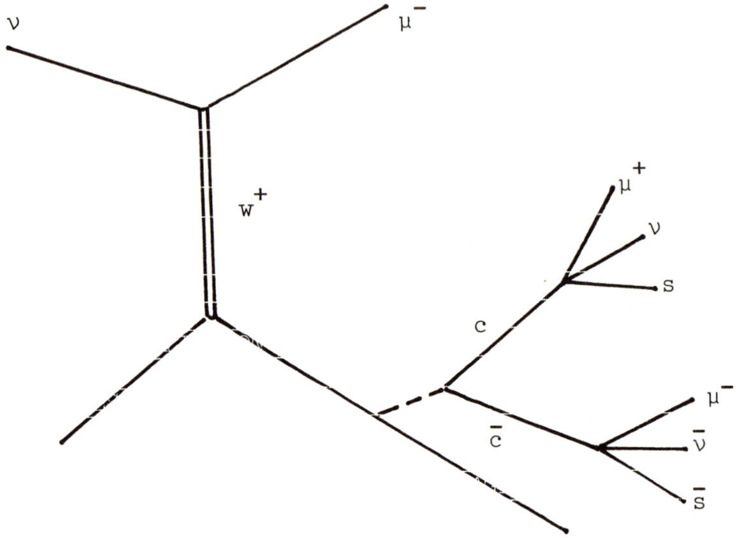

d) more exotic possibilities involving i) a heavy lepton cascade[25,26] ii) a quark cascade and iii) lepton-hadron models[27]

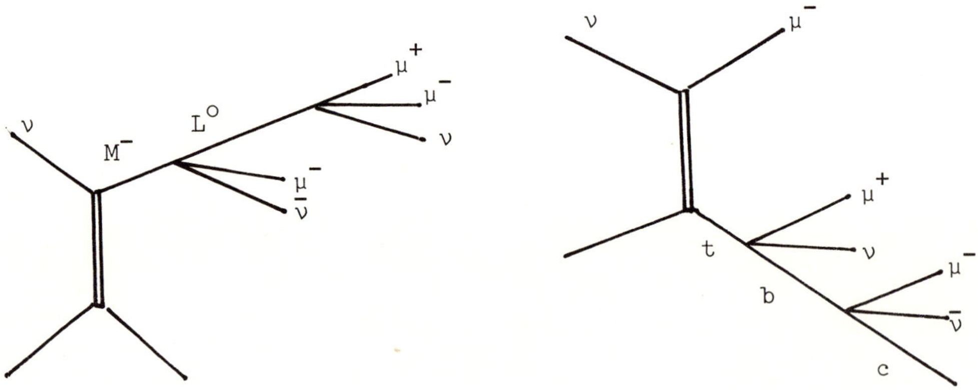

More complete references to theoretical work can be found in a review by Barger[28].

We then turn to experimental data. One example of a trimuon event $\nu\ Fe \rightarrow \mu^-\mu^-\mu^+X$ in the CDHS detector exposed to the wide-band beam is shown in fig. 13 in three 120° views.

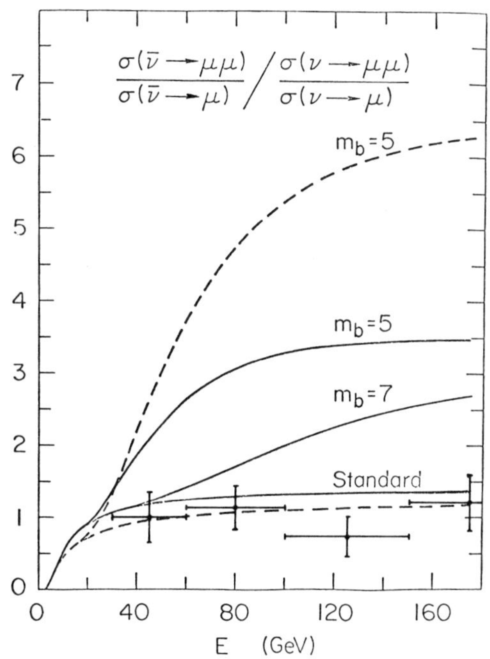

Fig. 12 Double ratio of dimuon to charged-current ratio for antineutrinos relative to neutrinos. Solid (dashed) curves include (exclude) asymptotic freedom corrections. Two lower curves correspond to the four-quark model, the three upper ones to models with an $(ub)_R$ current assuming m_b = 5 or 7 GeV.

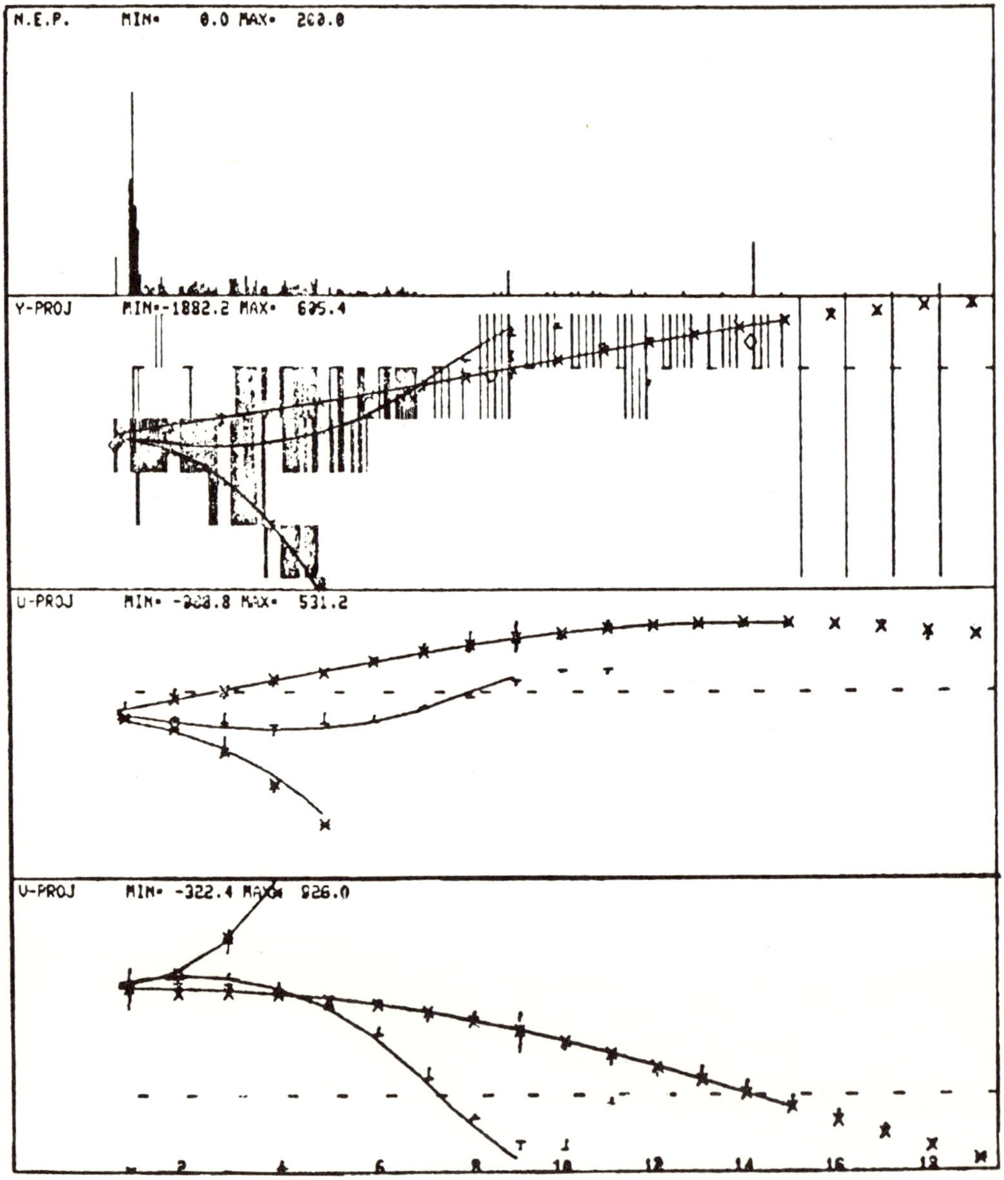

Fig. 13 Trimuon event display. On top, pulse heights. Next lower, horizontal drift chamber wires and scintillator hits. Botton, drift chambers at $\pm$ 60° to horizontal.

From the measured vector momenta of all three muons, the hadron energy E_h, and the neutrino direction we can reconstruct the kinematics of the event. In particular the total visible energy is $E = E_{\mu 1} + E_{\mu 2} + E_{\mu 3} + E_h$. A plot of trimuon event numbers vs. E is shown in fig. 14 for three different exposures together with the corresponding dimuon- and charged-current rates. The measured ratio $3\mu/CC$ ranges between 2×10^{-5} and 8×10^{-5} in this energy region, which seems lower than the figure 5×10^{-4} quoted earlier in Refs. 20 and 29.

The calculated rates for electromagnetic processes[22,23] are around 3×10^{-5} including experimental cuts, and the charm-anti-charm production is estimated[24] to be less than 10^{-6}. A simple model on hadronic μ pair production was made using experimental data[30] on dimuon production in $\pi - N$ reactions and assuming a resemblence between the interaction of the virtual W boson and this $\pi - N$ interaction. The resulting trimuon rate is 2×10^{-5}.

We see that from the rate alone a reliable choice between different production mechanisms is difficult to make. We therefore turn to kinematical properties of the trimuon events. Since there are two μ^- and one μ^+ in these events, we have to choose the leading μ^- amongst the two. In analogy to the dimuon events where the leading muon nearly always has a larger four-momentum transfer $Q^2 = (k_\mu - k_\nu)^2$ than the second muon (with k_μ and k_ν being the four-momenta of μ and ν respectively), we define here as leading μ^- the one with the larger Q^2 compared to the other μ^-, and we call it μ_1. (Other choices are possible). Fig. 15 then illustrates a difference between the leading and non-leading μ^-. Here the invariant dimuon masses M_{--} and M_{-+} are plotted against the trimuon invariant mass M_{--+}. For M_{-+} there are two possible combinations; the one indicated by a cross is the one between μ^+ and μ_1^-, the one given by a dot is the combination of μ^+ with the non-leading μ_2^-. It appears from the data that there is little correlation between the leading μ_1^- and both μ_2^- and μ^+, while the invariant mass

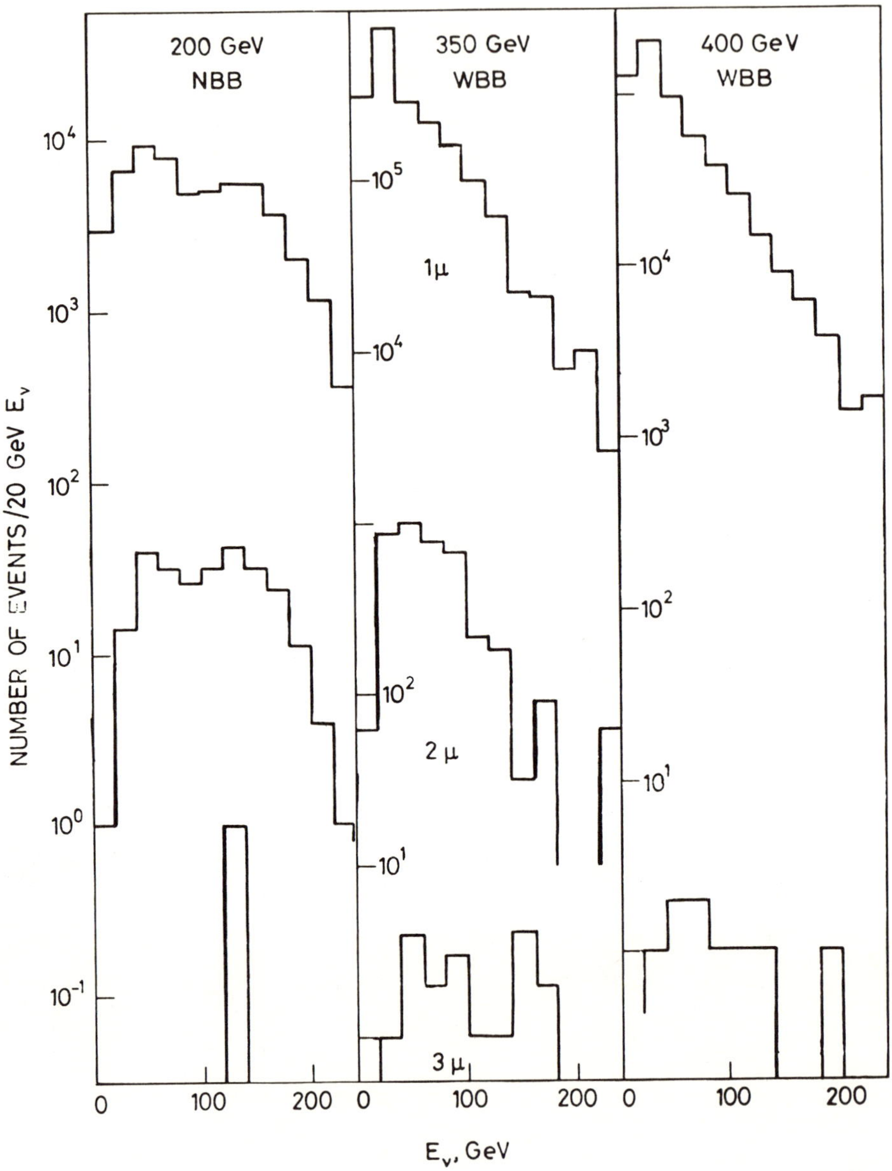

Fig. 14 Rate of 1μ, 2μ , and 3μ events versus neutrino energy for narrow-band (NBB) and wide-band (WBB) beam exposures.

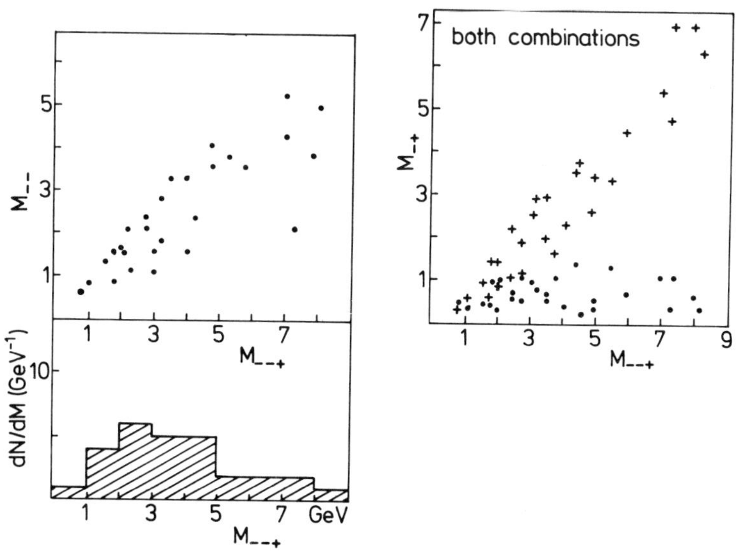

Fig. 15 Scatter plot of invariant 2μ and 3μ masses for trimuon events.

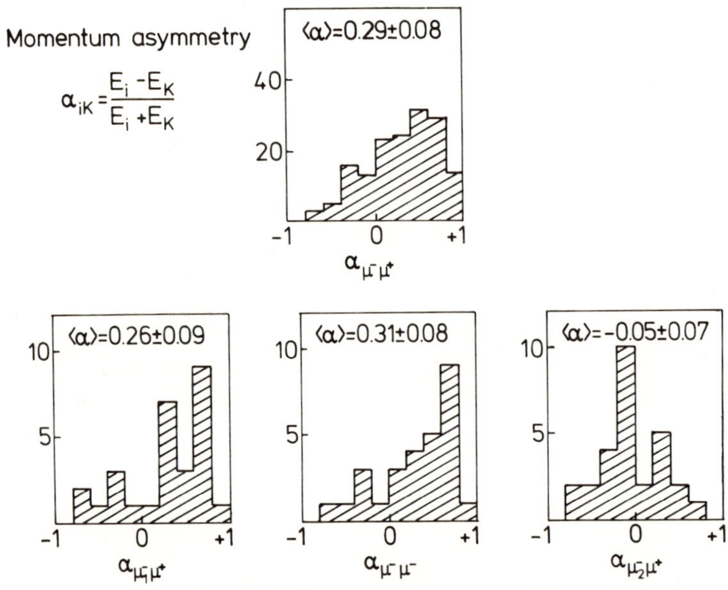

Fig. 16 Muon energy asymmetries α_{iK} for dimuon events and for the three two-muon combinations of trimuons.

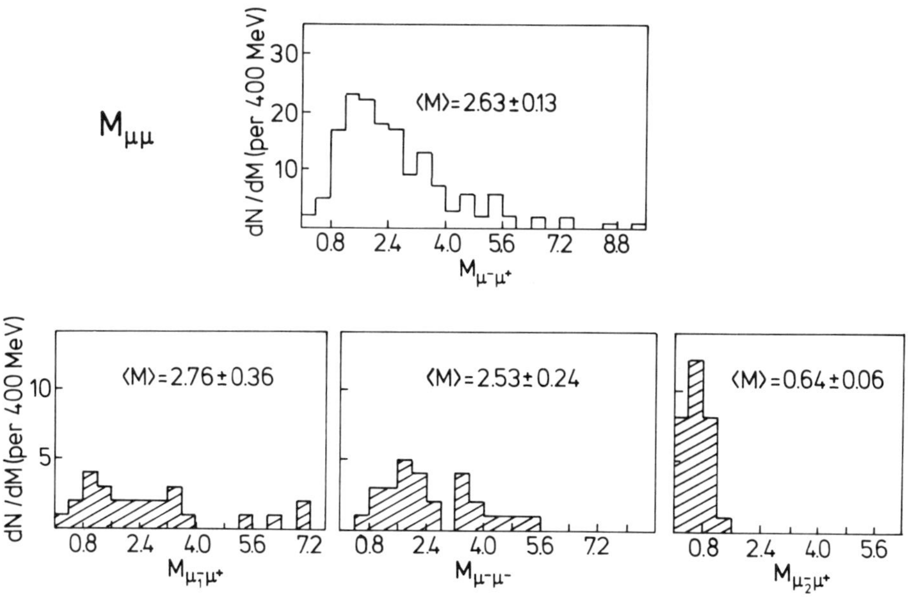

Fig. 17 Invariant two-muon masses for dimuon events and for the three two-muon combinations for trimuons.

of μ^+ with μ_2^- is bounded below 1.5 GeV, indicating a possible common origin of those two muons.

A similar conclusion can be drawn from the energy (or momentum) asymmetries $\alpha_{iK} = (E_i - E_K)/(E_i + E_K)$ of pairs (i, K) of muons shown in fig. 16. While the combinations (μ_1^- μ^+) and (μ_1^- μ_2^-) are as asymmetrical as the one from dimuons (top), the combination (μ_2^- μ^+) is symmetrical within the experimental error. The invariant masses corresponding to these combinations are displayed in fig. 17. They again show the striking difference between the (μ_2^- μ^+) pairing and other combinations.

The low values of m (μ_2^- μ^+) are of course reproduced well by the electromagnetic μ pair production models [22,23] and by the experimental data on hadronic μ pair production[30] which are dominated by ρ production. The leptonic cascade models[25,26] can only reproduce the data if the mass of the neutral heavy lepton L_o is around 1.5 GeV, which means a severe restriction on these models.

A more decisive test can, as in the case of dimuons, be done on the basis of the projections of the muon momenta on the plane perpendicular to the neutrino direction. If we then compute the angle ϕ in this plane between the leading μ_1^- and the vector sum of μ_1^- and μ_2^- and the vector sum of μ_2^- and μ^+ , we expect quite different distributions for the models in question: the electromagnetic models would yield a peak around $\phi = 0^o$ due to radiation from the μ_1^- and a second component around $\phi = 180^o$ due to quark radiation; the relative amount of these two contributions differ slightly between the models[22,23]. Lepton cascade models[25,26] yield a rather flat distribution in ϕ, while the hadronic production will occur predominantly around $\phi = 180°$. Fig 18 shows ϕ distributions according to several models and the data. The curves are normalized to the same event number. It appears that neither the electromagnetic nor the lepton cascade mechanism can be the dominating source of trimuons. On the other hand, the hadronic pair production model can reproduce the data quite well. A new physics mechanism is therefore

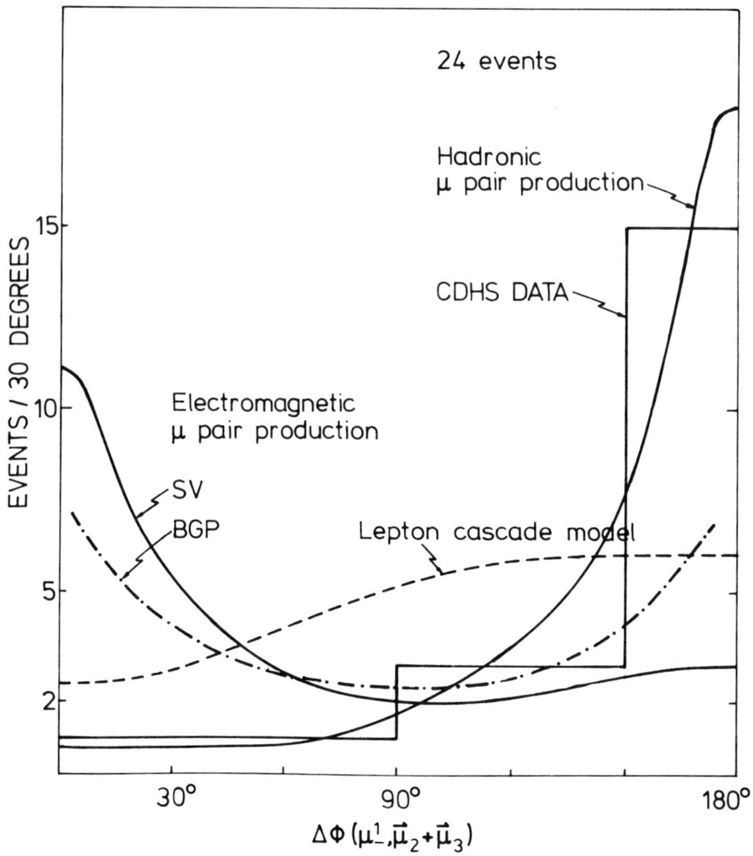

Fig. 18 Distribution in ϕ, the azimuthal angle in the plane transverse to the neutrino direction between the leading μ_1^- and the vector sum of μ_2^- and μ_3. SV (Ref. 23) and BGP (Ref. 22) are model calculations.

not required by these data.

5. THE TETRAMUON EVENT

In the wide-band neutrino beam exposure a first example of an event with four muons in the final state[8] has been found. The computer reconstruction of the event is shown in fig. 19. Since the number of charged-current events in the same exposure is estimated (on the basis of a sample) to be 3×10^6 events and the dimuon number $\sim 7 \times 10^3$, this tetramuon event corresponds to a rate of 1.4×10^{-4} relative to dimuons. The measured parameters of the event are contained in table I.

The largest background contribution comes from trimuons with a $\pi \to \mu$ or $K \to \mu$ decay in the hadron shower. For an everage[21] $E_h = 28$ GeV we expect $(1.2 \pm 0.4) \times 10^{-2}$ background tetramuon events from this source.

If we now consider possible origins of the event, three conventional processes can be invoked (fig. 20): 1) charm production together with electromagnetic μ pair production from the leading μ; 2) charm production together with μ pair production in the hadron shower; 3) charm-anticharm production by a charmed quark produced at the weak vertex. For the total contribution of processes 1) -3) we estimate a rate of 0.2 events using refs. 22,23, 24 and 30. Additional processes involving new particles could include a quark cascade with new flavours

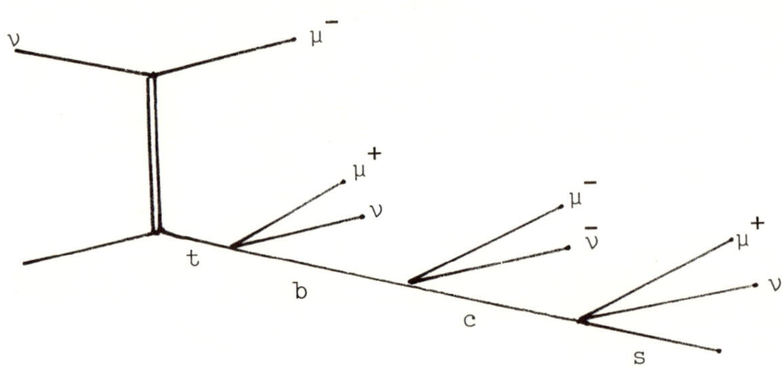

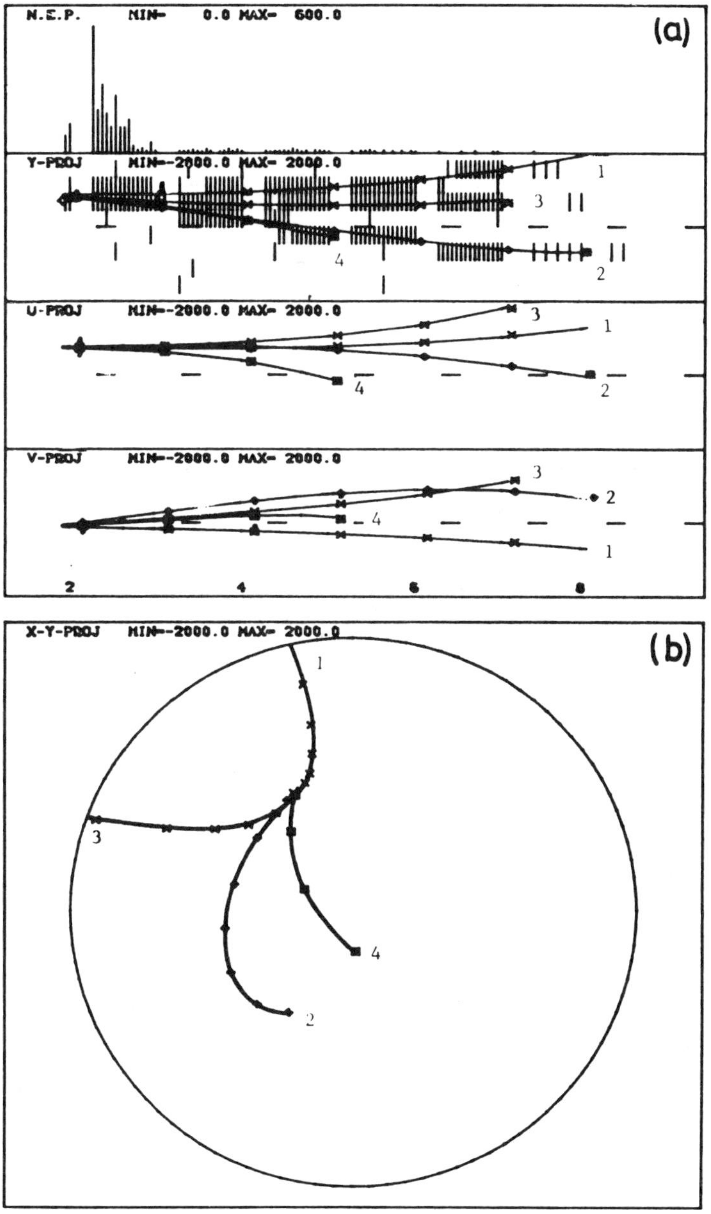

Fig. 19 Tetramuon event display a) On top, pulse heights; next lower, horizontal drift chamber hits and scintillator hits. Botton, drift chamber hits in $\pm$ 60° views to horizontal b) Drift chamber hits projected into the plane perpendicular to the neutrino direction.

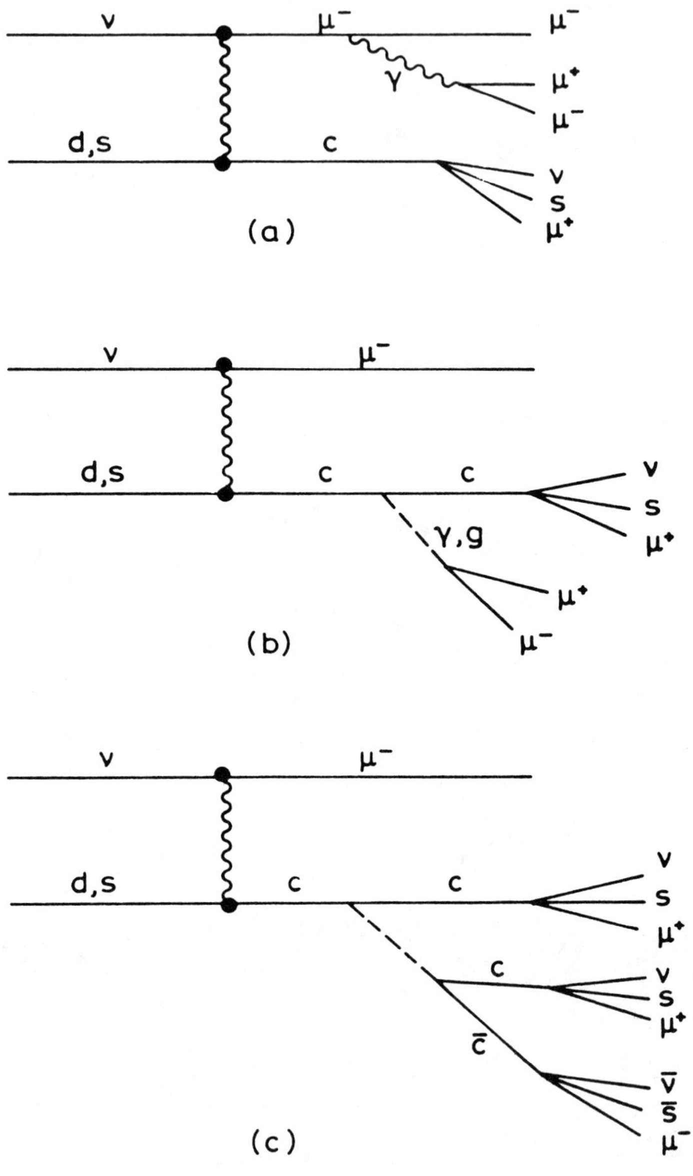

Fig. 20 Conventional sources of tetramuon events.

or a heavy lepton cascade together with charm production:

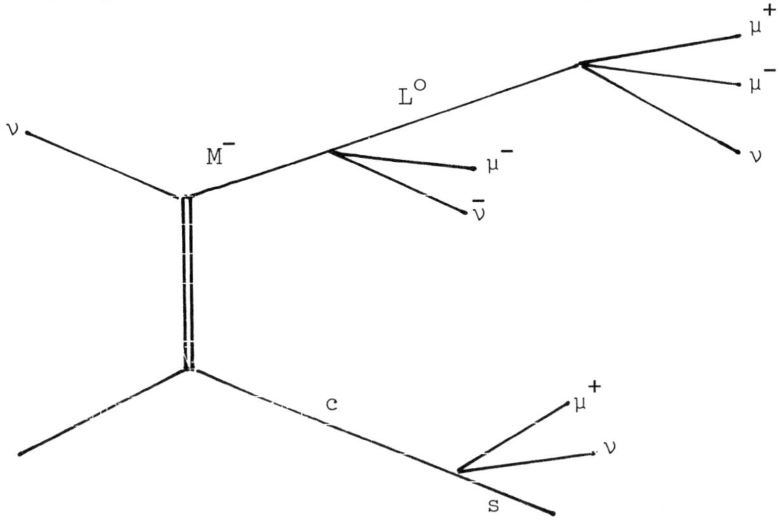

For the processes 2) and 3) and the quark cascade only one μ^- is associated with the lepton vertex, while the other three are produced in the hadron shower. For those processes we would then expect small transverse momenta of the three non-leading muons relative to the hadron shower axis. For the two possible choices of leading muon, these transverse momenta are given in table I. At least one of them is rather large in both cases. This gives slight preference to the other mechanisms listed, though it is premature to make a definite statement on the basis of this one event.

TABLE 1

Measured quantities of the four muon event

1.	Total observed energy	91.4 ± 7.3 GeV		
2.	Hadron shower energy	58 ± 7 GeV		
3.	Momenta of the muons (GeV/c) *)	P_x	P_y	P_z
	Muon 1 (+)	0.50	0.20	11.1
	Muon 2 (-)	-1.52	-1.69	8.4
	Muon 3 (+)	0.93	-1.12	9.0
	Muon 4 (-)	-0.24	-0.63	4.45
4.	Muon masses, pair-wise *)			
	Muons 1 and 2	3.0 GeV		
	Muons 1 and 3	2.1 GeV		
	Muons 1 and 4	1.3 GeV		
	Muons 2 and 3	0.95 GeV		
	Muons 2 and 4	0.87 GeV		
	Muons 3 and 4	0.40 GeV		
5.	Three muon masses *)			
	Muons 1, 2 and 3	3.8 GeV		
	Muons 1, 2 and 4	3.4 GeV		
	Muons 2, 3 and 4	1.3 GeV		
	Muons 1, 3 and 4	2.5 GeV		
6.	Four muon invariant mass	4.1 GeV		

---

*) The momenta and the masses have the following errors due to multiple scattering and measurement: $\Delta p/p = 0.12$; $\Delta M_{\mu\mu}/M_{\mu\mu} = 0.12$; $\Delta M_{\mu\mu\mu}/M_{\mu\mu\mu} = 0.16$; $\Delta M_{4\mu}/M_{4\mu} = 0.20$.

Transverse momenta with respect to the hadron shower direction for the three non-leading muons.

TABLE I (con't)

Leading Muon	$P_{TSH,1}$	$P_{TSH,2}$	$P_{TSH,3}$	$P_{TSH,4}$
2	0.29	--	1.71	0.80
4	0.48	2.33	1.52	--

REFERENCES

1. M. Holder et al., Nucl. Instr. Methods 148 (1978) 235
2. M. Holder et al., Phys. Lett 71B (1977) 222
3. M. Holder et al., Phys. Lett. 72B (1977) 254
4. M. Holder et al., Phys. Rev. Lett. 39 (1977) 433
5. M. Holder et al., Phys. Lett. 69B (1977) 377
6. M. Holder et al., Phys. Lett. 70B (1977) 396
7. M. Holder et al., Phys. Lett. 70B (1977) 393
8. M. Holder et al., Phys. Lett. 73B (1978) 105
9. A. Benvenuti et al., Phys. Rev. Lett. 34 (1975) 419
10. A. Benvenuti et al., Phys. Rev. Lett. 35 (1975) 119, 1203, 1249
11. B. C. Barish et al., Phys. Rev. Lett. 36 (1976) 939
12. K. Kleinknecht, in Proc. 1977 Int. Symp. on Lepton and Photon Interactions at High Energy, Hamburg 1977, p. 271
13. L. N. Chang et al., Phys. Rev. Lett. 35 (1975) 6
14. L. M. Seghal and P. M. Zerwas, Nucl. Phys. B108 (1976) 483
15. V. Barger et al., Phys. Rev. D16 (1977) 746
16. M. Barnett, Phys. Rev. Lett. 34 (1975) 41 and SLAC-PUB-1892
17. A. Benvenuti et al., Phys. Rev. Lett. 37 (1976) 189
18. B. Barish et al., Phys. Rev. Lett. 39 (1977) 981
19. B. C. Barish et al., Phys. Rev. Lett. 38 (1977) 577
20. A. Benvenuti et al., Phys. Rev. Lett. 38 (1977) 1110
21. M. Holder et al., Characteristics of trimuon events in neutrino interactions, Phys. Lett., to be published.
22. V. Barger, T. Gottschalk and R. J. N. Phillips, Electromagnetic muon pair contributions in neutrino trimuon production, preprint COO-881-9, Nov. 1977.
23. J. Smith and J. A. M. Vermaseren, Electromagnetic Backgrounds in Neutrino Produced Trimuon Events, preprint IPT-SB-77-66, and J. Smith, priv. comm.
24. B. L. Young, T. F. Walsh and T. C. Yang, Trimuons from Charm, preprint DESY 77/8/, Dec. 1977.
25. C. H. Albright et al., Phys. Rev. Lett. 38 (1977) 1187

26. V. Barger et al., Phys. Rev. Lett. 38 (1977) 1190.

27. D. Horn and G. Ross, Phys. Lett. 69B (1977) 364

28. V. Barger, Phenomenology of heavy quark and heavy lepton production by neutrinos, Ben Lee memorical Conf, Oct. 1977. Wisconsin preprint COO-881-7.

29. A. Benvenuti et al., Phys. Rev. Lett. 40 (1978) 488.

30. K. F. Anderson et al., Phys. Rev. Lett. 37 (1976) 799.

NEW (QUARK) FLAVORS

C. Quigg*

Fermi National Accelerator Laboratory

P.O. Box 500, Batavia, Illinois 60510

ABSTRACT

Some possible characteristics of the new quark suggested by the discovery of Υ(9.4) are surveyed. An inverse scattering approach to the interquark potential is summarized.

INTRODUCTION

The Moderators of today's sessions have persuaded me that there is social value in making a public display of my ignorance of the properties of the next quark flavor. Before doing so I shall briefly recall the well-known flavors of quarks and leptons and summarize the immediate experimental motivation for speculating on the nature of a fifth quark flavor: the Υ family of new heavy particles. Next I will survey a number of possible assignments for the putative fifth quark and indicate very briefly the consequences of each. I then turn to the problem of determining the force that binds quarkonium, and outline a novel application of the inverse scattering formalism. The strengths and weaknesses of this new approach are indicated.

---

*Alfred P. Sloan Foundation Fellow; also at Enrico Fermi Institute, University of Chicago, Chicago, Illinois 60637.

II. WHY MORE FLAVORS?

As Dr. Meshkov has indicated in his provocative introductory remarks,[1] our present theories do not deal with the questions of how many flavors of quarks and leptons exist, and of the mass spectrum of new flavors. At the moment, the primary impetus to consider new flavors comes from experiment. Beyond the well-established quarks: u, d, s, and c (each in three colors) and leptons: ν_e, e, ν_μ, μ, there are now very strong indications[2] for a new massive lepton $\tau(1.8\ \mathrm{GeV}/c^2)$ with a companion neutrino ν_τ. Amid the euphoria attending QCD and unified theories of the weak and electromagnetic interactions, I admit my concern at the proliferation of "fundamental" fields. A conservative count[3] includes 12 quarks, 4 leptons, 8 colored gluons, 1 photon, 3 intermediate bosons (W^+, W^-, Z^0), 1 Higgs scalar, and as a future consideration 1 graviton, for a total of 30, with the end not in sight!

The strongest suggestion[4] of a fifth quark comes from the discovery[5] by the Columbia-Fermilab-Stony Brook Collaboration of the upsilon family of massive vector mesons. These were found in measurements of the reaction

$$p + N \to (\mu^+ \mu^-) + \text{anything}$$

with 400 GeV/c incident protons. The most recent published data are shown in Fig. 1. Two prominent peaks are seen near 9.4 and 10.0 GeV/c^2. The data after subtraction of a smooth background are shown in Fig. 2. The curves represent two-(dashed line) and three-peak (solid line) fits to the data which correspond to the resonance parameters given in Table I. At least two resonances are observed, and their natural widths are consistent with zero.

Although I do not have time to discuss all the possible explanations of these structures, it is important to remark that many hypotheses are consistent with what is known experimentally. The

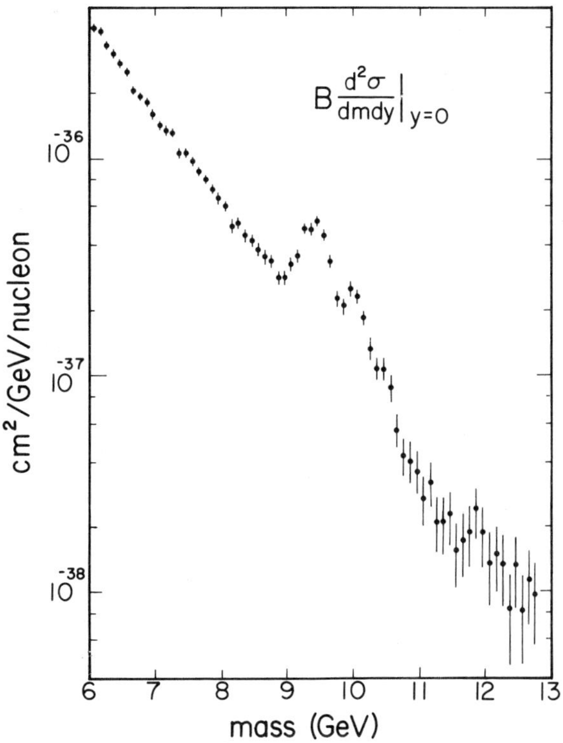

Fig. 1: Dimuon spectrum above 6 GeV, from Innes et al., Ref. 5

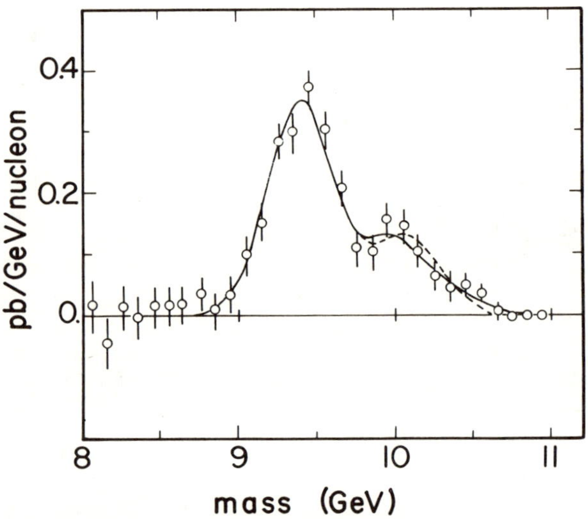

Fig. 2. Excess of the data over an exponential fit to the continuum. Errors shown are statistical only. The solid curve is the 3-peak fit; the dashed curve is the 2-peak fit.

TABLE I

Resonance Fit Parameters[a]

		2 peak	3 peak	
Υ	M_1	9.41 ± 0.013	9.40 ± 0.013	GeV/c^2
	$B\frac{d\sigma}{dy}\Big\vert_{y=0}$	0.18 ± 0.01	0.18 ± 0.01	pb
Υ'	M_2	10.06 ± 0.03	10.01 ± 0.04	GeV/c^2
	$B\frac{d\sigma}{dy}\Big\vert_{y=0}$	0.069 ± 0.006	0.065 ± 0.007	pb
Υ"	M_3	---	10.40 ± 0.12	GeV/c^2
	$B\frac{d\sigma}{dy}\Big\vert_{y=0}$	---	0.011 ± 0.007	pb
	X^2/DF	19.3/18	14.2/16	

[a]From Innes et al., Ref. 5. Errors are statistical only.

favorite, in light of the charmonium experience,[6] is to regard $\Upsilon(9.4\ \mathrm{GeV}/c^2)$ and $\Upsilon'(10.0)$ as the 1^3S_1 and 2^3S_1 states of a new $(Q\bar{Q})$ quarkonium family. It is this idea that we now pursue.

III. WHAT IS THE NEW FLAVOR?

If Υ signals the existence of a new quark, what is it? How does it participate in the weak interactions? What are the attributes of flavored $(Q\bar{q})$ states? Unlike our recent adventure with charm, wherein a single definite and attractive conjecture[7] could be elaborated[8] and confronted with experimental data, there is not a pressing theoretical need for a particular new quark. Several assignments have already been discussed in the context of specific gauge theories.[9] Here I shall catalog a few possibilities in terms of charged-current structures.

<u>If the new quark has charge $e_Q = -1/3$ (b-quark):</u>

1. Last year's favorite assignement was $\binom{u}{b}_R$, which provided a natural explanation for the celebrated high-y anomaly in $\bar{\nu}N$ scattering.[10] Although recent data from CERN[11] and Fermilab[12] speak against the spectacular effect first reported, a systematic difference between low-energy and high-energy results persists.[13] Whether that is to be understood as an experimental disagreement, an approach to scaling, "asymptotic freedom" violations of scaling, or new-flavor production is not entirely clear. Consequently I regard this assignment as unlikely but not impossible.

2. The assignment $\binom{c}{b}_R$ would imply charmed particles in decay products and a negligible effect upon neutrino physics. The lifetime of the flavored hadrons, estimated in a free quark model to be $\tau \sim 10^{-15}$ sec., might permit the detection of extremely short tracks at extremely high energies.

3. Coupling one new quark to another, as in $\binom{t}{b}_R$, leads to interesting alternatives. If $M_b > M_t$, where is the $(t\bar{t})$ family? It should have been seen more prominently than Υ in the CFS

experiment, so this ordering is ruled out. If $M_b < M_t$, the possibility of stable hadrons arises.[14] Depending upon mixing angles, the lifetime of the least massive flavored hadron could range from 10^{-15} sec. $\lesssim \tau \lesssim \infty$.

4. Weak currents mediated by new intermediate bosons would allow assignments such as $\binom{u}{b}_{L'}$, accompanied in the lepton sector by $\binom{\nu_\mu}{M^-}_{L'}$, for example. With the implied new leptons, detectable short tracks are likely. Quarks which are weak isosinglets may also be imagined.

<u>If the new quark has $e_Q = + 2/3$ (t-quark):</u>

1. The assignments $\binom{t}{b}_{R,L}$ would lead to stable particles if $M_t < M_b$. If $M_t > M_b$, we must ask whether a $(b\bar{b})$ quarkonium family could have been overlooked in the mass region between 4 and 10 GeV/c^2. Given the smaller branching ratio into lepton pairs for charge = - 1/3, I am not prepared to say this is impossible. I view it as unlikely, however.

2. A low-y anomaly in νN scattering is implied by the assignment $\binom{t}{d}_R$. At energies near the threshold, the effect would be an excess of events at medium values of y. The flavored particles would decay into ordinary hadrons, and would likely be too ephemeral to be observed as short tracks.

3. New weak currents are of course possible for this case as well.

<u>Exotica:</u>

Exotic quark charges, such as $e_Q = + 5/3, - 4/3, \ldots$ would yield spectacular signals at e^+e^- storage rings. We may also imagine stable quarks of any charge, or quarks which are distinguished by unconventional color properties.[15]

Compared to what lies ahead, finding charm experimentally was easy!

IV. APPROACHES TO THE INTERQUARK POTENTIAL

Let us now turn to the strong interaction properties of the new quarks. We hope[16] and believe[17] that the ideas of nonrelativistic quantum mechanics can fruitfully be applied to quarkonium systems. What is the interaction that binds heavy quarks together? A theoretical problem on which much work has been done is to solve QCD and compute the interquark potential.[18] This program is still underway.[19] Much remains to be learned experimentally about the Υ family. It is likely that the existence of additional narrow states can be settled at Fermilab, but the detailed spectroscopy and study of radiative transitions awaits new e^+e^- machines. At the level of phenomenology, a wide variety of studies has been undertaken. These include explicit potential models,[17] inferences from scaling rules,[20] general results from nonrelativistic quantum mechanics,[21] and general consequences of analyticity.[22]

Today I want to report on a new phenomenological approach, based on an approximate solution of the inverse scattering problem for quarkonium systems. I will merely state results, because a complete description has been given elsewhere.[23]

The inverse scattering program of Gel'fand, Levitan, Marchenko, and others[24] is basically a dispersion theory for the Schrödinger wavefunction. In one space dimension, a reflectionless potential[25] which supports N bound states (at $E_n = -\kappa_1^2, -\kappa_2^2, \ldots, -\kappa_N^2$) is given by

$$V(x) = -2\frac{d^2}{dx^2}\log(\text{Det } A) \quad , \tag{1}$$

where

$$A_{mn} = \delta_{mn} + \frac{c_m e^{-\kappa_m x} c_n e^{-\kappa_n x}}{\kappa_m + \kappa_n} \quad . \tag{2}$$

It is determined in general by the 2N parameters $\kappa_1, \ldots, \kappa_N$ and $c_1, \ldots, c_N$. The Schrödinger wavefunctions are themselves given by

$$\psi_n(x) = -\frac{1}{c_n e^{-\kappa_n x}} \frac{\mathrm{Det}A^{(n)}}{\mathrm{Det}A} , \tag{3}$$

where $A^{(n)}$ is the matrix obtained by replacing the n-th column of A by its derivative. For the special case of symmetric potential,

$$V(x) = V(-x) , \tag{4}$$

the c_n's can be eliminated[26] in favor of the bound-state pole positions

$$c_n^2/2\kappa_n = \prod_{m\neq n}^{N} \left|\frac{\kappa_m+\kappa_n}{\kappa_m-\kappa_n}\right| . \tag{5}$$

Hence a symmetric one-dimensional reflectionless potential which supports N bound states can be reconstructed from the N pole positions.

Our hope is that this formalism will provide us with a reliable local approximation to a confining potential (which supports an infinite number of discrete levels). We investigate whether a suitable approximation can be obtained using a small number of bound states. To do so, we study several examples numerically.

We first consider the harmonic oscillator potential

$$V(x) = x^2 , \tag{6}$$

which has bound states at

$$E_n = 2n - 1 , \quad n = 1,2,..., \tag{7}$$

for a reduced mass $\mu = 1/2$. To define the pole positions of the bound states in our reconstructed reflectionless potential, we must choose a zero of energy:

$$\kappa_n = [2\mu(E_0 - E_n)]^{\frac{1}{2}} . \tag{8}$$

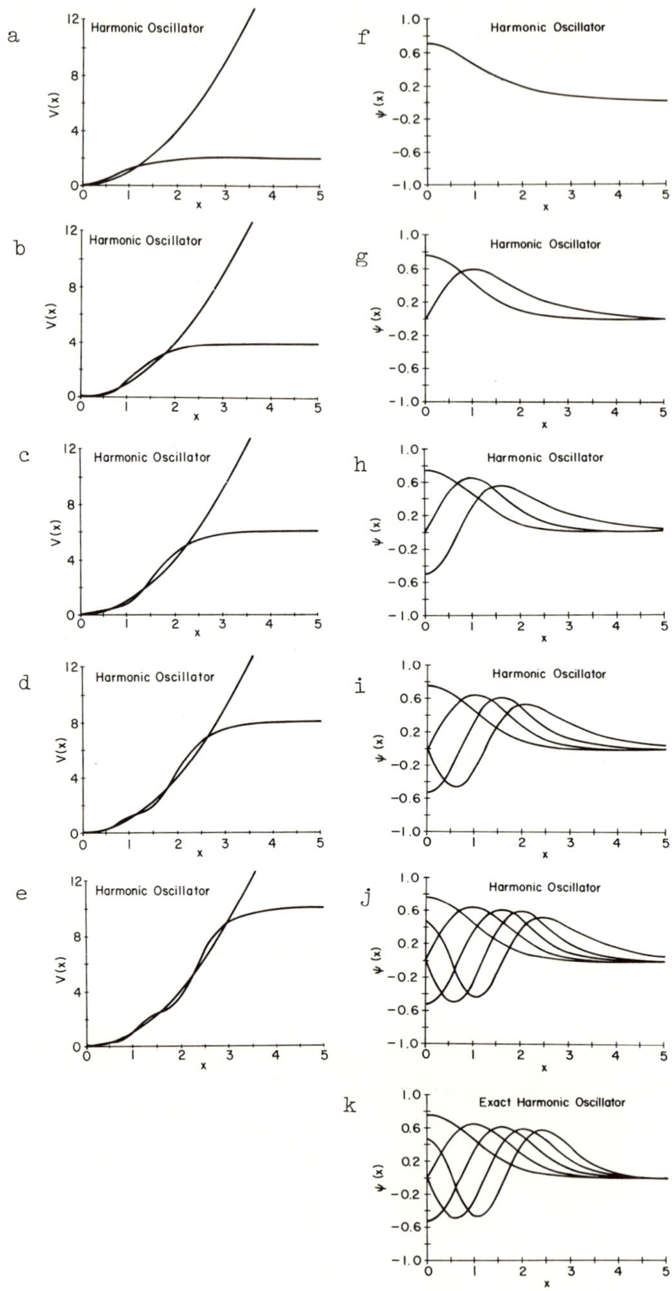

Fig. 3: Approximate reconstruction of the harmonic oscillator potential (a) - (e): N=1,2,3,4,5 approximations to the potential. The true potential is shown for comparison; (f) - (j): wave-functions obtained in the N=1,2,3,4,5 approximations; (k): Exact wave- functions.

The parameter E_0 is restricted to the interval $E_N < E_0 < E_{N+1}$ for the N-bound-state reconstruction. We find that excellent results follow from the choice

$$E_0 = \frac{1}{2}(E_N + E_{N+1}) \quad . \tag{9}$$

The N = 1, 2, 3, 4, 5 approximations to the harmonic oscillator potential (6) are shown in Fig. 3(a)-(e). The true potential is reproduced closely, up to the classical turning point of the last level included, after only a few bound states. I also show in Fig. 3(f) - (j) the resulting approximations to the Schrödinger wavefunctions. These may be compared with the exact wavefunctions in Fig. 3(k). The agreement is impressive.

Similar studies of the linear potential

$$V(x) = x \tag{10}$$

and the infinite square well

$$V(x) = \begin{cases} 0, & |x| < \pi/2 \\ \infty, & |x| > \pi/2 \end{cases} \tag{11}$$

are shown in Figs. 4 and 5. The results encourage the belief that this technique will lead to useful representations of a confining potential in the region of space probed by the low-lying levels.

The s-wave reduced radial equation in 3 space dimensions,

$$-\frac{u''(r)}{2\mu} + [V(r)-E]u(r) = 0 \quad , \tag{12}$$

has the same form as the 1-dimensional Schrödinger equation, but is supplemented by the boundary condition

$$u(0) = 0 \quad . \tag{13}$$

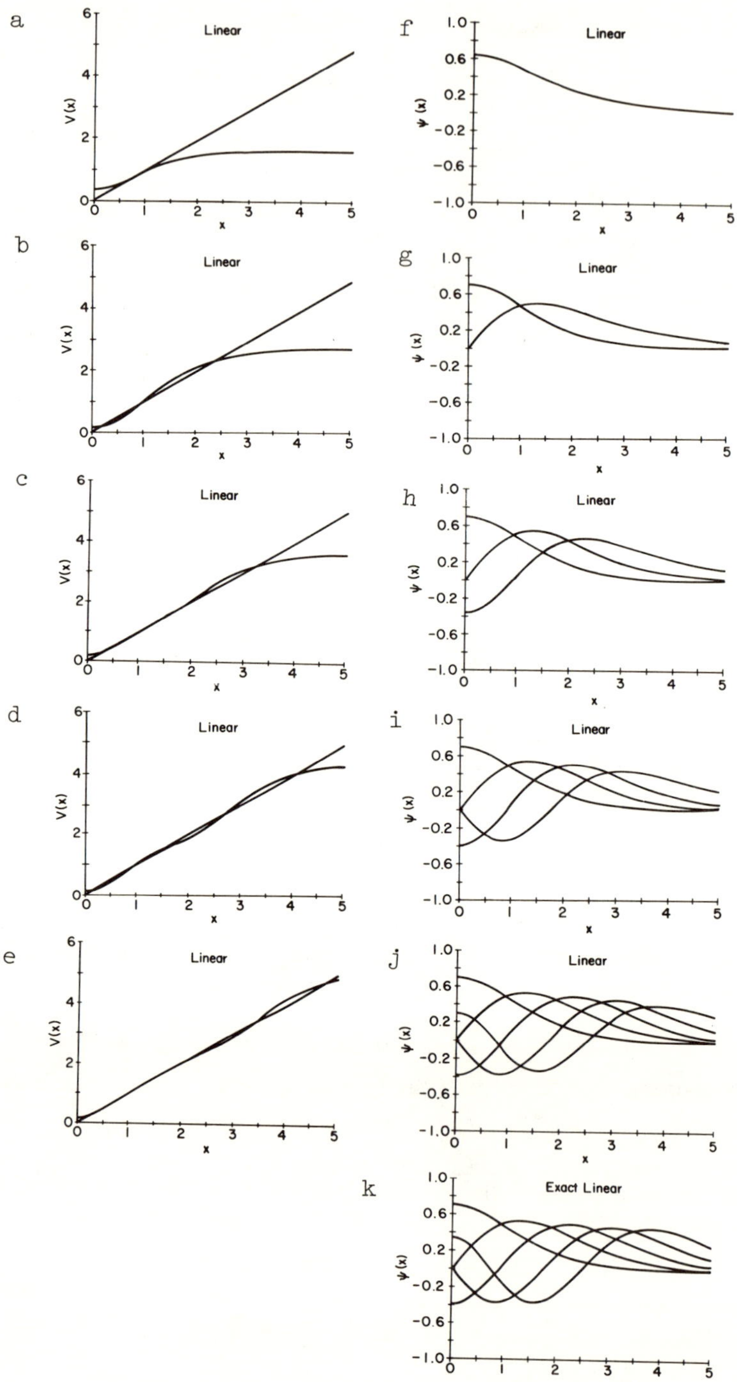

Fig. 4: Approximate reconstruction of the linear potential. See the caption to Fig. 3.

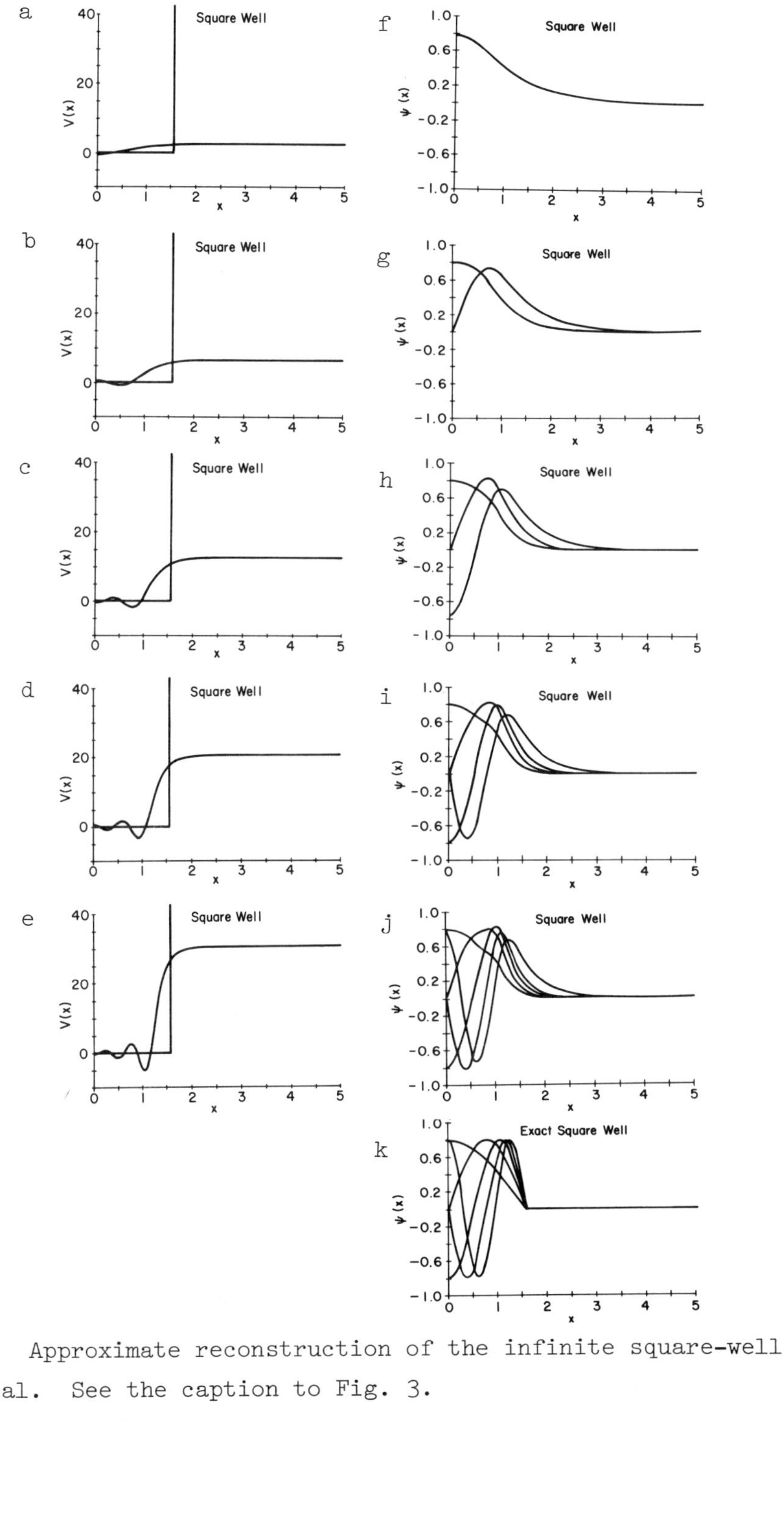

Fig. 5: Approximate reconstruction of the infinite square-well potential. See the caption to Fig. 3.

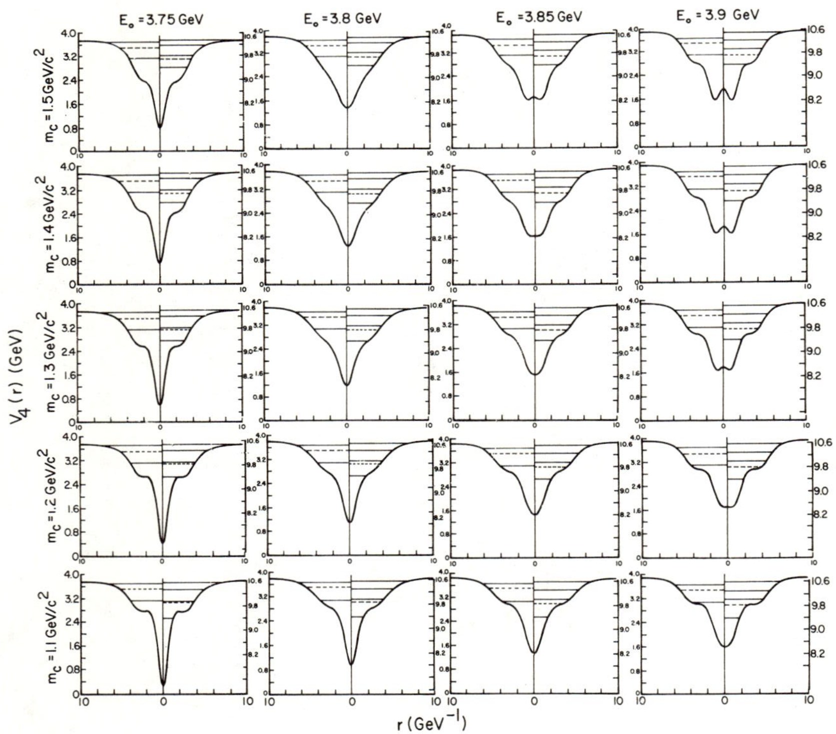

Fig. 6: Interquark potentials reconstructed from the masses and leptonic widths of $\psi(3.095)$ and $\psi'(3.684)$. The levels of charmonium are indicated on the left-hand side of each graph. Those of the upsilon family are shown on the right-hand side of each graph. The solid lines denote $^3S_1$ levels; dashed lines indicate the 2^3P_J levels. The twenty potentials depicted correspond to the choices E_0 = 3.75, 3.8, 3.85, 3.9 GeV and m_c = 1.1, 1.2, 1.3, 1.4, 1.5 GeV/c^2.

Therefore, only odd-parity one-dimensional solutions are admissible s-wave solutions in three dimensions. In the charmonium family, $\psi(3.095)$ and $\psi'(3.684)$ correspond to the n = 2 and n = 4 levels of the analogous 1-dimensional problem. We may use their positions to specify the parameters κ_2 and κ_4. The remaining parameters κ_1 and κ_3 are fixed by means of eq. (3) to reproduce the wavefunctions at the origin, which are measured[27] by the leptonic widths of ψ and ψ':

$$\Gamma(V \to e^+e^-) = 16\ \pi\alpha^2 e_Q^{\ 2}|\Psi(0)|^2/M_V^{\ 2} \quad . \tag{14}$$

Again we have found, through the study of simple examples, that reasonable approximations are reconstructed from the positions and leptonic widths of two states.

Our procedure, then, is to construct reflectionless potentials for which[28]

$$\begin{aligned} M(\psi) &= 3.095\ \text{GeV}/c^2 \quad , \\ M(\psi') &= 3.684\ \text{GeV}/c^2 \quad , \\ \Gamma(\psi \to e^+e^-) &= \underline{4.8} \pm 0.6\ \text{keV} \quad , \\ \Gamma(\psi' \to e^+e^-) &= \underline{2.1} \pm 0.3\ \text{keV} \quad . \end{aligned}$$

We encounter two sources of ambiguity: the zero of energy (E_0) discussed above, and the mass m_c of the charmed quark. We choose m_c = 1.1, 1.2, 1.3, 1.4, 1.5 GeV/c^2 and E_0 = 3.75, 3.8, 3.85, 3.9 GeV. The twenty resulting reflectionless potentials are shown in Fig. 6. All of these reproduce--by construction--the positions and leptonic widths of ψ and ψ'. The diversity of the potential shapes is noteworthy. To choose among these potentials, we may examine their implications for other observables. I shall cite but two examples.

First, we may solve the p-wave Schrödinger equation in the

reconstructed potentials for the position of the $2^3P(\chi_c)$ levels. These are indicated as the dashed lines on the left sides of the potentials in Fig. 6. Contours of the resulting χ_c masses in terms of the parameters E_0 and m_c are shown in Fig. 7. Experiment[29] favors the lower right-hand half of the plot.

Next, we may use these potentials to predict the properties of the Υ family. The only quantity now measured is the Υ-Υ' splitting, for which the predictions are shown as contours in Fig. 8. Again, experiment[5] favors the lower right-hand half of the display. Additional distinctions between the potentials are discussed at length in Ref. 23.

Let us conclude by assaying the strengths and weaknesses of this new approach to the interquark potential. It requires no assumptions on the behavior of the potential at $r = 0$ or $r = \infty$, but supplies no clues to the nature of the confining force. It provides a systematic display of the possible potential shapes, in the range of r relevant to the observed properties of the charmonium family. At the same time, because this intermediate range of r is least accessible to theoretical conjectures, the significance of inferences which can be drawn from the reconstructed potentials is not a priori clear. Only experimental properties of the $^3S_1$ states are used to determine the potentials. However, we are not guaranteed a pleasingly shaped monotonic potential. One important benefit of the exercise is that it illustrates the ambiguities inherent in extrapolating from the ψ family to the Υ family. Finally, I regard the inverse method as extremely promising when it can be applied to the Υ family, in which three or four narrow $^3S_1$ levels are to be expected.[30]

ACKNOWLEDGMENT

I thank my collaborators J.L. Rosner and H.B. Thacker for many discussions of the topics reported here.

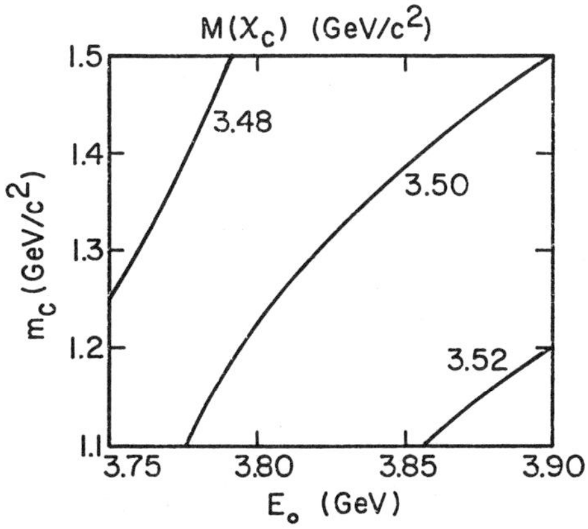

Fig. 7: Contours of the predicted mass of the $2^3P_J(\chi_c)$ level of the charmonium system as functions of the parameters E_0 and m_c.

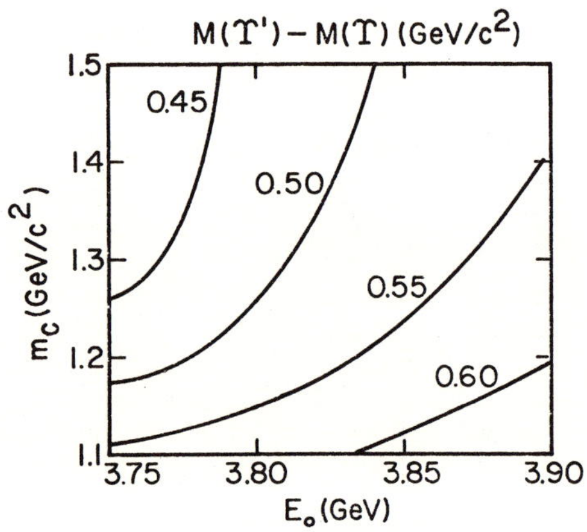

Fig. 8: Contours of the predicted Υ-Υ' level splitting as functions of the parameters E_0 and m_c.

REFERENCES

1. S. Meshkov, these proceedings.
2. An up-to-date summary is given by Martin Perl, in Unification of Elementary Forces and Gauge Theories, edited by D. Cline and F. Mills (Gordon and Breach, New York, 1978), p. 315.
3. Inclusion of the last two distinguished conservative from reactionary. See also M.A.B. Beg, these proceedings.
4. For earlier suggestions, see the summary by R.M. Barnett, in Deeper Pathways in High Energy Physics, edited by A. Perlmutter and L.F. Scott (Plenum, New York, 1977).
5. S.W. Herb et al., Phys. Rev. Lett. 39, 252 (1977); W.R. Innes et al., ibid., 39, 1240 (1977).
6. For recent reviews, see J.D. Jackson, invited paper at the 1977 European Conference on Particle Physics, Budapest, CERN Report No. TH.2351; K. Gottfried, invited paper at the 1977 International Symposium on Lepton and Photon Interactions at High Energies, Hamburg, Cornell Report No. CLNS-376; V.A. Novikov et al., ITEP-42,58,65,79-1977, to appear in Physics Reports.
7. B.J. Bjørken and S.L. Glashow, Phys. Lett. 11, 255 (1964); S.L. Glashow, J. Iliopoulos, and L. Maiani, Phys. Rev. D2, 1285 (1970).
8. M.K. Gaillard, B.W. Lee, and J.L. Rosner, Rev. Mod. Phys. 47, 277 (1975); for more recent references, see C. Quigg and J.L. Rosner, Phys. Rev. D17, (1978).
9. J. Ellis, M.K. Gaillard, D.V. Nanopoulos, and S. Rudaz, Nucl. Phys. B131, 285 (1977); R.M. Barnett, invited paper at the 1977 European Conference on Particle Physics, Budapest, SLAC-Pub-1961 and addendum; C.H. Albright, R.E. Shrock, and J. Smith, Fermilab-Pub-77/81-THY.
10. A. Benvenuti et al., Phys. Rev. Lett. 36, 1478 (1976), 37, 189 (1976).

11. M. Holder et al., Phys. Rev. Lett. 39, 433 (1977); K. Kleinknecht, these proceedings; P.C. Bosetti et al., Phys. Lett. 70B, 273 (1977).

12. B.C. Barish et al., Phys. Rev. Lett. 39, 1595 (1977).

13. For a review of Fermilab 15' Bubble Chamber results, see F. Nezrick, Fermilab-Conf-77/112-EXP. The Gargamelle results are given by D.H. Perkins, in Proceedings of the 1975 International Symposium on Lepton and Photon Interactions at High Energies, edited by W.T. Kirk (SLAC, Stanford, 1976), p. 571.

14. The stable hadron alternative has been discussed by R.N. Cahn, these proceedings, and Phys. Rev. Lett. 40, 80 (1978).

15. Unconventional color assignments have been discussed by E. Ma, Phys. Lett. 58B, 442 (1975); G. Karl, Phys. Rev. D14, 9 (1976); F. Wilczek and A. Zee, Phys. Rev. D16, 860 (1977); R. Giles and S.-H. H. Tye, Fermilab-Pub-77/96-THY; S.L. Glashow, remarks at the 1977 Irvine Conference.

16. T. Appelquist, and H.D. Politzer, Phys. Rev. Lett. 34, 43 (1975).

17. E. Eichten and K. Gottfried, Phys. Lett. 66B, 286 (1977); M. Machacek and Y. Tomozawa, "Leptonic decay of the new resonances and the nature of the confinement potential", Univ. of Michigan report UM HE 76-16, May, 1976 (unpublished); E. Eichten et al., Phys. Rev. Lett. 34, 369 (1975); Phys. Rev. Lett. 36, 500 (1976); J.F. Gunion and R.S. Willey, Phys. Rev. D12, 74 (1975); R. Barbieri, R. Kögerler, R. Gatto and Z. Kunszt, Nucl. Phys. B105, 125 (1976); Howard J. Schnitzer, Phys. Rev. Lett. 35, 1540 (1975); Phys. Rev. D13, 74 (1976); Phys. Lett. 65B, 239 (1976); A.B. Henriques, B.H. Kellet, and R.G. Moorhouse, Phys. Lett. 64B, 85 (1976); J.G. Wills, D.B. Lichtenberg and J.T. Kiehl, Phys. Rev. D15, 3358 (1977); D.B. Lichtenberg, J.G. Wills and J.T. Kiehl, Phys. Rev. Lett. 39, 1592 (1977); C. Quigg and J.L. Rosner, Phys. Lett. 71B, 153 (1977); T.-M. Yan, Cornell preprint CLNS-368 (1977, unpublished); William Celmaster

Howard Georgi, and Marie Machacek, Harvard University reports HUTP-77/A051 and HUTP-77/A060 (1977, unpublished).

18. F. Feinberg, Phys. Rev. Lett. 39, 316 (1977); T. Appelquist, M. Dine, and I.J. Muzinich, Phys. Lett. 69B, 231 (1977); W. Fischler, Nucl. Phys. B129, 157 (1977).
19. R. Dashen, these proceedings.
20. J.D. Jackson, in Proceedings of the 1976 Summer Institute on Particle Physics, edited by Martha C. Zipf (SLAC, Stanford, California, 1976), p. 147; C. Quigg and Jonathan L. Rosner, "Scaling the Schrödinger Equation," FERMILAB-Pub-77/90-THY, to appear in Comments on Nuclear and Particle Physics; J.S. Kang and Howard J. Schnitzer, Phys. Rev. D12, 841 (1975); J. Kandaswamy, J. Schechter, and M. Singer, Phys. Rev. Letters 38, 128; 1101(E) (1977).
21. C. Quigg and Jonathan L. Rosner, Phys. Lett. 72B, 462 (1978); M. Krammer and P. Leal Ferreira, Revista Brasileira de Fisica, 6, 7 (1976); C. Quigg and Jonathan L. Rosner, "Semiclassical Sum Rules," Fermilab-Pub-77/106-THY, November, 1977, submitted to Phys. Rev. D; K. Ishikawa and J.J. Sakurai, UCLA preprint, 1978.
22. A. Martin, Phys. Lett. 67B, 330 (1977); Phys. Lett. 70B, 192 (1977); H. Grosse, Phys. Lett. 68B, 343 (1977).
23. H.B. Thacker, C. Quigg, and J.L. Rosner, Fermilab-Pub-77/108, 109-THY, December, 1977, submitted to Phys. Rev. D.
24. F.J. Dyson, in Studies in Mathematical Physics, edited by E.H. Lieb, B. Simon, and A.S. Wightman (Princeton Univ. Press, Princeton, 1976), p. 151.
25. By reflectionless we mean a potential for which all phase shifts vanish in the continuum. The spectral function of such a potential consists entirely of bound-state poles. In reconstructing approximately a confining potential, it is natural to consider the reflectionless class.
26. This is most easily constructed using the remarkable connection

between soliton solutions to the Korteweg-deVries equation and Schrödinger potentials, as reviewed in the first paper of Ref. 23.

27. R. Van Royen and V.F. Weisskopf, Nuovo Cimento 50, 617 (1967); 51, 583 (1967).
28. P.A. Rapidis et al., Phys. Rev. Lett. 39, 526 (1977).
29. G.J. Feldman and M.L. Perl, "Recent Results in Electron-Positron Annihilation above 2 GeV," SLAC-Pub-1972, submitted to Physics Reports.
30. C. Quigg and J.L. Rosner, Phys. Lett. 72B, 462 (1978).

NEW ABSOLUTELY STABLE HADRONS*

Robert N. Cahn

University of Michigan

Ann Arbor, Michigan

The fundamental question which the discovery[1] of the Υ(9.4) raises is the relationship of the new quark, supposing that this is the proper interpretation of the resonance, to the four previously known quarks. The answer to this may come only when the new e^+e^- machines are completed and conduct detailed studies of the decay patterns of the new pseudoscalars (let us call them $\mathfrak{D}$'s) which are expected by analogy to the charmed system. Since such experiments are still some time off and since the identification of actual decay modes may well be further off, there is plenty of time for theorists to speculate without fear of being contradicted by real evidence.

For nearly all theorists, such speculation means considering some gauge group which, after spontaneous symmetry breaking, yields a unified theory of weak and electromagnetic interactions. After a gauge group is chosen, there is still much freedom in assigning the fermions of the theory to multiplets. These choices are constrained by our knowledge of weak interactions: the standard V-A phenomenology must be reproduced for the charged currents and what is known of the neutral currents must be accounted for as well.

*Work supported by the United States Department of Energy.

A primary constraint then on such models is observed Cabibbo universality: the strength of the weak interaction between hadrons is the same as it is among leptons, except for a rotation by θ_c, the Cabibbo angle, which mixes the s and d quarks slightly. J. Ellis, Gaillard, and Nanopoulos[2] in a recent analysis found that $\cos^2\theta_c + \sin^2\theta_c = 1.001 \pm 0.004$ where $\cos\theta_c$ is determined by comparing μ decay and nuclear β decay, and $\sin\theta_c$ is determined from semi-leptonic hyperon decays. The good agreement between the sum of the terms and the expected value of unity means that additional quarks cannot be mixed with the old ones with very large mixing angles. J. Ellis et al. have given estimates of these limits for SU(2) x U(1) models.[3] The conclusion, though, is more general than any specific gauge model.

The choices which are open are to have rigorously no mixing between the new quarks and the old, or to have some quite small mixing. In the absence of any data, the choice becomes one of aesthetics, and one can argue that small mixing angles are simply the reflection of large mass differences between the new quarks and the old ones (though the fifth quark is not enormously heavier than the fourth). Still, the possibility of zero mixing is not unattractive.

Zero mixing of the new quarks with the old ones can arise in gauge models when there are discrete symmetries present. An interesting example is the SU(3) x U(1) model of Lee and Weinberg[4] and Lee and Shrock[5]. In order to reproduce the standard charged current phenomenology, they find they must have a discrete symmetry, RU, under which the familiar fermions are even and the yet to be discovered fermions are odd.

Such discrete symmetries can be introduced as well in models based on other gauge groups and, indeed, may be desirable for producing an appropriate amount of CP violation. This has been stressed by Weinberg[6] for the case of SU(2) x U(1) and by Mohapatra and Sidhu[7] for $SU(2)_L \times SU(2)_R \times U(1)$.

If there are new quarks which are not connected to the old ones by the weak currents, the pseudoscalar mesons, $\mathcal{D}$, would be absolutely stable, except that the heavier member of the isodoublet would beta decay into the lighter[8]. This decay would be controlled by CVC and would be analogous to pion beta decay. On the basis of old quark phenomenology, we expect the pseudoscalars containing u or $\bar{u}$ quarks to be the stable ones. Thus, for example, if the new, unmixed quark has charge - 1/3 the truly stable pseudoscalar would be charged. A five MeV mass splitting of the isodoublet would yield a three second lifetime for the neutral pseudoscalar.

The ideal way to study these new particles would be in e^+e^- annihilation. Very near threshold for the pseudoscalar production, however, the new particles would be very elusive since they would have little kinetic energy and might be stopped in the beam pipe by dE/dx losses. Far enough above threshold their presence would be unmistakable: they would produce tracks showing high momentum but low velocity!

It might be possible to prove or disprove the existence of such stable pseudoscalars in the 5 GeV mass range by experiments at Fermilab. The difficulty with such experiments is that the expected production cross section is quite small, perhaps in the range of 10^{-35} - 10^{-36} cm^2. Nevertheless, this could be overcome by using a momentum-defined beam and Cerenkov counters designed to be sensitive only to particles with β less than a certain critical value (but greater than the value required to produce Cerenkov light) together with time-of-flight information.[9]

Should new stable pseudoscalars actually be found, many remarkable studies could be done. Perhaps the most obvious would be to measure the x-rays emitted by a captured $\mathcal{D}^-$. These would be analogous to muonic or pionic x-rays: after initial capture by an atom, the pseudoscalar would cascade down the E1 and Auger transitions. When the principal quantum number reached $(m_{\mathcal{D}}/m_e)^{\frac{1}{2}} \approx 100$, the pseudoscalar's orbit would lie inside the 1s electron. It would continue its cascade in an essentially classical manner, spiralling down towards the nuclear surface. Except for the lightest nuclei, the pseudoscalar would reach the nuclear surface before reaching the 1s state. There it would be expected to undergo an exothermic reaction if the ordinary quark present were in fact an anti-quark ($\bar{u}$), or to continue falling if it were a quark (d).

An exception to this would be capture by hydrogen. The 1s state would have the proton in orbit about the $\mathcal{D}^-$ with a Bohr radius of about 30 fermi. This bizarre object would not be infinitely stable since if the $\mathcal{D}^-$ were $d\bar{q}$ it would beta decay, while if it were $\bar{u}q$ there would be an annihilation by capture. Still, it might be quite long lived and conceivably could serve as a catalyst for fusion. Such a scheme was discussed twenty years ago when Alvarez and collaborators[10] observed catalysis of fusion by muons. The principle involved is simple.[11] The muonic hydrogen is a compact neutral object which can penetrate a deuterium atom and capture a deuteron to form muonic H D^+. This ion is smaller than its electronic counterpart by the ratio m_e/m_μ. Thus the overlap between the proton and the deuteron is large enough to generate a significant rate of fusion.

Muon-catalyzed fusion was limited by several effects. First, the muons have a limited lifetime of 2×10^{-6} sec. Second, catalysis could proceed only if, after each fusion, the muon which had bound the proton and deuteron were liberated to participate in another reaction. In fact, some fraction of the time, the muon was captured by the newly formed helium nucleus. Finally, the cross section for the capture of the deuteron by the muonic hydrogen was not great, leading to a low rate of catalysis.

Using stable pseudoscalars instead of muons would alleviate the first problem substantially, at least for the case in which $\mathcal{D}^- = d\bar{q}$. However, the cross section for capture of the deuteron would be reduced since the "demonic" hydrogen would be even more compact than muonic hydrogen. The dynamics of the three body p -d - $\mathcal{D}$ system are yet to be explored, so it is not possible to estimate the probability for the $\mathcal{D}^-$ being captured by the fused helium nucleus. This factor is likely to determine the number of fusions each pseudoscalar could catalyze. In any event, the low production cross section for the new pseudoscalars in either hadronic or electromagnetic processes makes their use as fusion catalysts economically worthless by many orders of magnitude.

It might be thought that if there were stable five GeV pseudoscalars they could not have evaded detection. Whether this is so depends on assumptions such as their production cross section in hadronic collisions. The observed cross section for the production of the T(9.4) times its branching ratio into muon pairs[1] is B $(d\sigma/dy) \approx 2 \times 10^{-37}$ cm^2. We estimate a production cross section for the pseudoscalars in the range $10^{-35} - 10^{-36}$ cm^2. This is slightly below the limits for stable particles established at Fermilab by Gustafson, et al.[12]

If the charge $\mathcal{D}$ is stable, there would be analogs of hydrogen with the proton replaced by a $\mathcal{D}$. A very interesting experiment was performed by Muller et al.[13] using the 88-inch cyclotron at Berkeley as a very powerful mass spectrometer. In this way, limits on the ratio of anomalous hydrogen to ordinary hydrogen of 2×10^{-19} were set. Now "demonic" hydrogen could have resulted either from primordial creation or from production by cosmic rays incident on the earth's atmosphere over a period of some billions of years. These processes have been considered by Zel'dovich, Okun, and Pikel'ner.[14] The cosmological arguments are very crude and involved. I cannot comment on them. The cosmic ray production estimates of Zel'dovich et al. seemed to have been based on an assumed production cross section of 0.01 mb, which is certainly much larger than the cross section for $\mathcal{D}$ production (Zel'dovich, et al. were actually considering quark production). More realistic estimates have been made recently by L. W. Jones,[15] who concludes that cosmic ray production would have led to concentrations less than 6×10^{-23}, so that the experiment of Muller et al. does not rule out our stable pseudoscalars.

It should be emphasized that new stable particles are not required by unified gauge theories of weak and electromagnetic interactions. On the other hand, they are not merely a logical possibility but a convenient way of reconciling the existence of more than four quarks with the observed universality of weak decays. They may result from discrete symmetries of the sort which are frequently introduced into gauge theories either to preserve universality or to prevent CP violations from becoming too large. Should they actually be found, it would be possible to study some rather bizarre phenomena including x-ray spectra, hypernuclei and possibly, catalyzed fusion.

REFERENCES

1. S. Herb et al., Phys. Rev. Lett. 39, 252 (1977).
2. J. Ellis, M. K. Gaillard, and D. V. Nanopoulos, Nucl. Phys. B 109, 213 (1976).
3. J. Ellis, M. K. Gaillard, D. V. Nanopoulos and S. Rudaz, Nucl. Phys. B131, 285 (1977).
4. B. W. Lee and S. Weinberg, Phys. Rev. Lett. 38, 1237 (1977).
5. B. W. Lee and R. Shrock, FERMILAB - Pub - 77/48.
6. S. Weinberg, "Unified Gauge Theories", invited talk at VIIth International Conference on High Energy Physics and Nuclear Structure, Zurich, Switzerland, Aug. 30, 1977.
7. R. N. Mohapatra and D. P. Sidhu, CCNY-HEP-77/9.
8. R. N. Cahn, Phys. Lett. 40, 80 (1978). The possibility of having new stable hadrons within the orthodoxy of fractionally charged quarks has long been apparent. See for example passing remarks in F. Wilczek, Phys. Letters 59B, 179 (1975) and A. De Rujula, H. Georgi, and S. L. Glashow, Phys. Rev. D12, 3589 (1975). There are probably other similar references of which I am unaware. Should stable hadrons actually be found, I am confident that many of the early references would be brought to our attention.
9. M. Longo, private communication. H. R. Gustafson et al., Fermilab proposal.
10. L. W. Alvarez et al., Phys. Rev. 105, 1127 (1957).
11. Actually, it is not really this simple. See the review by Ya. B. Zel'dovich and S. S. Gershtein, Usp. Fiz. Nauk 71, 581 (1960), trans. Sov. Physics Uspekhi, 3, 593 (1961).
12. H. R. Gustafson et al., Phys. Rev. Lett. 37, 474 (1976).
13. R. A. Muller, L. W. Alvarez, W. R. Holley, and E. J. Stephenson, Science 196, 521 (1977).
14. Ya. Zel'dovich, L. Okun and S. Pikel'ner, Usp. Fiz. Nauk, 87, 113 (1965), Sov. Phys. Usp. 8. 702 (1966).

15. L. W. Jones, private communication and UM HE 78-5, unpublished. Of course, if the neutral $\mathcal{D}$ were the stable form, no anomalous hydrogen would be found in any event.

QUANTUM FLAVORDYNAMICS: A STATUS REPORT[+]

M.A.B. Bég

The Rockefeller University, New York, New York

In Memoriam

BENJAMIN W. LEE

ABSTRACT

The problems posed and the challenges offered by QFD, at this time, are analyzed and discussed. A convenient starting point is afforded by a comparison of the different circumstances of QFD and QCD. Some features of the conventional methodology of QFD, which may be deemed to be questionable, are underlined. The prospects of the standard model are examined in the light of recent experiments, and some theoretical deficiencies of the model are noted. Following a brief comment about the status of some recently proposed models, it is suggested that one search for new physical principles which may be governing the structure of weak interactions in a hitherto unperceived way. Manifest left-right symmetry and time-reversal invariance, realized as symmetries of the Nambu-Goldstone type, are mentioned as possible examples

[+]Work supported in part by the U.S. Energy Research and Development Administration under grant number EY-76-C-02-2232B.*000

of such principles; additional motivation for these symmetries emerges in a discussion of recent work on the problem of natural suppression of strong T-violation. The report concludes with a brief appraisal of the present state of QFD.

1. INTRODUCTION

The conceptual framework governing our thinking about subnuclear phenomena has evolved considerably during the last few years. It is now universally recognized that Gauge Field Theory furnishes the most economical and elegant description of all elementary particle interactions. Quantum Electrodynamics (QED) the most successful of gauge theories - and the simplest in that it is based on the group U(1)--is no longer a solitary discipline, rather it is an integral part of Quantum Flavordynamics (QFD) which embraces both electromagnetic and weak interactions.[1] Strong interactions, which--in the not too distant past--inspired attempts to bypass field theory altogether, are now described by the $SU(3)_c$-based gauge theory known as Quantum Chromodynamics[2] (QCD).

The momentum generated by the weak-electromagnetic synthesis has led to a number of interesting proposals for a further synthesis[3]: the embedding of QCD and QFD in a guage theory based on some, suitably chosen, Lie group of rank ≥ 4. It is to be hoped that this trend will continue and that some convincing way will be found to bring gravitation into the picture, so that none of the fundamental interactions remains an island unto itself. At this time, however, the search for the "ultimate unifying theory" is akin to the quest for the Holy Grail; the nature of the elusive prize is not well defined and the precise direction to be taken is not too clear.

In this report I shall not get involved with any of the super unification schemes. I want to concentrate rather on a couple of narrower topics: (a) the present status of QFD, as a fundamental theory, and (b) the adequacy of some specific realizations of QFD, as descriptions of weak interactions, with respect to the criteria of completeness and naturality. I assume, of course, that QFD and QCD are generated by separate, commuting, gauge groups.

To put the state of QFD in perspective, a comparison with QCD is worthwhile. In QCD there is not a shadow of a doubt that we have the physically correct gauge group, namely $SU(3)_c$; this group is one of the more significant legacies from [4,5,6] the days of SU(6). Furthermore, there are many good reasons for, and almost none against, the view that this guage group is an <u>unbroken</u>[7] symmetry of nature. Thus, given the number of quark flavors, QCD is characterized by a well defined Lagrangian. In situations where this Lagrangian lends itself to reliable calculations[8], it yields results in accord with experiment. Also, with some cooperation from QFD, QCD can reproduce automatically all the known symmetries of strong interactions. Our optimistic appraisal of QCD should not, by any means, be taken to imply that all the important problems associated with it may be regarded as solved. We know little, for example, of the mechanism whereby chiral flavor groups such as $SU(2)_L \otimes SU(2)_R$ manage to choose the Nambu-Goldstone(NG) rather than the Wigner-Weyl (WW) realization, we know less about the problem of reconciling the NG-mechanism with another legacy from SU(6), the naive quark model. That the theory leads to color-confinement is still an article of faith, not an established fact. Indeed, to unlock the physical content of the theory in the low energy domain, it is necessary to develop non-perturbative methods and face the concomitant problems[2]; the gauge-fixing ambiguities, discovered by Gribov,[9] provide an interesting example of a conceptual point that could not be foreseen in perturbation theory. Nonetheless, QCD is a theory in which one does have a very well-defined starting point, and one may proceed with the analysis of field equations without worrying that the next bubble-chamber picture might force a revision of the basic Lagrangian.

In contrast, the situation in QFD is strikingly different. We are not yet quite sure if we have found the right gauge group, many of us have nagging doubts about the use of fundamental Higgs fields, and the enforcement of naturality is largely an ad-hoc game.[1] There is considerable room for speculation and it seems prudent to keep an eye on new theoretical ideas as well as experimental signals that might, conceivably, change the course of the subject.

This report is organized as follows: In Section 2, I review some questionable features of the present methodology of QFD and propose a classification scheme to bring out the distinction between true or natural naturality and ad-hoc or artificial naturality. Section 3 is devoted to a discussion of the prospects of the standard model. Section 4 is a comment on the status of some recent proposals. In Section 5, I consider the possibility that new, as yet unperceived, physical principles may be playing a significant role in determining the structure of weak interactions, choosing as examples some principles abstracted from recently proposed models. The present state of QFD is very briefly surveyed in Section 6.

2. QUESTIONS PERTAINING TO THE METHODOLOGY OF QFD

A feature of the canonical methodology that has been deemed to be unsatisfactory by many authors is the manner in which one implements the Higgs mechanism.[1] As every practitioner of the art knows, one breaks the symmetry of the vacuum by introducing a judicious assortment of spin-0 fields in the Lagrangian and contriving their couplings, in such a way as to yield the desired mass splittings for the fermions and the gauge fields but without getting stuck with any unacceptably light near Goldstone bosons. This spontaneous symmetry breaking is about as spontaneous as the spontaneous political demonstrations in some well known countries.

More seriously, with this strategy the essential simplicity of the gauge-theoretic approach is clouded by the complexity of the Higgs sector in any realistic model. To make sure that the physics comes out both correctly and naturally, in the technical sense, one is obliged to play mathematical games with the Higgs fields; however, no physical effect that may be regarded as stemming directly from the Higgs sector has yet been localized in nature. This latter circumstance has led to the now customary practice of simply discarding the contributions of physical Higgs fields in calculations of radiative corrections.[10]

There have been many speculations about the future of the Higgs sector, centered around the possibility that the whole picture may change when an understanding of the dynamical Higgs mechanism is in hand.[11] An example of such a mechanism has been known for a long time: electromagnetic waves propagating through a dense plasma acquire a longitudinal component, and mass.[12] However, even in the abelian case, no convincing field-theoretic formulation in 3+1 dimensional space has yet been found.

An article[11] published in 1974 summarized the situation as follows: "The ideal state of affairs would be to have no Higgs fields in the Lagrangian at all, and to be able to trigger the Higgs-Kibble mechanism in a dynamical way .... . A theory which generates its own Higgses as bound states may, conceivably, also manage to generate broken hadron symmetries in a natural way. One is held back, however, by the difficulty of doing any explicit calculations, or even achieving a precise formulation of the scheme in any but toy models. Until one has developed the technology for handling dynamical theories in full generality, one can not make a go of it; in the meantime, one can pretend that the scalar fields inserted by hand in the Lagrangian are phenomenological props to be discarded in the future - and thereby feel less unhappy about the proliferation of Higgses." In January 1978 the situation remains unchanged.

The next topic to which I want to address myself is the concept of naturality.[1] In QFD, a statement may be said to be natural if it follows from general group theoretic considerations rather than fiat, a relationship is natural if it follows from some - exact or softly broken - symmetry rather than manual adjustment of parameters. Alternatively, and more precisely, one may define a natural relationship as follows: If a and b are quantities such that each of them is subject to infinite renormalization, then a=b is natural provided $a_{renormalized} = b_{renormalized}$ + calculable and small finite corrections. (The corrections, which arise at the loop level, vanish if the symmetry underlying the relationship is an exact symmetry of the system).

The conventional formulation of the concept of naturality, outlined above, served QFD well during the early stages of its development. Thus the simple criterion that universality be implemented in a natural way was helpful in appraising some first-generation models. At this time, however, it seems advisable to refine the concept somewhat, and I propose to make a start in this direction by distinguishing between various types of naturality. By Type I naturality I shall mean naturality stemming directly from the gauge group and the particle assignments, Type II will refer to naturality predicated on continuous (ungauged) or discrete symmetries which have a well-defined physical meaning, and by Type III I indicate naturality forced by the injection of ad-hoc unphysical symmetries into the Lagrangian. Quite obviously, this classification reflects a value judgment on my part. A theory which makes essential use of artificial, or Type III, naturality can hardly be deemed to be satisfactory, a Type III-natural formula can not exactly be said to be well founded. Unfortunately, the present state of the art is such that we are often obliged to choose between this kind of naturality and no naturality at all. Since the source of the difficulty is the Higgs sector of the Lagrangian, it is tempting to speculate that the dynamical Higgs mechanism might just liberate us from this problem.

3. PROSPECTS OF THE STANDARD MODEL

By "the standard model" one usually means the specific realization of QFD based on the group $U(1) \otimes SU(2)_L$ with four leptons (ν_e, e; ν_μ, μ), four quarks (u, d; c, s) and one complex Higgs doublet.[1] To accommodate recent experimental findings,[13,14] and to make peace with the old CP-problem, a leptonic doublet (ν_τ, τ), a quark doublet (t, b) and extra Higgs multiplets[15] have been added on to the model. These "extended standard models" differ from each other in the manner in which the Higgs sector has been enlarged; the only extensions of interest are those in which all the Higgs multiplets are doublets, so that the good predictions of the model for neutrino induced processes are maintained.

In the following, I shall first review the status of the model in the light of some recent experimental results and then try to play advocatus diaboli by underlining, what appear to me to be, the principal shortcomings of the model.

(a) Experimental Status

I shall confine myself, by and large, to experimental results which were either announced or widely debated and discussed during the last year.

(i) Neutral current experiments with ν_μ beams. These experiments provide the major triumphs of the model to date. All data[16,17] are well fitted with $\sin^2\xi_{W.S.} \doteq \frac{1}{4}$, corresponding to $m_W \cong 74$ GeV and $m_Z \cong 87$ GeV. Direct observation of the intermediate bosons, at these masses, would be a spectacular confirmation of the model. The first results from ISABELLE are eagerly awaited.

(ii) The process $\mu\to e\gamma$. It is remarkable indeed that one vague rumor of a positive experimental signal touched off such a flurry of speculation, model building, press reports, etc., during the preceding year.[18] As of today, there is no evidence that the process goes, in accord with the null prediction of the standard model. (The isomorphic process $s\to d+\gamma$ would also be forbidden in the limit $m_u\to m_c$).

(iii) Neutrino induced trimuon events[19,20]. Such events with "unusual" characteristics could well pose a serious threat to the standard model. Understandably, therefore, the "sociological" reaction of the theorists to the first experimental reports[19] has been comparable to that elicited by $\mu\to e\gamma$. However, in this case, the experimentalists are not yet agreed among themselves, and the situation is far from clear. In the CERN experiment[20], the events seen with the wide-band beam correspond to

$$\frac{\sigma(\nu_\mu+N\to\mu^-\mu^+\mu^- + \ldots)}{\sigma(\nu_\mu+N\to\mu^- + \ldots)} \cong 5\times 10^{-5} \tag{3.1}$$

for $E_\nu>100$ GeV. Furthermore, the parameters of the events are such that just about all of them can be readily explained in terms of the process

$$\nu_\mu+N\to\mu^- + \text{"}\gamma\text{"} + \ldots, \qquad \text{"}\gamma\text{"}\to\mu^+\mu^-$$

"γ" being a virtual photon. In the Fermilab experiment, the absolute trimuon rate is substantially larger. More interesting than the absolute rate are the so-called "super" events[21], reported by the Fermilab group, events in which the muons seem to be carrying off almost all of the energy of the incident neutrino and which appear to defy any conventional explanation. Such events may be present in the CERN sample[22] as well, at a branching ratio of about 10^{-6}. Until the dust settles down, one can not assess the magnitude of the threat - if any - to the standard model from the trimuon sector.

(iv) The high-y anomaly. The preceding year witnessed the fall[23] of the high-y anomaly, which was often cited in the past[24] as an argument against the standard model. There is still some disagreement among the experimentalists; however, for all practical purposes the story seems to be over. The high-y anomaly, if any, is not of a magnitude such as to warrant a QFD-based explanation; QCD effects[24] appear to be quite adequate. The same holds true for the energy-dependence of $\sigma_{cc}(\bar{\nu}_\mu)/\sigma_{cc}(\nu_\mu)$. In other words, there is no evidence for a non-standard coupling of the form: $\frac{G_F}{\sqrt{2}}\bar{\nu}_\mu\gamma_\lambda(1-\gamma_5)\mu^-\cdot\bar{b}\gamma^\lambda(1+\gamma_5)u$, b being the so-called bottom quark with charge $-\frac{1}{3}$.

(v) Parity violation in Atomic Physics[25,26]. In September 1977 it seemed clear to everyone that experiments on parity-violation in atomic Bismuth had dealt the death blow to the standard model. The measured value of the optical rotation in Bismuth,[26] consistent with zero, seemed to be at least ten times smaller than the predictions of the model. Very recently, however, many people have taken a second look at the atomic physics calculations[27] which are needed to express the observable quantity in terms of the parameters of the model. Enough doubt has been cast on the validity of these calculations to give the model a fresh lease on life.[28] Perhaps the only way to clear the matter would be to do an experiment on deuterium.[29] [Hydrogen seems unsuitable in the light of the new value for the Weinberg-Salam angle, derived from the neutrino data; at $\sin^2\xi_{W.S.}=0.25$ there is no (measurable) parity violation in the electron-proton system. A null result in Hydrogen would, therefore, convey no information at all.] Also, one might look for parity-violation effects in deep-inelastic electron-nucleon scattering.[30]

(b) Theoretical Shortcomings

To avoid unnecessary repetition of points already made earlier, let me proceed on the assumption that we accept the present methodology of QFD. Even within the framework of this conventional wisdom, several features of the standard model appear to be somewhat less than satisfactory. In my view, the following observations are worthy of serious consideration.

(i) The model does not provide a "true" unification of weak and electromagnetic interactions.

The point, raised by Faddeev[31] and more recently by Bjorken[32], is that one has merely traded two old coupling constants for two new ones; to have true unification one should start with a simple - or a well-chosen semi-simple - gauge group. A simple group will also be helpful in understanding charge quantization.

(ii) The model does not yield any insight into the nature of CP-violation.

Actually, the standard model can not even accommodate CP-violation[15]; the extended standard models can, but it has to be put in by hand.

(iii) The model does not furnish us with any natural basis for new flavors.

It is evident that the model does not require new flavors such as (ν_τ, τ) and (t, b) in the same way that it required the c-quark to get off the ground. One can, of course, throw in any new flavors that come along; to have a natural or group-theoretic raison d'etre for the flavors, however, one would have to enlarge the gauge group.

(iv) Isotopic spin is not a natural symmetry in the model.

Even in the limit in which one ignores all effects associated with gauge field couplings, the iso-spin currents are only partially conserved; this means, among other things, that all $\Delta I=1$ mass differences are incalculable.[10]

(c) Outlook

It is obvious from our discussion that the standard model is not in any serious trouble with experiment at this time. However, there is reasonable cause for concern about the neutrino induced trimuon events, of the "super" variety, and there is the distinct possibility that the question of parity violation in atomic physics may yet prove to be the Achilles' heel of the model. For these reasons alone, it seems prudent to keep one's gauge-model options open. Independent of the experimental considerations are the theoretical deficiencies of the model; these provide excellent reasons indeed for forging ahead with the quest for a better model.

4. ON SOME RECENT PROPOSALS

The preceding year has been one of great productivity in the model building industry[33]. Unfortunately, most of the proposed models attempted to follow the "shifting winds of experimental trends", and interest in them has, perhaps justifiably, waned. Some, however, contained the germs of interesting new ideas, ideas which might well survive the models-in which they have hitherto been embedded-to emerge, in their own right, as viable physical principles. To the extent that the 1977 crop of models lends itself to abstraction of new principles, which may be helpful in pinpointing the structure of QFD, the labor that went into it has not been in vain.

5. THE SEARCH FOR NEW PHYSICAL PRINCIPLES

While it is evident that there is a legitimate need for alternatives to-and, hopefully, improvements on-the currently available realizations of QFD, the precise direction to be taken is not as clear at this time. Perhaps it is just a matter of trial and error, and we have not made enough trials and enough errors. Perhaps we are missing some important ingredient that may be revealed by an experimental signal which is not as evanescent as the ones that caused such confusion in the recent past. Or perhaps we must wait until the formidable mathematical problem posed by the non-abelian dynamical Higgs mechanism in a 3 + 1 dimensional world has been brought under control, and the basic methodology has undergone a radical change. There are pathways to suit the taste of every investigator trying to solve the QFD puzzle.

A relatively non-controversial strategy at this time would be to look for new general physical principles that may be governing the weak interactions in a hitherto unperceived way, and to see how these principles fit into the gauge theoretic framework. Needless to say, such principles can not in general be inferred directly from contemporary experiments; they must be freely postulated.

Explicit examples of novel physical principles, that may well be playing a decisive role in determining the structure of weak interactions, are afforded by a couple of suggestions that were recently made in the literature[34,35]. A brief discussion of these ideas, and possible extensions thereof, appears to be in order.

(a) The Principle of Manifest Left-Right Symmetry[34]

It has been suggested that, in common with all other known interactions, weak interactions are characterized by manifest invariance under left-right conjugation, and that the observed pattern of parity nonconservation is a consequence of the Nambu-Goldstone nature of the symmetry-realization. The merger of Wigner-Weyl and Nambu-Goldstone realizations, at small distances,[36] makes this hypothesis amenable to direct tests in high energy experiments[37].

The full content of this postulate is expressed through the equations:

$$P\, H(\vec{x}, t) P^{-1} = H(-\vec{x}, t) \quad , \tag{5.1}$$

$$P\, |0\rangle \neq |0\rangle \quad , \tag{5.2}$$

$$J_R^{\text{Physical}} = J_L^{\text{Physical}}(\gamma_5 \to -\gamma_5) + \underline{\text{Finite}} \text{ loop-level corrections of } O(\alpha) \; . \quad (5.3)$$

Here H is the Hamiltonian density and P is the parity operator. Eq. (5.3) serves to distinguish this formulation of left-right symmetry from other, non-manifest, formulations in which

$$J_R^{\text{Physical}} = J_L^{\text{Physical}}(\gamma_5 \to -\gamma_5) + \text{Tree-level corrections of } O(1) \; . \quad (5.4)$$

The principal advantage of the non-manifest realization of left-right symmetry is that one need not enlarge the gauge group beyond $U(1) \otimes SU(2)_{L+R}$; fermion mass splittings suffice to uphold the V-A structure of weak interactions at low frequencies[38]. Unfortunately, however, all $U(1) \otimes SU(2)_{L+R}$ based "vector-like" models - in the form in which they were originally proposed - have the additional feature that the leptonic and hadronic neutral currents are both pure vector, contrary to the findings in neutrino experiments[16,17].

To implement manifest left-right symmetry the gauge group must be enlarged to[39] $U(1) \otimes SU(2)_L \otimes SU(2)_R$, or to some G such that[35] $G \supset U(1) \otimes SU(2)_L \otimes SU(2)_R$. [With the enlarged gauge group, the non-manifest option is not closed; it is simply less elegant.] Many of the consequences of this symmetry are, however, independent of the choice of G or the details of the Higgsology. For example:

(i) Parity violation in Atomic Physics is controllable. (With judicious choice of the Higgs sector, it can be made zero at the tree level in a technically natural way[40].)

(ii) There is no high-y anomaly, or any energy dependence in $\sigma_{cc}(\bar{\nu}N)/\sigma_{cc}(\nu N)$, from the QFD sector[34].

(iii) All parity-violating amplitudes become negligibly small at energies which probe distances[34] of $O[(10^3 \text{GeV})^{-1}]$.

Indeed it is possible that the trend towards parity conservation in weak interactions may become visible at ISABELLE energies.

(b) Time Reversal as a Nambu-Goldstone Symmetry

If one accepts the view that space-inversion and charge-conjugation are symmetries of the Lagrangian, realized in the Nambu-Goldstone way, it seems natural to put time-reversal on the same footing. Actually, the notion that time-reversal invariance is broken spontaneously is not new[41]; what makes a reconsideration of it timely is the current emphasis on spontaneous P-violation, together with the need to secure a natural suppression of strong CP-violation[42].

Strong CP-violation emanates from the QCD sector via the interaction:

$$L = \frac{\theta}{64\pi^2} \cdot \mathrm{Tr} \cdot \varepsilon^{\mu\nu\rho\sigma} G_{\mu\nu} G_{\rho\sigma} , \tag{5.5}$$

G being the curl of the gluon field matrix. Despite the fact that this interaction is a total divergence, its physical effects need not vanish[42]. Also, it conserves strangeness and violates both P and T; consequently, it can not be the source of the CP-violation effects observed in nature[43]. The physical T-violating effects presumably reside in the QFD sector, and it is their presence that leads to potential difficulties with the above interaction. (If either T or P were an exact Wigner-Weyl symmetry, θ would vanish to all orders.)

Several options are available for resolving the problem; in making the following list, I have assumed-for the sake of definiteness - that the QFD sector is described by one of the "extended standard models" mentioned earlier.

(i) Adjust θ by hand, setting it equal to zero or some other desired value. This is the unnatural solution.

(ii) Set m_u, the mass of the up-quark, equal to zero[44]. This permits us to implement chiral transformations which, apart from redefining the quark fields, do nothing more than change the value of θ in the Lagrangian; in particular, we can rotate θ to zero. Whether this solution leads to serious trouble with mass formulae is not fully clear[45].

(iii) Enforce a global U(1) symmetry-of the Nambu-Goldstone genus on the Lagrangian to transform θ away[46]. The concomitant pseudoscalar Goldstone boson can, however, acquire a mass from instanton effects, which lead to explicit but very soft symmetry breaking. (The boson is a pseudo-Goldstone boson, in the usual jargon.) While the mass is, in principle, calculable, the best one can do at this time is to make order of magnitude estimates; these indicate that the boson may be quite light, in the 100 KeV to 1 MeV range. This "axion" alternative appears to be in very serious difficulty with existing experiments[47], if the axion mass is taken to be less than 1.02 MeV.

(iv) Introduce T-violation into the theory in such a way that $\theta_{bare} = 0$ and $\theta_{renormalized}$ is a calculable finite number of the right order of magnitude $\sim 10^{-10}$. While spontaneous violation of either P or T is adequate to guarantee $\theta_{bare}=0$ and $\theta_{ren.}<\infty$, getting the right order of magnitude for $\theta_{ren.}$ appears to be non-trivial.

(c) Nature of Lepton-Hadron Symmetry

Lepton-hadron symmetry has hitherto been taken to imply that to each lepton assigned to a given representation of the gauge group there correspond three quarks, one for each color degree of freedom, belonging to representations of the same dimensionality[48]. A recent paper[35] introduces in a natural way an alternate formulation of lepton-hadron symmetry in which there is a one to one correspondence between leptons and quarks, and the lepton spectrum is characterized by a Han-Nambu like structure. Since the number of possible quark flavors is limited[8], $n_Q \leq 16$, the question of whether the number of leptonic flavors is n_Q or $3n_Q$ is experimentally decidable. Unfortunately, our lack of understanding of the mass-splitting pattern in flavor space prevents us from helping the experimentalist, with specific predictions, in the search for new flavors.

6. CONCLUDING REMARKS

Having underlined several unsatisfactory features of the present state of QFD, it is important that we do not lose sight of the positive aspects. When all is said and done, the basic gauge-theoretic approach embodied in QFD remains unsurpassed in elegance and in the promise it holds for unravelling the structure of weak interactions. As stressed earlier, we have not yet reached the point where it can be said that the QFD Lagrangian has been pinpointed; however, let us not forget that the problems which stand in our way may also be regarded as opportunities for probing deeper into the nature of the fundamental interactions. The specific topics, on which we have dwelt - to wit, the questionable features of present day methodology, the shortcomings of the standard model, the quest for new physical principles that may be playing an important but as yet unperceived role in weak interactions, the need for a theory of "soft" T-violation etc. - provide concrete and worthwhile challenges to workers in QFD. The path to the Lagrangian may be clearer after some of these challenges have been met.

ACKNOWLEDGMENTS

It is a pleasure to thank G. Feinberg, M. Gell-Mann, R. Jackiw, W. Marciano, S. Meshkov, H. Pagels, H. Quinn, A. Sirlin and H.-S. Tsao for comments and observations relevant to the material of this report.

REFERENCES AND FOOTNOTES

1. For reviews of QFD, written before the discipline was so christened by M. Gell-Mann, see: E.S. Abers and B.W. Lee, Phys. Rep. C9, 1(1973); M.A.B. Bég and A. Sirlin, Ann. Rev. Nucl. Sci. 24, 379 (1974).
2. For a review of QCD, see: W.J. Marciano and H.R. Pagels, Phys. Rep. C (to be published).
3. For example, F. Gürsey and P. Sikivie, Phys. Rev. Lett. 36, 775 (1976).
4. F. Gürsey and L.A. Radicati, Phys. Rev. Lett. 13, 173 (1964).
5. M.A.B. Bég, B.W. Lee and A. Pais, Phys. Rev. Lett. 13, 643 (1964); M.A.B. Bég and V. Singh, Phys. Rev. Lett. 13, 418 (1964).
6. O. W. Greenberg, Phys. Rev. Lett. 13, 598 (1964).
7. Symmetry breaking with the conventional Higgs mechanism jeopardizes the status of $SU(3)_c$, as a classification symmetry, as well as the asymptotic freedom property of the theory [Ref. 8, below].
8. H.D. Politzer, Phys. Rev. Lett. 30, 1346 (1973); D.J. Gross and F. Wilczek, ibid. 1343 (1973)
9. V. N. Gribov, Lecture at the 12th Winter School of the Leningrad Nuclear Physics Institute (1977).
10. A. Sirlin, Rev. Mod. Phys. (to be published).
11. M.A.B. Bég and A. Sirlin, Ref. 1.
12. P.W. Anderson, Phys. Rev. 112, 1900 (1958)
13. M.L. Perl et al., Phys. Rev. Lett. 35, 1489 (1975), and Phys. Lett. 63B, 466 (1976).
14. S.W. Herb et al., Phys. Rev. Lett. 39, 252 (1977).
15. T.D. Lee, Phys. Rep. C9, 148 (1974); P. Sikivie, Phys. Lett. 65B, 141 (1976); S. Weinberg, Phys. Rev. Lett. 37, 657 (1976).
16. P. Wanderer et al., Phys. Rev. D (to be published) and references cited therein.
17. M. Holder et al., Phys. Lett. 72B, 254 (1977) and references

cited therein.

18. For a recent review of the theoretical and the experimental situation, see W.J. Marciano's report to "Orbis Scientiae 1978"; for an extensive list of references, see: B. Humpert, "$\mu \to e\gamma$; Why so interesting?", SLAC Preprint (1977).
19. A. Benvenuti et al., Phys. Rev. Lett. 38, 1110, 1185 (1977).
20. K. Kleinknecht, Report to "Orbis Scientiae 1978"; J. Steinberger, Report to the Irvine Conference (1977).
21. For information about the new super events seen by the HPWF group, I am indebted to V. Barger (Private communication).
22. K. Kleinknecht, Private Communication.
23. M. Holder et al., Phys. Rev. Lett. 39, 433 (1977).
24. See, for example, R.M. Barnett, H. Georgi and H.D. Politzer, Phys. Rev. Lett. 37, 1313 (1976).
25. M.A. Bouchiat and C.C. Bouchiat, Phys. Lett. 48B, 111 (1974); Le Journal de Physique 35, 899 (1974) and 36, 493 (1975); G. Feinberg and M.Y. Chen, Phys. Rev. D10, 3789 (1974); M.A.B. Bég and G. Feinberg, Phys. Rev. Lett. 33, 606 (1974); 35, 130 (E) (1975).
26. L.L. Lewis et al., Phys. Rev. Lett. 39, 795 (1977); P. Baird et al., ibid. 39, 798 (1977).
27. C.E. Loving and P. Sandars, Oxford University Preprint (1977).
28. G. Feinberg, "Parity Nonconservation in Atoms", Columbia University preprint (1977); S. Meshkov & S.P.Rosen, unpublished.
29. R.R. Lewis and W.L. Williams, Phys. Lett. 59B, 70 (1975); R. Cahn and G. Kane, ibid. 71B, 348 (1977).
30. M.A.B. Bég and G. Feinberg, Ref. 25.
31. L.D. Faddeev, Dokl. Akad. Nauk SSSR 210, 807 (1973) [English translation: Sov. Phys.-Doklady 18, 382 (1973)].
32. J.D. Bjorken, SLAC Preprint (1977).
33. The reader is referred to recent volumes of the standard journals, particularly "Physics Letters" and "Physical Review Letters", for a reasonably complete bibliography.

34. M.A.B. Bég, R. Budny, R. Mohapatra and A. Sirlin, Phys. Rev. Lett. 38, 1252 (1977).

35. M.A.B. Bég, R. Mohapatra, A. Sirlin and H.-S. Tsao, Phys. Rev. Lett. 39, 1054 (1977).

36. M.A.B. Bég and S.-S. Shei, Phys. Rev. D12, 3092 (1975).

37. Earlier formulations of left-right symmetry, and suggestions to the effect that parity may be good at high energies, may be found in the following papers (listed in chronological order): E.M. Lipmanov, Yad. Fiz. 6, 541 (1967) [Sov. J. Nucl. Phys. 6, 395 (1968)]; M.A.B. Bég and A. Zee, Phys. Rev. Lett. 30, 675 (1973) and Phys. Rev. D8, 1460 (1973); P. Fayet, Nucl. Phys. B78, 14 (1974); J. Pati and A. Salam, Phys. Rev. D10, 275 (1974); H. Fritzsch and P. Minkowski, Nucl. Phys. B103, 61 (1976); R.N. Mohapatra and D. P. Sidhu, Phys. Rev. Lett. 38, 667 (1977); A. de Rujula, H. Georgi and S.L. Glashow, Harvard Preprint (1977).

38. The first "vector-like" model was constructed by M.A.B. Bég and A. Zee, Ref. 37; for other references, see the report of H. Georgi in the Proceedings of "Orbis Scientiae, 1977".

39. R.N. Mohapatra and D.P. Sidhu, Ref. 37.

40. R.N. Mohapatra, F.E.Paige and D.P. Sidhu, BNL Preprint (1977). Parity violation effects arising at the one-loop level have been calculated by W.J. Marciano and A.I. Sanda, Rockefeller University Report No. COO-2232B-142. (to be published in Phys. Rev. D)

41. The first discussion, in the gauge theoretic contest, is that of T.D. Lee Ref. 15.

42. R. Peccei and H. Quinn, Phys. Rev. Lett. 38, 1440 (1977); Phys. Rev. D16, 1791 (1977). H. Quinn, Report to "Orbis Scientiae 1978".

43. If one chooses θ so as to reproduce the correct order of magnitude for $K_L \to 2\pi$, one obtains a value for the electric dipole moment of the neutron which is about six orders of magnitude larger than the experimental upper limit.

[See, for example, K. Kleinknecht, Ann. Rev. of Nuc. Sci. 26, 1 (1976) and references cited therein]

44. This course is favored by M. Gell-Mann (Private communication)

45. Arguments against $m_u = 0$ [see, for example, S. Weinberg, Ref. 46], based on the application of chiral perturbation theory to the K^0-K^+ mass difference, are not very convincing.

46. S. Weinberg, Phys. Rev. Lett. 40, 223 (1978); F. Wilczek, ibid. 40, 279 (1977).

47. C. Baltay, G. Feinberg and M. Goldhaber (Unpublished); G. Feinberg (Private communication).

48. And corresponding to the same eigenvalues for the Casimir operators of the group. (This qualification is, of course, redundant for representations of low dimensionality).

STATUS OF MUON-NUMBER CONSERVATION AND COMMENTS ON PARITY VIOLATION IN ATOMS +

William J. Marciano

The Rockefeller University

New York, New York 10021

INTRODUCTION

Last year two of the hottest topics of conversation at this conference were[1]: 1) the rumor that muon-number violation might have been observed in the decay $\mu^+ \rightarrow e^+ + \gamma$ and 2) the failure of bismuth experiments to detect parity violation at the level predicted by the standard Weinberg-Salam model[2]. During the intervening time between then and now, several significant theoretical and experimental developments regarding these topics have occurred.

We have learned that unified gauge theories of weak and electromagnetic interactions could, if necessary, easily accommodate small violations of muon-number conversation. Furthermore, within the framework of renormalizable gauge theories, the rates for processes such as $\mu^+ \rightarrow e^+ + \gamma$, $\mu^+ \rightarrow e^+ + e^+$, μ^- + Nucleus $\rightarrow e^-$ + Nucleus, etc. are finite and calculable, even if they arise as higher order effects; a feature missing from earlier considerations[3]. Because of this important property, we witnessed the birth of a plethora of models which were designed to account for an anticipated $\mu^+ \rightarrow e^+ + \gamma$ branching ratio of about $\sim 5 \times 10^{-9}$. However, the validity of these models has experienced serious experimental setbacks. The muon-number violating decay $\mu^+ \rightarrow e^+ + \gamma$ was not observed[4,5] and the bound on its branching ratio has been lowered to $<1.1 \times 10^{-9}$. In addition, a recent experimental search for muon conversion, $\mu^- + \text{Nucleus} \rightarrow e^- + \text{Nucleus}$, found no such events[6] and thereby lowered the bound on this process considerably (see section III for a complete discussion of this reaction). This finding is extremely important because the rate for muon conversion was predicted to be rather sizeable in most of the proposed muon-number violating models[7].

As far as parity violation in atoms is concerned, the bismuth experiments are still consistent with no parity violation and the errors on their findings have been lowered considerably. The published results for the optical rotation of polarized light in bismuth due parity violating effects are now[8]

$$R_{876nm} = (-0.7 \pm 3.2) \times 10^{-8} \qquad \text{(Washington Exp.)}$$

$$R_{648nm} = (2.7 \pm 4.7) \times 10^{-8} \qquad \text{(Oxford Exp.)}$$

$$(1.1)$$

To account for these essentially null results, several theoretical alternatives to the standard Weinberg-Salam model have been suggested. The most popular idea is to make the electron's neutral current coupling pure vector (vector-like models), so as to avoid a large coherent enhancement effect in heavy atoms such as bismuth. A second interesting possibility, which is easily realized within the framework of left-right symmetric models such as $SU_L(2) \times SU_R(2) \times U(1)$ is to have all neutral currents naturally conserve parity at the tree level[9]. Both possibilities accommodate the results in (1.1) and await similar experiments on ordinary hydrogen and deuterium along with other neutral current tests to validate or invalidate them.

In addition to creating a model building boom, the results in (1.1) have motivated a reexamination of the atomic physics calculations used to obtain theoretical predictions and recently some doubt has been cast on their credibility. For example, including the effects of electronic shielding, Loving and Sandars[10] have reduced their theoretical predictions by a factor of .55. When this revision is taken into account and the experimental value $\sin^2\theta_W \simeq .24$ is used, the standard Weinberg-Salam model predicts[10-12]

$$R^{W-S}_{876nm} = -9.2 \times 10^{-8}$$

$$(\text{for } \sin^2\theta_W = .24) \qquad (1.2)$$

$$R^{W-S}_{648nm} = -12 \times 10^{-8}$$

(Considerations of the atomic physics by other authors predict larger values[12].) Even these considerably reduced predictions disagree with (1.1) by about three standard deviations and seem to pose a serious problem for the Weinberg-Salam model. However, before any definite conclusions can be reached, more reliable atomic physics calculations should be carried out and the results of similar experiments on ordinary hydrogen, deuterium, cesium and thallium which are already underway should be analyzed[12]. Independent of any theoretical model, it would be extremely interesting if a positive signal for parity violation were observed in any of these experiments.

With these introductory remarks out of the way, let me briefly outline the remainder of this talk. In section II the present status of muon-number conservation is reviewed, experimental bounds are given and the outlook for the immediate future is described. In section III a detailed description of muon conversion along with a review of theoretical calculations and predictions for this process are presented. Finally, in section IV some recent results on parity violation in atoms obtained by A.I. Sanda and myself are reported[13]. We find that even models which avoid very large parity violation in bismuth by making the electron's neutral current pure vector or by having parity conserving neutral currents at the tree level may still predict sizeable effects due to radiative corrections. For example, a particular SU(2) x U(1) vector-like model in which the electron's neutral current coupling is pure vector predicts, as a result of one loop contributions, an optical rotation in bismuth which is about 11% as large as that predicted by the standard Weinberg-Salam model and opposite in sign, a surprisingly large result with testable experimental consequences[13].

II. STATUS OF MUON-NUMBER CONSERVATION

To begin this discussion, let us recall the experimental status of some muon-number nonconserving processes prior to last year (t<1977).

$$R_{e\gamma} \equiv \frac{\Gamma(\mu^+ \to e^+ + \gamma)}{\Gamma(\mu^+ \to e^+ + \nu_e + \bar{\nu}_\mu)} < 2.2 \times 10^{-8} \quad (1964 \text{ Exp.})^{14} \tag{2.1}$$

$$R_{3e} \equiv \frac{\Gamma(\mu^+ \to e^+ + e^- + e^+)}{\Gamma(\mu^+ \to e^+ + \nu_e + \bar{\nu}_\mu)} < 1.9 \times 10^{-9} \quad (1976 \text{ Exp.})^{15} \tag{2.2}$$

$$R_{eN} \equiv \frac{\omega(\mu^- + N \to e^- + N)}{\omega(\mu^- + N \to \nu_\mu + N')} < 1.6 \times 10^{-8} \quad (1972 \text{ Exp.})^{16}, N=Cu \tag{2.3}$$

The first two branching ratios are relative to ordinary muon decay. Experiments designed to search for these exotic decays employ μ^+, rather than μ^- beams, because they are easier to obtain with high intensities and they do not undergo muon capture when stopped[3]. The ratio R_{eN} measures the rate for muon conversion, $\mu^- +$ Nucleus $\rightarrow e^- +$ Nucleus, relative or ordinary muon capture. (a detailed description of this process is given in section III.) Searches for these three exotic processes presently represent the best tests of muon-number conservation; so they will be the only ones we will discuss[17,18].

The advent of high intensity meson factories has given experimentalists the opportunity to significantly lower the bounds in (2.1) - (2.3) or perhaps even discover one of these exotic reactions if it exists. With such high stakes and considering the fact that these are good sound basic experiments, it is not surprising that three large meson laboratories (SIN, TRIUMF, and LAMPF) have undertaken some of these experiments. Let us review what results have been obtained during the last year and the prospects for the near future. (t> 1977)

1) $\underline{\mu^+ \rightarrow e^+ + \gamma}$: Groups at both TRIUMF and SIN now have reported bounds on this process which are below the rumored value of $R_{e\gamma} \sim 5 \times 10^{-9}$ that was circulating last year. The published bounds from these two independent experiments are[4,5]

$$R_{e\gamma} < 3.6 \times 10^{-9} \quad (\text{TRIUMF})^4 \tag{2.4a}$$

$$R_{e\gamma} < 1.1 \times 10^{-9} \quad (\text{SIN})^5 \tag{2.4b}$$

TRIUMF is still analyzing data; we can probably expect their bound to also go down to $\sim 1 \times 10^{-9}$. In addition, a high intensity experiment at LAMPF has been running since this summer[19]. This experiment is designed to detect $\mu \rightarrow e + \gamma$ if it occurs with a branching ratio of at least $R_{e\gamma} \sim 1 \times 10^{-10}$ or to yield a bound of about $R_{e\gamma} < 5 \times 10^{-11}$. Should this experiment come close to achieving its goals, it will provide a significant advancement.

(2) $\mu^+ \to e^+ + e^- + e^+$: To my knowledge, there has been no improvement in the bound of (2.2); however, that number comes from an experiment in the Soviet Union which may have been revived. I believe that a proposal to search for this decay at LAMPF has also been submitted[19].

(3) μ^- + Nucleus $\to e^-$ + Nucleus : Recently a group at SIN reported a null result for their search to find muon conversion in a sulfur target. Their experiment produced the bound[6]

$$R_{eN} < 4 \times 10^{-10} \quad \text{(SIN), N = Sulfur} \tag{2.5}$$

We note that this is a very significant improvement over the bound in (2.3); however, these two experiments should be distinguished in that (2.5) comes from a sulfur target while (2.3) employed a copper target. (We comment further on this distinction in section III)

The SIN experiment is very sophisticated and may be able to lower their bound to $\sim 1 \times 10^{-10}$. In addition they can repeat their experiment using other targets, a worthwhile endeavor. A muon conversion experiment has also been proposed[20] at TRIUMF and cancelled at LAMPF.

In light of some recent theoretical work[7], muon conversion experiments appear to be an extremely good test of muon-number conservation. For many of the gauge theory models recently considered, the bound in (2.5) turns out to be a much more stringent constraint then (2.4). To amplify this point, we now present a somewhat detailed discussion of muon conversion.

III. MUON CONVERSION: μ^- + Nucleus $\to e^-$ + Nucleus

Since this exotic process is unfamiliar to many people, we begin with a brief description.

Muon Conversion: 1) Negative muons are captured in atomic orbits of a stopping material such as Cu or S. 2) Each muon quickly transitions its way down to a 1s orbit by electromagnetic cascades.

(The time scale for these cascades is much smaller than the muon lifetime.) 3) While in the 1s state the following fates are available to the muon, i) ordinary capture $\mu^- + N(A,Z) \rightarrow \nu_\mu + N'(A,Z-1)$, ii) decay $\mu^- \rightarrow e^- + \bar{\nu}_e + \nu_\mu$ or perhaps iii) conversion $\mu^- + N \rightarrow e^- + N$.

The experimental signature for conversion is very clear; a monoenergetic electron with energy

$$E_e \simeq m_\mu - m_\mu^2/M_N \simeq 105 \text{ MeV} \qquad (3.1)$$

should be observed in the final state. (Remember that in ordinary muon decay (at rest) the <u>maximum</u> energy of the final state electron is $E_e^{max} \simeq m_\mu/2$; so it does not present a background problem.) Because only a single particle need be detected in muon conversion, it is an easier and cleaner experiment than either $\mu^+ \rightarrow e^+ + \gamma$ or $\mu^+ \rightarrow e^+ + e^- + e^+$.

Notice that the rate for conversion is compared to ordinary muon capture in (2.3) and (2.5). This is because capture in heavy atoms is much more probable than decay[21] (for $Z \geq 11$, capture exceeds decay). Primakoff's semi-phenomenological capture rate formula nicely illustrates this fact

$$\omega(\mu^- + (A,Z) \rightarrow \nu_\mu + (A,Z-1) \simeq Z_{eff}^4 (4.13\text{x}10^{-4}) \frac{G_F^2 m_\mu^5}{192\pi^3} (1-(3.15)\frac{A-Z}{2A}) \qquad (3.2)$$

where Z_{eff} is the effective nuclear charge, $G_F^2 m_\mu^5/(192\pi^3)$ is the rate for ordinary free muon decay and $(1 - (3.15)(A-Z)/2A)$ represents a suppression factor that results from the Fermi statistics of neutrons (i.e. protons converted into neutrons cannot go into states already occupied by other neutrons). So for example in the case of Cu, $Z = 29$ while $Z_{eff} \simeq 21 \rightarrow$ Capture $\sim$ 13 x Decay. The factor Z^4_{eff} which enhances capture also occurs in the muon conversion rate formula. (It arises primarily from the overlap wavefunction of the muon with the nucleus.) Furthermore, there is no suppression of muon conversion due to statistics; on the contrary, it is further enhanced by coherence effects. These general features alone point out why muon conversion was originally suggested to be a nice reaction to search for muon number non-conservation [22].

A detailed theoretical analysis of muon conversion was given in the pioneering work of Weinberg and Feinberg[22]. Their results provide convenient formulas for estimating the rate of muon conversion of any gauge theory, so we briefly outline some of them.

Because of Lorentz covariance and electromagnetic gauge invariance, the induced matrix element of the electromagnetic current between muon and electron states must be of the form (neglecting the electron's mass)

$$< e|J^{el}_{\lambda}(0)|\mu> = -ie\bar{u}_e \{[f_{E0}(q^2) + \gamma_5 f_{M0}(q^2)]\gamma^{\alpha}(g_{\alpha\lambda} - \frac{q_{\alpha}q_{\lambda}}{q^2})$$

$$+i[f_{M1}(q^2) + \gamma_5 f_{E1}(q^2)]\sigma_{\lambda\alpha}q^{\alpha}/m_{\mu}\}u_{\mu} , \quad (3.3)$$

where $q = p_{\mu} - p_e$. Two of the four form factors in (3.3) must vanish at $q^2 = 0$

$$f_{E0}(0) = f_{M0}(0) = 0 , \quad (3.4)$$

so they cannot contribute to $\mu \rightarrow e + \gamma$, the rate for which is given by

$$\Gamma(\mu \rightarrow e+\gamma) = \frac{\alpha m_{\mu}}{2} [|f_{E1}(0)|^2 + |f_{M1}(0)|^2] . \quad (3.5)$$

On the other hand, all four form factors can contribute to $\mu^+ \rightarrow e^+ + e^- + e^+$ and $\mu^- + N \rightarrow e^- + N$, since these processes would occur at $q^2 \neq 0$. Weinberg and Feinberg, using (3.2) in their denominator, estimated that in terms of the form factors in (3.3)

$$R_{eN} \simeq \frac{|f_{E0}(-m_\mu^2) + f_{M1}(-m_\mu^2)|^2 + |f_{M0}(-m_\mu^2) + f_{E1}(-m_\mu^2)|^2}{8G_F^2 m_\mu^4} \tag{3.6}$$

for nuclei with mass number A > 60.[7,22,23,24] Note that sulfur which has Z = 16, A = 32 does not satisfy this criterion. If we merely extrapolate Weinberg and Feinberg's analysis to sulfur, we find the result in (3.6) multiplied by ∿ .6, however this factor should not be taken too seriously, since some of their approximations are not valid for light atoms[22]. Estimates for R_{eN} could be improved using current phenomenological imput; however, unless this exotic process is observed, (3.6) would seem to be a satisfactory approximation.

In renormalizable gauge theories of weak and electromagnetic interactions, the form factors in (3.3) are finite and calculable. However, f_{E0} and f_{M0} may be gauge dependent (under nonabelian gauge transformations), a reminder that in such theories there may be other contributions to processes such as muon conversion in addition to those involving photons. The easiest way to include all of these additional effects is to replace f_{E0} and f_{M0} in (3.6) by "effective" form factors which include photonic as well as non-photonic contributions[7]

$$f_{E0}^{\text{effective}} = f_{E0}^{\text{photonic}} + f_{E0}^{\text{non-photonic}},$$
$$f_{M0}^{\text{effective}} = f_{M0}^{\text{photonic}} + f_{M0}^{\text{non-photonic}}. \tag{3.7}$$

The non-photonic contributions are easily obtained from the relevant Feynman diagrams in the following way: Writing the muon conversion amplitude as

$$M^{\text{non-photonic}} = i\,[\bar{e}\gamma_\lambda(a^u+b^u\gamma_5)\mu\bar{u}\gamma^\lambda u + \bar{e}\gamma_\lambda(a^d + b^d\gamma_5)\mu\bar{d}\gamma^\lambda d]\ , \tag{3.8}$$

where u and d are up and downs quarks (we present the sum of up and down quark amplitudes and only the hadronic vector part since this part contributes coherently, thereby dominating in heavy atoms), one finds

$$f_{EO}^{\text{non-photonic}} = \frac{m_\mu^2}{e^2}\,[a^u\,\frac{Z+A}{Z} + a^d\,\frac{2A-Z}{Z}]\ ,$$

$$f_{MO}^{\text{non-photonic}} = -\,\frac{m_\mu^2}{e^2}[b^u\,\frac{Z+A}{Z} + b^d\,\frac{2A-Z}{Z}]\ . \tag{3.9}$$

Using this approach, it was found[7] that R_{eN} turns out to be much larger than one might naively anticipate in most of the models which were designed primarily to accommodate muon-number violation in $\mu \to e + \gamma$. To illustrate the magnitude of these finding, I have listed in table I the ratio of R_{eN} to $R_{e\gamma}$ in a variety of SU(2) x U(1) models which are identified by their leptonic representations. As a means of comparison, the predictions for $R_{3e}/R_{e\gamma}$ are also given.[7,25-31]

Models	$R_{eN}/R_{e\gamma}$	$R_{3e}/R_{e\gamma}$
A	140	2
B	5	.1
C	50	.05
D	140	7

A	$\begin{pmatrix}\nu_e\\ e\end{pmatrix}_L \begin{pmatrix}\nu_\mu\\ \mu\end{pmatrix}_L \quad N_{1_L} \quad N_{2_L} \begin{pmatrix}N_1c+N_2s\\ e\end{pmatrix}_R \begin{pmatrix}N_2c-N_1s\\ \mu\end{pmatrix}_R$	$c \equiv \cos\phi$ $s \equiv \sin\phi$
B	$\begin{pmatrix}\nu+(m_e/m_1)N_1c+(m_e/m_2)N_2s\\ e\end{pmatrix}_L \begin{pmatrix}\nu'-(m_\mu/m_1)N_1s+(m_\mu/m_2)N_2c\\ \mu\end{pmatrix}_L$	The rest like A
C	$\begin{pmatrix}\nu_1\\ e\end{pmatrix}_L \begin{pmatrix}\nu_2\\ \mu\end{pmatrix}_L \begin{pmatrix}\nu_3\\ L\end{pmatrix}_L \quad e_R \ \mu_R \ L_R \ \nu_{L_R}$	$\nu_i \equiv$ orthogonal combinations of ν_e, ν_μ, ν_L
D	$\begin{pmatrix}\nu_e\\ e\\ hc+ks\end{pmatrix}_L \begin{pmatrix}\nu_\mu\\ \mu\\ kc-hs\end{pmatrix}_L \quad e_R, \mu_R, h_R, k_R$	

TABLE I. Estimates of the ratios $R_{eN}/R_{e\gamma}$ and $R_{3e}/R_{e\gamma}$ in a variety of SU(2) x U(1) models which are designated by their leptonic representations. Values quoted were found using eq. (3.6) and the results of ref. 7 along with $\sin^2\theta_W = .24$, $\ln(M_W^2/m_\ell^2)=5$ for models A, B, & D, $\ln(M_W^2/m_L^2) = 3.5$ for model C. Models A, B, C, and D are presented in Refs. 25, 26, 27, and 28 respectively.

The sizeable magnitude of these results requires some explanation, since naively one would expect $R_{eN}/R_{e\gamma} \sim \alpha/\pi$. In the first three models listed, the dominant contributions to $\mu^- + N \rightarrow e^- + N$ come from non-photonic parts of

$f_{E0}^{\text{effective}}$ and $f_{M0}^{\text{effective}}$ induced by the box diagrams in fig. 1. The amplitudes from these diagrams are sizeable[7,29] because they have a large logarithm ∿ 3.5 - 5 and they depend on the weak coupling which is larger than the electric charge $g^2 = e^2/\sin^2\theta_W \simeq 4e^2$. In addition, when added coherently, they turn out to be sensitive to the number of quarks in the nucleus, 3A, rather than merely Z or A, the "quark coherence effect". All of these enhancements occur at the amplitude level, they are squared in going to the transition rate; hence the large values for $R_{eN}/R_{e\gamma}$ in the table[7].

The sizeable value for R_{eN} relative to $R_{e\gamma}$ in the fourth model results from the charge radius contributions shown in fig. 2. This makes f_{E0} and f_{M0} much larger than f_{E1} and f_{M1} at $q^2 = -m_\mu^2$. When such a condition arises, both R_{eN} and R_{3e} can both be larger than $R_{e\gamma}$ and then satisfy $R_{eN} \simeq 20R_{3e}$.[7,28,31] This illustrates another reason why $\mu^- + N \rightarrow e^- + N$ is a much better process to search for than $\mu^+ \rightarrow e^+ + e^- + e^+$.

In additions to these SU(2) x U(1) models, it can be shown for the SU(2) x U(1) x U(1) model of Bég and Sirlin[32] that $R_{eN} > R_{e\gamma}$. Therefore, we see that the bound in (2.5) is a more stringent constraint than those in (2.4) for a large class of models[33].

The moral of these considerations, as I see it, is the following: If muon number is not absolutely conserved, then muon conversion is potentially the best place to look for a violation. Its only drawback from a theoretical point of view is that it depends on the weak interactions of quarks which automatically implies some uncertainly in our estimates. In any case, with the meson facilities that are now available, all three processes $\mu^+ \rightarrow e^+ + \gamma$, $\mu^- + N \rightarrow e^- + N$ and $\mu^+ \rightarrow e^+ + e^- + e^+$ (along with other exotic reactions) should be searched for. Either the bounds will be further lowered, or a fundamental discovery will be made.

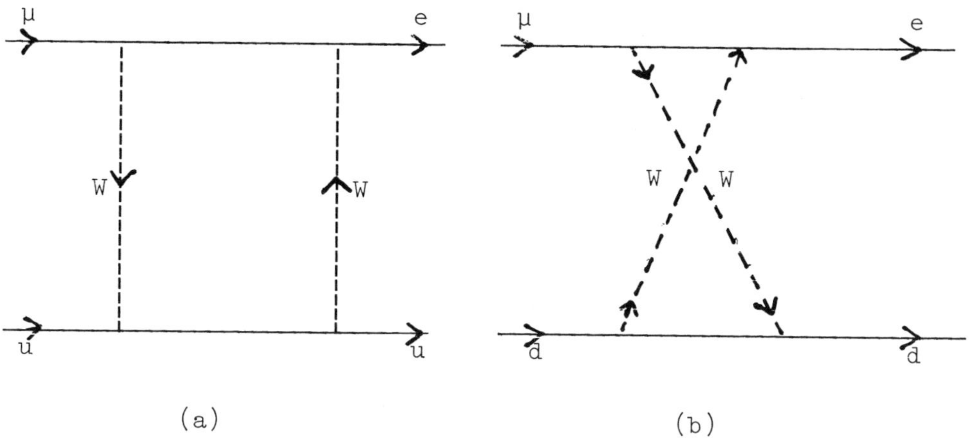

Box diagrams involving the exchange of two charged W bosons, which can give a large contribution to muon conversion.

Fig. 1

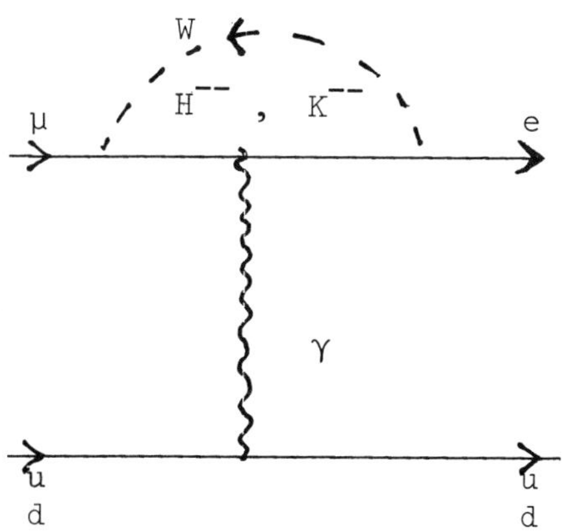

Charge radius diagrams which dominate muon conversion in the Wilczek-Zee model.

Fig. 2

IV. PARITY VIOLATION IN ATOMS

I would like to conclude this talk with some comments about gauge theory models which avoid large parity violation in bismuth either by making the electron's neutral current coupling pure vector (vector-like models) or by having all neutral currents naturally conserve parity at the tree level in a viable manner (e.g. the recent $SU_L(2) \times SU_R(2) \times U(1)$ model of ref. 9). These alternatives to the standard $SU(2) \times U(1)$ model certainly accommodate the essentially null results cited in (1.1); however, they must predict parity violating effects at some level due to radiative corrections.[13] Indeed, the box diagrams in fig. 3 which involve the exchange to two charged W bosons give rise to a parity violating neutral current effect in all models. Cognizant of the enhancements that may arise for such diagrams from our previous work[7] (see section III), A.I. Sanda and I decided to investigate the magnitude of such higher order effects for models in which they are the dominant source of parity violation.[13] The size of the effect that we found can be quite large and will probably surprise most of the people here. Because the details of our calculations already exist in preprint form,[13] I will describe only one subtle result and then illustrate our estimates for two specific models.[34]

Consider the following neutral current interaction Langrangian for the photon field, A_μ, and any massive neutral vector boson Z_μ

$$\mathcal{L}_{int.} = eA^\mu(-\bar{e}\gamma_\mu e+Q\bar{q}\gamma_\mu q)+ \frac{g}{2\sqrt{2}\cos\theta_W} Z^\mu(\bar{e}\gamma_\mu(a^e+b^e\gamma_5)e$$

$$+\bar{q}\gamma_\mu(a^q+b^q\gamma_5)q) \quad , \tag{4.1}$$

where q=u or d, Q=2/3 for the up quark, Q=-1/3 for the down quark, $g^2/8M_Z^2 \cos^2\theta_W=G_F/\sqrt{2}$ and a^e,b^e,a^q,b^q are, for the moment, arbitrary In terms of these couplings, the parity violating amplitude that results from the exchange of a single Z boson between an electron and quark (see fig. 4) is

$$M^{Z}_{P.V.} = -i \frac{G_F}{\sqrt{2}} [b^e a^q \bar{e}\gamma_\mu\gamma_5 e\bar{q}\gamma_\mu q + a^e b^q \bar{e}\gamma_\mu e\bar{q}\gamma^\mu\gamma_5 q] \quad . \tag{4.2}$$

For heavy atoms such as bismuth, only the first term in (4.2) (proportional to $b^e a^q$) is important because its effect is enhanced by the coherence of all quarks that make up the nucleus.[11,12,13] The magnitude of parity violation in such atoms is usually determined by the so called weak charge Q_W, which for the amplitude in (4.2) is given by [11,12,13]

$$^{(Z)}Q_W(Z,A) = 2b^e(a^u(Z+A)+a^d(2A-Z)] \; . \tag{4.3}$$

The terms proportional to $a^e b^q$ in (4.2) are neglected because they do not add coherently.

In vector-like models one makes $Q_W=0$ in lowest order by having $b^e=0$, i.e. the neutral current for the electron in (4.1) is pure vector $a^e \neq 0$, $b^e=0$. For example in SU(2) x U(1) models $b=\sqrt{2}\ (t_3^L - t_3^R)$ where t_3 is the third component of weak isospin; therefore b^e will be zero if the left and right-handed components of the electron field are put into SU(2) multiplets for which $t_3^L = t_3^R$. In this way large parity violating effects can be avoided in heavy atoms.

Now let us look at box diagrams involving a photon and a Z boson as pictured in fig. 5. Using the same interaction Lagrangian, (4.1), we find that these give rise to the parity violating amplitude[13]

$$M^{\gamma Z}_{P.V.} = -i \frac{G_F}{\sqrt{2}} \left(\frac{3\alpha}{2\pi} Q \ln \frac{M_Z^2}{m^2}\right) \{a^e b^q \bar{e}\gamma_\mu\gamma_5 e\bar{q}\gamma^\mu q + b^e a^q \bar{e}\gamma_\mu e\bar{q}\gamma^\mu\gamma_5 q\} \tag{4.4}$$

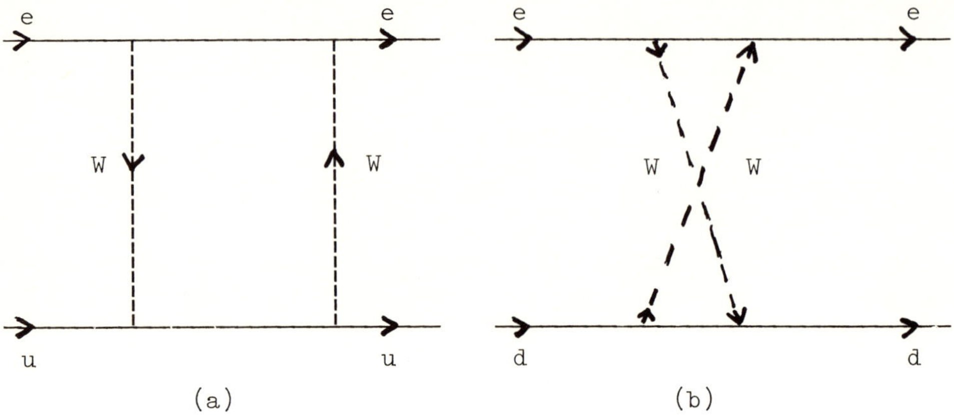

Box diagrams involving the exchange of two charged W bosons which induce neutral current parity violating effects.

Fig. 3

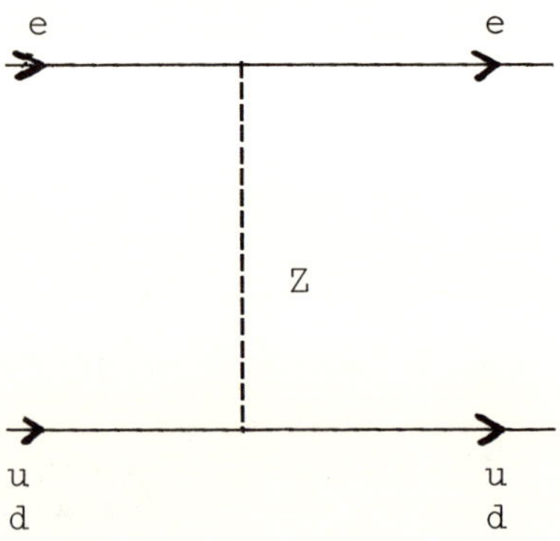

Lowest order parity violating diagrams in which a neutral vector boson is exchanged between an electron and up or down quark.

Fig. 4

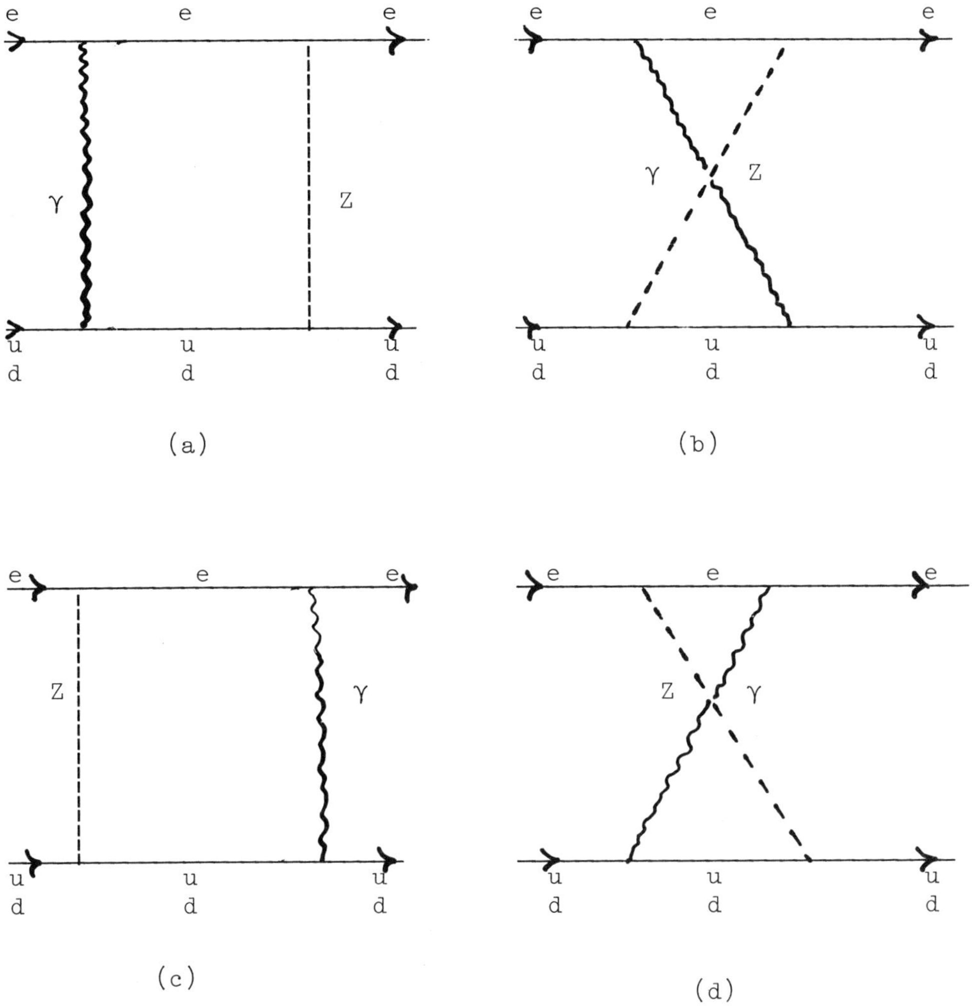

Box diagrams involving the exchange of a proton and a massive neutral vector boson between an electron and up or down quark.

Fig. 5

where m ∿ 1-2 GeV a typical hadronic mass scale. The most important point to notice is that the roles of $a^e b^q$ and $b^e a^q$ have been interchanged in going from (4.2) to (4.4). Now the terms proportional to $a^e b^q$ add coherently. Therefore, the contribution of this amplitude to the weak charge is

$$^{(\gamma Z)}Q_W(Z,A) = a^e \left(\frac{\alpha}{\pi} \ln \frac{M_Z^2}{m^2} \right) \{2b^u(Z+A) - b^d(2A-Z)\} \quad , \tag{4.5}$$

which is independent of b^e and thus non-vanishing in vector-like models. (Both b^u and b^d cannot be zero for theories with a single Z boson because of the observed inequality of ν and $\bar{\nu}$ cross-sections). The weak charge in (4.5) is of order α/π; but it is enhanced by several effects: The presence of a large logarithm $\ln(M_Z^2/m^2) \approx 8$ and the fact that b^u and b^d often have opposite signs so that (4.5) is sensitive to ∿3A the number of quarks in the nucleus rather than merely Z or A; we call this[13] the "quark coherence effect". For example, in the SU(2) x U(1) Vector-Like model with fermion field representations taken to be

$$\begin{pmatrix} \nu_e \\ e \end{pmatrix}_L \quad \begin{pmatrix} N_e \\ e \end{pmatrix}_R \qquad N_{e_L} \quad \begin{pmatrix} u \\ d_\theta \end{pmatrix}_L \begin{pmatrix} c \\ s_\theta \end{pmatrix}_L \qquad u_R \; d_R \; c_R \; s_R \text{ etc} \tag{4.6}$$

one has $t_3^{eL} = t_3^{e_R} = -1/2$, $t_3^{u_L} = 1/2$, $t_3^{d_L} = -1/2$, $t_3^{u_R} = t_3^{d_R} = 0$ and then

(4.5) becomes

$$^{(\gamma Z)}Q_W(A,Z) = \frac{\alpha}{\pi} \ln (M_Z^2/m^2) \cos 2\theta_W (4A+Z) . \quad (4.7)$$

Using $\ln (M_Z^2/m^2) \approx 8$, $\cos 2\theta_W \approx .52$ and A=209, Z=83 for Bi; this gives[35].

$$^{(\gamma Z)}Q_W(Bi) \approx 9 . \quad (4.8)$$

This higher order effect is to be compared with the prediction of the standard Weinberg-Salam model[13]

$$Q_W(Bi) = - 123, \text{ (W-S model with } \sin^2\theta_W \approx .24) . \quad (4.9)$$

The result in (4.8) is about 7% of (4.9) in magnitude and opposite in sign. Of course in a complete treatment other comparable sources of radiative corrections must be added to the results of the γZ diagrams.

To dramatize the effect of radiative corrections further, I have listed in table 2 some of our results for two models in which $Q_W=0$ to lowest order. The SU(2) x U(1) Vector-Like model (see (4.6)) gets contributions from box diagrams[36] involving two W's or two Z's in addition to the result of (4.8). All contributions have the same sign, so finally we wind up with a result that is about 11% of the standard model prediction and opposite in sign.[13]

Also illustrated in this table are our predictions for a recent $SU_L(2) \times SU_R(2) \times U(1)$ model[9] in which there are two weak neutral currents and both conserve parity in lowest order. (R. Mohapatra will talk about this model, so I won't give any details.) The largest parity violating radiative correction in this case is

Quantity	Experiment	Weinberg-Salam SU(2) x U(1)	Vector-Like SU(2) x U(1)	Left-Right
Q_W(Bi)	$-52 < Q_W < 33$ $-20 < Q_W < 74$	-123	13.9	4.6 - 14.4
R_{876nm}	(-0.7 ± 3.2) $\times10^{-8}$	-9.2×10^{-8}	1×10^{-8}	$(0.35-1.1)$ $\times10^{-8}$
R_{648nm}	(2.7 ± 4.7) $\times10^{-8}$	-12×10^{-8}	1.4×10^{-8}	$(0.47-1.5)$ $\times10^{-8}$

TABLE II. A comparison of the parity violation predictions of the Weinberg-Salam model, Vector-Like SU(2) x U(1) model of Eq. (4.6) and $SU_L(2) \times SU_R(2) \times U(1)$ model of Ref. 9 with the present experimental bounds on the weak charge Q_W and the values for the optical rotations in Bi. All estimates were obtained using the atomic physics calculations of Ref. 10 along with $\sin^2\theta_W = .24$ and $\ln(M_Z/m) = 4$. The range of values given for the Left-Right model correspond to $2.7 \leq M_R/M_L \leq 100$.

proportional to ln (M_R/M_L) where M_R and M_L are the masses of the charged right and left-handed intermediate vector bosons.[37,13] The range of values given in our table are for M_R/M_L between 2.7 and 100 (the 2.7 is a phenomenological lower bound[38] while the 100 is an arbitrary choice).

Included in this table are also the predictions for the optical rotations in bismuth. These estimates were arrived at by using the same atomic physics calculations that lead to (1.2). (They are directly proportional to Q_W (Bi)). Notice that the predictions that arise from radiative corrections are certainly within the reach of the bismuth experiments. Therefore, an order of magnitude increase in the precision of these experiments will be a nice test of these results.

In closing, I would like to reiterate that the observation of parity violation in any of the atomic physics experiments presently underway would be an extremely important finding independent of any model's predictions. Hopefully, our results will provide some further incentive to continue such experiments.

ACKNOWLEDGEMENTS

I would like to thank A.I. Sanda for a productive collaboration on questions of muon-number violation and parity violation in atoms. In addition, valuable conversations with M.A.B. Beg, J.D. Bowman, D. Bryman, G. Feinberg, H. Pagels, A. Pais, P. Sandars and A. Sirlin are gratefully acknowledged.

+Work supported in part by the U.S. Energy Research and Development Administration under grant number EY-76-C-02-2232B.*000.

REFERENCES AND FOOTNOTES

1. The status of these topics at this time last year can be found in talks by Ta-Pei Cheng and C. Bouchiat, Deeper Pathways in High-Energy Physics, edited by Perlmutter and Scott (Plenum Press, New York, 1977) p. 659 and p. 435
2. S. Weinberg, Phys. Rev. Lett. 19, 1264 (1967); A. Salam in Elementary Particle Physics ed. N. Svartholm (Almquist and Wiksells, Stockholm 1968) p.367.
3. For a pre-gauge-theory review of these processes, see S. Frankel, in Muon Physics, edited by C. S. Wu and V. W. Hughes (Academic Press, New York, 1975), Vol. II, p. 83 and references therein.
4. P. Depommier et al., Phys. Rev. Lett. 39, 1113 (1977).
5. H. Povel et al., Phys. Lett. 72B, 183 (1977).
6. A. Badertscher et al., Phys. Rev. Lett. 39, 1385 (1977).
7. W. J. Marciano and A. I. Sanda, Phys. Rev. Lett. 38, 1512 (1977); G. Altarelli, L. Baulieu, N. Cabibbo, L. Maiani and R. Petronzio, Nucl. Phys. B125, 285 (1977).
8. L. L. Lewis et al., Phys. Rev. Lett. 39, 795 (1977); P. Baird et al. ibid 39, 798 (1977).
9. R. Mohapatra, F. Paige and D. Sidhu, B.N.L. preprint 1977; and R. Mohapatra in his talk at this conference.
10. C. E. Loving and P. Sandars, Oxford University preprint 1977.
11. Theoretical background for these effects can be found in: M. A. Bouchiat and C. C. Bouchiat, Phys. Lett. 48B, 111 (1974), Le Journal De Physique 35, 899 (1974); 36, 493 (1975).
12. G. Feinberg, "Parity Non-Conservation in Atoms", Columbia University preprint 1977. This paper reviews the status and background of these calculations. It also describes the feasibility of performing similar experiments in other atoms.
13. W. J. Marciano and A. I. Sanda, "Parity Violation in Atoms Induced by Radiative Corrections" Rockefeller University preprint COO-2232B-142. Submitted by Phys. Rev. D.

14. S. Parker, H. Anderson and C. Rey, Phys. Rev. 133, B768 (1964).

15. S. M. Korenchenko et al. Zh. Eksp. Teor. Fiz. 70, 3 (1976) [Sov. Phys. JETP 43, 1 (1976)].

16. D. A. Bryman, M. Blecher, K. Gotow and R. J. Powers, Phys. Rev. Lett. 28, 1469 (1972).

17. A discussion of other muon-number non-conserving processes can be found in Ta-Pei Cheng and Ling-Fong Li, Phys. Rev. D16, 1425 (1977).

18. The best bounds on the possible decay $\nu_\mu \to \nu_e + \gamma$ come from Astrophysics, see M.A.B. Bég, W. J. Marciano and M. Ruderman, to be published in Phys. Rev. D and references therein.

19. J. D. Bowman, Private communication.

20. D. Bryman, private communication.

21. For a review of muon capture which contains many references to the literature, see N. C. Mukhopadhyay, Phys. Reports 30C, 1 (1977).

22. S. Weinberg and G. Feinberg, Phys. Rev. Lett. 3, 111, 244 (E) (1959).
This paper provides a clear analysis of muon conversion, and formulas which are easily applied to gauge theories.

23. F. J. Ernst, Phys. Rev. Lett. 5, 478 (1960).

24. If we discard inelastic processes (which experimentalists would probably not identify as conversions anyway), then the 8 in the donominator of (3.6) becomes ~10 and our numerical predictions are reduced by about a factor of .8 C.f. ref. 22.

25. T.-P. Cheng and L.-F. Li, Phys. Rev. Lett. 38, 381 (1977); S. M. Bilenky, S. T. Petcov, and B. Pontecorvo, Phys. Lett. 67B, 309 (1977).

26. J. D. Bjorken, K. Lane, and S. Weinberg, Phys. Rev. D16, 1474 (1977).

27. B. W. Lee, S. Pakvasa, R. E. Shrock and H. Sugawara, Phys. Rev. Lett. 38, 937, 1230 (E) (1977); B. W. Lee and R. Shrock, Phys. Rev. D16, 1444 (1977).

28. F. Wilczek and A. Zee, Phys. Rev. Lett. 38, 531 (1977).

29. S. B. Treiman, F. Wilczek and A. Zee, Phys. Rev. D16, 152 (1977).

30. S. M. Barr and S. Wandzura, Phys. Rev. D16, 707 (1977).

31. W. J. Marciano and A. I. Sanda, Phys. Lett. 67B, 303 (1977).

32. M. A. B. Bég and A. Sirlin, Phys. Rev. Lett. 38, 1113 (1977).

33. In a model proposed by J. D. Bjorken and S. Weinberg, Phys. Rev. Lett. 38, 622 (1977) muon-number violations result from the Higgs sector of the theory; therefore the magnitude of R_{eN} depends on the quark-Higgs couplings which are essentially free parameters.

34. The details of these calculations can be found in the appendix of ref. 13.

35. I have used the value $\sin^2\theta_W \simeq .24$ throughout this talk. C. f. M. Holder et al. "Study of Inclusive Neutral Current Interactions of Neutrinos and Antineutrinos, CERN preprint 1977.

36. The box diagrams involving two W exchanges contribute about 4%, while the diagrams with two Z's are almost insignificant for this model.

37. Q. Shafi and Ch. Wetterich, Phys. Lett. 69B 464 (1977). This paper presents a nice procedure for extracting this leading logarithm.

38. M.A.B. Bég, R. Budny, R. Mohapatra and A. Sirlin, Phys. Rev. Lett. 38, 1252 (1977).

WEAK INTERACTION MODELS WITH SPONTANEOUSLY BROKEN LEFT-RIGHT SYMMETRY*

Rabindra H. Mohapatra

Department of Physics, City College of City University

Of New York, New York, N. Y. 10031

ABSTRACT

We review the present status of weak interaction models with spontaneously broken left-right symmetry. The theoretical basis for asymptotic parity conservation, manifest left-right symmetry in charged current weak interactions, natural parity conservation in neutral currents and CP-violation in the context of $SU(2)_L \otimes SU(2)_R \otimes U(1)$ models are outlined in detail. Various directions for further research in the theoretical and experimental side are indicated.

I. INTRODUCTION

The V-A theory of weak interactions suggested by Marshak, Sudarshan[1] and subsequently by Feynman and Gell-Mann[2] twenty years ago provides an excellent phenomenological description of nearly all known charged current weak phenomena at low energies. Taken in conjunction with the standard parton model[3]

---

*Work supported in part by National Science Foundation grant No. Phy-77-03354 and the City University of New York PSC-BHE research award under grant No. 11687 and No. 11449.

assumptions about the hadron substructure, V-A theory also seems to provide an adequate description of all available high energy charged current data. The form in which this theory existed prior to the birth of modern gauge theories[4] was, however, mathematically quite unsatisfactory. It was beset with uncontrollable divergence difficulties.[5] With the advent of spontaneously broken gauge theories, this problem was cured. The $SU(2)_L \times U(1)$ gauge theory of Weinberg[6] and Salam[6] supplemented by the charm hypothesis to understand the suppression of $K_L \to \mu\bar{\mu}$ decay by Glashow, Illiopoulos and Maiani[7] was a milestone in the history of weak and electromagnetic interactions. The discovery of neutral currents and hadrons containing charmed particles have confirmed the basic ideas and some of the predictions of the $SU(2)_L \times U(1)$ model (to be called the "standard model" henceforth). There are, however, several unsatisfactory aspects of the model on both theoretical and phenomenological grounds. On the theoretical side

(a) the model, as it stands cannot support CP-violation[8] as observed in K^0-decays;

(b) observed parity violation in low energy weak interaction is put by hand into the theory. Indeed, it is perhaps a little odd that we are trying to unify the various forms of interactions, without in some way making their spacetime structures similar to start with.

On the phenomenological side,

(c) we have to accommodate more leptons[9] and more quark degrees of freedom[10]. This is of course easily done by adding more weak isopin doublets to the "standard" model.[11] The price we pay for this is that Cabibbo form of universality in such a model becomes highly unnatural owing to the proliferation mixing angles.

(d) More serious for the model and indeed for the gauge group

$SU(2)_L \otimes U(1)$ are the recent experimental findings[12] that in heavy atoms such as Bismuth no parity violating interactions appears to exist up to the level of $\sim(G_F/10)$. There is however a certain degree of theoretical uncertainty associated with the atomic calculations[13] that allowed us to draw these conclusions. But nevertheless, a potential threat to the standard model does indeed seem to exist.

A serious and elegant alternative to the standard $SU(2)_L \otimes U(1)$ model on the left-right symmetric weak gauge group $SU(2)_L \otimes SU(2)_R \otimes U(1)$ has recently been suggested by several of us.[14,15,16,17,18] It provides an interesting resolution of all the above shortcomings of the standard model and offers what appears to be an immensely more satisfactory framework for the description of weak and electromagnetic interactions.

An important distinguishing feature of this class of theories is that the weak interaction Lagrangian in common with strong and electromagnetic interactions has the property that it is invariant under space reflection. The observed parity non-invariance of low energy weak interactions in this class of theories is attributed to the idea that vacuum state of the theory is left-right asymmetric.[14,15] These theories, therefore, provide a conceptually different framework for the observed parity violation in low energy weak processes. It then follows that at very high energies, where all weak vector boson masses in the theory can be ignored, weak interactions should be parity invariant. The parity invariance would also reappear for a weakly interacting physical system in acute enviromental conditions such as high temperatures or high magnetic fields, undergoing a phase transition and returning to a symmetric phase. We will briefly touch on these questions at the end of this lecture.

At this point, we remark that the above concept of asymptotic parity conservation can be implemented via either an $SU(2) \times U(1)$ gauge group[19] or $SU(2)_L \otimes SU(2)_R \otimes U(1)$

gauge group or of course any generalization of the latter such as $SU(N)_L \otimes SU(N)_R \otimes G$. Coupled with the requirement of "natural" flavor conservation in weak neutral current interactions[20,21] to order G_F (excluding the Higgs boson interactions), "vector like" $SU(2) \otimes U(1)$ theories would predict pure vector neutral current interactions. This predicts the equality of the cross-section for neutrino and antineutrino deep inelastic inclusive scattering off nucleons ($\sigma^{\nu N}_{N.C.} = \sigma^{\bar{\nu}N}_{N.C.}$) as well as the elastic ν_μ and $\bar{\nu}_\mu$ scattering off electrons i.e. $\sigma^{\nu_\mu e} = \sigma^{\bar{\nu}_\mu e}$. The first relation is definitely in disagreement with the present data[22], while the situation with respect to the second one is not entirely clear.[23] Thus $SU(2) \otimes U(1)$ vector-like models are ruled out. We will soon see how deftly the $SU(2)_L \otimes SU(2)_R \otimes U(1)$ theory deals with this situation.

The $SU(2)_L \otimes SU(2)_R \otimes U(1)$ theory has two neutral Z-bosons, thereby allowing for a very rich structure for the neutral current interactions.[16,17,18] It turns out that in a certain limit (which can be implemented in a "natural" manner[24]), (i) the neutral current interactions are parity conserving[16,18] and (ii) the effective Hamiltonian for neutral current interactions of neutrinos has the same structure as predicted by the standard model.[16,18,25,26,27]

Another very interesting feature of this model is that it provides a link between the magnitude of CP-violation and P-violation.[15] CP-violation, therefore, does not remain any more as an isolated phenomenon as in most other theories.

Another advantage of this class of theories is their ability to generate (via discrete flavor symmetries) natural zeroth order relations[28] between the flavor mixing angles and quark masses, thus making it possible to predict magnitudes of Cabibbo angle and CP-phases using plausible values for quark masses.

This review is organized as follows: In Sec.(II), we discuss the general pattern of gauge symmetry breaking in left-right symmetric gauge theories and the physical manifestation of left-right symmetry in charged current weak interactions along with its phenomenological implications; in Sec. III, we discuss the general structure of neutral current interactions in these theories and the various interesting limiting cases; and a Higgs system that leads to a "naturally" parity-conserving neutral current structure and its implications; in Sec. IV, we discuss CP-violation in left-right symmetric models with four quarks and show how one is led to the concept of absolutely stable hadrons; in Sec. V, we discuss possible avenues of further research into left-right symmetry by investigating extended gauge groups such as $SU(2)_L \otimes SU(2)_R \otimes U(1)_L \otimes U(1)_R$ and $SU(3)_L \otimes SU(3)_R$, and superunifying groups that incorporate the strong interaction gauge group, i.e. color SU(3), and the left-right symmetric $SU(2)_L \otimes SU(2)_R \otimes U(1)$ group of weak and electromagnetic interactions are discussed. The groups touched on are SO(10) and $[SU(4)]^4$. In Sec VI, we present a very brief list of possible experiments to test first the general idea of spontaneously broken left-right symmetry and secondly, the hypothesis of parity conserving neutral currents that arises for certain choices of symmetry breaking patterns in these theories. In Sec. VII, we conclude with some additional general remarks.

II. SYMMETRY BREAKING, FERMION MASSES AND CHARGED CURRENT WEAK INTERACTIONS

In this section, we outline the general theoretical framework of left-right symmetric models, which hold for arbitrary even numbers of quarks and lepton flavors. As stressed earlier, one appealing feature of these theories is that they provide a conceptually different approach to the V-A theory of weak interactions in the sense that the basic weak interaction Lagrangian is left-right symmetric, while the vacuum state is not. The observed left-right asymmetry in low-energy weak interactions owes its origin to the parity-asymmetric vacuum. At high enough energies, where this vacuum asymmetry is not important or in the high temperature phase (like the early stages of the universe), where owing to a phase transition vacuum symmetry is restored, weak interactions ought to exhibit no trace of parity violation. In this section, we will explicitly construct the Higgs mechanism that will enable us to achieve the desired breakdown of symmetry and will delineate the various natural alternative paths the theory will take subsequent to spontaneous breakdown. Examples of the various alternatives will be presented for four, six and eight quark flavors. We stress that the discussions of this section apply to general left-right symmetric groups of the form $G_L \otimes G_R \otimes U(1)$ with slight modification. We will however restrict ourselves to the case of the $SU(2)_L \times SU(2)_R \times U(1)$ gauge group.

Let us suppose that there are 2N quark flavors; N of them (called p_i) have electric charge $Q = +\frac{2}{3}$ and the remaining N have electric charge $Q = -\frac{1}{3}$. The fields p_i and n_i are weak eigenstates. We assign them to doublets under the group $SU(2)_L \otimes SU(2)_R \otimes U(1)$ in a left-right symmetric manner as follows:

$$\psi_{iL} = \begin{pmatrix} p_{iL} \\ n_{iL} \end{pmatrix} i = 1, 2, \ldots\ldots N \quad \left(\frac{1}{2}, 0, \frac{1}{3}\right)$$

$$\psi_{iR} = \begin{pmatrix} p_{iR} \\ n_{iR} \end{pmatrix} i = 1, 2, \ldots\ldots N \quad \left(0, \frac{1}{2}, \frac{1}{3}\right) \qquad \text{(II.1)}$$

The numbers in the parenthesis (I_L, I_R, γ) stand for representation labels with respect to the groups $SU(2)_L \times SU(2)_R \times U(1)$ respectively. We will ignore the leptons for a while except for the remark that they are also assigned to doublets in an identical fashion, with however a different U(1) quantum number ($\gamma = 1$).*

Secondly we note the left-right symmetry of the fermion multiplets i.e.

$$L \leftrightarrow R: \quad \psi_{iL} \leftrightarrow \psi_{iR} \tag{II.2a}$$

We further let the gauge bosons transform under left-right conjugation as follows:

$$\vec{W}_L \leftrightarrow \vec{W}_R \tag{II.2b}$$

We then require the gauge invariant Lagrangian to be invariant under this discrete $L \leftrightarrow R$ transformation. This then leads to the equality of the two SU(2) gauge couplings $g_L = g_R \equiv g$. The U(1) gauge coupling is denoted g'. Also, we note that the electric charge is expressed in terms of the group generators as follows:

$$Q = I_{3L} + I_{3R} + \frac{Y}{2} \tag{II.3}$$

where $I_{3L,R}$ stand for the third components of the weak isospins, $SU(2)_{L,R}$. The general weak gauge interaction of the quarks can be written as:

$$\mathcal{L}_{WK} = \frac{i}{2}\left[g \sum_{i=1,\dots.N} \left\{ \bar{\psi}_{iL} \gamma_\mu \vec{\tau}\cdot\vec{W}_{\mu L}\, \psi_{iL} + \bar{\psi}_{iR} \gamma_\mu \vec{\tau}\cdot\vec{W}_{\mu R}\, \psi_{iR} \right\} + \frac{g'}{3} \sum_i \bar{\psi}_i \gamma_\mu \psi_i B_\mu \right] \tag{II.4}$$

---

* $\begin{pmatrix}\nu \\ e^-\end{pmatrix}_L$, $\begin{pmatrix}\nu' \\ \mu^-\end{pmatrix}_L$ transform as (1/2, 0, -1) and $\begin{pmatrix}\nu \\ e^-\end{pmatrix}_R$, $\begin{pmatrix}\nu' \\ \mu^-\end{pmatrix}_R$ as (0, 1/2, -1). This implies that neutrinos have arbitrary mass.

Prior to spontaneous breakdown of the gauge symmetry, the gauge bosons are massless. It is therefore clear the $\mathcal{L}_{\omega k}$ in eqn. (II.4) conserves parity prior to spontaneous breakdown. Also, since ψ_{iL} and ψ_{iR} transform as doublets under different groups, all fermions are massless at this stage. We will now trigger the spontaneous breakdown. First, we discuss the Higgs multiplets that generate the mass of the fermions and the resulting structure for the charged current weak interaction.[24] Next we will discuss the Higgs particles responsible for the structure of the neutral currents.

IIa. Manifest Left-Right Symmetry of Charged Current Weak Interactions[29]

The Higgs multiplet that generates fermion masses belongs to a representation $(\frac{1}{2}, \frac{1}{2}, 0)$ under the gauge symmetry. We take only one such multiplet and denote it by Φ. Under left-right conjugation

$$\Phi \leftrightarrow \Phi^\dagger \qquad \text{(II.5)}$$

We hasten to add that it is possible to define other left-right transformations of Φ than the one in Eqn. (II.5). We will call these "unorthodox" $L \leftrightarrow R$ transformations and discuss them separately. Associated with Φ is of course the complex conjugate representation

$$\tilde{\Phi} = \tau_2 \Phi^* \tau_2$$

where τ_2 is the Pauli matrix. The most general gague invariant fermion Yukawa coupling involving ϕ and $\tilde{\phi}$ is:

$$\mathcal{L}_Y = \sum_{i,j=1,\dots N} \left(f_{ij} \bar{\psi}_{iL} \Phi \psi_{jR} + h_{ij} \bar{\psi}_{iL} \tilde{\Phi} \psi_{jR} \right) + \text{h.c.} \qquad \text{(II.6)}$$

We choose f_{ij} and h_{ij} to be real. We now require $\mathcal{L}_Y$ to be left-right symmetric, which then leads to:

$$f_{ij} = f_{ji} \text{ and } h_{ij} = h_{ji} \tag{II.7}$$

Spontaneous symmetry breaking takes place when Φ develops non-zero vacuum expectation values:

$$\langle \Phi \rangle = \begin{pmatrix} z & o \\ o & z' \end{pmatrix} \tag{II.8}$$

In general, z and z' are complex for a certain range of parameters in the potential (i.e. $z = \kappa e^{i\alpha}$, $z' = \kappa' e^{i\alpha'}$, where κ, κ', α, α' are real) and are real for another range (i.e. $\alpha = \alpha' = 0$). The mass matrix for $Q = +\frac{2}{3}$ quarks (p_i) denoted by $M^{(+)}$ (corresponding to weak $I_{3L,R} = +\frac{1}{2}$) and that for $Q = \frac{1}{3}$ quarks (n_i) denoted by $M^{(-)}$ (corresponding to weak $I_{3L,R} = -\frac{1}{2}$) are given by:

$$M^{(+)} = f_{ij} z + h_{ij} z'^{*}$$

$$\text{and} \quad M^{(-)} = f_{ij} z' + h_{ij} z^{*} \tag{II.9}$$

From Eqn. (II.7), it follows that $M^{\pm}$ are symmetric, i.e.

$$M^{(\pm)} = M^{(\pm)^T} \tag{II.10}$$

So for the case where z and z' are real, $M^{\pm}$ are real symmetric matrices. They are therefore diagonalized by a single orthogonal transformation $O^{(\pm)}$ on both left- and right-handed quarks[29a] i.e.

$$O^{(\pm)} M^{(\pm)} O^{(\pm)^T} = M_d^{(\pm)} \tag{II.11}$$

The physical quark states which are eigenstates of the mass matrix are then given by

$$p_{iL,R} = O^{(+)} \begin{pmatrix} u \\ c \\ t \\ \vdots \end{pmatrix} \tag{II.12}$$

and similarly for the n_i sector. The important point to note is that the left and right-handed mixing angles are the same. Thus, at the tree level, the physical left- and right- handed charged currents have identical flavor content or in general:

$$J^+_{\mu,L} = J^+_{\mu,R}\ (\gamma_5 \to -\gamma_5) + \text{finite, computable corrections } O(\alpha). \tag{II.13}$$

Eqn. (II.13) represents the full content of the idea of manifest left-right symmetry as expounded in ref. 29. We thus see that due to "manifest left-right symmetry" being natural, the number of fermion mixing angles gets reduced by half. Thus for four quarks, even though we have both left and right-handed charged currents, there is only one mixing angle, the Cabibbo angle, and the extended GIM[7] structure for currents of both chirality. Explicitly, for the case of four quarks, the charged current quark weak interaction would look like:

$$\mathcal{L}^{c.c.}_{WK} = \frac{ig}{\sqrt{2}} W^+_{\mu_L} \{\bar{U}_L \gamma_\mu (d_L \cos\theta_c + S_L \sin\theta_c) + \bar{C}_L \gamma_\mu (-d_L \sin\theta_c + S_L \cos\theta_c)\}$$
$$+ \frac{ig}{\sqrt{2}} W^+_{\mu R} \{\bar{U}_R \gamma_\mu (d_R \cos\theta_c + S_R \sin\theta_c) + \bar{C}_R \gamma_\mu (-d_R \sin\theta_c + S_R \cos\theta_c)\} + \text{h.c.} \tag{II.14}$$

Finally, we would like to note that, if f_{ij} and h_{ij} were complex, and z and z' were real, the left-right symmetry would require $f_{ij} = f^*_{ji}$ and $h_{ij} = h^*_{ji}$; as a result of this, the mass matrices would be hermitean, in which case we have $U_L = U_R$, leading again to manifest left-right symmetry. It is however doubtful that this case will be "natural" since complex Yukawa couplings in general will tend to induce complex ϕ^4 type couplings which in turn would keep z and z' from being real.

IIb. Case of Complex <Φ> and Pseudo-manifest Left-Right Symmetry

The mass matrix in this case can be written as:

$$M_{ij} = f_{ij}\, Z + h_{ij}\, Z'^{*} \quad \text{(II.15)}$$

Therefore, we still have, $M_{ij} = M_{ji}$; however, M is now complex. In general M is diagonalized by a bi-unitary transformation:

$$M_D = U_L M U_R^{-1} \quad \text{(II.16)}$$

Since, $M_D = M_D^T$ and $M = M^T$, it follows that,

$$U_L = U_R^{*} K \quad \text{(II.17)}$$

where K is a diagonal unitary matrix. This implies that when K can be absorbed by redefinition of fields, $U_{L,R}$ are parametrized by real angles $\theta^i_{L,R}$(i=1....) and phases $\delta^i_{L,R}$(i=1....) and one gets

$$\theta^i_L = \theta^i_R \text{ and } \delta^i_L = -\delta^i_R \quad \text{(II.18)}$$

Thus, even for complex <φ>, there is almost manifest left-right symmetry in the charged currents. We call this "pseudo-manifest" left-right symmetry. The situation does not change even when we include several φ-type multiplets to generate the fermion masses as long as their left-right transformation property is the same as in Eq. (II.4).

Thus, we see that, if the Lagrangian conserves CP prior to spontaneous breakdown of the gauge symmetry and has only one complex multiplet $\Phi \equiv (\frac{1}{2}, \frac{1}{2}, 0)$, then under the conventional definition of left-right symmetry for Φ i.e.

$$\Phi \leftrightarrow \Phi^{+} \quad \text{(II.19)}$$

one always ends up with equal left and right flavor-mixing angles. It thus appears that with a single Φ - multiplet it is impossible to have a "naturally" left-right symmetric theory with left and right mixing angles different, provided the theory is CP-conserving prior to spontaneous breakdown.

IIc. Unorthodox Left-Right Symmetry and Relation Between Up and Down Quark Masses

As mentioned before, we will work with a single ϕ and real coupling constants. However, we will choose an unorthodox definition of left-right symmetry for the Higgs sector i.e.

$$\Phi \leftrightarrow \tilde{\Phi}^{+} \tag{II.20}$$

Then, left-right symmetry of the fermion Yukawa couplings demands that

$$f_{ij} = h_{ji} \tag{II.21}$$

It then follows using Eqs. (II.9), and (II.21), that for real $<\phi>$,

$$M^{(+)^T} = M^{(-)} \tag{II.22}$$

i.e. the mass matrices for the $I_3 = +\frac{1}{2}$ and $I_3 = -\frac{1}{2}$ fermions are related. Eq. (II.22) leads to the following relation between the up and down fermion masses i.e.

$$(M^{(+)}\, M^{(+)^+})^T = M^{(-)^+}\, M^{(-)} \tag{II.23}$$

from which it follows that the eigenvalues of the $I_3 = +\frac{1}{2}$ and $I_3 = -\frac{1}{2}$ sectors are equal i.e.

$$m_u = m_d;\; m_c = m_s;\ldots. \tag{II.24}$$

This is physically unacceptable. Thus, we prove that with a single fermion mass generating Higgs boson ϕ, and with real coupling constants, three distinct physical cases result subsequent to spontaneous breakdown:

(i) Manifest left-right symmetry i.e. $U_L = U_R$;

(ii) Pseudo-manifest left-right symmetry: $U_L = U_R^* e^{i\theta}$

(iii) $m_u = m_d$, $m_c = m_s$, ...

We reject the third possibility and find the second case to be most interesting since it can also lead to non-trivial CP-violation even in the four quark case.

IId. Non-manifest Left-Right Symmetric Case

In order to obtain a charged current weak interaction Lagrangian which does not have manifest left-right symmetry, i.e. $\theta_L \neq \theta_R$, we have to make the fermion mass-matrix non-symmetric and non-hermitean. This can be done in two ways:

(a) Complex Yukawa Coupling Constants and Complex $\langle\phi\rangle$. If we choose the Yukawa couplings constants f_{ij} and h_{ij} to be complex, then for the Lagrangian to be left-right symmetric, we must have

$$f_{ij} = f^*_{ji}$$

$$h_{ij} = h^*_{ji} \qquad \text{(II.25)}$$

The fermion mass-matrix with complex $\langle\phi\rangle$ turns out in this case to be neither symmetric nor Hermitian. Thus, the left and right unitary transformations that diagonalize this are in general not related and thus, it is possible to have the left-right symmetry completely disappear from charged weak-interactions. It is important to note that in this case, departure from manifest left-right symmetry is proportional to the magnitude of CP-violation.

(b) Generalized Left-Right Symmetry and more than one Φ. Another way to make the left and right flavor mixing angles different is to have at least two complex multiplets with representation content $(\frac{1}{2}, \frac{1}{2}, 0)$ i.e. Φ_1 and Φ_2. Define

$$\tilde{\Phi}_m = \tau_2 \Phi_m^* \tau_2 \quad m = 1, 2.$$

The most general fermion Yukawa coupling to ϕ_i's is

$$\mathcal{L}_Y = \sum_{i,j;m} \left(f_{ij}^{(m)} \bar{\psi}_{iL} \Phi_m \psi_{jR} + h_{ij}^{(m)} \bar{\psi}_{iL} \tilde{\Phi}_m \psi_{jR} \right) + \text{h.c.} \qquad \text{(II.26)}$$

Let us assume that under left-right conjugation, the ϕ_m's transform as follows:

$$\Phi_1 \leftrightarrow \tilde{\Phi}_2^{\dagger}$$

$$\Phi_2 \leftrightarrow \tilde{\Phi}_1^{\dagger} \qquad \text{(II.27)}$$

The condition for $\mathcal{L}_Y$ to be invariant under this transformation is:

$$f_{ij}^{(1)} = h_{ji}^{(2)}; \; f_{ij}^{(2)} = h_{ji}^{(1)} \qquad \text{(II.28)}$$

After spontaneous breakdown, let the ϕ_m's acquire the following vacuum expectation values:

$$\langle \phi_m \rangle = \begin{pmatrix} Z_m & o \\ o & Z_m' \end{pmatrix} \qquad \text{(II.29)}$$

Z_m and Z_m^1 could be real or complex. The fermion mass matrices are these: for up-quarks:

$$M_{ij}^{(+)} = \sum_{m=1,2} \left(f_{ij}^{(m)} Z_m + h_{ij}^{(m)} Z_m'^{*} \right)$$

for down-quarks:

$$M_{ij}^{(-)} = \sum_{m=1,2} \left(f_{ij}^{(m)} Z_m' + h_{ij}^{(m)} Z_m^{*} \right) \qquad \text{(II.30)}$$

Thus, we see that, the mass matrices for up and down-like fermions are independent and non-symmetric. Thus, in general, one will need biorthogonal transformations of type (for real $\langle\phi\rangle$)

$$O_L M O_R^{-1} = M_d \qquad \text{(II.31)}$$

to make it diagonal. This will lead to different left and right flavor mixing angles.

From the above discussion, it is clear that there are two distinct realizations of left-right symmetry in charged current weak interactions; one where the symmetry is manifest and the other where left and right-handed fermion mixing angles are free. These two cases have distinct physical implications, some of which are: (a) in the case of manifest left-right symmetry, there cannot be right-handed charged currents connecting the light and heavy quarks (e.g. $(\bar{u}b)_R$ type current) with significant strength since observations in the low energy regime prevent such currents in the left-handed charged currents. Thus, the so called "high-y anomaly" dies a "natural" death in this model! (b) in models with manifest left-right symmetry, low energy data requires the right-handed charged W-bosons ($W_R^{\pm}$) to be much heavier than the left-handed charged W-bosons. On the other hand, in non-manifest theories, by adjusting the right-handed fermion mixing angles, one could have the right-handed charged currents connect only the light-heavy pairs (such as $(\bar{u}b)_R, (\bar{t}d)_R, \ldots$ where u,d,s,c are taken to denote the four light quark types and t,b,gh....to denote the heavier quark flavors). In that case, low energy weak interactions impose no restriction on the mass of the $W_R^{\pm}$. As we will discuss later, $m_{W_L^{\pm}}$ will be equal to, bigger than or smaller than the $m_{W_R^{\pm}}$ depending whether the parity violation in atomic physics to order G_F is zero or non-zero.

We now make the following observation that, at this state the gauge symmetry has been broken down to the level $U(1)_{L+R} \otimes U(1)$ from the original $SU(2)_L \otimes SU(2)_R \otimes U(1)$ as can be seen by computing the gauge boson masses. In the case of a single ϕ-multiplet for the charged W's we have

$$\begin{array}{c c} & \begin{array}{cc} W_L^+ & W_R^+ \end{array} \\ \begin{array}{c} W_L^- \\ \\ W_R^- \end{array} & \begin{pmatrix} \frac{g^2}{2}(\kappa^2+\kappa'^2) & -g^2\kappa\kappa' e^{i\beta} \\ & \\ -g^2\kappa\kappa' e^{-i\beta} & \frac{g^2}{2}(\kappa^2+\kappa'^2) \end{pmatrix} \end{array} \qquad (II.32)$$

where $\beta = -\alpha + \alpha'$. Out of all the neutral gauge bosons, the one linear combination i.e. $Z_A = \frac{1}{\sqrt{2}}(W_{3L}-W_{3R})$ picks up a mass $\frac{g^2}{2}(\kappa^2+\kappa'^2)$. The two parity even eigenstates involving gauge bosons i.e. $(W_{3L}+W_{3R})/\sqrt{2}$ and B remain massless resulting in an as yet unbroken U(1) X U(1) symmetry. We also note that, since in the case of manifest left-right symmetry, we must have $m_{W_R^\pm} \gg m_{W_L^\pm}$, we must add Higgs multiplets other than ϕ; because, (II.32) leads to (for $\beta=0$), eigenstates,

$$W_1^+ = \frac{1}{\sqrt{2}}(W_L^+ + W_R^+); \; m_{W_1}^2 = \frac{g^2}{2}(\kappa-\kappa')^2$$

$$W_2^+ = \frac{1}{\sqrt{2}}(W_L^+ - W_R^+); \; m_{W_2}^2 = \frac{g^2}{2}(\kappa+\kappa')^2 \qquad (II.33)$$

A little thought can convince the reader that, eq. (11.33) would lead to conflict with known low energy phenomenology. To avoid this situation, we introduce a pair of Higgs multiplets $\delta_L \equiv (1, 0, 0)$ and δ_R (0, 1, 0) such that, $\langle\delta_L\rangle = 0$ and $\langle\delta_R\rangle = \binom{0}{b}$. The charged boson mass matrix in the presence of δ takes the following form: (for $\beta=0$)

$$\begin{array}{c c} & \begin{array}{cc} W_L^+ & W_R^+ \end{array} \\ \begin{array}{c} W_L^- \\ \\ W_R^- \end{array} & \begin{pmatrix} \frac{g^2}{2}(\kappa^2+\kappa'^2) & -g^2\kappa\kappa' \\ & \\ -g^2\kappa\kappa' & \frac{g^2}{2}(\kappa^2+\kappa'^2+8b^2) \end{pmatrix} \end{array} \qquad (II.34)$$

The eigenstates of this matrix are:

$$W_1^+ = \text{Cos}\zeta\, W_L^+ + \text{Sin}\zeta\, W_R^+$$

$$W_2^+ = -\text{Sin}\zeta\, W_L^+ + \text{Cos}\zeta\, W_R^+ \qquad \text{(II.35)}$$

where

$$\tan 2\xi \simeq \frac{2\kappa\kappa'}{8b^2}$$

and

$$m_{W_1}^2 \simeq \frac{g^2}{2}(\kappa^2+\kappa'^2); \; m_{W_2}^2 \simeq \frac{g^2}{2}(\kappa^2+\kappa'^2+8b^2) \qquad \text{(II.36)}$$

At this stage, choosing $b \gg \kappa, \kappa'$, we get $m_{W_2} \gg m_{W_1}$ and also a small mixing angle.

IIe. Manifest Left-Right Symmetry and Limits on the Mass of $W_R^\pm$

At this stage, it is apparent that the low energy effective charged current weak Hamiltonian in the case of manifest left-right symmetry can be parameterized in terms of two parameters:[29] tan ζ and m_{W_2}/m_{W_1}.

$$\mathcal{L}_{WK}^{c.c.} = \frac{ig}{\sqrt{g}}[(V-A)_\mu W_{L_\mu}^+ + (V+A)_\mu W_{R\mu}^+] + h.c. \qquad \text{(II.37)}$$

and using Eq.(II.35), we can write, for low q^2 the effective Hamiltonian:

$$H_{WK}^{c.c.} = \frac{G_F}{\sqrt{2}}[V_\mu^+ V_\mu + \eta_{AA} A_\mu^+ A_\mu + \eta_{AV}(V_\mu^+ A_\mu + A_\mu^+ V_\mu)] \qquad \text{(II.38)}$$

where,

$$\frac{G_F}{2} = \frac{g^2}{8}\left[\frac{(\cos\zeta+\sin\zeta)^2}{m_{W_1}^2} + \frac{(\cos\zeta-\sin\zeta)^2}{m_{W_2}^2}\right] \qquad (II.39)$$

$$\eta_{AA} = \frac{\varepsilon^2 m_{W_2}^2 + m_{W_1}^2}{m_{W_2}^2 + \varepsilon^2 m_{W_1}^2}$$

$$\eta_{VA} = \frac{-\varepsilon(m_{W_2}^2 - m_{W_1}^2)}{m_{W_2}^2 + \varepsilon^2 m_{W_1}^2}$$

$$\text{and} \qquad \varepsilon = \frac{1-\tan\zeta}{1+\tan\zeta} \qquad (II.40)$$

The standard low energy parameters describing β- decay and μ-decay can be expressed in terms of η_{AA} and η_{AV} (see ref. 29). Note that in the V-A limit, $\eta_{AA} = +1$ and $\eta_{AV} = -1$. So, any deviation from these values will indicate the presence of V+A currents in addition to the conventional V-A ones. The present experimental values and the errors allow one[29] to set bounds on the parameters η_{AA} and η_{AV}, which in turn can be translated into bounds on parameters tan ζ and $(m_{W_2}/m_{W_1})^2$. We get using the experimental values of the Michel parameter and the electron polarization in Gamow-Teller transitions the following bounds:

$$\left(\frac{m_{W_2}}{m_{W_1}}\right) \geqslant 2\cdot 76$$

and

$$|\tan \zeta| \leq .06 \qquad \text{(II.40a)}$$

Clearly, determination of the low energy parameters such as ρ, $p_\pi(\mu^+)$, etc. with higher precision will be of decisive significance for theories with manifest left-right symmetry. For a somewhat modified bound, see ref. 30.

IIf. Example of a Theory With Non-Manifest Left-Right Symmetry

In this section, we would like to briefly touch on a model, where the left and right-handed mixing angles are completely different. We would like to maintain lepton-quark symmetry. The most interesting constraints on model building arises when the $W_L^\pm$ and $W_R^\pm$ are nearly degenerate in mass. In this case, the right-handed currents can couple only light and heavy fermions in order not to conflict with low energy phenomenology. It is, then, clear that one must have at least eight leptons with the following assignment for left and right-handed doublets:

$$\begin{pmatrix}\nu\\ e^-\end{pmatrix}_L \begin{pmatrix}\nu'\\ \mu^-\end{pmatrix}_L \begin{pmatrix}\nu_\tau\\ \tau^-\end{pmatrix}_L \begin{pmatrix}E^0\\ E^-\end{pmatrix}_L$$

$$\begin{pmatrix}\nu\\ E^-\end{pmatrix}_R \begin{pmatrix}\nu'\\ \tau^-\end{pmatrix}_R \begin{pmatrix}\nu'_\tau\\ \mu^-\end{pmatrix}_R \begin{pmatrix}E'\\ e^-_0\end{pmatrix}_R \qquad \text{(II.41)}$$

The τ^- could be identified as the heavy lepton observed at SLAC[9]. Arbitrary mixings between E^- and τ^- or ν'_τ and $E^{0'}$ are of course allowed in the right-hand sector. For example a mixing of the type $\nu'_\tau = \nu_\tau \cos\beta + E^{0'} \sin\beta$ and $E^0 = -\nu_\tau \sin\beta + E^0 \cos\beta$ could give rise to trimuon events[30a] in neutrino interactions.

As far as the quark sector is concerned, lepton-quark symmetry would dictate that we choose eight quark flavors, called u, d, s, c (light flavors) t, b, g, h (heavy flavors). The assignment to left-handed doublets are as follows:

$$\begin{pmatrix} u \\ d(\theta_c) \end{pmatrix}_L \begin{pmatrix} c \\ s(\theta_c) \end{pmatrix}_L \begin{pmatrix} t \\ b(\alpha) \end{pmatrix}_L \begin{pmatrix} h \\ g(\alpha) \end{pmatrix}_L \tag{II.42}$$

Of course, one could have chosen a more general mixing scheme in which the heavy quarks mix with light quarks. But, Carribo universality requires that light-heavy mixing angles in the left-handed sector must be very small. Coming to the right-handed quarks, we may choose their multiplet structure as follows:

$$\begin{pmatrix} u \\ b \end{pmatrix}_R \begin{pmatrix} c \\ s' \end{pmatrix}_R \begin{pmatrix} t \\ d' \end{pmatrix}_R \begin{pmatrix} h \\ g' \end{pmatrix}_R \tag{II.43}$$

where s', d' and g' are linear combinations of s, d, g. Other allowed mixing schemes could also be chosen in place of the one displayed in Eq. (II.43). They must, however, be such as to exclude $(\bar{u}d)_R$ type current.

Let us, now, look at the mass of the charged W-bosons in such a theory. As already mentioned, since $W_R^{\pm}$ need not be heavy in this case, we can dispose of the Higgs multiplets $\delta_L(1, 0, 0)$ and $\delta_R(0, 1, 0)$ multiplets. To decide the $W^{\pm}$ masses, we only have to look at Eq. (II.32). To get almost degenerate masses for $W_L^{\pm}$ and $W_R^{\pm}$, we must set $\kappa' \ll \kappa$ as is obvious from Eq. (II.33). For $\kappa'=0$, we get

$$m_{W_L^{\pm}} = m_{W_R^{\pm}} \tag{II.44}$$

It is then easy to convince oneself that there is no left-right mixing in the effective Hamiltonian, H_W. This, for example, implies that terms like $(\bar{\nu}\mu-)_L(\bar{b}u)_R$ which could give rise to the so called "high y-anomaly" do not exist in the weak Hamiltonian. Thus, in

Ψ

such models, the b-quark mass could be chosen to be very light (≈5 GeV).

III. GENERAL STRUCTURE OF NEUTRAL CURRENTS IN LEFT-RIGHT SYMMETRIC GAUGE THEORIES

In the last section, enough Higgs mesons were introduced so as to break the gauge group $SU(2)_L \otimes SU(2)_R \otimes U(1)$ down to the level of $U(1)_{L+R} \otimes U(1)$. Of the three neutral gauge bosons of the theory W_{3L}, W_{3R} and B, only one linear combination $Z_A = \frac{1}{\sqrt{2}} (W_{3L} - W_{3R})$ had acquired mass due to the ϕ-multiplet. Two other combinations $(W_{3L} + W_{3R}) \frac{1}{\sqrt{2}}$ and B (which incidentally are parity-even eigenstates) remained massless. In this section, we will introduce the rest of the symmetry breaking so that we are ultimately left with only one massless gauge boson corresponding to electromagnetism. There are two most economical ways of achieving this. The resulting physics is of course different in both cases. We will therefore discuss them separately. In the first case, one introduces[15, 16] weak doublets: $\chi_L(\frac{1}{2}, 0, 1)$ and $\chi_R(0, \frac{1}{2}, 1)$ and give non-zero vacuum expectation values to each of them. In the second case,[24] one takes a multiplet $\rho \equiv (\frac{1}{2}, \frac{1}{2}, -2)$ and breaks the rest of the unbroken symmetry. In the first case, in the tree level, there is an arbitrary amount of parity violation in the neutral current weak Hamiltonian, whereas in the second case, at the tree level no parity violation is permissible in the neutral current Hamiltonian. In the latter case, we say that weak neutral currents conserve parity "naturally", while in the former case, they do not. The advantage of having natural parity conservation is that parity violating effects can in that case be predicted[31]. We will discuss this and compare the predictions of left-right symmetric models for atomic P-violation with other models.

IIIa. Symmetry Breaking by Doublet Higgs

In this subsection, we discuss symmetry breaking by means of the $\chi_L(\frac{1}{2}, 0, 1)$ and $\chi_R(0, \frac{1}{2}, 1)$ multiplets. We choose

$$<\chi_L> = \begin{pmatrix} 0 \\ \lambda_L \end{pmatrix}; \; <\chi_R> = \begin{pmatrix} 0 \\ \lambda_R \end{pmatrix} \qquad \text{(III.1)}$$

Including the effect of $\phi(\frac{1}{2}, \frac{1}{2}, 0)$ discussed in the previous section, we can write the neutral boson mass matrix as follows:

$$\begin{array}{c} \\ W_{3L} \\ W_{3R} \\ B \end{array} \begin{array}{c} \begin{array}{ccc} W_{3L} & W_{3R} & B \end{array} \\ \begin{pmatrix} \frac{g^2}{2}(\kappa^2+\kappa'^2+\lambda_L^2) & -\frac{g^2}{2}(\kappa^2+\kappa'^2) & -\frac{1}{2}gg'\lambda_L^2 \\ -\frac{g^2}{2}(\kappa^2+\kappa'^2) & \frac{g^2}{2}(\kappa^2+\kappa'^2+\lambda_R^2) & -\frac{1}{2}gg'\lambda_R^2 \\ -\frac{gg'}{2}\lambda_L^2 & -\frac{1}{2}gg'\lambda_R^2 & \frac{g'^2}{2}(\lambda_L^2+\lambda_R^2) \end{pmatrix} \end{array}$$

To express the eigenstates and eigenvalues of this matrix in a tractable fashion, we define

$$\frac{1}{\sqrt{2}}\tan\theta = g'/g$$

$$\varepsilon = \frac{\kappa^2+\kappa'^2}{\kappa^2+\kappa'^2+\frac{\lambda_L^2+\lambda_R^2}{2}}$$

$$\sigma = \left(\frac{\lambda_L^2-\lambda_R^2}{2}\right)\left|\kappa^2+\kappa'^2+\frac{\lambda_L^2+\lambda_R^2}{2}\right|^{-1} \qquad \text{(III.2)}$$

$$A = \frac{\sin\theta}{\sqrt{2}}(W_{3L}+W_{3R})+\cos\theta\; B$$

$$Z_V = \frac{\cos\theta}{\sqrt{2}}(W_{3L}+W_{3R})-\sin\theta\; B$$

$$Z_A = \frac{1}{\sqrt{2}}(W_{3L}-W_{3R}) \qquad \text{(III.3)}$$

The eigenstates are

$$Z_1 = -\sin\phi\ Z_V + \cos\phi\ Z_A$$

$$Z_2 = \cos\phi\ Z_V + \sin\phi\ Z_A \qquad \text{(III.4)}$$

where $\tan 2\phi = \dfrac{-2\sigma \sec\theta}{(1-\varepsilon)\sec^2\theta - (1+\varepsilon)}$ (III.5)

The eigenvalues are:

$$m^2_{2,1} = \frac{g^2}{4}\left(\kappa^2+\kappa'^2+\frac{\lambda_L^2+\lambda_R^2}{2}\right)\left[\sec^2\theta(1-\varepsilon)+(1+\varepsilon)\pm\sqrt{\{(1+\varepsilon)-\sec^2\theta(1-\varepsilon)\}^2+\frac{4\sigma^2}{\cos^2\theta}}\right] \qquad \text{(III.6)}$$

For the sake of comparison, we present the mass of the $W_L^{\pm}$ for the case of manifest left-right symmetry,

$$m^2_{W_L^{\pm}} \simeq \frac{g^2}{2}(\kappa^2+\kappa'^2+\lambda_L^2) \qquad \text{(III.7)}$$

Using (III.7), (III.6) can be rewritten as

$$m^2_{2,1} = \frac{m^2_{W_L^+}}{2(1+\sigma)}\left[(1-\varepsilon)\sec^2\theta+(1+\varepsilon)\pm\sqrt{\{(1-\varepsilon)\sec^2\theta+(1+\varepsilon)\}^2+4\sigma^2\sec}\,\right] \qquad \text{(III.8)}$$

The neutral current interaction Lagrangian $\mathcal{L}$ in this case can be written as:

$$\mathcal{L}^{N.C.}_{WK} = -\frac{ig}{\sqrt{2}\cos\theta}(\cos\phi\ Z_{2\mu} - \sin\phi\ Z_{1\mu})\sum_{i=1,\cdots N}\bar{\psi}_i\gamma_\mu\{I^{i'}_{3_L} - I^{i'}_{3_R} - Q^i\sin^2\theta\}\psi_i$$

$$-\frac{ig}{\sqrt{2}}(\sin\phi Z_{2\mu}+\cos\phi Z_{1\mu})\sum_{i=1,\cdots N}\bar{\psi}_i\gamma_\mu(I^{i'}_{3L}-I^{i'}_{3R})\psi_i$$

where

$$I'^{i}_{3{L \atop R}} \equiv I^{i}_{3{L \atop R}}(1\pm\gamma_5)\cdot\frac{1}{2} \qquad \text{(III.9)}$$

Several things are clear from Eq. (III.9).

(i) The effective neutral current Hamiltonian violates parity for general ϕ.

(ii) For the case $\lambda_L=\lambda_R$, Eq. (III.5) yield $\phi=0$ in which case, Z_V and Z_A become exact eigenstates; and neutral currents conserve parity. This case is so interesting that, we will study it in detail later. Before we do that, let us study the general case in somewhat greater depth. The effective weak neutral current Hamiltonian can be written as:

$$H^{N.C.}_{WK} = H^{(+)} + H^{(-)} \qquad \text{(III.10)}$$

where

$$H^{(+)} = \frac{g^2}{2\cos^2\theta}\left(\frac{\cos^2\phi}{m_2^2}+\frac{\sin^2\phi}{m_1^2}\right)V_\mu V_\mu$$

$$+ \frac{g^2}{2}\left(\frac{\sin^2\phi}{m_2^2}+\frac{\cos^2\phi}{m_1^2}\right)A_\mu A_\mu \qquad \text{(III.11)}$$

$$H^{(-)} = \frac{g^2\cos\phi\sin\phi}{2\cos\theta}\left(\frac{1}{m_2^2} - \frac{1}{m_1^2}\right)(V_\mu A_\mu+A_\mu V_\mu) \qquad \text{(III.12)}$$

where

$$V_\mu = \sum_{i=1,\ldots N}\bar{\Psi}_i\gamma_\mu\{\tfrac{1}{2}(I^i_{3L}+I^i_{3R}) + \frac{\gamma_5}{2}(I^i_{3L}-I^i_{3R})-Q^i\sin^2\theta\}\Psi_i$$

$$A_\mu = \frac{1}{2}\sum_{i=1,\ldots N}\bar{\Psi}_i\gamma_\mu\{(I^i_{3L}-I^i_{3R}) + \gamma_5(I^i_{3L}+I^i_{3R})\}\Psi_i \qquad \text{(III.13)}$$

$H^{(+)}$ is the parity conserving part and $H^{(-)}$ is the parity violating part.

$$\text{for } I^i_{3L} = I^i_{3R}$$

IIIb. Parity Conserving Neutral Currents

As observed in the previous page, in the limit $\lambda_L = \lambda_R$, we get $\phi = 0$. Then the eigenstates of the neutral gauge boson mass matrix are Z_V and Z_A of Eq. (III.3) and their masses are:

$$m^2_{Z_V} = m^2_{W_L^+}(1-\varepsilon)\sec^2\theta$$

and

$$m^2_{Z_A} = m^2_{W_L^+}(1+\varepsilon) \qquad \text{(III.14)}$$

where the parameter ε satisfies the following inequality.

$$0 \le \varepsilon \le 1$$

and

$$\varepsilon = (m^2_{Z_A} - m^2_{W_L^+})/m^2_{W_L^+} \qquad \text{(III.15)}$$

It therefore follows from Eq.(III.15) $m^2_{Z_A} \geq m^2_{W_L^+}$. The $\mathcal{H}^{N.C.}_{WK}$ is then parity conserving and has the VV + AA form:

$$\mathcal{H}^{N.C.}_{WK} = \frac{g^2}{2\cos^2\theta m^2_{Z_V}} V_\mu V_\mu + \frac{g^2}{2m^2_{Z_A}} A_\mu A_\mu \qquad \text{(III.16)}$$

It therefore follows that, to order G_F, there will be no parity violation in atomic physics. However, as we will show now, $\sigma^{\nu n}$ and $\sigma^{\bar{\nu} n}$ will <u>not</u> in general be equal to each other. The reason for this is that $\sigma^{\nu n}$ gets contribution from two graphs; one mediated by Z_V and another by Z_A. Both have opposite charge conjugation property. Since $\sigma^{\bar{\nu} n}$ is obtained from $\sigma^{\nu n}$ by crossing symmetry, we get

$$\sigma^{\nu N} = C|M_V + M_A|^2$$

$$\sigma^{\bar{\nu} N} = C|M_V - M_A|^2 \qquad \text{(III.17)}$$

Thus, unless either M_V or M_A is zero, we will always have $\sigma^n \neq \sigma^n$. Furthermore, since all laboratory experiments are conducted with predominantly left-handed neutrinos, the effective Hamiltonian for neutrino interaction looks like:

$$H^{N.C.}_{WK,\nu} = \frac{g^2}{8}\,\bar{\nu}\gamma_\mu(1+\gamma_5)\nu\left[\frac{1}{\cos^2\theta m^2_{Z_V}} V_\mu + \frac{1}{m^2_{Z_A}} A_\mu\right]$$

$$= \frac{G_F}{\sqrt{2}}\,\bar{\nu}\gamma_\mu(1+\gamma_5)\,\nu\left[\frac{V_\mu}{1-\varepsilon} + \frac{A_\mu}{1+\varepsilon}\right] \qquad \text{(III.18)}$$

We see that, in the limit $\varepsilon \to 0$, using Eq. (III.13), we can rewrite Eq. (III.18) as follows:

$$H^{N.C.}_{WK,\nu} = \frac{G_F}{\sqrt{2}}\,\bar{\nu}\gamma_\mu(1+\gamma_5)\nu \sum_{i=1,\ldots N}\bar{\psi}_i\gamma_\mu\left[(1+\gamma_5)I^i_{3L} - Q^i\sin^2\theta\right]\psi_i \qquad \text{(III.19)}$$

Here, we have displayed the $(1+\gamma_5)$ explicitly. Note that, this is precisely the structure of the neutral current[15,18,32] in the Weinberg-Salam model if we write $\sin^2\theta = 2\sin^2\theta_W$. Thus, we see that our model has the interesting features that it reproduces the results of Weinberg-Salam (W-S) model for all fermion masses going to zero (since $\varepsilon \to 0$ means κ, $\kappa' \to 0$ separately). Since the W-S model appears to be in agreement with experiments for neutrino neutral currents, so does our model. However, the difference lies in the electron neutral current effects. Whereas, in our model, the electron neutral current conserves parity, the same is not true for the W-S model. This will have experimental implications that we will discuss in a subsequent section. This interesting feature however has an embarrassing aspect to it i.e. it requires all fermions to be nearly massless. If heavier quark-antiquark bound states (such as J/ψ (3.1)) keep getting discovered, the limit $\varepsilon \to 0$ will not represent a credible situation.

But there does indeed exist an alternative method, which avoids any possible embarrassments and yet yields the **Weinberg-Salem** structure for neutral current Hamiltonian.

IIIc. Weinberg-Salam Limit for Neutrino Neutral Current Interaction

In the previous section, we observed that structure of neutrino neutral current Hamiltonian discussed in Sec. IIIb (i.e. $\lambda_L=\lambda_R$) becomes similar in structure to that of the Weinberg-Salam model in the limit of $\varepsilon\to 0$ (i.e. all fermion masses vanishing). In this subsection, we show that this interesting property of the left-right symmetric gauge theories holds even in the limit when $\lambda_L\neq\lambda_R$, but only in the limit $\varepsilon\to 0$. To show that, we study Eq. (III.11) and (III.12) in the case when $\phi\neq 0$. Let us set $\varepsilon=0$ from the beginning. Then, we observe that,

$$H^{N.C.}_{WK,\nu}=\frac{G_F}{\sqrt{2}}\bar{\nu}\gamma_\mu(1+\gamma_5)\nu\left[a\ V_\mu+b\ A_\mu\right]$$

where

$$a=\frac{m^2_{W^+_L}}{\cos^2\theta}\left(\frac{\cos^2\phi}{m_2^2}+\frac{\sin^2\phi}{m_1^2}\right)+\frac{m^2_{W^+_L}\cos\phi\sin\phi}{\cos\theta}\left(\frac{1}{m_2^2}-\frac{1}{m_1^2}\right)$$

and

$$b=m^2_{W^+_L}\left(\frac{\cos^2\phi}{m_1^2}+\frac{\sin^2\phi}{m_2^2}\right)+\frac{m^2_{W^+_L}\cos\phi\sin\phi}{\cos\theta}\left(\frac{1}{m_2^2}-\frac{1}{m_1^2}\right) \tag{III.20}$$

For $\varepsilon\to 0$, we get,

$$m^2_{2,1}=\frac{m^2_{W^+_L}}{2(1+\sigma)}\left[(\sec^2\theta+1)\pm\sec 2\phi\tan^2\theta\right]$$

and

$$\mathrm{Sec}2\phi=\sqrt{1+4\sigma^2\cot^2\theta\ \mathrm{cosec}^2\theta} \tag{III.21}$$

Using these, one gets,

$$a = b = \frac{4(1-\sigma^2)}{\cos^2\theta} \frac{1}{[(\sec^2\theta+1)^2-\sec^2 2\phi\tan^4\theta]} \quad \text{(III.22)}$$

As a result, we get,

$$H^{N.C.}_{WK,\nu} = \frac{G_F}{\sqrt{2}} \bar{\nu}\gamma_\mu(1+\gamma_5)\nu \sum_{i=1,\ldots N} \bar{\psi}_i\gamma_\mu[I^i_{3L}(1+\gamma_5)-\sin^2\theta Q^i]\psi_i \quad \text{(III.23)}$$

Apart from the scale factor, a, the structure is again same as in W-S model provided we identify $\sin^2\theta=2\sin^2\theta_W$. This proves our assertion. Indeed, Georgi and Weinberg[26] have extended this result to any gauge group of the form $SU(2)_L \otimes G \otimes U(1)$.

IIId. 'Natural' Parity Conservation in Neutral Current Interaction

In section IIIb., we saw that if we set the angle $\phi=0$, the neutral current conserves parity. However, since for left-handed neutrinos, it goes to the W-S model in the limit for $\varepsilon\to0$, there is no conflict with the available neutral current data. At the same time, it can accommodate no parity violation in atoms to order G_F. As we stressed there, since θ is set equal to zero arbitrarily, the situation is not natural and Z_A-Z_V mixing can be induced in higher orders.

Suppose it turns out that atomic experiments show parity violating effects at a level $\leq G_F\alpha$, then it would be desirable to understand this result in terms of some kind of a selection rule, which forbids parity violation in neutral currents at the tree level and generates it at the one-loop level. In this section, we present an alternative set of Higgs multiplets which by its group theoretical structure enables us to attain this goal.

We suggest[24] instead of the two doublets χ_L and χ_R, to choose $\rho \equiv (\frac{1}{2}, \frac{1}{2}, -2)$, which has the following charge structure:

$$\rho \equiv \begin{pmatrix} \rho_1^{+} & \rho_3^{0} \\ \rho_3^{++} & \rho_2^{+} \end{pmatrix} \qquad \text{(III.24)}$$

We give it the following vacuum expectation values

$$\langle\rho\rangle = \begin{pmatrix} 0 & c \\ 0 & 0 \end{pmatrix} \qquad \text{(III.25)}$$

We also note that, under left-right conjugation

$$\rho \leftrightarrow \tilde{\rho}^{\dagger} \qquad \text{(III.26)}$$

The neutral boson mass matrix resulting from $\langle\rho\rangle$ and $\langle\phi\rangle$ is:

$$\begin{array}{c|ccc} & W_L^3 & W_R^3 & B \\ \hline W_L^3 & g^2(x+y) & -g^2(x-y) & -2gg'y \\ W_R^3 & -g^2(x-y) & g^2(x+y) & -2gg'y \\ B & -2g'gy & -2\,gg'y & 4g'^2y \end{array} \qquad \text{(III.27)}$$

where

$$x = \frac{1}{2}(\kappa^2+\kappa'^2)$$

$$y = \frac{1}{2}c^2$$

The eigenstates of this matrix are the parity eigenstates defined in Eq. (III.3) and this happens without having to adjust any parameter of the theory arbitrarily. Thus, <u>the parity conservation of the neutral current Hamiltonian is technically natural and any parity violating effects in such models are computable</u> (see ref. 31 for a computation of this effect). The eigenvalues of the matrix (III.27)are

$$m_A^2 = 0; \; m_{ZA}^2 = \frac{4e^2x}{\sin^2\theta}; \; m_{Z_V}^2 = \frac{16e^2y}{\sin^2 2\theta} \qquad \text{(III.28)}$$

It has also been pointed out by Sidhu and this author[27] that this yields the neutral current structure of the W-S model in a rather interesting limit. To see this, we note that we can cast the $H^{N.C.}_{wk,\nu}$ in this case exactly in the form given in Eq. (III.18) with the difference that ε is defined now as

$$\varepsilon = (m^2_{Z_A} - m^2_{W_L^+})/m^2_{W_L^+} \equiv \frac{x-y}{x+y} \qquad \text{(III.29)}$$

so the range of values accessible to ε is now different from that given in Eq. (III.15):

$$-1 \leq \varepsilon \leq 1 \qquad \text{(III.30)}$$

[Incidentally we see that, unlike the case $\langle\chi_L\rangle = \langle\chi_R\rangle$, here Z_A can be lighter than the charged W_L^+ boson.] More interestingly the W-S limit is attained for $\varepsilon = 0$, which is satisfied when $x = y$ or when

$$\langle T_\gamma \phi^\dagger \phi\rangle \qquad \langle T_\gamma \rho^\dagger \rho\rangle \qquad \text{(III.31)}$$

In this case, κ and κ' are non-zero and therefore can lead to a realistic quark mass spectrum, even the left-right symmetric theory in the limit $\varepsilon \to 0$ describe the real world.

Finally, we would like to formally state the conditions[24] which a left-right symmetric theory (with $g_L = g_R$) must satisfy in order to have "naturally" parity conserving neutral currents.

(i) All fermions must have the save value for I_{3L} and I_{3R} i.e.

$$I_{3L}(\psi_i) = I_{3R}(\psi_i) \qquad \text{(III.32)}$$

(ii) The Higgs system must be so as to yield the Z_V, Z_A eigenstates "naturally".

Finally, we would like to comment that this attractive "naturalness" feature is not shared by the model discussed in ref. 14 even when left-right symmetry is made "natural" in these models and this is a clear distinguishing feature between the $SU(2)_L \times SU(2)_R \times U(1)$ models[15,16,17] and the $SU(2)_L \times SU(2)_R \times SU(4')$ model[14].

IIIe. Natural Flavor Conservation in Neutral Current Interaction

One of the interesting features of the W-S model with GIM quark structure is the property that neutral currents conserve strangeness and charm "naturally" and thus lead to an understanding of why $K_L \to \mu\bar{\mu}$ decay occurs only in level $G_F\alpha^2$. <u>One of the conditions</u> for this to happen in any SU(2) gauge theory is that quark flavors of a given charge must <u>all</u> have the same value for I_{3L}. When translated to the language of $SU(2)_L \times SU(2)_R \times U(1)$ type of theories, this implies that quark flavors of equal charge must <u>all</u> have the same value for I_{3L} and I_{3R} for the left and right-handed helicities respectively. From Eq. (II.1), we see that, this condition is indeed satisfied. This implies that the neutral current gauge interactions are flavor diagonal. We are, of course, aware that true "natural" flavor conservation also implies stringent constraints on the Higgs sector which cannot be met in a realistic $SU(2)_L \times SU(2)_R \times U(1)$ model [31,35]. One way to make the theory have true natural flavor conservation in neutral current interaction is to let only ϕ couple to the fermions and not $\tilde{\phi}$, because then, diagonalization of the mass matrix diagonalizes all neutral Higgs interactions involving left-over Higgs mesons. It is easy to check that in this case, one has the interesting mass relations

$$\frac{m_u}{m_d} = \frac{m_c}{m_s} = \frac{m_t}{m_b} = \ldots . \qquad \text{(III.33)}$$

However the difficulty in this case is that the Cabibbo angle in this case also turns out to be zero and thus makes the model unacceptable.

IIIf. Properties of Neutral Z-Bosons

The mass and decay characteristics[33] of the two neutral heavy vector bosons Z_1, Z_2 (or Z_A and Z_V in the parity conserving case) in our model depends on the two parameters ε and $\sin\theta$ and $\sin\phi$ (or ε and $\sin\theta$ in the parity conserving case). They, in turn, are determined by the neutrino neutral current fits. Recently, Sehgal[36] has extracted the various neutral current parameters using single-pion inclusive production data in neutral current couplings; for the parity conserving case, this allows:

$$\varepsilon \approx 0 \text{ to } \pm.1 \text{ and } \sin^2\theta = .5 \qquad \text{(III.34)}$$

This gives the following masses for the charged and neutral gauge bosons:

$$m_{W_L^+} \simeq 76 \text{ Gev}$$

$$m_{Z_A} \simeq 71 \text{ to } 81 \text{ Gev}$$

$$m_{Z_V} \simeq 112 \text{ to } 102 \text{ Gev} \qquad \text{(III.35)}$$

Substituting the value of $\sin^2\theta \simeq .5$, the decay widths to various hadronic and leptonic channels can be calculated (see Table II). It is interesting that for this value of θ, Z_V has vanishing decay width to leptonic channels unlike the Z_A. Therefore, only Z_A is likely to show up in the dilepton mass distribution in pp-collision at ISABELLE energies. Moreover, unlike the Z-boson of the Weinberg-Salam models (Z_W), there will be no parity violating asymmetries associated with our Z_A-boson. The Z_W of course has a different mass ($m_{Z_W} \simeq 88$ Gev). One important point to note is that, the neutral bosons in our model are not very different from the W-boson mass and are therefore unlikely to show up as resonances at PETRA or PEP energies.

The situation for the parity violating case i.e. $\phi \neq 0$ is of course very different from the parity conserving case. This case is easier to analyze for the case of $\varepsilon \to 0$, since the neutrino neutral current analysis is here similar to the W-S case apart from an overall scale factor, a (see Eq. III.23). In the analysis below, we set a=1 and the neutral current fit in that case requires $\sin^2\theta = .5$.

The masses of the two bosons Z_1, Z_2 are related to the W_L^+ by the following formula:

$$\frac{m_{2,1}^2}{m_{W_L^+}^2} = \sqrt{2} \left(\frac{3 \pm \sec 2\phi}{2\sqrt{2} + \tan 2\phi} \right) \qquad \text{(III.36)}$$

where $0 < \phi < \frac{\pi}{2}$. In this case one of the Z bosons can be very much lighter than W_L^+ (see fig. 1). For example, for $\phi = 30^0$

$$m_1^2 \simeq .78 m_{W_L^+}^2$$

This gives $m_1 \approx 68$ Gev, where $m_{W_L^+} \approx 75$ Gev. We point out however that, in this case, one has a large parity violation in atomic physics. We will express the atomic effects as a function of ϕ in a subsequent section.

The left-right symmetric models also lead to several inequalities involving neutral current cross-sections[37,33], which distinguish them from the SU(2) ⊗ U(1) type theories with a single Z-boson.[38,39] In particular, in ref. 33, it has been observed that certain type of neutral current relations are valid in both 1-Z and 2-Z boson theories. A test of those relations could provide important information on these models.[40]

IIIg. Atomic Parity Violation and Higher Order Effects

In subsection IIIb and IIId, we saw that it is possible to

get natural parity conservation for the neutral current Hamiltonian in the tree approximation.[24] However since the charge current weak Hamiltonian violates parity in higher orders, it will induce parity violating neutral current effects. This has been discussed in the literature by several authors.[31] First of all, we note that, if a Z_A-Z_V mixing exists in the tree approximation, that would lead to an effective parity violating Hamiltonian (relevant for the discussion of heavy atoms) of the following type:

$$H^{atom}_{P.V.} = \frac{e^2 \sin 2\phi}{8\sin^2\theta\cos\theta} \left(\frac{1}{m_2^2} - \frac{1}{m_1^2}\right) \bar{e}\gamma_\mu\gamma_5 e$$

$$\times \left[\frac{1}{3}(4\cos^2\theta - 1)\bar{u}\gamma_\mu u - \left(\frac{2\cos^2\theta+1}{3}\right)\bar{d}\gamma_\mu d\right] \qquad \text{(III.37)}$$

From this, in the limit $\varepsilon\to 0$, we get for the weak charge Q_W defined in terms of the non-relativistic atomic Hamiltonian[41] as follows:

$$H^{(-)}_{WK} \simeq \frac{G_F}{\sqrt{2}} Q_W \left\{\frac{\vec{\sigma}\cdot\vec{\rho}}{2m_e}, \delta^3(\gamma)\right\} \qquad \text{(III.38)}$$

$$Q_W = -2\sigma\,[2Z\cos^2\theta - A] \qquad \text{(III.39)}$$

For $\cos^2\theta = \frac{1}{2}$, for Bi, we get $Q_W = 252\sigma$. The present experimental limits on Q_W are

$$Q_W^{Bi} = 40 \pm 32 \quad \text{(Oxford)}^{42}$$
$$= -42 \pm 16 \quad \text{(Washington)}^{42}$$

This yields $|\sigma| < .15$.

In models such as those discussed by Paige, Sidhu and this author[24] where $Q_W = 0$ in the tree approximation, the higher order effects induce parity violating effects. Keeping only the gauge boson contribution, the relevant Feynman diagrams are listed in fig. 1. For the models, where $m_{W_R} \gg m_{W_L}$, the dominant contribution is given by (these results are taken from the paper of G. Senjanovic and A. Sokorac[31]):

$$Q_W = \frac{5\alpha}{\pi} Z \ln \left(\frac{m^2_{W_R}}{m^2_{W_L}}\right) + Q_W^{Higgs} \tag{III.40}$$

It is clear from the above expression that $m_{W_R^+}$ cannot be chosen arbitrarily larger than $m_{W_L^+}$ since that will lead to large atomic parity violating effects. If we choose $m_{W_R^+} = 3m_{W_L^+} \simeq 225$ Gev, then Eq. (III.40) leads to a value for Q_W for Bismuth

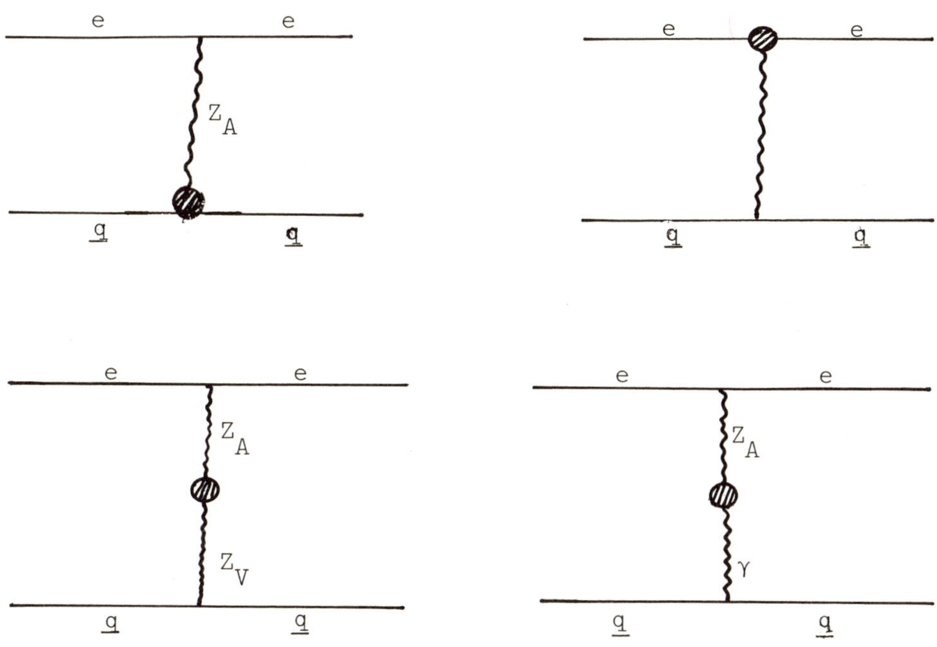

Fig. 1. Given here are the Feynman diagrams, which contribute to parity violating effects in atoms in the leading log approximation. The blobs include all possible contributions such as $W^+_{L,R}$ bosons, Higgs bosons as well as fermions.

$$Q_W^{Bi} \simeq 2.1 + Q_W^{Higgs} \quad \text{(III.41)}$$

In the above expression Q_W^{Higgs} stand for the contribution of the Higgs bosons to Q_W. These will be significant if $m_H >> m_{W_L}$; however, for $m_H \approx$ 10-20 Gev, these are negligible.

There is however another possibility, where $m_{W_L^+} = m_{W_R^+}$ but the right-handed currents mix only light and heavy fermions as in the model discussed in sec. IIf. In this case, however, the dominant contribution is proportional to the quark-mass differences and is proportional to $G_F^2(\Delta m_q)^2$. This would lead to a much smaller value for Q_W.

IV. CP-VIOLATION, HEAVY FERMIONS AND ABSOLUTELY STABLE HADRONS

IVa. CP-Violation with Four Flavors

One of the principal drawbacks of the standard WSGIM model [i.e. $SU(2)_L \otimes U(1)$ theory with 4-quark flavors, 4-leptons and one Higgs doublet] is that it cannot support CP-violation in the weak interaction sector. It turns out that, in this case, all non-trivial CP-phases can be removed by redefinition of the phases of the fermion fields in the theory. The same also happens to be true if we extend the group to $SU(2) \otimes U(1) \otimes U(1)$, provided all fermions having the same chirality and the same charge have the same quantum numbers each U(1). Let us state this as a theorem:

Theorem: For a gauge theory (denoted by S) based on the gauge group $SU(2)X[U(1)]^n$ [Y_i(i=1,...n) denote the U(1) quantum numbers] with four massive quarks, all CP-violating phases can be removed from the gauge interactions; therefore, if we ignore the effect of Higgs bosons, we have

$$\eta_{00} = \eta_{+-} = 0 \quad \text{(IV.1)}$$

(where η_{ij} are the standard CP-violation parameters) provided

$Y_i(\psi_Q) = C_i$ for each chirality and charge.

To apply this theorem, let us define a general gauge theory as $G(n,\varepsilon)$, where n stands for the number of fermions, and ε stands for a parameter in the theory.

$$\text{Let} \quad G(n,\varepsilon) \xrightarrow[\varepsilon\to 0]{} S \tag{IV.2}$$

It follows from Eq. (IV.1) that in such a theory,

$$\eta_{+-} = C\varepsilon \tag{IV.3}$$

where C is a constant.

Now let us apply this to the case of left-right symmetric theories with four quarks[15]. To generate the CP-violation phases, we choose the minimum of the Higgs potential with complex $\langle\phi\rangle$ (see Eq. (II.14)) and real Yukawa couplings.[43] In this case, the origin of CP-violation is also spontaneous. To show how many independent phases exist in this case, we may proceed as follows. We have a general complex symmetric matrix both for $I_{3L}=+\frac{1}{2}$ and $I_{3L}=-\frac{1}{2}$ quarks i.e.

$$M^{(\pm)} = M^{(\pm)^T} \tag{IV.4}$$

$$\text{Let} \qquad U_L^{(\pm)} M^{(\pm)} U_R^{(\pm)^{-1}} = M_d^{(\pm)} \tag{IV.5}$$

From Eq. (II.17), we realize that,

$$U_R^{(\pm)} = K^{(\pm)} U_L^{(\pm)^*} \tag{IV.6}$$

where $K^{(\pm)}$ are arbitrary diagonal unitary matrices. Let us concentrate on the $M^{(+)}$ sector. U's are general 2x2 unitary matrices:

$$U_L^{(\pm)} = e^{i\gamma_L} \begin{pmatrix} \cos\theta_L e^{i\alpha_L} & \sin\theta_L e^{i\beta_L} \\ -\sin\theta_L e^{-i\beta_L} & \cos\theta_L e^{-i\alpha_L} \end{pmatrix} \tag{IV.7}$$

Using Eq. (IV.6) with an appropriate choice of $K^{(\pm)}$, it is possible to write the weak fermion doublets in terms of the flavor eigenstates as follows:

$$\begin{pmatrix} u_L \\ d_L\cos\theta_c + e^{i\delta} s_L \sin\theta_c \end{pmatrix} \begin{pmatrix} c_L \\ -d_L \sin\theta_c e^{-i\delta} + s_L \cos\theta_c \end{pmatrix}$$

$$\begin{pmatrix} u_R \\ d_R \mathrm{Cos}\theta_c + s_R \sin\theta_c e^{-i\delta} \end{pmatrix} \begin{pmatrix} c_R \\ -d_R \sin\theta_c e^{i\delta} + s_R \cos\theta_c \end{pmatrix} \tag{IV.8}$$

In the limit of $\delta \to 0$, CP-violation vanishes. We furthermore note that, for $m_{W_L^+}/m_{W_R^+} \to 0$, $SU(2)_L \otimes SU(2)_R \otimes U(1)$ theory reduces to $SU(2)_L \otimes U(1)_L \otimes U(1)_R$ theory that satisfies the conditions of the theorem stated above. It then follows that in this case $\varepsilon = (m_{W_L^+}/m_{W_R^+})^2$.

$$\eta_{+-} = c \sin 2\delta \left(\frac{m_{W_L^+}}{m_{W_R^+}}\right)^2 \tag{IV.9}$$

Detailed calculation[15,43a] also yields this same formula with $c \approx 1$. We thus see that, in left-right symmetric models with four quarks, there is a deep connection between CP and P-violation.[15] Turning the Eq. (IV.9) around, we provide an upper bound on the mass of the W_R^+:

$$2.76 \leq \left(\frac{m_{W_R^+}}{m_{W_L^+}}\right) \leq 22 \tag{IV.10}$$

Furthermore, the CP-violating weak Hamiltonian in this model satisfies the "iso-conjugate" relations[44] i.e.

$$[I_3, H_{WK}^{(-)}] = iaH_{WK}^{(+)} \tag{IV.10a}$$

As a result of this, one immediately obtains the relation

$$\eta_{+-} = \eta_{00} \tag{IV.10b}$$

without any further dynamical assumptions. We of course assume that the contribution of the Higgs-boson parts are negligible, which results if we assume large masses for Higgs bosons. Incidentally, such large masses for the relevant Higgs boson are required in the model to supress their contribution to $K_L^0 \to \mu\bar{\mu}$ decay.

Even though this model of CP-violation simulates the superweak[45] model in predicting Eq. (IV.10b), this is a milliweak model and will lead to milliweak CP-violating effects in other non-leptonic decays of charmed and uncharmed particles. Another difference between this and the superweak model is in its prediction of neutron dipole moment. We predict,

$$10^{-29}\text{ecm} < d_n^e < 10^{-25}\text{ecm}$$

whereas superweak type models predict $d_n^e \lesssim 10^{-29}$ecm.

IVb. Heavy Fermions and Quark Islands

We have noted that the discussions of Sec. II and III are valid for an arbitrary even number of quarks and leptons. The simple prescription is to add more and more left-right symmetric doublets as new fermions or their bound states are discovered. The discussions of the charged and neutral current interaction can be applied to these cases without any modification. The only point to note is that, in the case of non-manifest realization (with $m_{W_L^+} = m_{W_R^+}$), left-right symmetry and natural parity (or flavor) conservation in neutral currents requires that

$$n_F \geq 8 \tag{IV.11}$$

where n_F is the number of flavors.

To study the physical implications of weak interactions models with more than four flavors, let us first look at the left-handed charged currents. Since in general all flavors of some electric charge mix, the number of flavor mixing angles will increase with increasing number of doublets (for n-doublets, the number of real mixing angles in the left-handed sector becomes $\frac{n(n-1)}{2}$ and in the right-handed sector, we have another $\frac{n(n-1)}{2}$ leading to a total of n(n-1) angles.) For the case of manifest left-right symmetry, the same number is $\frac{n(n-1)}{2}$. So a general weak doublet can be written as:

$$\begin{pmatrix} u_i \\ \sum_j a_{ij} d_j \end{pmatrix}_L \quad i,j=1,\cdots n \tag{IV.12}$$

where u_i is the ith flavor eigenstate of $Q = +\frac{2}{3}$ and d_i with $Q = -\frac{1}{3}$ and a_{ij} are mixing parameters which satisfy the equation $\sum_{j=1} a_{ij}^2 = 1$. If we choose $u_1 = u$, $u_2 = c, \ldots$ $d_1 = d$; $d_2 = s, \ldots,$ we know that

$$a_{11}^2 + a_{12}^2 \simeq 1.001 \tag{IV.13}$$

thus, the other mixing parameters i.e. a_{13}, $a_{14}, \ldots$ must be arbitrarily chosen to be close to zero. This makes Cabibbo universality look very unnatural.

Moreover, we know that, for $SU(2)_L \times U(1)$ model with six or more quarks, the theorem stated in Sec. IVa does not hold anymore. Therefore, in general, the relation between CP and P-vilation does not hold. On the other hand, the attractiveness of the relation certainly makes one search for constraints on hadron models that would enable us to maintain this relation. We are aware of two ways by which these apparently unattractive features can be avoided.

(A) Quark Islands. One obvious way to achieve both the objectives i.e. maintaining exact Cabibbo universality as well as to maintain the CP and P- relation, is to forbid all heavier quarks from mixing with the four light quarks u, d, s, c. It may even be interesting to speculate that, there are several decoupled islands of flavor with four flavors in each island i.e.

$$[u, d, s, c]; [t, b, g, h]; [\qquad]; \ldots$$

This decoupling can be achieved naturally by requiring the heavier quarks to have one extra quantum number. If the total number of flavors is sixteen then, there may be four such islands with the nth island having a discrete symmetry

$$\psi_n \rightarrow e^{i\pi n/2} \psi_n, \ \psi_1 \equiv [u, d, s, c] \qquad (IV.14)$$

$$\cdot \ \cdot \ \cdot \ \cdot$$

One immediate consequence of this idea is that the lightest members of any two islands may combine to form pseudo-scalar bound states (e.g. $\bar{u}b$) or baryonic bound states (e.g. uub,...), which are stable[43]. Some experimental implications of this idea have been recently discussed by Cahn[46].

(B) Larger Groups. Another way to accommodate the desirable features of the four quark models without having to postulate stable high mass hadrons is to go to a larger group such as $SU(4)_L \otimes SU(4)_R \otimes U(1)$.[47] The larger groups can accommodate more flavors within fewer multiplets, thus reducing the number of mixing angles. For example, in the model considered in ref. 46, we had eight quark flavors assigned to two multiplets leading to only one fermion mixing angle for both heavy and light quarks, the Cabibbo angle. The CP and P relation can also be preserved in these models.

V. LEFT-RIGHT SYMMETRIC THEORIES BASED ON HIGHER GROUPS

So far we have concentrated our attention on the simplest example of a left-right symmetric gauge theory based on the group

$SU(2)_L \otimes SU(2)_R \otimes U(1)$. Most features of this model such as asymptotic parity conservation, manifest left-right symmetry and parity conserving neutral currents generalize to higher groups in a very straightforward manner, provided the group is of the form $G_L \otimes G_R \otimes U(1)_{L+R}$. In this section, we will discuss extension of this model to two types of higher groups. The first group we will consider is $SU(2)_L \otimes SU(2)_R \otimes U(1)_L \otimes U(1)_R$ and the next higher group we will consider is $SU(3)_L \otimes SU(3)_R$. The latter is a non-abelian group and therefore explains electric charge quantization. Possible unifying groups that unify this with strong interactions will also be commented on.

Va. $SU(2)_L \otimes SU(2)_R \otimes U(1)_L \otimes U(1)_R$ Gauge Theory[47a,48]

Let us denote the gauge bosons by $\vec{W}_L$, $\vec{W}_R$, B_L and B_R in obvious correspondence with the gauge group. Imposing the left-right discrete symmetry, we are left only with two gauge couplings: $g_L = g_R \equiv g$ and $g'_L = g'_R \equiv g'$. Let us denote $g'/g = \tan\theta$. The quarks and leptons are assigned to the group as follows:

Quarks:

$$\psi_{iL} \equiv \binom{p_i}{n_i}_L \quad i=1,\dots \quad N \quad (\tfrac{1}{2}, 0, \tfrac{1}{3}, 0)$$

$$\psi_{iR} \equiv \binom{p_i}{n_i}_R \quad i=1,\dots \quad N \quad (0, \tfrac{1}{2}, 0, \tfrac{1}{3})$$

Leptons:

$$L_i \equiv \binom{\nu_i}{E_i^-}_R \quad i=1,\dots \quad N \quad (0, \tfrac{1}{2}, 0, -1)$$

$$R_i \equiv \binom{\nu_i}{E_i^-}_R \quad i=1,\dots \quad N \quad (0, \tfrac{1}{2}, 0, -1) \qquad \text{(V.1)}$$

(where for leptons i = 1 stands for electron type, i = 2 stands for muon type, i = 3... stands for heavier lepton type fields). The following Higgs bosons are chosen to give mass to leptons and

and quarks respectively:

$$\phi_1^{(m)} = (\frac{1}{2}, \frac{1}{2}, -1, 1) \quad m=1,2,\ldots$$

$$\phi_2^{(m)} = (\frac{1}{2}, \frac{1}{2}, \frac{1}{3}, \frac{-1}{3}) \quad m=1,2,\ldots \qquad (V.2)$$

The following multiplets enable us to make $m_{W_R^+} >> m_{W_L^+}$;

$$\delta_L = (1, 0, 0, 0)$$

$$\delta_R = (0, 1, 0, 0) \qquad (V.3)$$

The neutral current structure is determined by the following doublets:

$$\chi_L = (\frac{1}{2}, 0, 1, 0)$$

$$\chi_R = (0, \frac{1}{2}, 0, 1) \qquad (V.4)$$

We choose two other Higgs multiplets which enable us to pass to two interesting subgroups of $SU(2)_L \otimes SU(2)_R \otimes U(1)_L \otimes U(1)_R$:

$$\Sigma = (\frac{1}{2}, \frac{1}{2}, 0, 0)$$

$$\text{with} \quad \Sigma = \tau_2 \Sigma^* \tau_2 \qquad (V.5)$$

and

$$\Delta = (0, 0, 1, -1) \qquad (V.6)$$

The electric charge formula is given by:

$$Q = I_{3L} + I_{3R} + \frac{1}{2}(Y_L + Y_R) \qquad (V.7)$$

We give non-zero VEV's to the following fields:

$$\langle\phi_i^{(m)}\rangle = \begin{pmatrix} \kappa_i^{(m)} e^{i\alpha_i^{(m)}} & 0 \\ 0 & \kappa_i^{(m)'} e^{i\alpha_i^{(m)'}} \end{pmatrix}$$

$$\langle\chi_L\rangle = \langle\chi_R\rangle = \binom{0}{\lambda}$$

$$\langle\Sigma\rangle = \begin{pmatrix} a & o \\ o & a \end{pmatrix}$$

$$\langle\Delta\rangle = c \qquad \text{(V.8)}$$

First of all, let us see how the various lower symmetry groups arise from the above group:

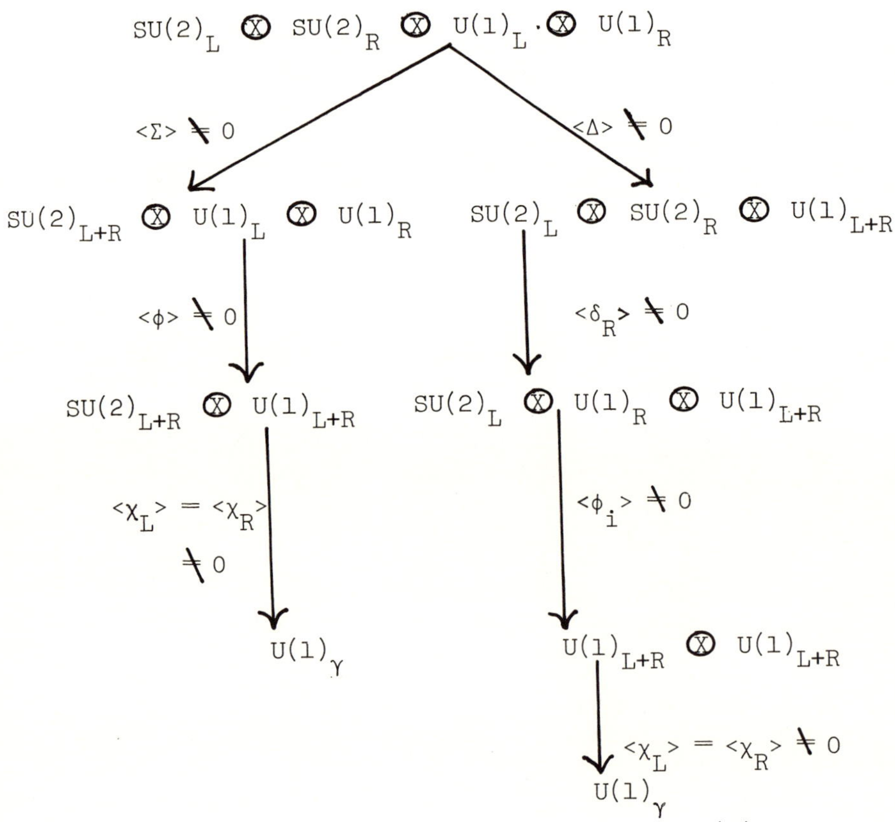

Case (B) Case (A)

We will consider the cases (A) and (B) separately.

Case (A): In this case, we can have manifestly left-right symmetric charged currents and we dispense with the Higgs multiplet Σ. Let us study the mass spectrum of the neutral gauge bosons and the structure of weak neutral currents in this theory. In terms of the parameters of spontaneous symmetry breaking given in Eq. (V.8), the eigenstates of the neutral boson mass matrix are given as follows:

$$A \equiv \frac{\sin\theta}{\sqrt{2}}(W_{3L}+W_{3R}) + \frac{\cos\theta}{\sqrt{2}}(B_L+B_R)$$

$$Z_V = \frac{\cos\theta}{\sqrt{2}}(W_{3L}+W_{3R}) - \frac{\sin\theta}{\sqrt{2}}(B_L+B_R)$$

$$Z_1 = \frac{\cos\eta}{\sqrt{2}}(W_{3L}-W_{3R}) + \frac{\sin\eta}{\sqrt{2}}(B_L-B_R)$$

$$Z_2 = -\frac{\sin\eta}{\sqrt{2}}(W_{3L}-W_{3R}) + \frac{\cos\eta}{\sqrt{2}}(B_L-B_R) \qquad \text{(V.9)}$$

where $\tan\theta = g'/g$ and

$$\tan\eta = \frac{-2\tan\theta(1-\frac{2}{3}\varepsilon_{2-}+2\varepsilon_{2-})}{1-\tan^2\theta\ (1+2Q_c) + 2\varepsilon_{2+}\ (1-\frac{\tan^2\theta}{9}) + 2\varepsilon_{2+}\ (1-\tan^2\theta)} \qquad \text{(V.10)}$$

with

$$\varepsilon_{i\pm} = \frac{\kappa_i^2 \pm \kappa_i'^2}{\lambda^2} \quad \text{(for m=1)}$$

$$Q_c = \frac{c^2}{\lambda^2} \qquad \text{(V.11)}$$

The mass spectrum looks as follows:

$$m_A = 0$$

$$m^2_{Z_V} = \frac{1}{2}(g^2 + g'^2)\lambda^2$$

$$m^2_{W_L^+} = \frac{g^2\lambda^2}{2}(1 + \varepsilon_{1+} + \varepsilon_{2+})$$

$$m^2_{Z_{1,2}} = \frac{m^2_{W_L^+}}{(1+\varepsilon_{1+}+\varepsilon_{2+})}\left[\frac{\sec^2\theta}{2} + \varepsilon_{1+}\sec^2\theta + \varepsilon_{2+}\left(1 + \frac{\tan^2\theta}{9}\right)\right.$$

$$\left. + Q_c\tan^2\theta\,\{1 \mp \sqrt{f(\theta, Q_c)}\}\right]$$

where

$$f(\theta, Q_c) = \left\{1 - \frac{1-\tan^2\theta}{2Q_c\tan^2\theta} - \frac{\varepsilon_{1+}(1-\tan^2\theta)+\varepsilon_{2+}\left(1 - \frac{\tan^2\theta}{9}\right)}{Q_c\tan^2\theta}\right\}^2 \tag{V.12}$$

$$+ \frac{\mathrm{Cot}^2\theta}{Q_c^2}\left(1+2\varepsilon_{1-} - \frac{2}{3}\varepsilon_{2-}\right)^2$$

First of all, we note that the neutral currents in this case are parity conserving. It follows from Eq. (V.12) that, in the limit of $Q_c \to \infty$, (and $\varepsilon_{i\pm} \to 0$)

$$m_{Z_1} \to \infty$$

$$m^2_{Z_2} \to \frac{g^2\lambda^2}{2} \tag{V.13}$$

This is the $SU(2)_L \otimes SU(2)_R \otimes U(1)$ limit discussed in Section III. However, for Q_c finite, we find that

$$m_{Z_2} < m_{W_L^+} \qquad (V.14)$$

We thus see that, unlike the $SU(2)_L \otimes SU(2)_R \otimes U(1)$ theory case, even in the case of parity conserving neutral currents, one of the neutral Z-bosons can be very light and could therefore be looked for in the PEP and PETRA energy ranges. We note that,

for $SU(2)_L \otimes U(1)$ $\quad m_Z > m_{W_L^+}$

for $SU(2)_L \otimes SU(2)_R \otimes U(1)$ $\quad m_{Z_1} \approx m_{W_L^+}$

(with parity conserving neutral currents) $\quad m_{Z_V} > m_{W_L^+}$

for $SU(2)_L \otimes SU(2)_R \otimes U(1)_L \otimes U(1)_R$, $\quad m_{Z_2} < m_{W_L^+}$

(parity conserving neutral currents) $\quad m_{Z_1}, m_{Z_V} > m_{W_L^+}$

Thus, if no parity violation is detected in atoms in order G_F and a Z-boson lighter than $m_{W_L^+}$ is found, the minimal gauge group which will be most attractive will be the $SU(2)_L \otimes SU(2)_R \otimes U(1)_L \otimes U(1)_R$ extension of our $SU(2)_L \otimes SU(2)_R \otimes U(1)_{L+R}$ group. The neutral current Lagrangian in this case looks like:

$$\mathcal{L}^{N.C.}_{WK} = -\frac{ie}{\sin 2\theta} Z_{V\mu} \sum_i \overline{\Psi}_i \gamma_\mu [(I^i_{3L}+I^i_{3R})+\gamma_5(I^i_{3L}-I^i_{3R})-2Q^i \sin^2\theta]\Psi_i$$

$$-\frac{ie}{2\sin\theta} Z_{1\mu} \sum_i \overline{\Psi}_i \gamma_\mu [\cos\eta\{(I^i_{3L}-I^i_{3R})+\gamma_5(I^i_{3L}+I^i_{3R})\}$$

$$+\frac{1}{2}\tan\theta \sin\eta \;\{(Y^i_L-Y^i_R)+\gamma_5(Y^i_L+Y^i_R)\}]\Psi_i$$

$$-\frac{ie}{2\sin\theta} Z_{2\mu} \sum_i \overline{\Psi}_i \gamma_\mu [-\sin\eta\{(I^i_{3L}-I^i_{3R})+\gamma_5(I^i_{3L}+I^i_{3R})\}$$

$$+\frac{1}{2}\tan\theta\cos\eta\{(Y^i_L-Y^i_R)+\gamma_5(Y^i_L+Y^i_R)\}]\Psi_i \qquad (V.14)$$

Case (B): In this case,[48] by choosing $\langle\Sigma\rangle$ large, the gauge group is reduced to $SU(2)_{L+R} \otimes U(1)_L \otimes U(1)_R$. Therefore, low energy phenomenology requires that the charged currents be completely non-manifest with right-handed currents connecting only light and heavy flavors (see II.f). The δ_L and δ_R multiplets are not required in this case. The charged boson mass spectrum in this case looks like:

$$W_1^+ \simeq \frac{1}{\sqrt{2}}(W_L^+-W_R^+);\; m^2_{W_1} \simeq \frac{g^2a^2}{2}$$

$$W_2^+ \simeq \frac{1}{\sqrt{2}}(W_L^++W_R^+);\; m^2_{W_2} \simeq \frac{g^2\lambda^2}{2} \qquad (V.15)$$

We of course choose $a \gg \lambda \rightarrow m_{W_1^+} \gg m_{W_2^+}$.

The eigenstates corresponding to the neutral bosons are:

$$Z_V = \frac{\cos\theta}{\sqrt{2}}(W_{3L}+W_{3R}) - \frac{\sin\theta}{\sqrt{2}}(B_L+B_R)$$

$$Z_A \simeq \frac{1}{\sqrt{2}}(W_{3L}-W_{3R})$$

$$X_A \approx \frac{1}{\sqrt{2}}(B_L-B_R) \qquad (V.16)$$

Their masses are: (for $\langle\phi_i\rangle \simeq 0$)

$$m^2_{Z_V} \simeq (g^2+g'^2)\frac{\lambda^2}{2}$$

$$m^2_{Z_A} \simeq \frac{g^2a^2}{2}$$

$$m^2_{X_A} \simeq \frac{g'^2\lambda^2}{2} \tag{V.17}$$

Again $m_{Z_A} >> m_{X_A}$.

The structure of the neutral current Lagrangian effective at low energies is parity conserving and looks like:

$$\mathcal{L}^{N.C.}_{WK} = \frac{-ie}{\sin 2\theta} Z_{V\mu} \sum_i \bar{\Psi}_i \gamma_\mu [(I^i_{3L} + I^i_{3R}) + \gamma_5 (I^i_{3L} - I^i_{3R}) - 2Q^i \sin^2\theta]\psi_i$$

$$\frac{-ie}{4\cos\theta} X_{A_\mu} \sum_i \bar{\Psi}_i \gamma_\mu \{(Y^i_L - Y^i_R) + \gamma_5 (Y^i_L + Y^i_R)\}\Psi_i \tag{V.18}$$

This kind of neutral current is not in conflict with present neutrino data[48].

Vb. Some Remarks on $SU(3)_L \otimes SU(3)_R$ Group

If we want to understand electric charge quantization and yet maintain the gross features of the previously described left-right symmetric models, $SU(3)_L \otimes SU(3)_R$[49] lends itself as a natural choice for the weak gauge group. Coupled with color $SU(3)_C$ group of strong interactions, $SU(3)_L \otimes SU(3)_R \otimes SU(3)_C$ could well provide a starting point for the description of all elementary particle interactions. Moreover, this group is also a subgroup of the exceptional unifying group[50] E_6:

$$E_6 = SU(3) \otimes SU(3') \otimes SU(3)_C \tag{V.19}$$

In fact, the basic 27-dimensional representation of E_6 has a decomposition under $SU(3) \otimes SU(3') \otimes SU(3)_C$ like the following:

$$27 = (3, \bar{3}, 1) \oplus (3, 1, \bar{3}) \oplus (1, \bar{3}, 3) \tag{V.20}$$

The proper assignment of fermions to this group and the question of whether this group really leads to asymptotic parity conservation are presently under investigation. We may, however, consider an alternative assignment, where leptons are assigned to octets whereas quarks are assigned to triplets (see Chakrabarti, ref. 49):

Quarks:

$$\begin{pmatrix} u_L \\ d_L \\ b_L \end{pmatrix} \quad \begin{pmatrix} c_L \\ s_L \\ g_L \end{pmatrix} \qquad (3, 1)$$

$$\begin{pmatrix} u_R \\ d_R \\ b_R \end{pmatrix} \quad \begin{pmatrix} c_R \\ s_R \\ g_R \end{pmatrix} \qquad (1, 3) \qquad \text{(V.21)}$$

To cancel anomalies, we must have six other heavier quarks belonging to the multiplets (3*, 1) and (1, 3*)

Leptons: (L, R)

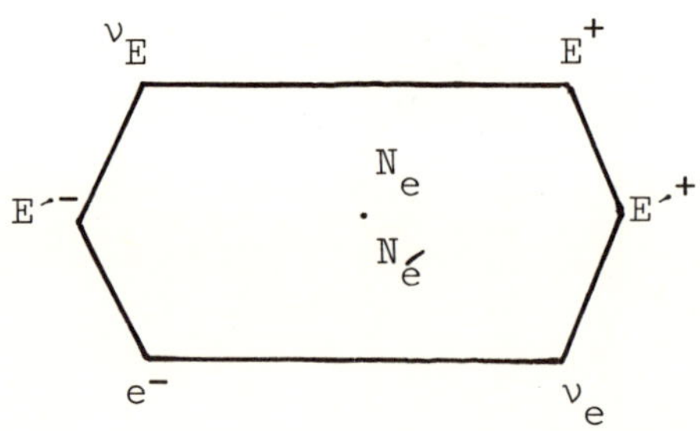

(V.22)

The detailed discussion of the Higgs system for this model has been carried out in ref. 49; we refer the interested reader to this paper.

Vc. Higher Unifying Groups such as SO(10) and $[SU(4)]^4$

Various possible ways of incorporating $SU(2)_L \otimes SU(2)_R \otimes U(1)$

gauge group within a higher unifying group that includes strong interactions have been suggested. Two most interesting groups are $SO(10)$[51] and $[SU(4)]^4$ [52]. Both have very different characteristics several of which are:

(a) The proton is necessarily unstable in the SO(10) model whereas in the $[SU(4)]^4$ unifying group the proton is stable as long as quarks are assumed to have fractional charges.

(b) The $SU(2)_L \otimes SU(2)_R \otimes U(1)_L \otimes U(1)_R$ is not a subgroup of SO(10) whereas it is one of $[SU(4)]^4$. Thus, observation of more than two neutral Z-type bosons would be evidence against the SO(10) group.

(c) Furthermore, $[SU(4)]^4$ leads to chiral quantum chromodynamics, whereas SO(10) does not.

Now let us briefly recound some of the salient features of both these unifying models:

SO(10): This is a group of rank 5 with $SU(4') \otimes SU(2) \otimes SU(2')$ as its subgroups. The decomposition of various representations of SO(10) in terms of this subgroup are:

$$\begin{aligned}
10 &= (6, 1, 1) \oplus (1, 2, 2) \\
16 &= (4^*, 1, 2) \oplus (4^*, 2, 1) \\
45 &= (6, 2, 2) \oplus (15, 1, 1) \oplus (1, 3, 1) \oplus (1, 1, 3) \\
120 &= (10, 1, 1) \oplus (\overline{10}, 1, 1) \oplus (\overline{6}, 3, 1) (6, 1, 3) \oplus (15, 2, 2) \\
&\quad \oplus (1, 2, 2) \\
126 &= (\overline{6}, 1, 1) \oplus (10, 3, 1) \oplus (\overline{10}, 1, \overline{3}) \oplus (15, 2, 2) \\
144 &= (20, 2, 1) \oplus (20, 1, 2) \oplus (4, 2, 3) \oplus (4, 3, 2) \oplus \\
&\quad (4, 2, 1) \oplus (4, 1, 2) \\
210 &= (50, 1, 1) \oplus (20, 2, 2) \oplus (6, 3, 3) \oplus (1, 4, 4) \oplus \\
&\quad (6, 1, 1) \oplus (1, 2, 2)
\end{aligned} \qquad (V.24)$$

Furthermore,

$$\begin{aligned}
16 \otimes \overline{16} &= 1 \oplus 45 \oplus 210 \\
16 \otimes 16 &= 10 \oplus 120 \oplus 126
\end{aligned} \qquad (V.25)$$

The fermions in this model are assigned to the 16-dimensional representation of SO(10). If we want to accommodate 2N quark flavors, we must have N 16-dim. representations. Besides, to give mass to the fermions, the Higgs particles must belong to the 10, 120 and 126 dimensional representation of the gauge group. Also, we must include a 45-dimensional representation to split the W_L and W_R masses. Since leptons and quarks belong to the same multiplet, there will be gauge bosons (X) connecting quarks and leptons when such gauge bosons will mix with the W-bosons that will give rise to proton decay. We must also include a 16-dimensional Higgs boson to generate the desired neutral current structure. Other than these sketchy remarks, we will refrain from giving any details.

<u>$[SU(4)]^4$:</u> This group is a very straightforward generalization of $SU(2)_L \otimes SU(2)_R \times (SU(4')_{L+R}$. Four SU(4)'s must be chosen in order to have only one overall gauge coupling describing all interactions. The fermions will transform as $(4, \bar{4}, \bar{4}, 4)$ of the gauge group.

As far as the gauge symmetry breaking is concerned, various possibilities exist, depending on one's taste. We can let the breaking proceed through different stages corresponding to a certain hierarchy of the gauge boson masses:

<u>Stage I:</u> At this stage, we break the gauge symmetry down as follows:

$$[SU(4)]^4 \to SU(4)_L \otimes SU(4)_R \otimes SU(3')_L \otimes SU(3')_R \otimes U(1)_L \otimes U(1)_R \tag{V.26}$$

This can be achieved by the quartet of Higgs multiplets:

$\Sigma_1=(15, 1, 1, 1)$; $\Sigma_2=(1, 15, 1, 1)$; $\Sigma_3=(1, 1, 15, 1)$; $\Sigma_4=(1, 1, 1, 15)$

and by giving non-zero vacuum expectation values to Σ_3 and Σ_4 as follows:

$$\langle E_j \rangle = c_j \begin{pmatrix} 1 & & \\ & 1 & \\ & & 1 \\ & & & -3 \end{pmatrix}; \; j=3, 4. \tag{V.27}$$

Stage II: At this stage, one may decide to leave all of $SU(3')_L \otimes SU(3')_R$ unbroken. However, if one wants to preserve only vectorial color as an unbroken symmetry, we may achieve this by a choice of Higgs multiplets of type $\alpha_1=(4, \bar{4}, 1, 1)$ and $\alpha_2=(1, 1, \bar{4}, 4)$ with,

$$\langle\alpha_2\rangle = d \begin{pmatrix} 1 & & \\ & 1 & \\ & & 1 \end{pmatrix} \qquad \text{(V.28)}$$

This gives rise to $SU(4)_L \otimes SU(4)_R \otimes SU(3')_{L+R} \otimes U(1)_L \otimes U(1)_R$ as the unbroken group at this stage.

Stage III: Next, we may concentrate on the breaking of the flavor gauge symmetry to reduce it to $SU(2)_L \otimes SU(2)_R \otimes U(1)_L \otimes U(1)_R \otimes SU(3')_{L+R}$. This may be achieved by using multiplets of type $(15, 1, 1, 1) \oplus (1, 15, 1, 1) \oplus \ldots$ type and following the techniques outlined in ref. 47.

Finally since $SU(2)_L \otimes SU(2)_R \otimes U(1)_L \otimes U(1)_R$ is a subgroup of this group, in these theories, there is the possibility of having light Z-bosons ($m_Z < m_{W_L^+}$) while at the same time having a parity conserving neutral current structure.

VI. SOME EXPERIMENTAL IMPLICATIONS

In the preceding sections, we have tried to outline some of the salient theoretical features of the left-right symmetric gauge theories based on the gauge group $SU(2)_L \otimes SU(2)_R \otimes U(1)$ and other left-right symmetric groups. In this section, we would like briefly to mention some of the experiments that may be used to test these theories.

(A) Higher precision low energy weak interaction experiments:

We find it rather amazing that, while all low energy experiments are consistent with a pure V-A theory of weak interaction, the best available data such as the Michel parameter and polarization measurements do allow for as large as 13% admixture of V+A current.

In our framework, this allows for as light a mass for W_R^+ as 200 Gev. Better measurements of these and other parameters[29] may serve either to confirm the presence of V+A currents at this level or may increase the lower bound on $m_{W_R^+}$. Both these possibilities hold exciting prospects for weak interactions.

(B) Restoration of left-right symmetry at high energies

An important implication of spontaneously broken left-right symmetry is that at sufficiently high energies where the spontaneous breaking can be ignored, weak interactions will conserve parity. If indeed $m_{W_R^+} \simeq 200$ Gev, then at momentum transfers $q^2 \gtrsim 200$ $(\mathrm{Gev})^2$, significant departures from V-A nature should be observed. In particular, if one looks at the deep inelastic polarized muon-capture cross-sections at such q^2- values, the following behaviour will be observed:

$$\frac{d\sigma_{\mu_L} - d\sigma_{\mu_R}}{d\sigma_{\mu_L} + d\sigma_{\mu_R}} \xrightarrow[q^2 \to \infty]{} \frac{1}{q^2} \tag{VI.1}$$

whereas in conventional V-A theories, the above asymmetry will go to a constant value.

(C) Symmetry restoration in high magnetic fields and high temperature

It has been suggested in the literature[53] that under "acute environmental" conditions such as high temperature or high magnetic fields, a phase transition to a symmetric phase may take place in gauge theories. In such a situation, all effects that owe their origin to spontaneous breakdown will disappear. It has also been suggested that such conditions may have been present in the early stages of the universe (the first few seconds), when the temperature $T \sim 10^{16}$ oK is what is required. Similarly a large magnetic field $\sim 10^{14}$ Gauss may be found in nuclei. Further theoretical study of these questions is clearly of great significance.

(D) Massive neutrinos and their implications

A necessary consequence of left-right symmetric models is the presence of massive neutrinos with both left as well as right-handed helicity states. However, in the conventional weak interactions, only left-handed neutrinos will be produced and therefore, all neutrino induced experiments in the laboratory will have similar characteristics as the V-A theory would predict. However, in cosmic rays, due to Drell-Yan type mechanisms, both left and right-handed neutrinos and antineutrinos will be produced if left-right symmetric theories describe their interactions. The reason for this is that, in our model, the neutral gauge bosons couple to both left and right-handed neutrinos with comparable strength. The production mechanism is depicted in fig. 2.

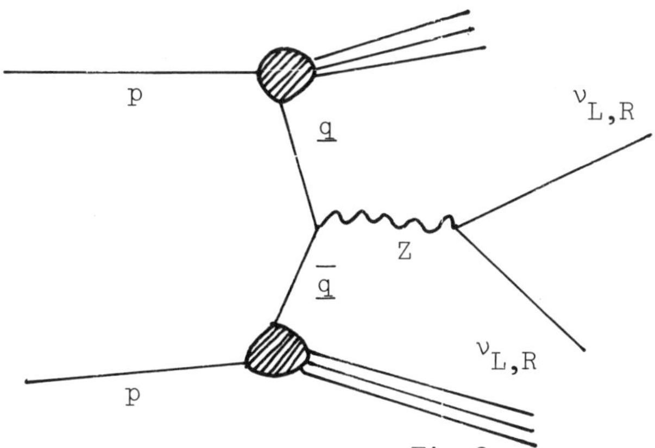

Fig.2

Therefore, experiments done with extraterrestrial neutrinos would exhibit different characteristics. For example,

$$\sigma^{\nu}_{N.C.}\Big|_{\text{cosmic }\bar{\nu}} > \sigma^{\nu}_{N.C.}\Big|_{\text{Lab}} \tag{VI.2}$$

So far we discussed possible phenomenological implications of left-right symmetric theories, which are independent of the detailed nature of the gauge symmetry breaking. Coming to the more specific case where the symmetry breaking is such that the resulting neutral currents are parity conserving, several tests may be proposed.

(E) Atomic parity violation experiments:

If the neutral currents are parity conserving, the atomic experiments ought not to show any parity violation to order G_F. It, however, turns out that, in the Weinberg-Salam model with $\sin^2\theta_W \simeq \frac{1}{4}$, the electron-current becomes pure axial vector and the vector part of the hadronic neutral current is such that for the hydrogen atom, it does not give any parity violation. Thus, experiments with Hydrogen[54] will be unable to distinguish between left-right symmetric models and the standard WSGIM model. However, experiments with the Deuterium atom can and so can experiments with heavier atoms such as Bismuth, Thallium, etc. since in these atoms, the standard model will lead to large parity violating effects.

(F) Deep inelastic polarized electron scattering experiments:

A promising experiment now in progress by a SLAC-Yale Bielefeld collaboration[55] which studies deep inelastic scattering of polarized electrons of nucleons can distinguish between our model and the standard WSGIM model. Here, one studies the following asymmetry parameter[56]:

$$A = \frac{d\sigma_L - d\sigma_R}{d\sigma_L + d\sigma_R} \tag{VI.3}$$

A=0 for models with parity conserving neutral currents.
Since the dominant contribution to the denominator comes from photon exchange graphs, for WSGIM model or any model with P-violating neutral current,

$$A \sim \frac{G_F q^2}{4\pi\alpha} \tag{VI.4}$$

In high energy experiments, $q^2 \sim 1\text{-}2\ (\mathrm{Gev})^2$. Thus, $A \sim 10^{-5}$. The above mentioned experiment[55] is apparently designed to reach this level of accuracy.

(G) Search for Neutral Z-bosons in e^+e^- Annihilation:

One distinguishing feature between the standard model and the left-right symmetric gauge models is the mass spectrum for neutral weak bosons in these theories. For example, if a weak neutral boson Z of $m_Z < m_{W_L^+}$ is found, one consistent interpretation of this in terms of left-right symmetric theories is that either (a) there is necessarily parity violation in order G_F in atoms or (b) if neutral current interactions conserve parity, the gauge group has to be at least $SU(2)_L \otimes SU(2)_R \otimes U(1)_L \otimes U(1)_R$. Needless to say this rules out the standard model. Thus, the search for weak boson resonances in e^+e^- annihilation is clearly a very crucial test of the ideas presented in this review.

VII. CONCLUDING REMARKS AND OUTLOOK

In this review, we have tried to outline the theoretical and experimental implications of a new approach to weak interactions where weak-interactions conserve parity at high energies. This concept implemented within an $SU(2)_L \otimes SU(2)_R \otimes U(1)$ model leads to several interesting consequences: (a) Manifest left-right symmetry of charged current weak interactions and the death of the high-y anomaly

(b) Parity conserving neutral currents of VV+AA type leading to neutrino neutral current interactions identical to the WS-GIM model as well as accommodating the null results in optical rotation experiments in high Z-atoms

(c) It gives an interesting relation between CP and P-violation and predicts $\eta_{+-} = \eta_{oo}$ although it is a milliweak model of CP-violation.

(d) We also suggest that an apparently attractive way to include more quarks and leptons without making the property of universality unnatural is to postulate the existence of stable or quasi-stable high mass hadrons.

We have tried to mention higher groups of which this group is a subgroup and also superunifying groups that lead to this $SU(2)_L \otimes SU(2)_R \otimes U(1)$ subgroup in the low energy domain. Possible experimental tests are suggested, first of the general idea of spontaneously broken left-right symmetry and then of one of its elegant outcomes - the parity conserving structure for neutral currents.

There are many things we would have liked to include in the review but could not due to length: among them are many interesting details, for which we refer the reader to original literature. At many points, we have simply advanced suggestions which ought to be pursued and some of which we are pursuing. Worthy of special mention among the things that we could not discuss is the recent interesting work on determining Cabibbo angle and CP-violating phases in terms of quark masses in $SU(2)_L \otimes SU(2)_R \otimes U(1)$ theories. Also, we have skipped many other types of vector-like models such as those based on $SU(3) \otimes U(1)$ groups[57], $SU(4) \otimes SU(4')$ groups[58] etc. which certainly are not yet ruled out by experiments.

Finally, for the skeptic, who wil ask, do we really need to go beyond the standard $SU(2)_L \otimes U(1)$ model, we once again reiterate the following: (i) here we have a very minimal extension of the standard model that provides an immensely more satisfactory basis for observed parity violation at low energies, yet preserves the nice properties of the standard model such as "natural" flavor conservation by the neutral gauge interactions, quark-lepton symmetry etc.; (ii) our model becomes, we believe, inevitable if indeed the search for parity violating effects of order G_F in atoms such as Bismuth, Thallium, Hydrogen and Deuterium fail. (iii) It is often alleged that the massive neutrino makes our models unattractive--we believe that a small neutrino mass in this model could very well owe its origin to radiative corrections.

The author would like to thank Prof. Robert E. Marshak for various discussions and comments on left-right symmetric models.

FOOTNOTES AND REFERENCES

1. E.C.G. Sudarshan and R. E. Marshak, Proceedings of the Padua-Venice Conference, 1957 and Phys. Rev. 109, 1860 (1958)
2. R. Feynman and M. Gell-Mann, Phys. Rev. 109, 193 (1958) See also J. J. Sakurai, Nuovo Cimento 7, 649 (1958)
3. R. Feynman, Photon Hadron Interactions, Benjamin (1972) J. D. Bjorken and E. A. Paschos, Phys. Rev. 185, 1975 (1969)
4. J. C. Taylor, "Gauge Theories of Weak Interactions", Cambridge Univ. Press (1975) M. A. B. Bég and A. Sirlin, Ann. Rev. of Nuc. Sc. 24, 379 (1973) H. Harari, Proc. of the Summer Institute on Particle Physics, SLAC Report No. 204 (1977) ed. by M. Zipf.
5. R. E. Marshak, Riazuddin and C. P. Ryan, "Theory of Weak Interactions in Particle Physics", John Wiley (1969)
6. S. Weinberg, Phys. Rev. Lett. 19, 1264 (1967) A. Salam, in Elementary Particle Theory: Relativistic Groups and Analyticity (Nobel Symposium No. 8), Ed. N. Svartholm (Almquist and Wikskells, Stockholm, 1968) p. 367
7. S. L. Glashow, J. Illiopoulos and L. Maiani, Phys. Rev. D2, 1285 (1970)
8. R. N. Mohapatra, "CP-Violation in Gauge Theories" - Particles and Fields - APS-DPF - '74; ed. by C. E. Carlson, p. 127 L. Wolfenstein, "Open Problems in Weak Interaction Theory" (to be published in Proc. of Aachen Conf.) (1976) A. Pais, "Five Decades of Weak Interactions" ed. by N. P. Chang (1977), p. 58
9. M. L. Perl et al, Phys. Rev. Lett. 35, 1489 (1975) G. J. Feldman et al, Phys. Rev. Lett. 38, 117 (1977)
10. S. W. Herb et al, Phys. Rev. Lett. 39, 252 (1977)
11. M. Kobayashi and S. Maskawa, Prog. of Theoretical Physics 49, 652 (1973) S. Pakvasa and H. Sugawara, Phys. Rev. D14 305 (1976)

L. Maiani, Phys. Lett. 62B, 183 (1976)

12. D. Soreide et al, Nature 264, 528 (1976)
P. Sandars, Review talk at the Ben Lee Memorial Conference (1977)
L. L. Lewis et al, Phys. Rev. Lett. 39, 795 (1977)
P. G. E. Baird et al, Phys. Rev. Lett. 39, 798 (1977)

13. M. A. Bouchiat and C. C. Bouchiat, Phys. Lett. 48B, 111 (1974); J. Physique 35, 899 (1974); 36, 493 (1975)
I. Khriplovich, JETP Lett. 20, 315 (1974)
E. M. Henley and L. Wilets, Phys. Rev. A14, 1411 (1976)
R. Lewis and R. Williams, Phys. Lett. 59B, 70 (1975)
M. Brinicombe, Loving and P. Sandars, J. Phys. B9, L237 (1976)
M. Grant, Pyper and P. Sandars, Oxford Preprint (1977)
R. Cowan, S. Meshkov and S. P. Rosen, NBA preprint (1977)
E. M. Henley, M. Klapisch and L. Wilets, Phys. Rev. Lett. 39, 994 (1977)
For a review, see G. Feinberg CU_TP-111 to be published in the Proc. of the Ben Lee Memorial Conference (1977)

14. J. C. Pati and A. Salam, Phys. Rev. D10, 275 (1974)
This model does not have "natural" left-right symmetry but however, introduces the left-right symmetric gauge structure as well as the fermion content.

15. R. N. Mohapatra and J. C. Pati, Phys. Rev. D11, 566, 2558 (1975) and G. Senjanovic and R. N. Mohapatra, Phys. Rev. D12, 1502 (1975 (1975). The first "natural" left-right symmetric model based on the gauge group $SU(2)_L \times SU(2)_R \times U(1)$ which is closer in spirit to the models of ref. 16, 17, 18 was constructed in these papers.

16. R. N. Mohapatra and D. P. Sidhu, Phys. Rev. Lett. 38, 667 (1977) and Phys. Rev. D16, 2843 (1977). "Naturally" left-right symmetric theories with neutral current structure substantially different from the standard model and relevant for the discussion of atomic physics experiments (e.g. parity conserving neutral

currents) were first suggested in this paper for the case of four and six quark flavors.

17. A. DeRujula, H. Georgi and S. L. Glashow, Annals of Phys. (N.Y.) 109, 242(1977). This paper discussed (independently of and simultaneously with the work of ref. 16) $SU(2)_L \otimes SU(2)_R \otimes U(1)$ models with six quarks and parity violating neutral current structure with two Z-bosons. Due to the presence of singlet leptons, this model does not have "natural" left-right symmetry but has massless neutrinoes unlike ref. 15 and 16.

18. H. Fritzsch and P. Minkowski, Nuc. Phys. B103, 61 (1976). The parity conserving neutral current structure that was studied in full generality in ref. 16 was also observed in this paper for massless quarks.

18a. The idea of left and right-handed charged currents coupling to different charged W-bosons was discussed in a purely phenomenological context by E. Lipmanov, Sov. Journal of Nuc. Phys. 6, 395 (1968).
The gauge group $SU(2)_L \otimes SU(2)_R \otimes U(1)$ was first used by S. Weinberg, Phys. Rev. Lett. 29, 388 (1972) to study approximate isospin invariance. The model had neither left-right symmetry in gauge coupling nor a left-right symmetric fermion assignment.

19. M. A. B. Bég and A. Zee, Phys. Rev. Lett. 30, 675 (1973)
A. de Rujula, H. Georgi and S. L. Glashow, Phys. Rev. D12, 3589 (1975)
R. Kingsley, S. Treiman, F. Wilczek and A. Zee, Phys. Rev. D12, 2768 (1975)
H. Fritzsch, M. Gell-Mann and P. Minkowski, Phys. Lett. 59B, 256 (1975)
S. Pakvasa, W. Simmons and S. F. Tuan, Phys. Rev. Lett. 35, 702, (1975)
G. Granco, T. Hagiwara and R. N. Mohapatra, Phys. Rev. D13,

104 (1976

R. N. Mohapatra, Phys. Rev. D13, 113 (1976)

E. Golowich and B. Hostein, Phys. Rev. Lett. 35, 831 (1975)

G. Branco and R. N. Mohapatra, Phys. Rev. Lett. 36, 926 (1976)

20. S. L. Glashow and S. Weinberg, Phys. Rev. D15, 1958 (1977)

21. E. A. Paschos, Phys. Rev. D15, 1966 (1977)

T. L. Trueman, F. E. Paige and E. A. Paschos, D15, 3416 (1977)

M. Chanowitz, J. Ellis and M. Gaillard, CERN Preprint (1977)

J. Kim and K. Kang, Phys. Lett. 64B, 93 (1976)

H. Harari, Phys. Lett. 57B, 265 (1975)

22. F. Hassert et al, Phys. Lett. 46B, 138 (1973)

J. Bliteschau *et al*, Nuc. Phys. B118, 218 (1977)

P. Wanderer *et al*, Brookhaven Preprint (1977), HPWF 77/1.

L. Stutte, Proc. of Int'l. conf. on New Quantum Numbers, ed. by D. Cline and J. Kolonko (1976)

M. Holder *et al*, CDHS Collaboration (1977)

23. F. Reines, H. Gurr and H. Sobel, Phys. Rev. Lett. 37, 315 (1976)

H. Faissner *et al*, Aachen Preprint - PITHA - (1977) - No. 91

24. R. N. Mohapatra, Frank E. Paige and d. P. Sidhu, Brookhaven Preprint, 23186 (1977)

25. J. C. Pati, S. Rajpoot and A. Salam, Phys. Rev. D (to be published)

26. See H. Georgi and S. Weinberg, Harvard Preprint HUTP-77/A052 for a generalization.

27. R. N. Mohapatra and D. P. Sidhu, Brookhaven Preprint (1977)

28. A. deRujala, H. Georgi and S. L. Glashow, Ann. of Phys. (N.Y.) 109, 258 (1977)

F. Wilczek and A. Zee, Phys. Lett. 70B, 418 (1977)

S. Weinberg, to be published in Festschrift in honor of I. I. Rabi (1977)

H. Fritzsch, Phys. Lett. 70B, 436 (1977)

R. N. Mohapatra and G. Senjanovic, Phys. Lett. B (to be published)

G. Bronco, Univ. of Bonn, Preprint (1977)

T. Hagiwara, K. Kitazoe, G. Mainland and K. Tanaka, Ohio Univ. Preprint, (1977)

29. M. A. B. Bég, R. V. Budny, R. N. Mohapatra and A Sirlin, Phys. Rev. Lett. 38 1252 (1977). The idea of manifest left-right symmetry was abstracted from the asymptotically parity conserving $SU(2)_L \otimes SU(2)_R \otimes U(1)$ model of ref. 15 and 16 and a detailed phenomenological analysis of charged current weak interaction for these models was done in this paper.

29a. Note a general mass matrix can be diagonalized by a bi-unitary transformation as $U_L M U_R^{-1} = M_d$ where U_L and U_R are determined as follows:

$$U_L M M^+ U_L^{-1} = M_d^2$$

and

$$U_R M^+ M U_R^{-1} = M_d^2$$

U_L and U_R give rise to the left and right-handed fermion mixing angles and phases.

30. R. Budny, Erice Lectures, (1977)

B. Holstein and S. Treiman, Phys. Rev. D16, 2369 (1977)

30a. For a review of trimuon phenomenology, see, V. Barger, talk presented at the Ben Lee Memorial Conference, 1977.

C. Albright, J. Smith and J. Vermaseren, Phys. Rev. D16, 3170 (1977)

V. Barger, T. Gottschalk, D. Nanopoulos and R. Phillips, Phys. Rev. D16, 2141 (1977)

31. Q. Shafi and C. Wetterich, Phys. Lett. 69B, 464 (1977)

G. Senjanovic and A. Sokorac, CCNY-HEP-1978

W. Marciano and A. I. Sanda, Rockefeller Preprint, COO-2232B-142 (1978)

32. J. C. Pati, S. Rajpoot and A. Salam, Phys. Rev. D (to be published)

33. R. N. Mohapatra and D. P. Sidhu, BNL preprint No. (1978)

34. M. Chanowitz, J. Ellis and M. K. Gaillard, CERN preprint TH.2312 (1977)

35. R. N. Mohapatra and D. P. Sidhu, Phys. Rev. D16, 2843 (1977)

36. L. M. Sehgal, Phys. Lett. 71B, 99 (1977)

37. V. Gupta and P. Roy, Tata Institute, preprint TIFR-TH-77-36 (1977)

38. P. Hung and J. Sakurai, Phys. Lett. 69B, 323 (1977)
L. Wolfenstein, in Proc. of the 1975 International Conference on Lepton-Photon at High Energies, (ed. W. T. Kirk) p. 613

39. J. Bernabeau and C. Jarkskog, CERN preprint (1977)

40. For some discussions of mixing angle θ in this model, see V. Elias, Phys. Rev. D16, 2284 (1977)

41. For any general parity violating Hamiltonian of the form,

$$H_{wk}^{(-)} = \frac{G_F}{\sqrt{2}} \bar{e}\gamma_\mu\gamma_5 e[g_V^u \bar{u}_{\gamma_\mu} u + g_V^d \bar{d}_{\gamma_\mu} d]$$

we have

$$Q_W = 2[g_V^u (Z+A) + g_V^d (2A-Z)]$$

42. E. N. Fortson et al and P. Sandars et al (see ref. 12)

43. T. D. Lee, Phys. Rep. C9, 148 (1974). Spontaneous CP-violation with many Higgs multiplets were first studied in this paper. In our case, however, we can get spontaneous CP-violation with just one Higgs multiplet ϕ.

43a. R N. Mohapatra and D. P. Sidhu, Phys. Rev. D17, (1978)

44. R. N. Mohapatra and J. C. Pati, Phys. Rev. D8, 2317 (1973)

45. L. Wolfenstein, Phys. Rev. Lett 13, 562 (1964)
For gauge theoretic realizations of this idea see:
R. N. Mohapatra, J. C. Pati and L. Wolfenstein, Phys. Rev. D11, 3319 (1975)
P. Sikivie, Phys. Lett. 65B, 141 (1976)

46. R. Cahn, these proceedings

47. M. A. B. Bég, R. N. Mohapatra, A. Sirlin and H. Tsao, Phys.

Rev. Lett. 39, 1054 (1977)

47a. Q. Shafi and C. Wetterich, Phys. Lett. B (to be published) The Higgs-boson sector and Z-boson mass matrix of this paper is in error and has been corrected in ref. 48. V. Elias, J.C. Pati and A. Salam, U. of Maryland, Technical Report (1977)

48. R.N. Mohapatra and D.P. Sidhu to appear, Brookhaven preprint (1978). Much of the details of this subsection Va are borrowed from this paper.
See also D.P. Sidhu, to appear as Brookhaven preprint (1978)

49. This group has considered in connection with left-right symmetry by:
Y. Achiman, Phys. Lett. 70B, 187 (1977)
J. Chakrabarti, CCNY-HEP-77/11
P. Minkowski, Univ. of Bern Preprint, 1977.

50. F. Gursey, P. Sikivie and P. Ramond, Phys. Lett. 60B, 177 (1976)

51. H. Fritzsch and P. Minkowski, Annals of Phys. (N. Y.) 93, 193 (1974)
H. Georgi and S. L. Glashow, "Particles and Fields" - 74 page, 575 (1974).
M. Chanowitz, J. Ellis and M. Gaillard, Ref. 21

52. J. C. Pati and A. Salam, D10, 275 (1974)
R. N. Mohapatra and J. C. Pati, D11, 2558 (1975)
V. Elias, J. C. Pati and A. Salam, ref. 47.

53. D. Khirtznis and A. Linde, P. N. Levedev Inst. Preprint (1975); Phys. Lett. 42B, 471 (1972)
S. Weinberg, Phys. Rev. D9, 3357 (1974)
R. Dolan and R. Jackiw, Phys. Rev. D9, 3320 (1974)
A. Salam and J. Strathdee, Nature 252, 569 (1974)

54. R. Cahn and G. Kane, Phys. Lett. 71B, 348 (1977)

55. C. Prescott, C. Sinclaire et al - quoted in ref. 56

56. R. Cahn and F. Gilman, SLAC-PUB-2002 (1977)

57. G. Segre and J. Weyers, Phys. Lett. 65B, 243 (1976)

P. Langacker, G. Segre and M. Golshani, U. of Penn Preprint (1977)

58. R. N. Mohapatra, Phys. Rev. D13, 113 (1976)

F. J. Yndurain, Nuc. Phys. B115, 293 (1976).

THE "NEW" U(1) PROBLEM*

Helen R. Quinn

Stanford University

Stanford, California 94305

Beford accepting the invitation to talk at this conference I pointed out that I had had no part in the most recent developments in the subject. However I agreed to present a review of those developments, along with some discussion of the work which prompted them, which I did about a year ago in collaboration with Roberto Peccei.[1]

The problem to which we addressed ourselves was that of CP invariance of the strong interactions. In general a theory will be CP invariant provided all masses and couplings which appear in the Lagrangian can be made real by a suitable redefinition of the fields in the theory. In the case of non-Abelian gauge theories there is an additional phase which appears and is related to the choice of vacuum. In such a theory, because the Hamiltonian and all observables commute with the generators of all gauge transformations, we know that the Hilbert space of the theory factorizes into subspaces labelled by their transformation properties under gauge transformations. The Θ parameter of the instanton picture labels such a property. In calculating any physical amplitude we find the term from the sector with topological charge q has a phase $e^{iq\theta}$. Hence unless $\theta \equiv 0$ there will in general be CP non-invariance in the theory.

Since we are talking of a strong interaction theory, for which I do not trust the dilute instanton gas approximation, I cannot actually calculate at present how big these CP violating effects would be. Naively however I estimate that they are of order $\frac{m_u}{m_H}\sin\theta$ where m_u=mass of the up-quark and m_H is a typical hadronic mass. The appearance of the quark mass factor will shortly be explained. Clearly θ must be a very small angle for the current limit on the neutron dipole to be in agreement with such a theory.

Given this fact, Peccei and I asked in what class of theories is θ naturally zero in the sense that it can be set to zero by a redefinition of the fields in the theory. This can clearly be done by making a chiral rotation of the fermion fields. Since when $\psi \to e^{i\gamma_5\alpha}\psi$ we have

$$\delta S = i\alpha\int d^4x\,\partial^\mu j^5_\mu \equiv -i\alpha(g^2/16\pi^2)\int d^4x F\tilde{F} + i\alpha\int d^4x\,\delta\mathcal{L} = -2i\alpha q + i\alpha\int d^4x\,\delta\mathcal{L}$$

we see that under such a rotation $\theta \to \theta - 2\alpha$. However if the fermion in question is massive this rotation also takes $m \to me^{2i\alpha}$ and hence introduces phases elsewhere in the theory. However if $\delta\mathcal{L} = 0$ under the rotation, that is if the Lagrangian in question possesses a chiral U(1) symmetry, then the only effect of this rotation is to redefine θ and the theory has natural strong CP invariance.

As I have already mentioned, if there is any quark in the theory which is truly massless then the Lagrangian possesses such a symmetry. The lightest quark, and hence the most likely canidate for masslessness is the u-quark. Current algebra calculations seem to prefer a non-zero value for its mass by about two standard deviations. As Heinz Pagels has mentioned earlier, setting m_u to zero would require enhancing the electromagnetic contributions to the mass differences by about a factor of four above the accepted values, a factor that is rather hard to come by in the context of those calculations. One point which has occurred to me is that it would be perhaps worthwhile to relax the standard SU(3) assumption that $\langle\bar{u}u\rangle = \langle\bar{d}d\rangle = \langle\bar{s}s\rangle$. However I doubt that any reasonable values for the ratios of these quanities would change the calculations sufficiently to give $m_u = 0$ as the preferred solution.

If we exclude this possibility how can we achieve the desired U(1)? Peccei and I considered the standard SU(2) × U(1) type Weinberg theory, and noticed that in this theory the chiral U(1) can be achieved provided the quarks get masses via the Higgs mechanism and there are at least two Higgs doublets in the theory. To see that we need two instead of the usual one doublet consider the two types of Yukawa couplings required to give mass to both up and down type quarks-

$$g\bar{\psi}_L\phi\psi_{u_R} + g'\bar{\psi}_L\tilde{\phi}\psi_{d_R} \qquad ,$$

where

$$\psi_L = \begin{pmatrix} \psi_{u_L} \\ \psi_{d_L} \end{pmatrix} \qquad \phi = \begin{pmatrix} \phi_0 \\ \phi_- \end{pmatrix} \qquad \tilde{\phi} = \begin{pmatrix} \phi^+ \\ \phi_0^* \end{pmatrix} \quad .$$

Clearly if the first term is to be invariant under $\psi_{u_R} \to e^{i\alpha}\psi_{u_R}$, $\psi_{u_L} \to e^{-i\alpha}\psi_{u_L}$ we must have $\phi_0 \to e^{2i\alpha}\phi_0$. Invariance of the second term would then require $\psi_{d_R} \to e^{-i\alpha}\psi_{d_R}$, $\psi_{d_L} \to e^{+i\alpha}\psi_{d_L}$.

This is certainly a U(1) symmetry, but it has net chirality zero. Such a rotation makes no net change to θ since under it $\theta \to \theta - 2\alpha_u - 2\alpha_d$. However the argument above gave $\alpha_d = -\alpha_u$, so that θ is unchanged by such a rotation.

Clearly this can be avoided by making the scalar doublet

coupled to ψ_{d_R} a different one from that coupled to ψ_{u_R}. One can impose a chiral U(1) symmetry under which $\psi \to e^{i\gamma_5 \alpha}\psi$ for all fermions but $\phi_1 \to e^{2i\alpha}\phi_1$ and $\phi_2 \to e^{-2i\alpha}\phi_2$. Clearly this leaves the Yukawa coupling terms invariant. However it places a constraint upon the scalar polynomial, namely that terms, such as $c(\phi_1^* \phi_2)^2$ + h.c., which are not invariant under the opposite rotations of ϕ_1 and ϕ_2 may not appear in $\mathcal{L}$. This form can readily be generalized to include Yukawa couplings to further quarks and to leptons. The U(1) symmetry then involves simultaneous rotations of all fermions.

As has been recently observed by Steve Weinberg[2] and independently by Frank Wilczek[3] the imposition of this constraint on the scalar polynomial has a rather drastic consequence-a new light pseudoscalar particle called an Axion (or sometimes Higglet) is predicted. To see this one need only observe that the additional U(1) we have imposed on $\mathcal{L}$ is spontaneously broken, so that one naively predicts the existence of a second neutral Goldstone particle in addition to that which is eaten by the neutral vector boson when it acquires its mass. However the particle is not a true Goldstone boson, because this U(1) is <u>not</u> a symmetry of the full theory in the Θ vacuum. Instanton effects induce additional terms in the effective scalar polynomial which violate the U(1) symmetry. This is exactly the effect observed by 't Hooft to be the resolution of the "old" U(1) problem.

The properties of the Higglet can be calculated by standard current algebra methods. This has been discussed by Weinberg[2] and by Bardeen and Tye[4]. One finds that the particle has a mass of about 50 ke V and a lifetime of the order of a second. The mass eigenstates are an admixture of $q\bar{q}$ and elementary scalar terms with a mixing angle proportional to f_π/f_{Higgs}. In the simple model discussed above $f_{Higgs} = (\lambda_1^2 + \lambda_2^2)^{\frac{1}{2}}$ where λ_1 and λ_2 are the vacuum expectation values of the scalar fields. This same combination of parameters appears in the vector meson masses, so that we find $f_{Higgs} = [\sqrt{2}G_F]^{\frac{1}{2}}$. The particle can thus be radiated by strong interactions through the $q\bar{q}$ admixture in its wave function, in particular the light quark admixture dominates production at present energies. There is some freedom to adjust the relative fractions of π^o and η that are mixed with the Higglet. I will shortly discuss the experimental constraints on this model, before I do so however I would like to comment that I believe the phenomenology is quite general, within the context of the standard SU(2) × U(1) model. For example I tried to construct a theory where the Higglet couples only to heavy quarks. To do so I had first to add additional scalar doublets, so that light quarks and heavy quarks obtained their masses from couplings to different doublets. This unfortunately introduced yet another Goldstone boson, which I could only avoid by gauging it away, at the price of a new neutral current. Even so when all was said and done I found I had achieved very little-the instanton terms effectively mix the light quarks back into the Higglet state, so that all I achieved was the result $f_{Higgs} \leq [\sqrt{2}G_F]^{\frac{1}{2}}$ replacing the equality of the simpler model. I could thus make the particle heavier, but only at the cost of increasing its strongly interacting component, not to mention the other uglinesses of my model. Another option, coupling more than one Higgs multiplet to the same quarks, could also modify the result, but in general it suffers the disease that the Higglet then couples to a flavor-changing current, a situation which is clearly untenable experimentally. However I have not explored models other than the standard SU(2) × U(1). Since, as we shall see, it is very unlikely that a particle such as I have described exists, it becomes very interesting to explore other alternatives. In particular one should investigate whether other weak gauge groups, or different representation content will allow the imposition of a chiral U(1) symmetry without the doubling up of Higgs multiplets and consequent additional (pseudo)-Goldstone particle.

Now to the experimental situation. Here I will be quoting

calculations made by Bjorken and his graduate student, Sinan Kaptanoglu[5], as well as the calculations discussed by Weinberg and also (in a private communication) by Gary Feinberg[6]. Let me begin first with the extreme cases. Kaptanoglu became interested in cosmological and astrophysical implications of the particle, an amusing game though it turned out not to produce very strong bounds. The first question he raised was the cosmological one-if the Higglet were stable enough to exist in the early universe long enough that it were out of equilibrium with the other particles before it decayed this would be a disaster. The subsequent decays would produce a photon distribution which would entirely destroy the 4° background radiation spectrum. However the constraint that the particle must decay before it goes out of equilibrium with the other particles produces only the very weak bound that it must be heavier than a few eV. A stronger, but more model dependent bound comes from the stability of red giant stars. A weakly coupled light particle provides a new energy transport mechanism which could affect the stability properties of the core of the red stars. Now we do not see the cores of these stars, we see only the hot gaseous envelope. Astrophysicists then make models of what they think the core must be like in order to maintain this hot envelope. The models have considerable variability in their estimates of the size and temperature of the core and the bound on Higglet mass depends sensitively upon these parameters. However Kaptanoglu (and Bjorken informs me Schramm has also made the calculation) estimated that this yielded a lower bound about a kilovolt on the particle mass.

At the other extreme is the hole-in-the-hill experiment of Schwartz et. al.[7] at SLAC. A detector was installed in a hole in the ground behind the beam dump area at SLAC and they looked to see what happened there. Not very much did. Bjorken has made some crude estimates of what they should see from Higglets Compton radiated by the SLAC electron beam as it is stopped in the dump and scattered in the active area of their detector. He estimates an upper bound on the Higglet mass of a few MeV on the basis of this experiment.

That leaves the interesting range where our calculations predicted the particle should be. In this range very severe constraints are imposed by the reactor experiments of Reines et al.[8] Weinberg and Feinberg have estimated that the ratio of Higglet production to photon production in the reactor should be of the order of 10^{-6} Higglets/photon. This then can be translated into a prediction of about 10^5 deuteron dissociation events/day due to Higglets and about 10^5 pulses/day in the NaI detector surrounding Reines $\bar{\nu}_e$ e scattering experiment. [Feinberg gives somewhat higher numbers, I quote here from Weinberg's reprint.] Since the observed rates are -2.9 ± 7.2 deuteron dissociation events/day and -160 ± 260 pulses/day in the NaI we seem to have a real problem. The predictions could be brought into agreement with these rates only by adjusting the Higglet mass so far that the bounds discussed previously come into play. The predictions can be somewhat reduced if the isospin one part of the quark admixture in the Higglet is reduced, so that the processes concerned must proceed chiefly through the isospin zero component, which couples less strongly to nucleons.

Roberto Peccei has raised another question about these predictions which I will now discuss, but I doubt that it could account for a discrepancy as large as 10^{3-4}. Peccei points out that both experiments have a threshold, the deuteron dissociation intrinsically, and in the NaI detector a cut-off at $E_\gamma \geq 1.5$ MeV was used. Further he observes that parity considerations show that nuclear transitions which are E-type for photons cannot proceed via Higglet emission, while M-type transitions may do so. Furthermore he notes that the spectrum of photons from the two types of transition is quite different, and in particular that many of the more energetic transitions are E2 transitions. Thus he states that, although 10^{-6} Higglet/photon may be the correct ratio when averaging over all energies, it may be a gross overestimate for the energy range relevant to these experiments. To make his observation quantitative rather than qualitative is a very grubby job, when one thinks of the number of different nuclear processes occurring inside the reaction. It seems unlikely to me that it would change the prediction by as much as 10^{-4}, but I mention it as a caution about the kind of theoretical uncertainties involved in this calculation.

All in all it appears now very unlikely that the predicted particle exists. I have already mentioned one option which might be possible, namely that there are theories with a chiral U(1) symmetry but no additional pseudo-Goldstone particles from the Higgs sector. The other possible way out of the contradiction is to abandon the Higgs mechanism and declare the fermion masses appear dynamically. In this case the original Lagrangian involves only massless quarks and the "new" U(1) is the old U(1) of chiral SU(N) × SU(N)- for which there is no longer any problem. At present no one can realistically calculate anything about dynamical mass generation but we should not forget that the possibility has not been ruled out. Should one be unable to find models with Higgs scalars that do not involve a Higglet, and should the experimental situation continue to mitigate against such a particle we may be forced to turn to this option if we are to achieve a theory which has natural strong CP conservation.

---

* Work supported by the Department of Energy and A. P. Sloan Foundation.

REFERENCES

1. R. D. Peccei and H. R. Quinn, Phys. Rev. Lett. 38, 1440 (1977); Phys. Rev. D16 , 1791 (1977).
2. S. Weinberg, "A New Light Boson?", Phys. Rev. Lett 40, 223 (1978).
3. F. Wilczek, "Problem of Strong P and T Invariance in the Presence of Instantons," Institute for Advanced Studies, Princeton Preprint, to be published in Phys. Rev. Letters.
4. W. A. Bardeen and S.-H. H. Tye, "Current Algebra Applied to the Properties of the Light Higgs Boson," Fermilab Pub. 77/110-THY.
5. J. D. Bjorken and S. Kaptanoglu, private communication
6. G. Feinberg, private communication
7. A. Rothenberg, Thesis, Stanford University, SLAC Report 147.
8. F. Reines, H. S. Gurr, and H. W. Sobel, Phys. Rev. Lett. 37, 315 (1976).
   H. S. Gurr, F. Reines and H. W. Sobel, Phys. Rev. Lett. 33, 179 (1974).
9. R. D. Peccei, private communication.

SUPERSYMMETRIC THEORIES OF PARTICLES*

P. Fayet**
California Institute of Technology

Pasadena, California

ABSTRACT

We show how spontaneously broken supersymmetric theories of weak, electromagnetic and strong interactions can be constructed.

A new class of particles is distinguished by means of a new quantum number, R. It includes fermions (new neutrinos such as the Goldstone neutrino, or heavy leptons) as well as bosons (heavy scalars associated with ordinary leptons and quarks). Supersymmetry also relates physical Higgs scalars with massive gauge bosons: they now sit in the same representations of supersymmetry, together with heavy leptons. The production and properties of the new states are discussed.

To describe gravitation, we also introduce the graviton and its spin $\frac{3}{2}$ partner, the gravitino; the latter becomes massive, while the Goldstone neutrino is eliminated. However, gravitational interactions of the massive gravitino do not vanish in the limit of small coupling constant κ, but give back the weak interactions of the Goldstone neutrino.

*Work supported in part by the U.S. Department of Energy under Contract No. EY76-C-03-0068.

**On leave of absence from Laboratoire de Phys. Theor. de l'Ecole Normale Superieure, Paris, France.

I. INTRODUCTION

a) THE SUPERSYMMETRY ALGEBRA

In supersymmetric theories[1], bosons and fermions are associated in multiplets. The supersymmetry generator Q_α transforms boson fields into fermion fields, and conversely:

$$[Q_\alpha, \text{boson field}] = \text{fermion field}$$
$$\{Q_\alpha, \text{fermion field}\} = \text{boson field} \tag{1}$$

Q_α is a Majorana spinor (with four real components in a Majorana representation where the γ matrices are real) satisfying the (anti) commutation relations:

$$[Q_\alpha, P^\mu] = 0 \tag{2}$$
$$\{Q_\alpha, \bar{Q}_\beta\} = 2(\gamma_\mu)_{\alpha\beta}P^\mu \tag{3}$$

Since $P^\mu P_\mu$ commutes with all other operators in the algebra, all particles in a given representation have the same mass; therefore, supersymmetry has to be broken. A spontaneous breaking generates a massless spin $\frac{1}{2}$ Goldstone particle, referred to as the "*Goldstone neutrino*"[2].

Moreover, in a gauge theory, supersymmetry associates to any gauge vector boson a massless spin $\frac{1}{2}$ particle. An interesting object will be the fermionic partner of the photon ("*photino*"). The question naturally arises whether these neutrinos can be identified with ν_e, ν_μ, or are indeed new neutrinos, and what are their roles in the theory of electromagnetic and weak interactions.

b) DESCRIBING THE FOUR INTERACTIONS

More generally we want to describe in a supersymmetric theory all interactions, strong, electromagnetic, weak and gravitational.

The point of view we shall adopt here is that gravitation can be turned off, while we still keep a theory for strong, electromagnetic and weak interactions; in other words, that there is a smooth limit $\kappa \to 0$, for which we can forget about gravitational effects.

The boson-fermion mass-splittings we shall consider will be finite in this limit*. We have to break both supersymmetry and weak-and-electromagnetic gauge invariance, and we shall get for both of them a single mass^2 breaking scale, namely m_W^2**.

Supersymmetry will lead us to introduce new types of fields; they will be distinguished by a non-zero value of a new quantum number, R. Some of these fields are heavy, some others stay massless, but their couplings involve heavy particles. Among the latter are the Goldstone neutrino and the photino already mentioned.

But supersymmetry also gives us relations between particles already used in the construction of gauge theories: Higgs scalars now join gauge vector bosons in massive multiplets of supersymmetry, together with heavy leptons.

c) MORE UNIFIED THEORIES?

Starting from theories with N = 1 spinorial generator, we can take pairs of multiplets[3] to form multiplets of a larger algebra with N = 2 spinorial generators[4], and again to obtain supersymmetry with N = 4 spinorial generators[5].

This will give us a still more unified theory; but the more spinorial generators we have, the more difficult it is to get, first a spontaneous breaking, and later a physically acceptable theory. The present situation for spontaneous breaking is summarized in the following table:

---

*In contrast with the possibility of mass-splittings of the order of the Planck mass ($\sim \kappa^{-1}$).

**Throughout the paper, we shall always describe the simplest possibility, although other ones do sometimes exist.

Number of spinorial generators	Spontaneous breaking of gauge invariance	Spontaneous breaking of supersymmetry
N = 1	Yes	Yes
N = 2	Yes	Yes
N = 4	Yes	No

Although we shall limit ourselves to theories with only one spinorial generator, it is quite possible that the multiplets we consider appear later as parts of larger multiplets, for an extended supersymmetry algebra.

II. SPONTANEOUS SUMMETRY BREAKING AND CONSERVATION LAWS

a) THE MULTIPLETS

Ordinary gauge theories involve three types of fields: vectors spinors and scalars. With supersymmetry, we use two types of superfields only:

- gauge superfields describe a vector and a Majorana spinor (which can be rewritten as a two-component Dirac spinor) both massless.
- chiral superfields describe a two-component Dirac spinor and a complex scalar.

Moreover, spontaneous breaking of gauge invariance will lead to massive gauge superfields[6], describing a vector, a Dirac spinor and a real scalar.

b) SPONTANEOUS BREAKING OF GAUGE INVARIANCE

The spontaneous breaking of an internal symmetry generates a massless Goldstone boson; if the symmetry is realized locally, the gauge vector boson acquires a mass by the Englert-Brout-Higgs-Kibble mechanism.

Moreover, in a supersymmetric theory, the Goldstone boson belongs to a massless chiral multiplet; the gauge boson, to a massless gauge multiplet.

If gauge invariance is spontaneously broken while supersymmetry remains conserved, the gauge multiplet becomes massive by joining the chiral multiplet containing the would-be Goldstone boson. Then it describes a vector, a Dirac spinor and a real scalar, with equal masses.

d) SPONTANEOUS BREAKING OF SUPERSYMMETRY

Using the anticommutation relation (3), we can express the Hamiltonian as

$$H = \frac{1}{4} \sum_{\alpha=1}^{4} Q_\alpha^2 . \qquad (4)$$

This has very important consequences for spontaneous breaking of supersymmetry, as can be understood from a naive argument: let $|\Omega>$, supersymmetric, and $|\Omega'>$, which is not, be two candidates for the vacuum state. The energy of $|\Omega>$ vanishes, whereas the energy (density) of $|\Omega'>$ is positive. Therefore, the supersymmetric state $|\Omega>$ is always stable.

This makes spontaneous supersymmetry breaking difficult, but not impossible. But it is necessary for that that no $|\Omega>$ state exists at all; then we have spontaneous supersymmetry breaking. Fig. (1), which represents in a schematic way the potential of scalar fields, shows the very special features of this breaking.

Spontaneous breaking of supersymmetry generates a massless Goldstone spinor.

The amplitude for emitting a Goldstone neutrino in a physical process satisfies low-energy theorems. Experimental data on β-

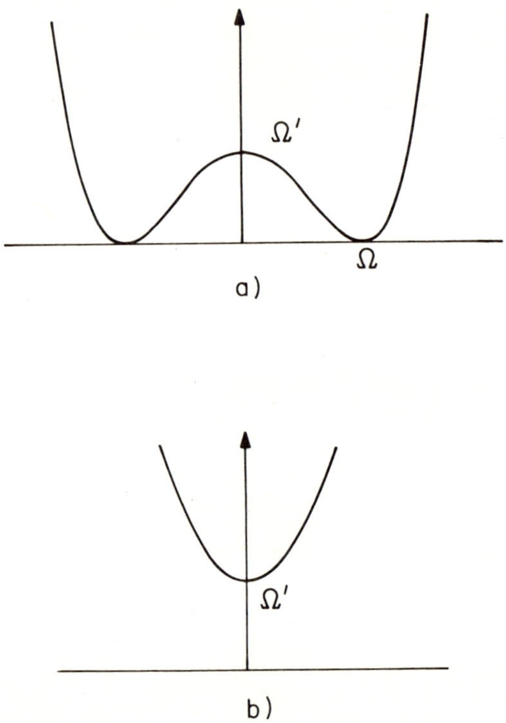

Figure 1. Representation of the potential in supersymmetric theories:
a) Ω, supersymmetric, is stable; Ω' is unstable: supersymmetry is conserved.
b) No Ω state exists; Ω', nonsupersymmetric, is stable: supersymmetry is spontaneously broken.

decay prevent to identify the Goldstone neutrino with the electron neutrino. On the other hand, an identification with the muon neutrino would make e-μ universality hard to understand. Therefore, the most natural idea is that the Goldstone neutrino is, indeed, a new neutrino.

e) R-INVARIANCE: A NEW QUANTUM NUMBER

We have to explain why this Goldstone neutrino remained unobserved up to now. It is, in fact, a member of a new class of particles for which we shall still use the word "leptons" (although some of them may have somewhat different properties).

The new class of leptons has its own conserved quantum number, R, which is carried by the spinorial generator in the supersymmetry algebra[7]. Inside a multiplet, bosons and fermions have values of R differing by one unit. Both the Goldstone neutrino and the photino have R = 1.

R-transformations act as a set of phase transformations for scalar fields, phase-and-γ_5 transformations for spinor fields. Such transformations have already been used for a long time in the construction of gauge theories, to restrict the Lagrangian density in a natural way[8]. They appeared to be very useful in the study of spontaneously broken supersymmetric theories.

R is now interpreted as the conserved quantum number associated with the new class of particles. This explains, for example, why a single new neutrino cannot be produced in a β-decay or a μ-decay. We shall come back in section III to the phenomenology of these new particles.

OTHER CONSERVED QUANTUM NUMBERS: INTRODUCING NEW SCALARS

Since we have only one spinorial generator, we can have, at most, one *internal* quantum number, R, carried by the spinorial generator in the supersymmetry algebra.

Other quantum numbers (electronic, muonic, baryonic,---) have to be defined as *external* quantum numbers; they have the same values for all particles in a multiplet.

To the ordinary leptons and quarks, we shall associate new scalar fields; let s and t denote the complex scalars associated with a left-handed or a right-handed Dirac spinor, respectively. The particle content for the electron and neutrino multiplets is given in the following table[9]:

Electron neutrino and associated scalars	Electron and associated scalars	Spin	R-Number	Electron Number
ν_{eL}	e_-	1/2	0	1
s_{eo}	s_{e-}	0	1	1
	t_{e-}	0	-1	1

We have a similar table for the muon sector, the quark sector, etc. Of course, these spinors and scalars are degenerated in mass as long as supersymmetry is conserved. The main difficulty is to obtain a symmetry breaking mechanism leading to large positive mass^2 for all these scalars: the state which one would like to choose as vacuum (i.e., with the desired symmetry properties: spontaneous breaking of supersymmetry, gauge invariance reduced to U(1) x color, conserved quantum numbers) is in general unstable, the scalar-boson mass^2 satisfying linear relations making some of them negative; but the corresponding translation is forbidden since it would break essestial invariances such as charge and color.

The solution to this problem gives constraints on the gauge group, the representations and the way they are coupled[9].

III. SPONTENEOUSLY BROKEN SUPERSYMMETRIC GAUGE THEORIES OF WEAK, ELECTROMAGNETIC AND STRONG INTERACTIONS

a) AN EXAMPLE

We now give an example for which the proposed program can be effectively realized. The gauge group is

$$SU(2) \times U'(1) \times U''(1) \times G_{color} \tag{5}$$

with $\vec{V}$, V', V'' and V_a as the gauge superfields.

In a first step, we realize a spontaneous breaking of gauge invariance, supersymmetry being conserved. For this we use two doublets of Higgs chiral superfields,

$$\begin{pmatrix} S_o \\ S_- \end{pmatrix} \quad \text{and} \quad \begin{pmatrix} T_o \\ T_- \end{pmatrix} \tag{6}$$

left-handed and right-handed, respectively. Their neutral scalar components acquire non-vanishing v.e.v, thus breaking the gauge group into a $U(1) \times G_{color}$ subgroup. These four Higgs chiral superfields are eliminated whereas four of the gauge superfields become massive, as explained in subsection IIb.

Then the model describes:

- massless color gauge multiplets with gluons and their spinorial partners ("*gluinos*")
- a massless gauge multiplet with the photon and the *photino*.
- four massive gauge multiplets with the heavy vector bosons W_-, Z', Z'', the physical Higgs scalars w_-, z', z'' and *heavy leptons* ℓ_-, $L-$, ℓ'_o, ℓ''_o.
- massless and massive matter multiplets, with ordinary leptons and quarks and *new scalars*.

In a second step (which makes use of one extra chiral super-field) supersymmetry is spontaneously broken. The Goldstone neutrino is a new neutrino, distinct from the photino. All scalar fields become heavy; such a result, almost trival for a nonsupersymmetric theory, is here very hard to obtain.

b) GENERAL FEATURES OF SUPERSYMMETRIC GAUGE THEORIES

Whatever the precise model may be, we can summarize the particle content of the theory as follows*:

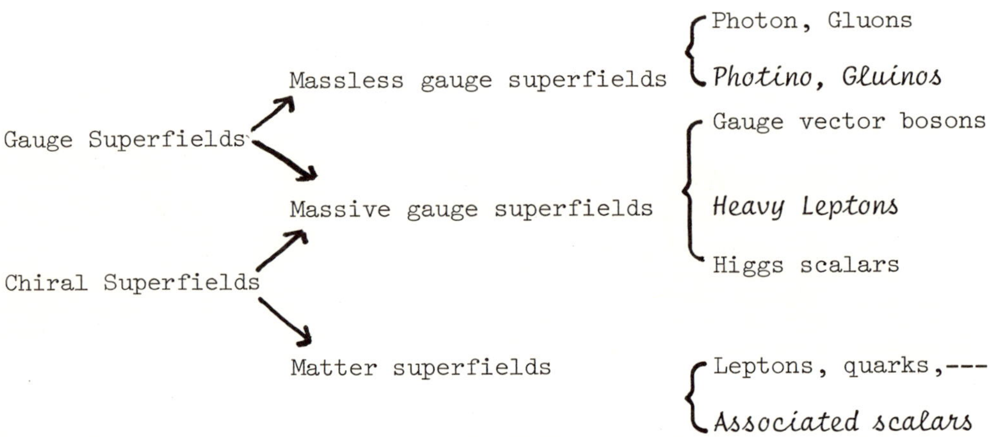

Italic characters have been used for the new particles carrying R.

---

*The Goldstone neutrino, which has not been indicated, appears, after spontaneous supersymmetry breaking, as a mixing of R = 1 - carrying neutral spinor fields.

c) USUAL WEAK, ELECTROMAGNETIC AND STRONG INTERACTIONS

Usual fermions, leptons and quarks, do not carry the new conserved quantum number R. Among the many scalars of the theory, these fermions can only exchange, at lowest order, scalars with R = 0; i.e., precisely the physical Higgs scalars (w_-, z' and z" in the previous example), which have very small couplings with light fermions.

For particles which do not carry R, one recovers weak, electromagnetic and strong interactions, mainly due to the exchange of gauge vector bosons, although the theory involves also many scalar bosons.

In the next paragraphs, we shall be interested in the particles which do carry R, fermions (Goldstone neutrino, photino, ---) as well as bosons (heavy partners of leptons and quarks, ---). They can be produced in scatterings, at least in pairs owing to R-conservation.

d) PRODUCTION AND INTERACTIONS OF NEW NEUTRINOS

They can be produced in pairs. Once produced, they have effective neutral current interactions with matter. Scattering with electrons is illustrated in Fig. 2.

e) PRODUCTION AND DECAYS OF NEW SCALARS

The new scalars associated with quarks and leptons can also be produced in pairs, provided the energy is high enough (see Fig. 3).

These heavy scalars have a very short lifetime: they can decay into the corresponding fermion, together with one of the new neutrinos (see Fig. 4)

Altogether, the creation and decay of a pair of s particles in e_+e_- annihilation

$$e_+e_- \to s_{e-}\bar{s}_{e-} \to e_+e_- + \text{Pair of new neutrinos}$$

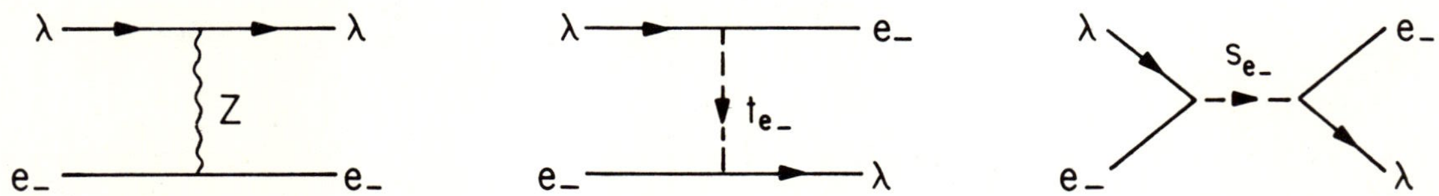

Figure 2. Scattering of new neutrinos (collectively denoted by λ) on electrons. The arrows show R-number conservation.

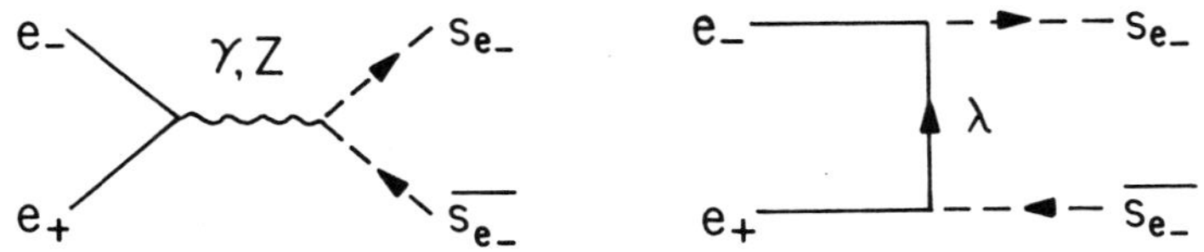

Figure 3. Production of a pair of new scalars $s_{e-}\overline{s_{e-}}$ in e_+e_- scattering. λ denotes any of the new neutrinos.

Figure 4. Possible decay modes of the new scalar bosons.

may manifest itself by the production of a non-collinear e_+e_- pair (or as well, $\mu^+ \mu^-$, quark-antiquark, etc), with missing energy carried away by neutrinos.

f) PRODUCTION AND DECAYS OF NEW HADRONIC STATES

Supersymmetry associates to the color octet of the gluons an octet of Majorana spinors (*gluinos*), which carry one unit of the R-quantum number. They are also massless and can be used to construct new color-singlet hadronic states carrying the R-leptonic number; for example, a quark, an antiquark and a gluino. One can ask whether such bound states should exist and what their masses may be. In any case, they would be unstable and decay into ordinary hadrons by emitting a new neutrino, in agreement with R-conservation.

An alternative to the possible existence of low-lying new hadronic states might be that the gluinos are indeed massive. In this prospect, theories which have no more a continuous R-symmetry but only a discrete one (R-parity, $(-1)^R$, still allows to distinguish between two classes of particles, but no more forbids gluinos mass terms) may be of interest.

IV. GRAVITATION

We shall now introduce gravitational interactions, which have a very small coupling constant κ (with dimension mass^{-1}). Then supersymmetry is realized locally[10] and associates to the massless spin-2 graviton a massless spin $\frac{3}{2}$ particle, the *gravitino*. The latter is coupled, with strength ~ κ, to the conserved vector - spinor current of supersymmetry.

a) GLOBAL SUPERSYMMETRY: INTERACTIONS OF THE GOLDSTONE NEUTRINO

A first important result, which will appear later on as connected with gravitation, can be derived in a spontaneously broken supersymmetric theory. Then we have a massless Goldstone neutrino, and masses of bosons and fermions in a multiplet are no more equal.

Both facts can be related: the coupling of the Goldstone neutrino with a boson-fermion pair can be derived from the boson-fermion mass splitting. More precisely, the mass spectrum of the theory (i.e., masses and mixing angles) will determine the interactions of the Goldstone neutrino[11].

b) LOCAL SUPERSYMMETRY: THE SUPER-HIGGS MECHANISM

Let us now suppose that supersymmetry is realized locally (the cosmological term coming from the non-vanishing vacuum energy density ε, in the global theory, being cancelled, as in ref. 12; a natural cancellation mechanism is still to be found). Then the gravitino becomes massive through the super-Higgs mechanism; it acquires new degrees of freedom, while the Goldstone neutrino is eliminated.

If we consider, as explained in the introduction, boson-fermion mass splittings small with respect to the Planck mass, then the gravitino mass

$$m_{3/2} \sim \kappa\sqrt{\varepsilon} \tag{7}$$

is very small. A numerical estimate may be

$$m_{3/2} \sim \frac{\kappa e}{G_{Fermi}} \sim 10^{-6} \text{ or } 10^{-5} \text{ eV}/c^2. \tag{8}$$

The Goldstone neutrino has been eliminated; nevertheless, we shall show that, since the gravitino is very light, its $\pm\frac{1}{2}$ helicity states behave as those of a Goldstone neutrino. In other words, we can forget about the super-Higgs effect for practical purposes of today's physics[11].

c) INTERACTIONS OF THE MASSIVE GRAVITINO

Let us study the amplitude for emitting a massive gravitino

with mass $m_{3/2}$, momentum $k^{\mu} >> m_{3/2}$, and helicity $\pm \frac{1}{2}$, in a physical process.

It is a gravitational effect: the gravitino field couples to the supersymmetry current with strength $\sim \kappa$. But a large factor ~ $\frac{k^{\mu}}{m_{3/2}}$ also comes from the gravitino wave function, with $m_{3/2}$ proportional to κ so that the gravitational coupling constant κ cancels out. Finally, we get back nothing but the amplitude for emitting the Goldstone neutrino of global supersymmetry in a weak interaction process such as those studied in section III.

Weak interactions of the Goldstone neutrino can be recovered as the limit of gravitational interactions of the massive gravitino, when κ gets small. This is, of course, connected with the fact that the interactions of the Goldstone neutrino depend only on the mass spectrum: they are reminiscent from gravitational interactions.

In conclusion, the super-Higgs mechanism leads to a mixing between gravitational and weak interactions. It gives to the gravitino, and to gravitation effects in particle physics, their chance to be detected.

REFERENCES

1. For review articles see: B. Zumino, Proc. of the 17th Int. Conf. on High Energy Physics, London 1974, ed. J. R. Smith (Rutherford Lab., Chilton, Didcot, Oxon, U.K.); P. Fayet and S. Ferrara, Physics Reports 32C, 249 (1977).
2. P. Fayet and J. Iliopoulos, Phys. Letters 51B, 461 (1974).
3. S. Ferrara and B. Zumino, Nucl. Phys. B79, 413 (1974); A. Salam and J. Strathdee, Phys. Letters 51B, 353 (1974).
4. P. Fayet, Nucl. Phys. B113, 135 (1976).
5. P. Fayet (1976), unpublished; J. Schwarz, these proceedings.
6. P. Fayet, Nuovo Cimento 31A, 626 (1976).
7. P. Fayet, Nucl. Phys. B90, 104 (1975).
8. P. Fayet, Nucl. Phys. B78, 14 (1974).
9. P. Fayet, Phys. Letters 69B, 489 (1977).
10. S. Ferrara, D. Z. Freedman and P. Van Nieuwenhuizen, Phys. Rev. D13, 3214 (1976); S. Deser and B. Zumino, Phys. Letters 62B, 335 (1976).
11. P. Fayet, Phys. Letters 70B, 461 (1977).
12. S. Deser and B. Zumino, Phys. Rev. Letters 38, 1433 (1977).

SPINNING STRING THEORY FROM A MODERN PERSPECTIVE[+]

John H. Schwarz

California Institute of Technology

Pasadena, California

ABSTRACT

The spinning string theory can be regarded as a locally supersymmetric theory in a superspace with ten Bose and sixteen Fermi dimensions. By making six of the Bose dimensions compact in a space of suitable topology, a four-dimensional theory with N=4 extended supersymmetry is obtained. The possibility that such a scheme could provide a unified description of all interactions is discussed.

INTRODUCTION

The preferred interpretation of the spinning string theory[1,2] has undergone a substantial evolution since its initial formulation in 1971*. The changing physical identification of the particles has been most striking. At first the motivation was to obtain a realistic theory of the hadrons.

+Work supported in part by the U.S. Department of Energy, San Francisco, under Contract EY76-C-03-0068.

*This has been reflected in an evolving name: dual pion model, Neveu-Schwarz model, dual fermion model, 10-dimensional string model, etc.

This restricted focus was a natural consequence of the historical origins in Regge theory and the Veneziano model[3]. In fact this viewpoint met with a certain degree of success. By identifying the ground state with the pion we were able to deduce the Lovelace-Shapiro $\pi\pi \to \pi\pi$ amplitude[4] and the Veneziano $\pi\pi \to \pi\omega$[3]. It was obvious from the beginning, however, that this approach had a number of serious shortcomings. For instance the π was a tachyon, the ρ was massless, the extension to SU(3) was unsatisfactory, and the fermions bore no resemblance with baryons.** Later we also learned that ten dimensions (9 space, 1 time) are required[6]. One subsequent suggestion, that already represented a substantial revision in thinking, was to identify the particles as quarks and gluons[7]. In this proposal (which also proved to have its shortcomings), six of the dimensions would be used to provide color degrees of freedom.

The possibility to be considered here, developed originally in collaboration with J. Scherk[8], involves identifying the (initially) massless particles of the string theory with the fundamental fields of nature-quarks, leptons, gluons, graviton, gauge bosons, etc. In this approach the dimensionless dual model coupling constant is essentially the electric charge, as in unified models of the Weinberg-Salam type, and the universal Regge slope is determined by Newton's constant (or equivalently, the Planck mass). Some recent theoretical developments[9)] (discussed below) make this suggestion even more attractive than when first proposed.

**Despite all these shortcomings the formulas can be modified to give a rather nice phenomenological description of the nonstrange mesons[5]. When doing this, one gives up all pretense of constructing a fundamental theory.

The free-particle spectrum of the spinning-string theory is free from ghosts (states of negative norm) for $D \leq 10$, (see Ref. 6), but only in exactly ten dimensions (9 space and 1 time) do multi-fermion trees and twisted loops have satisfactory duality properties required for unitarity. The spectrum consists of two distinct types of particles: open and closed strings. The states of each sector lie on linear, parallel Regge trajectories (in lowest order of perturbation theory), and the slope of closed-string trajectories is one-half that of the open-string trajectories. Each sector contains both bosons and fermions. Consistent subtheories are obtained by considering meson strings only, closed strings only, or closed meson strings only. However, the most interesting possibility involves none of these restrictions, but another one instead.

The open-string meson amplitudes allow the identification of a multiplicitively conserved quantum number (identified as "G parity" in the original paper) with respect to which the ground-state is odd. This state was identified as the "pion." It's mass is given by $\alpha' m^2 = -1/2$. It is the only tachyon in the open-string sector of the theory. (There is an analogous state in the closed string sector). It was clear from the beginning that one could restrict the model to the subspace of even "G parity" particles, which is free from tachyons. This restriction was not advocated partly because of our commitment to a hadronic interpretation, partly because of other hopes for eliminating the tachyon*, and

---

*Many people felt that the tachyon indicated that the perturbation theory was being formulated about an unstable vacuum and that a stable one should exist (as in the Higgs mechanism). The main problem with this approach was that the required off-mass-shell extrapolations proved extremely difficult to formulate.

partly because of the concern that fermionic couplings would restore the odd-G states through duality. The last point was resolved when Gliozzi, Scherk, and Olive (GSO)[9] demonstrated that a corresponding restriction could be made in the fermionic sector. Specifically, by requiring that the ground-state open-string fermion be simultaneously Majorana and Weyl (i.e., real and of definite handedness), they were able to show that exchange of even-G bosons in fermion-fermion scattering is dual to the exchange of even-G bosons in the crossed channel. They also pointed out that ten is the lowest number of dimensions in which the Majorana and Weyl conditions can be simultaneously realized. (The next one is eighteen). A spinor in 10 dimensions has $2^5=32$ components. However, once the Majorana and Weyl conditions are imposed this is reduced to sixteen real components describing eight physically distinct states of polarization.

The complete description of the string theory requires choosing an internal symmetry group for the open strings. The method of Chan and Paton[10] is evidently the only one possible. It requires that a U(n) group or product of U(n) groups be used* and that the particles be assigned to the adjoint representation. The origin of these restrictions is that on associating representation matrices $\lambda(1),\lambda(2),\ldots$ with the various lines, the duality-unitarity requirement becomes

$$\sum_i \mathrm{tr}[\lambda(1)\lambda(2)\ldots\lambda(M)\lambda_i]\mathrm{tr}[\lambda_i\lambda(M+1)\ldots\lambda(N)]$$

$$\propto \mathrm{tr}[\lambda(1)\lambda(2)\ldots\lambda(N)] \; . \qquad (1)$$

*The one exception is that a single SU(2) is allowed. The special feature of SU(n) for n=2 is the absence of d couplings.

Thus the intermediate set of matrices λ_i must be a completed set, and in particular there must be $\Sigma\, n_i^2$ of them. It should be emphasized that even when more than one U(n) factor group is included, only a single common coupling constant is allowed. This fact first becomes evident when the duality-unitarity properties of twisted-loop diagrams are considered. An equivalent way to understand this restriction is to note that the closed string states, which are necessarily singlets of the internal symmetry group, must couple universally (like gravity).

When the restrictions suggested by GSO are imposed the 10-dimensional spinning string theory is free from ghosts and tachyons. Moreover it then possesses local supersymmetry in ten dimensions! As far as I know this fact has not been fully proved, but there are several convincing indications. First, one can show that there are the same number of boson and fermion states at each mass level. For mass m given by $\alpha' m^2 = p$, the number of open-string boson states is the coefficient of q^{2p} in the "partition function"

$$f_B(q) = d(G)\,\frac{1}{2q}\prod_{n=1}^{\infty}\{(1+q^{2n-1})^8 - (1-q^{2n-1})^8\}\prod_{n=1}^{\infty}(1-q^{2n})^{-8}$$

$$= d(G)[8 + 128\, q^2 + \ldots]\ , \qquad (2)$$

while the number of fermions is given similarly by

$$f_F(q) = d(G)\ 8\prod_{n=1}^{\infty}(1+q^{2n})^8\,(1-q^{2n})^{-8}$$

$$= d(G)\ [8 + 128\, q^2 + \ldots]\ . \qquad (3)$$

The factor d(G) represents the dimension of the adjoint representation of the internal symmetry group G. As noted by GSO, these two expressions were proved equal in 1829 by Jacobi[11]. There are similar formulas for the closed-string degeneracies [without the d(G) factor]*. A second indication of the local supersymmetry of the full string theory is that of the ten-dimensional local field theory corresponding to the low-energy limit. (This will be discussed further in the next section.)

The spinning string theory appears to be renormalizable in a perturbation-theoretic sense. First of all, momentum integrations in loops are always Gaussian because of the Regge structure of amplitudes, and therefore there are no ultraviolet problems-even in ten dimensions. The theory does contain massless particles so one should also consider the infrared end. In ten dimensions there are no problems, of course, but when one plays the tricks (described below) required for a four-dimensional interpretation, the singularities characteristic of four dimensions appear. The behavior is in fact then exactly the same as in ordinary four-dimensional Yang-Mills and gravity theories and the same hopes that are customarily invoked should apply. The most troublesome divergences have no analog in ordinary field theory. They originate from summing the contributions of an infinite number of different internal particles in loop diagrams. There is a procedure of constructing counterterms that amounts to a coupling constant renormalization[12]. It would be sufficient to eliminate the infinities in fewer than ten dimensions, but we insist that D=10 and so a logarithmic divergence survives. It has been proved at the one-loop level (for the string model without intrinsic spin) that this remaining divergence can be absorbed in a slope-parameter renormalization[13].

---

*The coefficients in the series are the squares of those in the open-string partition function. Thus there are 8^2 bosons and 8^2 femions with $\alpha' m^2=0$, $(128)^2$ bosons and $(128)^2$ fermions with $\alpha' m^2=2$, and so forth.

Furthermore, some promising indications have been given that the proof can be extended to all orders, although as far as I am aware the argument has not yet been made airtight. The self interaction of the massless particles of the string theory may be consistently isolated in the low-energy limit ($\alpha' E^2 << 1$)*. In this limit the massless open-string states are described by the ten-dimensional supersymmetric Yang-Mills theory[9,14] :

$$\mathcal{L}_{10}^{YM} = -\frac{1}{4} F_{\mu\nu}F^{\mu\nu} + i\,\bar{\lambda}\Gamma\cdot D\lambda \quad . \tag{4}$$

The group index has been suppressed, and D represents a covariant derivative. $F_{\mu\nu}$, which is formed out of a 10-vector potential A_μ in the usual way, describes 8d(G) physical degrees of freedom since a massless vector has eight possible transverse polarizations. λ is a 32-component Majorana-Weyl spinor that also describes 8d(G) physical polarization states. The action is invariant under the supersymmetry transformations

$$\delta A_\mu = i\,\bar{\alpha}\,\Gamma_\mu\,\lambda \tag{5a}$$

$$\delta\lambda = \Sigma^{\mu\nu}F_{\mu\nu}\,\alpha \quad , \tag{5b}$$

where Γ_μ and $\Sigma_{\mu\nu}$ are 32 x 32 Dirac matrices (analogous to γ_μ and $\sigma_{\mu\nu}$ and α is an infinitesimal anticommuting 32-component Majorana-Weyl spinor. The theory, eq. (4), taken by itself is very singular in the ultraviolet since the high mass states that conspire to produce Regge behavior in the complete string theory are not included.

---

*One used to speak of the zero-slope limit, which is mathematically equivalent, but physically less clear.

Therefore when saying that it describes the low-energy limit well, we really only mean it in the tree approximation.

The corresponding low-energy limit of the closed-string sector also gives a supersymmetric ten-dimensional field theory - supergravity in ten dimensions[9]. The Lagrangian for this theory has not been fully worked out, but it is known to involve the following fields: (1) a "10-bien: V_μ^r describing a ten-dimensional graviton with 35 distinct polarization states, (2) a Rarita-Schwinger field Ψ_μ that is a ten-dimensional Majorana-Weyl spinor and gives 56 distinct polarization states, (3) an antisymmetric tensor field $B_{\mu\nu}$ giving 28 states, (4) a Majorana-Weyl spinor χ with eight polarization states, and (5) a scalar. Altogether these fields describe 64 Bose and 64 Fermi states that form an irreducible representation of supersymmetry in ten dimensions. When the interaction between open and closed strings is also considered, the Lagrangian in eq. (4) becomes covariantized in the usual way and some additional terms required by local supersymmetry also appear. The Lagrangian formed in this way is assured to reproduce the low-energy limit of the on-mass-shell string theory amplitudes in tree approximation.

The algebra of supersymmetry in ten dimensions contains the 55 generators of the Poincaré algebra and sixteen supersymmetries. The supersymmetry generators can be described by a Majorana-Weyl 32-component spinor S_a. They satisfy the anticommutation relation

$$\{S_a, S_b\} = i \quad [(\frac{1+\Gamma_{11}}{2})\Gamma \cdot P]_{ab} \qquad , \qquad (6)$$

where a right-handed projection operator appears as a consequence of the Weyl condition. An important fact about the 71-parameter superalgebra of ten-dimensional supersymmetry is that (unlike four-dimensional supersymmetry algebras) it cannot be obtained as a group contraction of a simple superalgebra[15]. This fact implies that one cannot add a cosmological term to ten-dimensional supergravity theory by embedding the ten-dimensional deSitter group [SO(10,1) or SO(9,2)] in a simple superalgebra. We also remark that a superspace formulation of these theories would require ten commuting and sixteen anticommuting dimensions. Presumably ten-dimensional supergravity can be understood as a geometrical theory of such a space.

In order to make contact with the real world one needs an interpretation in terms of four-dimensional spacetime. A method for achieving one has been extensively discussed in the literature[8,16,17]. The basic idea is that six of the spatial dimensions should form a compact space, whose size is sufficiently small (much less than a fermi) that it would not yet have been detected. This compactification can occur in a large number of topologically distinct ways of which only two will be discussed.* The most symmetrical choice is a six-dimensional sphere of radius L. In this case the ten-dimensional Lorentz-group symmetry is reduced to a direct product of the four-dimensional Lorentz group and SO(6)=SU(4), which is interpreted as an additional internal symmetry group. This choice is probably physically unacceptable because when the curvature scalar of the ten-dimensional supergravity Lagrangian is reinterpreted in four dimensions it gives rise, among other things, to the four-dimensional curvature scalar and a cosmological constant of the form const./L^2. We have already indicated that there can be no cosmological constant present in the initial ten-dimensional theory to cancel this effect. Thus one induces a deSitter radius for spacetime that is about the same size as the "internal" six-dimensional space. This appears to be a fatal objection to this topology, but the

---

*Cremmer and Scherk[17] have presented arguments indicating that the compactification may occur spontaneously. This raises the hope that the number of dimensions that become compactified and the topology they assume may perhaps be derivable from first principles.

question deserves further investigation. An alternative topology that seems more satisfactory consists of a direct product of six circles of circumference $L_1, L_2, \ldots, L_6$ (generalized torus). In this case the only remnants of the ten-dimensional Lorentz group besides the four-dimensional one are possible discrete transformations if any of the six radii should happen to be equal. A space of this type does not have any intrinsic curvature and therefore an induced cosmological term in four dimensions can be avoided. An additional technical advantage of this choice is that the normal mode expansion just involves multiple Fourier series, which helps to keep the mathematics simple. In fact this is the only case for which it has been demonstrated that the compactification does not destroy the desirable features of the loop amplitudes of the string theory[16] (although no problems are anticipated in other cases).

A four-dimensional Lagrangian is obtained from the ten-dimensional one (in the toroidal case) by forming

$$\mathcal{L}_4 = \frac{1}{L_1}\int_0^{L_1} dy_1 \cdots \frac{1}{L_6}\int_0^{L_6} dy_6 \mathcal{L}_{10} . \tag{7}$$

The fields $\phi(x,y)$ appearing inside $\mathcal{L}_{10}$ are expanded in Fourier series to give an infinite set of four-dimensional fields

$$\phi(x,y) = \sum_{\{n\}} \phi_{\{n\}}(x) \exp\left(2\pi i \Sigma \frac{n_i y_i}{L_i}\right) . \tag{8}$$

The four-dimensional mass of the $\phi_{\{n\}}$ quanta is then given by

$$M^2_{\{n\}} = \sum_i \left(\frac{2\pi n_i}{L_i}\right)^2 , \tag{9}$$

assuming that the ten-dimensional mass was zero. If one extends the notion of low-energy limit to include low compared to all these masses [as well as $(\alpha')^{-1/2}$], then we can restrict attention to the $n_i = 0$ mode. In this limit the resulting theory will display full O(6) or SU(4) symmetry since the shape of the compact space becomes irrelevant. Of course even so the symmetry group is smaller than the ten-dimensional one since rotations between the four spacetime and six internal-space dimensions no longer are symmetries. In fact the resulting subalgebra is precisely N=4 extended supersymmetry in four dimensions. When this reduction is performed on the Lagrangian in eq. (4) one obtains the N=4 supersymmetric Yang-Mills theory[9,14]. The gravity theory has not been analyzed as explicitly, but one knows what to expect from counting states. The ten-dimensional supergravity multiplet gives a reducible representation of the four-dimensional algebra. It consists of one N=4 supergravity multiplet and six N=4 matter multiplets. This system of closed string states is self-coupled with gravitational strength. If g denotes the dimensionless coupling constant of the open string Yang-Mills multiplet, then Newton's constant is identified as $G \sim g^4 \alpha'$, aside from factors of 2π[8,16]. This implies that if g is essentially the electric charge, the masses of neglected Regge recurrences are on the order of 10^{17} GeV.

From the point of view of the string theory it is not very meaningful to calculate beyond the tree approximation in the effective low-energy supergravity field theory, since neglected high mass states are expected to play an important role. On the other hand, higher order quantum effects of the Yang-Mills matter theory should be more important than effects of high mass states at sufficiently low energies.

N=4 supersymmetry has a number of special features in addition to its relation with the spinning string theory. For other choices of N (when N does not coincide with the dimension of ordinary spacetime) the internal symmetry part of the superalgebra is U(N), but for the N=4 case the internal symmetry is just SU(4). This is important for the string theory inasmuch as the internal symmetry originates from rotation in extra dimensions and SU(4) is the same as SO(6). Since no other U(N) is a rotation group this is another reason why the spinning-string "had" to be ten-dimensional. N=4 supersymmetry is the largest one that admits "matter" representations (defined to consist of particles with spin not exceeding J=1). Therefore if one is reconciled to the fact that nature cannot be described by a single gravity multiplet (based on N=8, for example), N=4 supersymmetry is the largest possible extension of the Poincare algebra*. On the other hand, if it is true that there is no renormalizable field theory in which gravity couples to matter not belonging to the same irreducible representation as the graviton, then I know of no alternative to the spinning string theory for getting a finite theory[19]. Another striking feature of N=4 supersymmetry concerns the charge renormalization function (called ψ or β) for the Yang-Mills theory. Since this function was known to vanish in the one-loop approximation, M. Gell-Mann and I conjectured that it may continue to do so to all orders. This was subsequently verified in the two-loop approximation[20]. This is the only four-dimensional theory known to have this property. It significance is not yet entirely clear.

Since the string theory provides the only known example of a renormalizable theory in which gravity couples to matter, a serious and systematic investigation of the possibilities for incorporating all the fundamental fields and interactions of nature seems warranted.

*Theories based on the "superconformal" algebra[18] may also be possible if certain fundamental issues can be resolved.

Here I will only raise some of the most obvious issues that bear on the problem and encourage others to help pursue them further.

Let us begin by summarizing the basic picture that emerges from the discussion in the preceding sections. We have a theory of strings, but the strings are so small - essentially the Planck length - that for all practical purposes they may be approximated by point particles. All that is lost by this approximation is the infinite set of super-heavy Regge recurrences that provide the cutoff ensuring the renormalizability of the full theory. If the theory is to describe nature, then it is also necessary that six dimensions form a small compact space. The size of this space provides the mass scale for a second infinite family of heavy particles. At energies low compared to this mass scale the only remaining fields are described by N=4 supersymmetric Yang-Mills theory (open strings) coupled to N=4 supergravity and six extra multiplets (closed strings).

The low-energy effective theory of all interactions other than gravity must therefore be given by N=4 supersymmetric Yang-Mills theory. The only allowed gauge group is a product of one or more SU(n)'s with a single common coupling constant. [Any U(1) factors would only couple with gravitational strength and are therefore ignored.] The SU(4) part of the supersymmetry algebra is not gauged. Therefore gauged color and flavor symmetries must be embedded in the Chan-Paton SU(n)'s. The most promising possibility is to use a single SU(n) group only. If there were more than one SU(n) factor particles associated with one factor would only interact with ones from the other with gravitational strength. It is also unacceptable to factor the flavor and color groups since we know that these are particles (namely the quarks) that are not singlets of either group. If we considered SU(5), for example, and suppose it breaks down to SU(3) × SU(2) × U(1) in a first stage of symmetry breaking so that the adjoint representation decomposes according to

$$24=(8,1)+(1,3)+(3,2)+(\bar{3},2)+(1,1) \quad , \qquad (10)$$

then we see that the representations ordinarily employed for gluons, weak gauge bosons, and quarks are present. However, leptons could only be accommodated as flavor singlets and triplets. Therefore $n \geq 6$ is probably a better choice.

Whatever group we choose, the N=4 supersymmetry implies that for each vector particle there are four Majorana (or Weyl) spinors, and six spin 0 fields. Perhaps the multiplicity of spinors can be employed to explain the observed reoccurrences of flavor representations [e.g. (e,ν_e), (μ,ν_μ), (τ,ν_τ), etc.], provided of course that the requisite symmetry breaking can be achieved. This would imply that there is still a fourth lepton family to be discovered.

The problem of symmetry breaking is a very serious one, especially as the string theory has so much symmetry and is so rigidly defined. One is simply not free to add new Higgs scalars, or interactions, or whatever "by hand". The theory has to do all its own dirty work. What causes the supersymmetry to be broken? If the compact space is not spherical that implies a breaking, but only at high energies. Another possibility may be that certain additional phases along the lines suggested by Cremmer and Scherk[16] can be incorporated in expansions of the type given in eq. (8). This would induce various masses and break the symmetry. One must then hope that explicit scalar self couplings or others induced by some dynamical mechanism are sufficient to break the gauge group symmetry in an appropriate fashion.

There are clearly a large number of difficult questions remaining to be straightened out. If nothing else, I hope that the reader is convinced that they at least merit investigation.

REFERENCES

1. P. Ramond, Phys. Rev. D3 (1971) 2415;
A. Neveu and J. H. Schwarz, Nucl. Phys. B31 (1971) 86;
Phys. Rev. D4 (1971) 1109;
A. Neveu, J. H. Schwarz, and C. B. Thorn, Phys. Letters 35B (1971) 529.

2. "Dual Theory," edited by M. Jacob, North Holland Pub. Co. (1974);
J. Scherk, Rev. Mod. Phys. 47 (1975) 123.

3. G. Veneziano, Nuovo Cimento 57A (1968) 190.

4. C. Lovelace, Phys. Letters 28B (1968) 265;
J. Shapiro, Phys. Rev. 179 (1969) 1345.

5. J. H. Schwarz, Phys. Rev. D5 (1972) 886.

6. P. Goddard and C. B. Thorn, Phys. Letters 40B (1972) 235;
J. H. Schwarz, Nucl. Phys. B46 (1972) 61;
R. C. Brower and K. A. Friedman, Phys. Rev. D7 (1973) 535.

7. J. H. Schwarz, Nucl. Phys. B68 (1974) 221.

8. J. Scherk and J. H. Schwarz, Nucl. Phys. B81 (1974) 118;
Phys. Letters 57B (1975) 463.

9. F. Gliozzi, J. Scherk, and D. Olive, Nucl. Phys. B122 (1977) 253;
Phys. Letters 65B (1976) 282.

10. Chan Hong-Mo and J. E. Paton, Nucl. Phys. B10 (1969) 519.

11. C. G. J. Jacobi, Fundamenta, Königsberg (1829).

12. A. Neveu and J. Scherk, Phys. Rev. D1 (1970) 2355;
D. J. Gross, A. Neveu, J. Scherk, and J. H. Schwarz, Phys. Rev. D2 (1970) 697.

13. M. Ademollo, A. D'Adda, R. D'Auria, F. Gliozzi, E. Napolitano, S. Sciuto, and P. Di Vecchia, Nucl. Phys. B94 (1975) 221;
J. A. Shapiro, Phys. Rev. D11 (1975) 2937;
R. Roth, Phys. Rev. D13 (1976) 1614.

14. L. Brink, J. H. Schwarz, and J. Scherk, Nucl. Phys. B121 (1977) 77.

15. This result was established in conversations with P. G. O. Freund.

16. E. Cremmer and J. Scherk, Nucl. Phys. B103 (1976) 399.

17. E. Cremmer and J. Scherk, Nucl. Phys. B108 (1976) 409; Nucl. Phys. B118 (1977) 61.

18. M. Kaku, P. K. Townsend, and P. van Nieuwenhuizen, Phys. Letters 69B (1977) 304; Phys. Letters 39 (1977) 1109.
S. Ferrara and B. Zumino, CERN Preprint TH.2418 (1977).

19. J. Scherk and J. H. Schwarz, Caltech preprint CALT-68-488 submitted to the 1975 Essay Competition of the Gravity Research Foundation.

20. E. G. Poggio and H. N. Pendleton, Brandeis Preprint, 1977; D. R. T. Joans, CERN Preprint TH. 2408, (1977).

COVARIANT QUANTIZATION OF SUPERGRAVITY

P. van Nieuwenhuizen*

CERN, Geneva, Switzerland

ABSTRACT

Unitarity requires a modification of the Faddeev-Popov quantization rules for Poincaré supergravity in the second-order formalism. In particular a four-ghost coupling is needed. By introducing a scalar, a pseudoscalar and an axial vector auxiliary field, the gauge algebra closes and ordinary Faddeev-Popov quantization is applicable. Elimination of the auxiliary fields <u>after</u> quantization reproduces the four-ghost coupling.

The status of the renormalizability of Poincaré supergravity is reviewed. The crucial question of three-loop finiteness is still unresolved. An explicit calculation of photon-photon scattering in the O(8) model is presented.

The gauge action of conformal supergravity is discussed. It arises naturally from gauging the graded conformal group, provided three constraints are imposed. These constraints turn out to be model-independent and to have a geometrical meaning. The gauge

*On leave from the Institute of Theoretical Physics, State University of New York, Stony Brook, N.Y. 11790

algebra closes and a connection between Poincaré and conformal supergravity arises.

1. INTRODUCTION

Supergravity is by now two years old, old enough to allow an assessment of its successes and shortcomings. From the beginning it was hoped that supergravity would bring significant progress to two important problems of theoretical physics:

i) the unification of all interactions, gravity included;

ii) the construction of a satisfactory quantum theory of gravity.

Models of supergravity can, in principle, unify the interactions of nature because they unify all particles of a given model into one (or sometimes two or more) superparticles and because all interactions arise by exchange of particles. To draw an analogy, consider nuclear physics where a proton and neutron form one nucleon. Exchange of the nucleon unifies the p-p, p-n and n-n forces into an isoscalar interaction. In supergravity, one superparticle can be a graviton, a photon, a gluon, an intermediate vector, an electron or a quark, etc., depending on the orientation of its "arrow in Fermi-Bose space", and the exchange of this superparticle leads at once to all interactions.

Supergravity is the gauge theory of supersymmetry (the latter is also sometimes called Fermi-Bose symmetry). Global supersymmetry models can be constructed in flat space-time and have nothing to do with gravity, but local supersymmetry requires curved space-time (see below) and is thus a theory of gravity. The symmetry between bosons and fermions gives all particles which constitute a superparticle the same mass. Thus symmetry breaking is needed, and recently progress was made here[1]. (Specifically: in the spin $(2, 3/2) + (\frac{1}{2}, 0^+, 0^-)$ system one can choose a potential such that there is no cosmological term and the spin 3/2 acquires a mass through the Higgs mechanism.) However, much remains to be done before phenomenology is possible. In particular, even the largest of all models with only one superparticle (the so-called O(8) model) has not enough internal symmetry to accommodate an $SU_3 \times SU_2 \times U_1$ group, and does not contain W^+, W^-, μ^- particles[2]. On the other hand, it is the only theory which predicts quark charges of 2/3 and -1/3.

The problem of quantum gravity is its dimensional coupling constant. In the relativistic extension of Newton's theory (general relativity), gravitons couple to energy instead of mass, so that the effective coupling constant in gravity, and in gravity alone, depends itself on energy. Thus in a one-loop Feynman diagram one has as always the divergences due to a summation over all virtual energies, but in quantum gravity the integrand itself blows up with energy. Thus one has worse divergences than in, say, Yang-Mills theory, and true renormalizability can never occur, but only the two extremes: finiteness of the S matrix or non-renormalizability. Finiteness is only possible due to "miraculous" cancellations which follow from some symmetry, and there is a possibility that local supersymmetry is that symmetry, as we shall see.

2. BASICS OF SUPERSYMMETRY

Under a Fermi-Bose rotation a boson B rotates into a fermion F as follows:

$$\delta B = F\varepsilon \quad .$$

It follows that ε must be anticommuting and a half-integer spinor, in order to keep both sides of the equation the same (fermions must anticommute since otherwise their actions vanish identically). The simplest choice, ε a spin $\frac{1}{2}$ object, leads to supersymmetry. It follows that in supersymmetry the elementary building blocks are doublets of one fermion and one boson with adjacent spins $(J, J + \frac{1}{2})$ or $(J, J - \frac{1}{2})$.

Since the dimension of a boson is 1 and of a fermion 3/2, it follows that ε has dimension $-\frac{1}{2}$, so that in the inverse rotation $\delta F \sim B\varepsilon$, there is a gap of one unit of dimension. This gap can be filled by either a derivative, or the gravitational constant (or a mass, but since we are heading for a gauge theory we exclude masses). Thus

$$\delta F = \partial_\mu B\varepsilon \qquad \text{or} \qquad \frac{1}{\kappa}(\partial_\mu + B)\varepsilon \tag{1}$$

omitting gamma matrices, etc. The commutator yields*

$$[\delta(\varepsilon_1), \delta(\varepsilon_2)]B = 2(\bar{\varepsilon}_2\gamma^\mu\varepsilon_1)\partial_\mu B + \partial_\mu\left(\frac{\bar{\varepsilon}_1\varepsilon_2}{\kappa}\right) \tag{2}$$

so that space-time symmetries ($\partial_\mu = P_\mu$ is part of the Poincaré group) and internal symmetries (the last term is an Abelian gauge transformation with parameter $\bar{\varepsilon}_1\varepsilon_2/\kappa$) become unified thanks to the introduction of the notion of fermionic symmetry. In operator notation, with $\delta(\varepsilon)B = [\bar{\varepsilon}Q, B]$ one thus has

$$[\bar{\varepsilon}_1 Q, \bar{\varepsilon}_2 Q] = 2(\bar{\varepsilon}_2\gamma^\mu\varepsilon_1)P_\mu + \text{internal symmetries} \tag{3}$$

Suppose that one deals with local supersymmetry, so that $\varepsilon = \varepsilon(x)$. Then the right-hand side describes a translation P_μ over a space-time dependent distance $\xi^\mu(x) = 2\bar{\varepsilon}_2(x)\gamma^\mu\varepsilon_1(x)$, which is the notion of a general co-ordinate transformation. Thus local supersymmetry leads to gravity! The converse holds, too. If one has in curved space a constant ε, then, after a gravitational gauge transformation, ε will have picked up a dependence on space-time.

What are the gauge fields of local supersymmetry? Quite generally, gauge fields are obtained by adding four-vector index μ to the parameters. Hence, the supersymmetry gauge field is the vectorial spinor ψ_μ. Its spin content is

*More precisely: in O(n) models the Abelian parameter is $\bar{\varepsilon}_1^i\varepsilon_2^j\varepsilon_{ij}$. In models with only one ε, there is no Abelian parameter.

$$\text{spin } \psi_\mu = \frac{1}{2} \otimes (1+0) = \frac{3}{2} + \frac{1}{2} + \frac{1}{2} \quad . \tag{4}$$

The lower gauge components with spin $\frac{1}{2}$ ($\partial.\psi$ and $\gamma.\psi$) are needed in the action in order that the theory be consistent. One can formulate a theory with only spin $\frac{3}{2}$ in the action (and helicity $\pm\frac{3}{2}$ physical states only), but this theory is then not ordinary supergravity but conformal supergravity with higher derivatives (see below).

From a group-theoretical point of view, supergravity can be obtained by gauging the graded Poincaré group, whose generators are P_μ, $M_{\mu\nu}$ and Q_a, satisfying

$$[Q_a,P_\mu] = 0; \; [Q_a,M_{\mu\nu}] = (\sigma_{\mu\nu})_a{}^b Q_b; \; \{Q_a,Q_b\} = (\gamma^\mu e)_{ab} P_\mu \tag{5}$$

where C must be the charge conjugation matrix in order that the Jacobi identities are satisfied. Thus one expects <u>three</u> gauge fields: for P_μ, for $M_{\mu\nu}$ and for Q_a. The gauge field for P_μ is the vierbein field e_μ^a (which is the square root of $g_{\mu\nu}$, $e_\mu^a e_{a\nu} = g_{\mu\nu}$), that of $M_{\mu\nu}$ is the spin connection ω_μ^{ab} and that of Q_a is ψ_μ. Since e_μ^a and ψ_μ both have two physical states with helicities ±2 and $\pm3/2$, respectively, there are already equal numbers of boson and fermion states and $\omega_{\mu ab}$ must be nonphysical. In supergravity based on the graded Poincaré group (Poincaré supergravity, or, just supergravity) this follows from the field equation of $\omega_{\mu ab}$ (Palatini formalism), but in conformal supergravity it follows from a constraint (vanishing of the translational curvature, see below).

3. THE GAUGE ACTION NAIVELY QUANTIZED

The gauge action of Poincaré supergravity is the sum of the Einstein action and the Rarita-Schwinger action in curved space

$$L = -\frac{e}{2\kappa^2} R - \frac{e}{2}\bar{\psi}_\mu R^\mu \tag{6}$$

$$R = e_b^{\ \mu} e_a^{\ \nu} (\partial_\mu \omega_\nu^{\ ab} + \omega_\mu^{\ ac} \omega_{\nu c}^{\ \ \ b} - \mu \leftrightarrow \nu) \tag{7}$$

$$R^\mu = e^{-1} \varepsilon^{\mu\nu\rho\sigma} \gamma_5 \gamma_\nu (\partial_\rho + \tfrac{1}{2} \omega_{\rho ab} \sigma^{ab}) \psi_\sigma \quad . \tag{8}$$

It is thus linear in the spin 2 and spin 3/2 curvatures. The field equation for $\omega_{\mu ab}$ is nonpropagating and can be solved algebraically (an elementary exercise)

$$\omega_{\mu ab} = \omega_{\mu ab}(e) + \frac{\kappa^2}{4} (\bar{\psi}_\mu \gamma_a \psi_b - \bar{\psi}_\mu \gamma_b \psi_a + \bar{\psi}_a \gamma_\mu \psi_b) \tag{9}$$

where $\omega_{\mu ab}(e)$ follows from the metric postulate

$$\partial_\mu e^a_{\ \nu} + \omega_\mu^{\ ab}(e) e_{b\nu} - \Gamma^\alpha_{\mu\nu}(g) e^a_{\ \alpha} = 0 \quad . \tag{10}$$

Thus supergravity is a theory with torsion since $\omega_{\mu ab} \neq \omega_{\mu ab}(e)$.

Substituting the solution for $\omega_{\mu ab}$ back into the action, it follows that one never need consider variations of $\omega_{\mu ab}$ since they always are multiplied by zero

$$\frac{\delta I(e,\psi,\omega)}{\delta\omega} \left(\frac{\delta\omega}{\delta e} \delta e + \frac{\delta\omega}{\delta\psi} \delta\psi\right) = 0 \text{ since } \left.\frac{\delta I}{\delta\omega}\right|_{e,\psi} = 0 \quad . \tag{11}$$

This obvious but invaluable trick treats ω as if it were an independent field (with $\delta\omega = 0$), but one really is working all the time in a second order formalism with only e and ψ as independent fields (in order to satisfy $\delta I/\delta\omega = 0$). It combines the simple features of the second order formalism[3] (where the transformation rules δe and $\delta\psi$ are simple but a complicated four-fermion term is present) with those of the first order formalism[4] (where the complicated four-fermion term is replaced by the independence of the field ω but where $\delta\omega$ is complicated). It has been of incomparable help in constructing new models[5] and has been named the 1.5 order formalism[6].

The action is invariant under the following transformation rules

$$\delta e^a{}_\mu = \kappa\bar{\varepsilon}\gamma^a\psi_\mu \;;\quad \delta\psi_\mu = \frac{2}{\kappa}\left(\partial_\mu + \frac{1}{2}\,\omega_\mu{}^{ab}\sigma_{ab}\right)\varepsilon \tag{12}$$

where, we repeat, $\omega = \omega(e,\psi)$. The invariance is shown as follows. Varying (only 1.5 order!) the vierbein fields in the Einstein action yields the Einstein tensor. Variation of ψ_σ in $D_\rho\psi_\sigma$ yields the commutator $[D_\rho, D_\sigma]$ hence a Riemann curvature. Variation of $\bar{\psi}_\mu$ yields the same result after partial integration and all curvature terms cancel. In the process of partially integrating one picks up terms proportional to $\partial_\mu\gamma_\nu$ and to $\omega_{\mu ab}$, while variation of the vierbein in γ_ν yields a $\psi^3\varepsilon$ term. These last three terms sum up to the torsion field equation in (9) and cancel too.

The gauge action (6) is invariant under (12) and under general co-ordinate and local Lorentz transformations, given by

$$\delta_G\begin{pmatrix} e^a{}_\mu \\ \psi_\mu \end{pmatrix} = \xi^\alpha\partial_\alpha\begin{pmatrix} e^a{}_\mu \\ \psi_\mu \end{pmatrix} + \begin{pmatrix} e^a{}_\alpha \\ \psi_\alpha \end{pmatrix}\partial_\mu\xi^\alpha \tag{13}$$

$$\delta_L e^a{}_\mu = \lambda^a{}_b e^b{}_\mu \;;\quad \delta_L\psi_\mu = \frac{1}{2}\lambda_{ab}\sigma^{ab}\psi_\mu \;. \tag{14}$$

If one were to quantize naively, one would add gauge fixing terms; for example

$$L(\text{fix}) = -\frac{1}{4}(\partial_\mu\sqrt{g}g^{\mu\nu})^2 + \frac{1}{4}\,\bar{\psi}\cdot\gamma\not{\partial}\gamma\cdot\psi + (e_{a\mu} - e_{\mu a})^2 \,. \tag{15}$$

The ghost actions would then follow as usual, starting with

$$L(\text{ghost}) = \bar{C}^\nu\Box C^\nu - \bar{C}_\gamma{}^\mu\partial_\mu C - \bar{C}^{ab}\,C_{ab} \tag{16}$$

where the gravitational ghosts C^ν and C^{ab} anticommute but the supersymmetry ghost C (a four-component complex spin $\frac{1}{2}$ field) commutes. The necessity of these opposite statistics not only follows from general theoretical arguments, but also from a theoretical "experiment"[7]. Calculating the self-energy of a graviton due to a spin 3/2 loop, one finds that the conservation equation $p_\mu D_{\mu\nu,}$ p_σ only holds if the spin 3/2 ghost field is commuting (so that no minus sign for the loop is added).

At this point propagators and vertices are given and one can calculate loop processes to investigate the renormalizability of supergravity. However, it is clear that there is something unusual going on as far as unitarity is concerned. Usually the ghost fields remove the unphysical modes of the gauge fields from the quantum theory, so that one ends up with two modes for a massless field at the quantum level. But the field ψ_μ has 4 (4-component spinors) times 4 (index μ) divided by 2 (since fermions have half as many degrees of freedom as bosons) = 8 components, while C has 4 components (ghosts are complex!). Apparently, there are 8 - 4 = 4 quantum modes of ψ_μ. There is really no problem as we shall see, but new quantization rules are needed[8]. The root of the problem can be traced back to a birth defect of supergravity, the non-closure of the gauge algebra[9].

4. THE GAUGE ALGEBRA

The gauge algebra of supergravity in second order formalism with only e^a_μ and ψ_μ as independent fields, does not close. For the commutator of two local supersymmetry transformations of e^a_μ one finds[9]

$$[\delta(\varepsilon_1), \delta(\varepsilon_2)]\, e^a{}_\mu = 2\, \bar{\varepsilon}_2\gamma^a D_\mu\varepsilon_1 - (1 \leftrightarrow 2) =$$

$$(t^\lambda\partial_\lambda e^a{}_\mu + e^a{}_\lambda\partial_\mu t^\lambda) + (t^\lambda\omega_{\lambda ab}e^b{}_\mu) + (-\frac{\kappa^2}{2}\, t^\lambda\bar{\psi}_\lambda\gamma_a\psi_\mu) \tag{17}$$

i.e., a sum of a general co-ordinate transformation G with parameter $t^\lambda \equiv 2\,\bar{\varepsilon}_2\gamma^\lambda\varepsilon_1$, a local Lorentz transformation L with parameter $t^\lambda\omega_{\lambda ab}$ and a local supersymmetry transformation S with parameter $-\frac{\kappa}{2}t^\lambda\psi_\lambda$. But a tedious calculation for the same commutator acting on ψ_μ yields extra terms

$$[\delta(\varepsilon_1),\delta(\varepsilon_2)]\psi_\mu = [\delta_G(t^\lambda) + \delta_L(t^\lambda\omega_{\lambda ab}) + \delta_S(-\tfrac{\kappa}{2}t^\lambda\psi_\lambda)]\psi_\mu$$

$$+ (\bar{\varepsilon}_1\gamma^\alpha\varepsilon_2)[\tfrac{1}{4}\gamma_\alpha g_{\mu\rho} + \tfrac{1}{2}\varepsilon_{\mu\alpha\rho\tau}\gamma_5\gamma_\tau]R^\tau$$

$$+ (\bar{\varepsilon}_1\sigma^{\rho\sigma}\varepsilon_2)[\tfrac{1}{2}\sigma_{\rho\sigma}g_{\mu\tau} + g_{\mu\rho}g_{\sigma\tau} + \tfrac{e}{2}\varepsilon_{\rho\sigma\tau\mu}\gamma_5]R^\tau \tag{18}$$

where R^τ is the spin 3/2 field equation. Thus off-shell ($R^\mu \neq 0$) the gauge algebra does not close. For all other commutators one finds at the right-hand side a sum of gauge transformations with field-dependent parameters.

In more formal and general notation[10)], we have found the following. If a gauge transformation with parameter ξ^α of a gauge field ϕ^i is given by

$$\delta(\xi)\phi^i = R^i{}_\alpha(\phi)\xi^\alpha \tag{19}$$

then the commutator algebra is given by

$$[\delta(\eta),\ \delta(\xi)]\phi^i = R^i{}_{\alpha,j}\, R^j{}_\beta\, (\eta^\beta\xi^\alpha - \xi^\alpha\eta^\beta)$$

$$= R^i{}_\gamma f^\gamma{}_{\alpha\beta}(\phi)\xi^\alpha\eta^\beta + \eta^{ij}_{\alpha\beta}\, I,_j\xi^\alpha\eta^\beta \quad . \tag{20}$$

The parameter of the gauge transformations on the right-hand side is $f^\gamma_{\alpha\beta}(\phi)\xi^\alpha\eta^\beta$ and is, in general, field dependent since the structure constants of the local algebra $f^\gamma_{\alpha\beta}$ do in general depend on ϕ^i. The nonclosure function $\eta^{ij}_{\alpha\beta}$ is nonzero only if i,j are the indices of a gravitino field and α,β refer to local supersymmetry. Note that we define, $I,_j$ by right derivatives; if $\phi^j = \psi_\mu$ then $I,_j = \delta I/\delta\psi_\mu$ and

$$\delta I/\delta\psi_\mu = -D_\rho\bar{\psi}_\sigma\gamma_5\gamma_\nu\varepsilon^{\mu\nu\rho\sigma} = (CeR^\mu)^T \tag{21}$$

where C is the charge conjugation matrix ($C = \gamma_o$) (for a Majorana spinor one has $\lambda = C\bar{\lambda}^T$).

One can repeat the foregoing for supergravity in first order formalism. One finds then that $\eta^{ij}_{\alpha\beta}$ is nonzero for i,j any of the fields e^a_μ, ψ_μ or $\omega_{\mu ab}$ but α,β still referring to local supersymmetry only.

Nonclosure of the algebra arises in global supersymmetry if one eliminates auxiliary fields from the algebra by means of their nonpropagating field equations which are given by an action which is invariant. Consider, for example, the spin $(1, \frac{1}{2})$ global supersymmetry multiplet

$$L = -\frac{1}{4}(\partial_\mu V_\nu - \partial_\nu V_\mu)^2 - \frac{1}{2}\bar{\lambda}\not{\partial}\lambda + \frac{1}{2}D^2$$

$$\delta V_\mu = -\bar{\varepsilon}\gamma_\mu\lambda;\ \delta\lambda = F_{\alpha\beta}\sigma^{\alpha\beta}\varepsilon + i\gamma_5 D\varepsilon;\ \delta D = i\bar{\varepsilon}\gamma_5\not{\partial}\lambda\ . \tag{22}$$

With D present, one finds $[\delta(\varepsilon_1), \delta(\varepsilon_2)]\lambda = 2(\bar{\varepsilon}_2\gamma^\mu\varepsilon_1)\partial_\mu\lambda$, but if one first eliminates D from the algebra by inserting its field equations D = 0, one finds

$$[\delta(\varepsilon_1), \delta(\varepsilon_2)]\lambda = 2(\bar{\varepsilon}_2\gamma^\mu\varepsilon_1)\partial_\mu\lambda + \frac{1}{2}(\bar{\varepsilon}_1\gamma^\mu\varepsilon_2)\gamma_\mu\not{\partial}\lambda + (\bar{\varepsilon}_1\sigma^{\rho\sigma}\varepsilon_2)\sigma_{\rho\sigma}\not{\partial}\lambda\ . \tag{23}$$

Incidentally, since δ(boson) $\approx \bar{\varepsilon}$ fermion, but δ(fermion) $\approx \partial_\mu$(boson)ε, the algebra always closes on boson fields if there are no auxiliary fields[11], since a propagating boson field equation requires two derivatives. This explains why the gauge algebra of supergravity closes on e^a_μ in second order formalism.

Since the beginning of supergravity, there have been speculations that in supergravity the reverse may hold: by adding extra fields to the transformation laws of e^a_μ and ψ_μ, one may obtain a closed algebra. What remains after finding these fields is to find an action containing these auxiliary as well as the physical fields, such that upon elimination of the auxiliary fields one regains the supergravity gauge action in second order formalism.

This conjecture has recently been verified[12]. The auxiliary fields are: a scalar S, a pseudoscalar P and an axial vector A_μ. The gauge transformations for local supersymmetry become

$$\begin{aligned}
\delta e^a{}_\mu &= \kappa\,\bar{\varepsilon}\,\gamma^a\psi_\mu \\
\delta\psi_\mu &= \frac{2}{\kappa}\,(D_\mu + \frac{i}{2}\,A_\mu\gamma_5)\varepsilon + \frac{1}{3}\,\gamma_\mu\,N\varepsilon \\
\delta A_\mu &= \frac{3i}{2}\,\bar{\varepsilon}\gamma_5\,(\hat{R}_\mu - \frac{1}{3}\gamma_\mu\gamma.\hat{R}) + \frac{i\kappa}{2}\,\bar{\varepsilon}\,N\gamma_5\psi_\mu \\
\delta S &= \frac{1}{2}\,\bar{\varepsilon}\gamma.\hat{R} - \frac{\kappa}{2}\,\bar{\varepsilon}\gamma^\sigma N\psi_\sigma \\
\delta P &= \frac{-i}{2}\,\bar{\varepsilon}\gamma_5\gamma.\hat{R} + \frac{i\kappa}{2}\,\bar{\varepsilon}\gamma_5\gamma^\sigma N\psi_\sigma \\
N &= S - i\gamma_5 P - i\not{A}\gamma_5 \\
e\hat{R}_\mu &= \varepsilon^{\mu\nu\rho\sigma}\,\gamma_5\gamma_\nu(D_\rho + \frac{i}{2}\,A_\rho\,\gamma_5)\psi_\sigma\ .
\end{aligned} \tag{24}$$

The action is even simpler

$$L = -\frac{e}{2\kappa^2}\,R - \frac{e}{2}\,\bar{\psi}_\mu\,R^\mu + \frac{e}{3}\,(A_\mu A^\mu - S^2 - P^2)\ . \tag{25}$$

Note that in the algebra one has chiral minimal coupling (except inside N), but not in the action, since there one finds R^μ instead of $\hat{R}^\mu$ as well as the mass term A_μ^2.

These results reveal an interesting connection between Poincaré and conformal supergravity. In conformal supergravity[13,14] there are two local supersymmetries associated with $Q \approx \sqrt{P}$ and $S \approx \sqrt{K}$, where K are conformal boosts. There are three gauge fields: e^a_μ, ψ_μ and an axial vector field A_μ! The transformation rules* are

*The normalization is A_μ [Ref. 13] = -2/3 A_μ (here).

$$\delta_Q e^a{}_\mu = \bar{\varepsilon}\gamma^a\psi_\mu \quad ; \quad \delta_Q\psi_\mu = 2(D_\mu + \frac{i}{2} A_\mu\gamma_5)\varepsilon$$

$$\delta_Q A_\mu = \frac{3i}{2} \bar{\varepsilon}\gamma_5 \, (\hat{R}_\mu - \frac{1}{3} \gamma_\mu \, \gamma.\hat{R}) \tag{26}$$

and

$$\delta_S e^a{}_\mu = 0; \; \delta_S\psi_\mu = - \gamma_\mu\varepsilon; \; \delta_S A_\mu = - \frac{3i}{2} \bar{\varepsilon}\gamma_5\psi_\mu \; . \tag{27}$$

Looking now at the Poincaré gauge algebra, we see that δe, $\delta\psi$ and δA are a linear combination of δ_Q and δ_S, with parameters ε and

$$\varepsilon_S = - \frac{1}{3} (S\text{-}iP\gamma_5\text{-}i\not{A}\gamma_5)\varepsilon$$

respectively. The kinetic terms of the conformal action are given by

$$L = e\kappa^2[(\frac{R_{\mu\nu}^2 - \frac{1}{3}R^2}{\kappa^2}) - \bar{\psi}_\mu(\Box g_{\mu\nu} - \partial_\mu\partial_\nu)(\hat{R}^\nu - \frac{1}{3}\gamma^\nu\gamma.\hat{R})] \tag{28}$$

$$- \frac{e}{3} F^2_{\mu\nu}(A) \; .$$

The Q- and S-invariance at the linearized level is obvious.

We now see what has happened. The fields e^a_μ, ψ_μ and A_μ have acquired two extra derivatives, so that A_μ becomes propagating. Thus e^a_μ describes two massless modes (since $\Box^2$ acts as twice $\Box$) and ψ_μ three massless modes (since $\Box\,\not{\partial}$ is equivalent to three times $\not{\partial}$)[14]. There are thus an equal number of bosonic and fermionic states (6), and this counting explains why the fields S and P are absent from the conformal supergravity gauge actions. The Q- and S-supersymmetry is needed to eliminate the $\gamma.\psi$ and $\partial.\psi$ spin $\frac{1}{2}$ parts of ψ_μ so that only three massless modes, each with helicity $\pm 3/2$, remain. It is intriguing that not only the transformation laws of e, ψ, A show a direct connection between Poincaré and conformal supergravity, but that also the fields S and P seem to know about S supersymmetry through N.

Also for various matter couplings, it has been shown that one can add auxiliary fields such that the algebra closes[15]. The auxiliary fields are of two kinds: either S, P, A_μ, or they are the auxiliary fields which are already needed for the matter system in order that the global supersymmetry algebra closes.

We stress that all these results hold only in second order formalism (or in 1.5 order formalism) where ω is not an independent field (but S, P, A_μ are independent).

5. UNITARITY IN SUPERGRAVITY

Suppose a classical action I is invariant under gauge transformation

$$\delta\phi^k = R^k{}_\alpha(\phi)\xi^\alpha \ . \tag{29}$$

One quantizes then, according to the Feynman-De Witt-Faddeev-Poppov-'t Hooft-Veltman procedure, by adding a gauge fixing and a ghost term

$$L^{qu} = L^d + \frac{1}{2} F_\alpha(\phi) a^{\alpha\beta} F_\beta(\phi) + \bar{C}^\alpha F_{\alpha,j} R^j{}_\beta C^\beta \tag{30}$$

where $F_{\alpha,j} = \delta F_\alpha(\phi)/\delta\phi^j$ may be field-dependent but $a^{\alpha\beta}$ should not depend on ϕ (otherwise unitarity is not satisfied). The (anti) ghost fields $(\bar{C}^\alpha)C^\beta$ are anticommuting for bosonic gauge invariances, but commute for local supersymmetry.

This is the usual quantization procedure for non-Abelian gauge fields and it is correct in all cases except in supergravity. In supergravity, one needs an extra <u>four-ghost coupling</u>[8,10,16] in order that the theory be unitary (by which one always means: a unitary <u>and</u> gauge-fixing-term independent S matrix). No six or higher ghosts seagulls are needed for unitarity.

The four-ghost coupling can be expressed in terms of the

nonclosure function η of the previous section by

$$L(4\text{-ghost}) = \frac{1}{4}\,\bar{C}^{\gamma}F_{\gamma,j}\;\bar{C}^{\delta}F_{\delta,i}\;\eta^{ij}_{\alpha\beta}\;C^{\alpha}C^{\beta} \quad . \tag{31}$$

For the gauge $F_{\alpha}(\phi) = -(\kappa/2)\gamma.\psi$ one finds from (18)

$$L(4\text{-ghost}) = \frac{5\kappa^2}{32}\,(\bar{C}\gamma_{\alpha}C\bar{C}^{T})(C^{T}C\gamma^{\alpha}C) \tag{32}$$

where we have chosen the normalization of F_{α} such that the kinetic term of the supersymmetry ghost is the usual $-\bar{C}\gamma^{\mu}D_{\mu}C$. Note that one cannot scale F_{α}, $\bar{C}$ and C such that all ghost terms are a density.

Before explaining these results, we show that these results are consistent. Suppose one would have quantized the gauge action of Poincaré supergravity with the auxiliary fields S, P and A_{μ}. Since its gauge algebra closes, $\eta^{ij}_{\alpha\beta} = 0$ and

$$L(\text{ghost}) = -\,\bar{C}\gamma^{\mu}[D_{\mu} + \frac{i\kappa}{2}A_{\mu}\gamma_5 + \frac{1}{6}\gamma_{\mu}\,(S - i\gamma_5 P - i\not{A}\gamma_5)]C \quad . \tag{33}$$

Eliminating S, P, A_{μ} from the quantum actions, one finds[16,24,12]

$$A_{\mu} = \frac{3i\kappa}{4}\,(\bar{C}\gamma_{\mu}\gamma_5 C);\; S = -\,\kappa(\bar{C}C);\; P = i\kappa(\bar{C}\gamma_5 C) \tag{34}$$

and, substituting these field equations back into the action, one indeed regains $L(4\text{-ghost})$ of Eq. (32).

These results can be obtained by requiring unitarity[16]. In order to prove unitarity one needs Ward identities, and in order to prove Ward identities one needs a local or global invariance of the quantum action. Although the classical gauge transformations no longer leave the quantum action invariant (for that purpose F_{α} was introduced!), there is an extension of the classical transformation rules under which the quantum action is invariant. This is BRS-invariance[17], and it is a global, nonlinear symmetry with a constant anticommuting parameter Λ. The anticommuting nature of Λ has led some authors to conjecture on a deep connection between supersymmetry and BRS-invariance, but the two have really nothing to do with each other: any gauge theory, including supergravity, is BRS invariant at the quantum level. The BRS transformation rules for any theory where the gauge algebra closes on-shell, read

$$\delta\phi^i = \delta\phi^i{}_{BRS} + \delta\phi^i{}_{\kappa} = R^i{}_{\alpha} C^{\alpha}\Lambda + \frac{1}{2}\bar{C}^{\gamma}F_{\gamma,j}\ \eta^{ji}_{\alpha\beta}\ C^{\alpha}C^{\beta}\Lambda$$

$$\delta C^{\beta} = \frac{1}{2} f^{\beta}{}_{\alpha\gamma} C^{\alpha}\Lambda C^{\gamma} \tag{35}$$

$$\delta\bar{C}^{\alpha} = \Lambda F_{\beta}\ a^{\beta\alpha}$$

They were found by R. Kallosh in an excellent paper[10] for F_{α} linear in ϕ*.

The invariance of the action under these extended BRS rules is shown as follows. The classical action is clearly invariant under $\delta\phi^i_{BRS}$. The variation $\delta\bar{C}^{\alpha}$ in L(2-ghost) cancels the variation $\delta\phi^i_{BRS}$ in L(fix). The variation $\delta\bar{C}^{\alpha}$ in L(4-ghost) cancels the variation $\delta\phi^i{}_{\kappa}$ in L(fix). The variation $\delta_{BRS}\phi^i$ in L(2-ghost) yields a commutator on ϕ^j since L(2-ghost) itself is proportional to $\delta_{BRS}\phi^j$. This is thus where the nonclosure function η comes in. One has now a large number of remaining terms, all containing η, and, miraculously, they cancel!

The Ward identities follow easily by noting that

$$\delta_{BRS} \langle\bar{C}^{\alpha}(x)\ \phi^k(y)\ \phi^{\eta_1}(z_1).....\phi^{\eta_\ell}(z_\ell)\rangle^{L(qu)} = 0 \quad . \tag{36}$$

This is because for any L varying both all external lines and the action from which the n point function is obtained, yields zero identically. But since our L(qu) is invariant under the extended BRS variations, one only needs to vary the external fields. For

*For nonlinear F_{α}, see P.K. Townsend and the author, to appear.

the unitarity relation we will only need all lines with on-shell momenta which means that one only needs the linear (pole) parts of $\delta_{BRS}\phi^i$ and $\delta_{BRS}\bar{C}^\alpha$, while $\delta_{BRS}C^\alpha = 0$. Moreover, we will only discuss here the case of one-loop untiarity (for n loop unitarity, see Ref. 16). In that case it is sufficient to put all fields $\phi^{n_i}(z_i)$ on-shell with physical momenta and polarizations. They are then invariant under linear BRS transformations, and play in the following the role of inert spectators. We will omit them from here on.

Thus one obtains the Ward identity

$$<\delta^{lin}_{BRS}\,\bar{C}^\alpha(x)\;\phi^k(y)> \;=\; -\;<\bar{C}^\alpha(x)\;\delta^{lin}_{BRS}\;\phi^k(y)> \quad . \tag{37}$$

For a supersymmetry gauge field ψ_μ,

$$\delta^{lin}_{BRS}\;\psi_\mu \;=\; \frac{2}{\kappa}\;\partial_\mu C\Lambda$$

and for the supersymmetric antighost field $\bar{C}$ one has

$$\delta^{lin}_{BRS}\;\bar{C} \;=\; -\;\frac{\kappa}{2}\;\Lambda\;\bar{\psi}\cdot\gamma \quad .$$

The resulting identity is shown in Fig. 1. Note in deriving that result, in (37) all external lines have propagators (for spin 3/2, see Eq. (38)). It says that saturating a gauge line, whose momentum is physical, at one end with one momentum factor flips the gauge line into a ghost line which at the other end emits the momentum of the ghost line. We have here only considered supersymmetry gauge and ghost fields, but the same holds for the other local symmetries. The other identity we need is the usual conservation relation in Fig. 1, which says that saturating all nonphysical lines by one momentum factor yields zero.

We come now to unitarity. Unitarity requires that in an N particle cut the sums of unphysical and of ghost modes cancel each

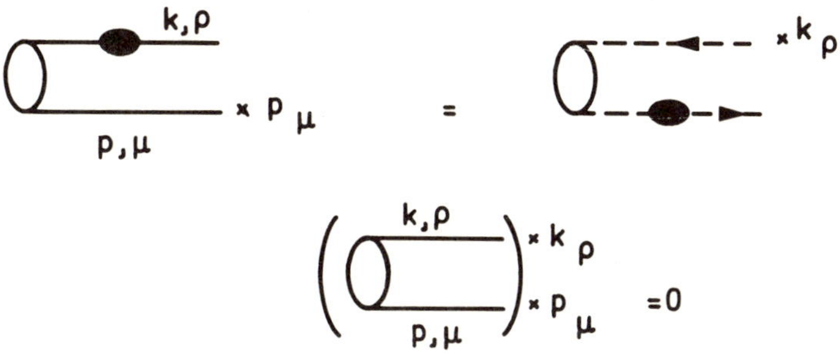

Fig. 1

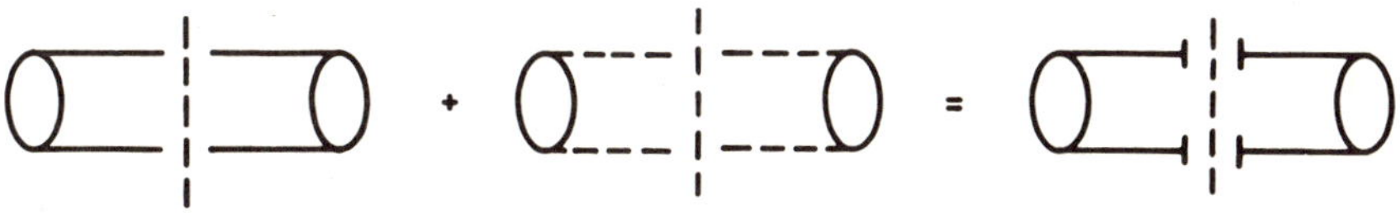

Fig. 2

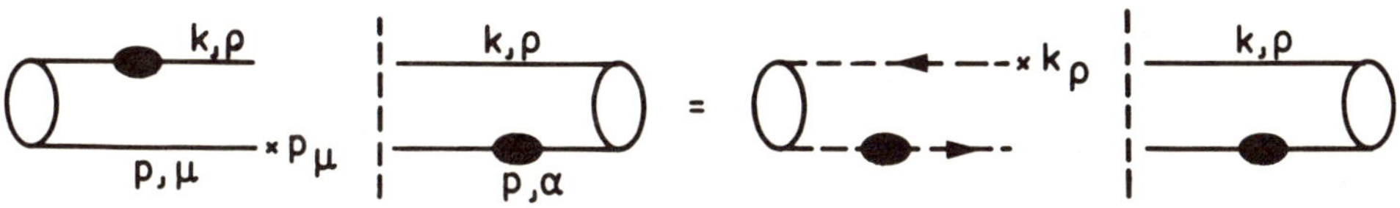

Fig. 3

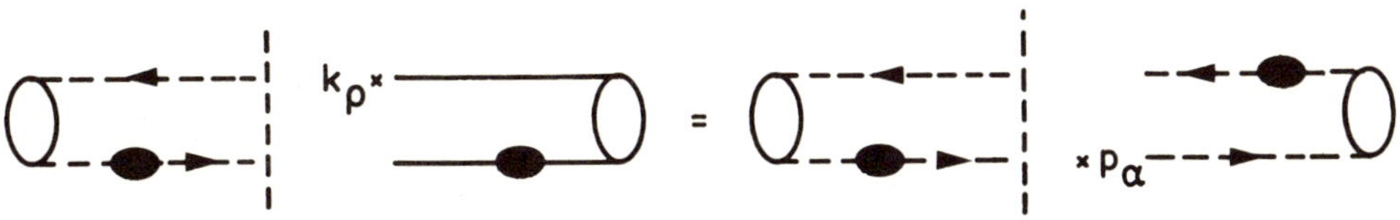

Fig. 4

other. For a two-particle cut (one-loop process) this implies Fig. 2.

In the first diagram, the propagators are the renormalizable propagators, obtained from L (cl) + L (fix) by inverting the field equations. For the spin 3/2 field one finds with Eq. (15)

$$P^{3/2}_{\mu\nu} = \langle\psi_\mu\bar{\psi}_\nu\rangle = \frac{1}{2}\,\gamma\nu\,\not{p}\,\gamma_\mu/p^2 \qquad . \tag{38}$$

It can be decomposed into its two physical modes with helicity ±3/2 plus gauge terms

$$P^{3/2}_{\mu\nu} = P^{phys}_{\mu\nu} + (p_\nu\bar{\not{p}}\,\not{p}\gamma_\mu + \gamma_\nu\not{p}\bar{\not{p}}\,p_\mu + 2p_\nu\bar{\not{p}}p_\mu)(2p.\bar{p})^{-1} \tag{39}$$

where $\bar{p} = (\vec{p}, -ip_4)$ is the time-reversed of $p = (\vec{p}, +ip_4)$ and where

$$P^{phys}_{\mu\nu} = \sum_{\lambda=\pm}(\varepsilon^\lambda_\mu u^\lambda)(\varepsilon^{\lambda *}_\nu \bar{u}_\nu) = \frac{1}{2}\,\bar{\delta}_{\mu\alpha}\,P^{3/2}_{\alpha\beta}\,\bar{\delta}_{\beta\nu} \qquad . \tag{40}$$

The symbol $\bar{\delta}_{\mu\nu}$ is the transverse photon projection operator

$$\bar{\delta}_{\mu\nu} = \delta_{\mu\nu} - (p_\mu\bar{p}_\nu + p_\nu\bar{p}_\mu)(p.\bar{p})^{-1} = \sum_{\lambda=\pm}(\varepsilon^\lambda_\mu \varepsilon^{\lambda *}_\nu) \qquad . \tag{41}$$

The propagators in the last diagram of Fig. 2 are $P^{phys}_{\mu\nu}$.

For definiteness we will consider two intermediate spin 3/2 lines. Consider the term with γ_ν in (39). It can be rewritten as

$$P_\mu\left(\frac{\bar{p}_\alpha}{p.\bar{p}}\right)\left(\frac{1}{2}\,\gamma_\nu\not{p}\gamma_\alpha\right) \tag{42}$$

hence as the product of the momentum of the cut line, times a scalar factor which we omit from now on for the time being, times the renormalizable propagator (index α acts on the left, ν on the right). Using the Ward identity once, one has Fig. 3. One can now use k_ρ on the right, and cycle once more with the Ward identity, cf. Fig. 4. The original factor $\bar{p}_\alpha(p.\bar{p})^{-1}$ contracted with the final factor p_α yields one, and since at no time have we added an extra minus sign for this loop, we end up with minus one times the oriented ghost loop (ghosts are complex!). Similarly the term with p_ν in (39) cancels the other orientation of the ghost loop. That leaves us with the last term in (39). Cycling p_μ once we get the factors k_ρ upstairs as before, but now we stop since the factors p_ν and k_ρ can be used in the conservation equation for the blob on the right to show that this term vanishes.

The conclusion is thus that the unphysical terms in the propagator cancel either with the ghosts, or by themselves, and unitarity holds. This argument has been extended to general N particle cuts (with Ward identities for cut-graphs) with gauge and ghost fields in the intermediate states. Also it has been shown[16] that starting from the requirement that the Ward identities hold (which is necessary for unitarity), one reconstructs the four-ghost interaction.

The four-ghost term is only effective from the two-loop level on. Thus all explicit one-loop calculations remain valid[18], while the proofs at the two-loop level[19] (which implicitly assumed that Ward identities on the external lines were satisfied) remain (or, rather, become) valid. One can also prove, using similar techniques, that the S matrix is independent of F_α, provided the Ward identities hold, hence provided the extra ghost term is added to the action.

6. RENORMALIZABILITY

The status of finiteness of the S matrix in pure and O(n) supergravity, supergravity with matter without extra O(n) symmetry, pure Einstein gravity and Einstein gravity plus matter is represented in the following table:

	O(n)	non-O(n)	Einstein	Einstein + matter
one-loop	finite	not finite	finite	not finite
two-loop	finite	not finite	(?)	not finite
three-loop	?	not finite	?	not finite

For a review, see last year's proceedings. New remarks are added.

At the two-loop level there is a proof of two years old by Grisaru, Wu and myself[20] that Einstein gravity is finite. Since however the proof uses helicity conservation which is only proven at the tree level, a half question mark has been noted down. At the three-loop level, however, no conclusion can be drawn at all for Einstein gravity. For supergravity, a proof of two-loop finiteness was first given[19] by noting that the only possible counterterm which does not vanish on-shell flips helicities, while supergravity (and Einstein gravity?) conserves helicity. Hence the coefficient of this term is zero. But anomalies might invalidate chirality conservation and hence two-loop finiteness. Also it is possible that supersymmetry anomalies affect the transformation rules under which the two-loop S matrix is invariant, so that proofs that there are no nonvanishing counter terms on-shell which are invariant under the classical rules tell us little about the actual counterterms if there are anomalies. The situation clearly deserves study.

An explicit calculation (see Fig. 5) of photon-photon scattering in the O(8) model[21] at the one-loop level yielded finite results. This was expected, although a proof of one-loop finiteness of the O(8) model is lacking -- but only because it requires some hairy algebra. The finiteness of this process makes it in my opinion almost certain that O(8) is one-loop finite. All divergences are again proportional to the square of the energy momentum tensor of the photon which again is a clear sign that the theory has a combined chirality-duality invariance. In fact, it probably has a global U(8) symmetry. Under each graph in Fig. 5 the coefficient of this divergence is written, and all divergences add up to zero.

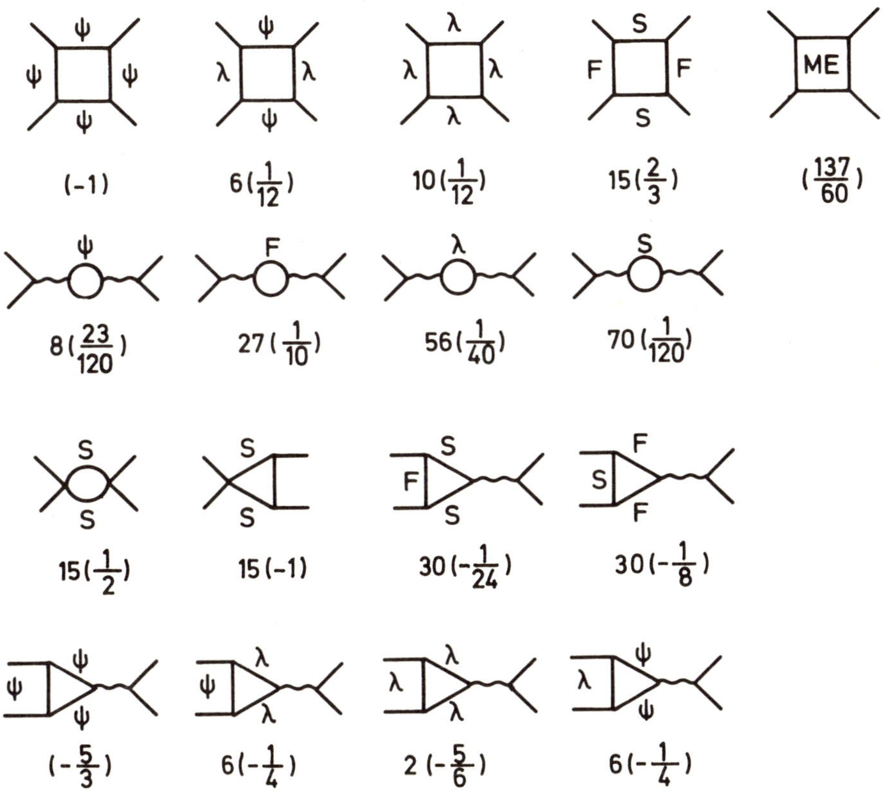

Fig. 5. Finiteness of photon-photon scattering in O(8) extended supergravity. The numbers are the coefficients of the standard divergence, ME refers to the Maxwell-Einstein system (with the graviton and one photon) while ψ = spin 3/2, λ = spin 1/2, F = spin 1 and S = spin 0.

This work was done with Mark Fischler of Stony Brook[22].

If anomalies do not spoil the two-loop finiteness of supergravity, then the finiteness at the three-loop level is the big question. It has been shown that there exists a possible counter term which is supersymmetric even off-shell[23]. Thus the only hope is that its coefficient vanishes. One possible approach is to consider extra symmetries (which do exist) which forbid this term. An explicit calculation seems impossible.

CONFORMAL SUPERGRAVITY

The conformal group has a special place in general relativity, and it seems interesting to see what Fermi-Bose gauge symmetry does with it. The superconformal group has 24 generators X_A: the 15 bosonic generators P_μ, $M_{\mu\nu}$, K_μ and D of the ordinary conformal group, to which are added one more bosonic generator A for chiral rotations, and the square roots Q_α and S_α satisfying

$$\{Q_\alpha,Q_\beta\}=-(\gamma^\mu C)_{\alpha\beta}P_\mu \;;\; \{S_\alpha,S_\beta\} = (\gamma^\mu C)_{\alpha\beta}\, K_\mu \quad . \tag{43}$$

Recently, the gauge action of the superconformal group was found[13]. It is a Yang-Mills type action because it is quadratic in curvatures $R^A_{\mu\nu}$, defined by

$$R^A_{\mu\nu} = \partial_\nu h^A_\mu - \partial_\mu h^A_\nu + f^A{}_{BC}\, h_\nu{}^B h_\mu{}^C \tag{44}$$

$$[X_A, X_B\} = f_{BA}{}^C X_C \quad . \tag{45}$$

The gauge fields h^A_μ assigned to the generators X_A are denoted by

$$\begin{array}{ccccccc} P_a & M_{ab} & K_a & D & Q_\alpha & S_\alpha & A \\ e^a{}_\mu & \omega_\mu{}^{ab} & f^a{}_\mu & b_\mu & \bar\psi^\alpha_\mu & \bar\phi^\alpha_\mu & A_\mu \end{array} \tag{46}$$

The action was found by writing down the most general action bilinear in curvatures, conserving parity and without dimensional constants. It reads

$$I = \int d^4x \, [\varepsilon^{\mu\nu\rho\sigma} \{R_{\mu\nu}(M)^{ab} \, R_{\rho\sigma}(M)^{cd} \, \varepsilon_{abcd} - 8 \, R_{\mu\nu}(Q)\gamma_5 \, \bar{R}_{\rho\sigma}(S)$$
$$+ 4i \, R_{\mu\nu}(A) \, R_{\rho\sigma}(D)\} - 8e \, R_{\mu\nu}(A) \, R^{\mu\nu}(A)] \quad . \tag{47}$$

It is invariant under all local gauge transformations (replacing P by general co-ordinate transformations)

$$\delta h_\mu^A = D_\mu \varepsilon^A \equiv \partial_\mu \varepsilon^A + f_{BC}{}^A h_\mu{}^B \varepsilon^C \tag{48}$$

provided three constraints are imposed. These constraints arise naturally, and the theory is invariant if and only if they hold.

i) Torsion: $R_{\mu\nu}(P) = 0$. This expresses the spin connection field $\omega_{\mu ab}$ in terms of the other fields. In Poincare supergravity it is a field equation, in conformal supergravity it is not. Hence it is a model-independent constraint. It tells one how a vector rotates under parallel transport depending on which fields are present, but not on which field equations these fields satisfy.

ii) Self-duality:

$$R_{\mu\nu}(Q) + \frac{e}{2} \gamma_5 \, \varepsilon_{\mu\nu\rho\sigma} \, R^{\rho\sigma}(Q) = 0 \quad .$$

Since $Q = \sqrt{P}$ one expects a constraint for Q, too. Pure vanishing, $R_{\mu\nu}(Q) = 0$, is too strong, as it would eliminate 24 field components, while there are only 32 fermion components (ψ_μ and ϕ_μ). Self-duality is the natural choice, and it corresponds in a sense to the square root of true vanishing. It gives 12 constraints, from which one can eliminate $\phi_\mu - \frac{1}{4} \gamma_\mu \, \gamma \cdot \phi$.

iii) The third constraint:

$$\sigma^{\mu\nu} \, R_{\mu\nu}(Q) = 0 \quad .$$

Its geometrical meaning is as yet unclear, but again it seems to have a deep meaning. It allows to eliminate $\gamma.\phi$ and leads to Q supersymmetry.

In addition, one finds that the field equation for the K_μ gauge field f^a_μ is nonpropagating, so that it can be eliminated by its field equation, while the D gauge field b_μ drops from the action altogether. The only remaining fields are: the vierbein, the gravitino and the chiral fields, whose two supersymmetry transformations are given in (26) and (27). For the solution of ϕ_μ in terms of these three fields, one finds

$$\phi_\mu = \frac{1}{3}\gamma^\nu (S_{\mu\nu} + \frac{e}{4}\gamma_5 \varepsilon_{\mu\nu\rho\sigma} S^{\rho\sigma})$$

$$S_{\mu\nu} = (\partial_\nu + \frac{1}{2}\omega_{\nu ab}\sigma^{ab} + \frac{1}{2} b_\nu + \frac{i}{2} A_\nu \gamma_5)\psi_\mu - (\mu\leftrightarrow\nu)$$

$$\omega_{\mu ab} = \omega_{\mu ab}(e) + \frac{1}{4}(\bar\psi_\mu\gamma_a\psi_b - \bar\psi_\mu\gamma_b\psi_a + \bar\psi_a\gamma_\mu\psi_b) + (e_{a\mu}b_b - e_{b\mu}b_a) \qquad (49)$$

$$\omega_{\mu ab}(e) = \frac{1}{2} e_a{}^\nu(e_{b\nu,\mu} - e_{b\mu,\nu}) + \frac{1}{2} e_a{}^\lambda e_b{}^\rho e_{c\lambda,\rho} e^c{}_\mu - (a\leftrightarrow b)$$

and

$$\delta_Q\phi_\mu = \gamma^a\varepsilon f_{a\mu} + \frac{i}{4}\gamma^\nu(\gamma_5 R_{\mu\nu}(A) + \frac{e}{2}\varepsilon_{\mu\nu\rho\sigma}R^{\rho\sigma}(A))\varepsilon$$

$$\delta_S\phi_\mu = (\partial_\mu + \frac{1}{2}\omega_{\mu ab}\sigma^{ab} - \frac{i}{2}A_\mu\gamma_5 - \frac{1}{2}b_\mu)\varepsilon \quad . \qquad (50)$$

For the proper conformal gauge field one finds

$$f_{\mu\nu} = -\frac{1}{4}(\hat R_{\nu\mu} - \frac{1}{6}g_{\mu\nu}\hat R) - \frac{1}{8}\bar\psi^\lambda\gamma_\nu\bar R_{\lambda\mu}(Q) - \frac{i}{16}R^{(A)}_{\rho\sigma}\varepsilon_{\mu\nu}{}^{\rho\sigma}$$

$$\hat R_{\nu\mu} = R_{\nu\mu} - \bar\psi_\nu\sigma_{\lambda\mu}\phi_\lambda + \bar\psi_\lambda\sigma_{\lambda\mu}\phi_\nu \qquad (51)$$

$$\bar R_{\lambda\mu}(Q) = (2D_\mu + \frac{1}{2}\omega_{\mu ab}\sigma^{ab} + \frac{1}{2}b_\mu + \frac{i}{2}A_\mu\gamma_5)\psi_\lambda - 2\gamma_\mu\phi_\lambda - (\lambda\leftrightarrow\mu)$$

$$R_{\rho\sigma}(A) = -F_{\rho\sigma}(A) - i\bar\psi_\rho\gamma_5\phi_\sigma + i\bar\psi_\sigma\gamma_5\phi_\rho \quad .$$

The complete action is complicated; for its kinetic part, see (28). Since it is a superconformal theory, there should be no lower gauge components. Indeed, S supersymmetry eliminates the $\gamma.\psi$ spin $\frac{1}{2}$ part of ψ_μ, while Q supersymmetry takes care of the $\partial.\psi$ spin $\frac{1}{2}$ part of ψ_μ. Weyl invariance (= local scale invariance) eliminates the scalar part (trace $g_{\mu\nu}$), while general coordinate invariance eliminates, as usual, the other scalar and the spin 1 part of $g_{\mu\nu}$. Chiral invariance gets rid of the spin 0 part of A_μ. Clearly, we do not need K_μ invariance, and indeed all fields are K-inert.

In some sense, the superconformal action can be viewed as the Poincaré action with two more derivatives on e_μ^a and ψ_μ. Proceeding in this way, one sees that A_μ should be one of the auxiliary fields in Poincaré supergravity because it is <u>here</u> propagating normally. And one also expects that the gauge algebra of superconformal gravity closes, since all nonpropagating fields of Poincaré supergravity which are effective in conformal supergravity now have a d'Alembertian. It indeed does! Thus, no four-ghost term is needed for conformal supergravity, which does not mean that there are no unitarity problems: it is lore (but only lore) that <u>all</u> higher derivative theories have ghosts at the classical level.

Conformal theories have no intrinsic scale (the fields ψ_μ have dimension $\frac{1}{2}$ instead of 3/2; therefore no explicit κ's are present). They do not reproduce the classical tests of general relativity, and for the time being it seems that their main advantage is that they clarify the structure of Poincaré supergravity. For example, the self-duality and third-constraint in Poincaré supergravity are just the spin 3/2 field equation, so that closure of the Poincaré gauge algebra modulo this field equation can be understood from this more general point of view. However, it is not excluded that future developments have surprises in petto for the superconformal group. For example, spontaneous symmetry breaking might lead to a term $\approx R$ in the action so that the classical tests would be satisfied, and let us not forget that (super) conformal theories are all-loop renormalizable! The parallel with Yang-Mills fields is exciting.

REFERENCES

1. J. Polonyi, Budapest preprint KFKI-1977-93. See also, A. Das, M. Fischler and M. Rocek, Phys. Letters 69B, 97 (1977) and Phys. Rev. D16, 3427 (1977).
2. See, for example, D.Z. Freedman, Proceedings American Physical Society meeting at Argonne Laboratories (1977).
3. D.Z. Freedman, P. van Nieuwenhuizen and S. Ferrara, Phys. Rev. D13, 3214 (1976).
4. S. Deser and B. Zumino, Phys. Letters 62B, 335 (1976).
5. E. Cremmer, J. Scherk and S. Ferrara, LPTENS-77-19, Phys. Letters to be published.
6. E.S. Fradkin and M. Vasiliev, Lebedev preprint. P.K. Townsend and P. van Nieuwenhuizen, Phys. Letters 67B, 439 (1977).
7. P. van Nieuwenhuizen and J.A.M. Vermaseren, Phys. Letters 65B, 263 (1976).
8. E.S. Fradkin and M. Vasiliev, Phys. Letters B72, 70 (1970).
9. D.Z. Freedman and P. van Nieuwenhuizen, Phys. Rev. D14, 912 (1976).
10. R. Kallosh, Zh. ETF Pis'ma v. Red. 26, 575 (1977).
11. E. Cremmer and J. Scherk, Nuclear Phys. B127, 259 (1977).
12. S. Ferrara and P. van Nieuwenhuizen, Phys. Letters B, to be published. K. Stelle and P.C. West, Imperial College preprint ICTP/77-78/6.
13. M. Kaku, P.K. Townsend and P. van Nieuwenhuizen, Phys. Letters 69B, 304 (1977); Phys. Rev. Letters 39, 1109 (1977); Phys. Rev. D. to be published; with S. Ferrara, Nuclear Phys. B129, 125 (1977).
14. S. Ferrara and B. Zumino, Nuclear Physics B, to be published.
15. S. Ferrara, M.T. Grisaru and P. van Nieuwenhuizen, CERN preprint TH.2476.
16. G. Sterman, P.K. Townsend and P. van Nieuwenhuizen, Phys. Rev. D, to be published.

17. C. Becchi, A. Rouet and R. Stora, Comm. Math. Phys. 142, 127 (1975).
18. For a review, see M.T. Grisaru and P. van Nieuwenhuizen, Proceedings Coral Gables Conference (1977).
19. M.T. Grisaru, Phys. Letters B66, 75 (1977).
20. M.T. Grisaru, P. van Nieuwenhuizen and C.C. Wu, Phys. Rev. D 12, 1563 (1975).
21. B. de Wit and D.Z. Freedman, Nuclear Phys. B130, 105 (1977).
22. M. Fischler and P. van Nieuwenhuizen, to be published.
23. B. Zumino, private communication and Ref. 14, Eq. (A.9). See also, R. Grimm, J. Wess and B. Zumino, Phys. Letters B, to be published. For the on-shell limit, see S. Deser, J. Kay and K. Stelle, Phys. Rev. Letters 38, 527 (1977).
24. B. de Wit and M.T. Grisaru, Phys. Letters to be published.

THE USE AND MISUSE OF THE PARTON MODEL*

Howard Georgi[†]

Harvard University

Cambridge, Massachusetts

ABSTRACT

Perturbation theory in quantum chromodynamics is used to discuss the parton model of semi-inclusive μP scattering and other processes.

[†]Research supported in part by the Science Foundation under Grant No. PHY77-22864, and the Sloan Foundation.

The parton model has been around for a long time. It is often said that Quantum Chromodynamics justifies the parton model because of asymptotic freedom. The truth is much more complicated. The connection between QCD and the parton model has been established only for inclusive e^+e^- annihilation and inclusive lepton-hadron scattering. For other processes, for example semi-inclusive e^+e^- annihilation, semi-inclusive lepton-hadron scattering, and the Drell-Yan process $P+P$ (or $\overline{P}$) $\to \mu^+ + \mu^-$ + anything, high p_T hadron production in hadron-hadron scattering, etc., the connection has not been demonstrated.

The first thing I want to do today is to try to make it obvious that the connection is not trivial. Except for inclusive e^+e^- annihilation, all the other processes involve momenta which are not large. The hadrons which are singled out and the partons out of which they are constructed have momenta p for which p^2 is small. So even in a process such as electroproduction or the Drell-Yan process where Q^2 of the virtual proton can be large, naive perturbation theory cannot be used. Even though the coupling constant

$$\alpha_s(Q^2) \propto \frac{1}{\ell n(Q^2/\Lambda^2)} \tag{1}$$

goes to zero as Q^2 gets large, perturbation theory is useless because large logarithms appear $-\ell n(Q^2/p^2)$ and

$$\alpha_s(Q^2)\ \ell n\ (Q^2/p^2) \simeq 1 \quad . \tag{2}$$

In view of this fundamental difficulty, it seems rather remarkable that the parton model works at all. We are all familiar with the ingenious way in which QCD inclusive lepton-hadron scattering evades this difficulty for the process. The large logarithms involving the target hadron and parton momenta can all be factored into a set of Q^2 dependent parton distribution functions which are related to the matrix elements of the twist-2 operators between target hadron states, with all other contributions suppressed by powers of p^2/Q^2. By an historical accident, we have learned that this property of factorization can be demonstrated by the use of the operator product expansion (OPE). But it is the factorization and not the OPE which is important.[2] In this talk I will review the factorization of the inclusive lepton-hadron scattering cross-section directly in parton language and argue that the obvious generalization of the factorization hypothesis to other processes offers the only hope for understanding the parton model in QCD perturbation theory.

WHAT FACTORS AND WHY?

In inclusive lepton-hadron scattering, it is the moments of the structure functions which factor. In parton language, the reason that it is the moments which factor is that the cross-section for inclusive lepton scattering of a specific hadron is obtained from the parton cross-sections and the parton distribution functions by means of a convolution integral which simplifies when moments are taken.

Explicitly, the inclusive lepton-hadron process is depicted in Fig. 1. $k_1 - k_2$ are the initial (final) lepton momenta, $q=k_1$, k_2 is the virtual photon (or intermediate vector boson) momentum, P is the target hadron momentum and p is the struck parton momentum. The hadron and parton momentum are assumed to be parallel--thus we ignore effects of the order $\langle p_\perp^2 \rangle/Q^2$ where $Q^2=-q^2$ and $\vec{p}_\perp$ is the parton tranverse momentum (we expect $\langle p_\perp^2\rangle \simeq (300\text{MeV})^2$ due to the nonperturbative confinement mechanism). ξ is the fraction of the target hadron momentum carried by the struck parton--loosely $p=\xi P$ (more precisely if $\vec{q}$ is in the positive Z direction $p^- = p^0 - p^Z = \xi P^-$). The scaling variables x_H of the hadron and x_p of the parton are defined by

$$x_H = -\frac{q^-}{P^-} \simeq \frac{Q^2}{2Pq} \qquad (3)$$

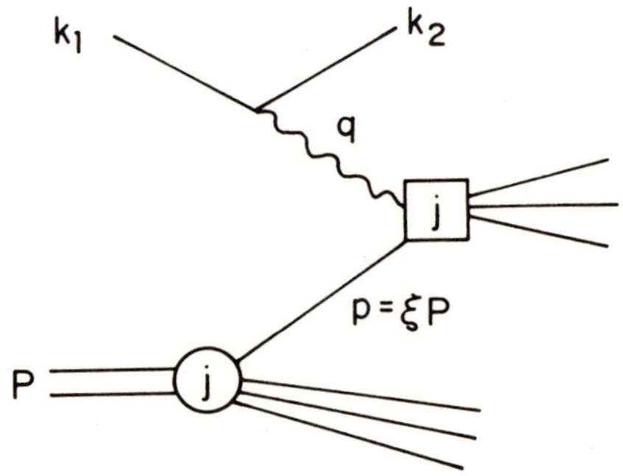

Figure 1. Inclusive lepton-hadron scattering in the parton model.

$$x_P = -\frac{q^-}{p^-} \sim \frac{Q^2}{2pq} ,$$

and finally $y=2Pq/2Pk_1$, is the fractional energy loss of the lepton. Then the hadron cross-section in the impulse approximation is given by

$$\frac{d\sigma_H}{dx_H dy} = \sum_j \int dx_p \, d\xi \, \delta(x_H - x_p \xi) \times$$

$$\times \frac{d\sigma_j}{dx_p dy} F_j(\xi) , \qquad (4)$$

where $F_j(\xi)$ is the j-parton distribution function in the hadron (the circle in Fig 1) and $d\sigma_j/dx_p dy$ is the cross-section for inclusive lepton-j-parton scattering (the square in Fig. 1). The sum on j runs over all partons--quarks, antiquarks and gluons.

Equation (4) is not complete. In view of the above discussion, we had better keep track of the p^2 dependence:

$$\frac{d d\sigma_H}{dx_H dy} = \sum_j \int dp^2 \, dx_p \, d\xi \, \delta(x_H - x_p \xi) \times$$

$$\times \frac{d\sigma_j}{dx_p dy}(Q^2, p^2, x_p, y) F_j(p^2, \xi) + O\left(\frac{p^2}{Q^2}\right) . \qquad (5)$$

As promised this simplifies when we take momenta. Define

$$\sigma^n = \int \frac{dx}{x} x^n \frac{d\sigma}{dxdy} \qquad (6)$$

$$F_j^{\ n} = \int \frac{d\xi}{\xi} \quad \xi^{\ n} \ F_j \quad . \tag{6}$$

The Eq. (5) becomes

$$\sigma_H^{\ n} = \sum_j \int dp^2 \quad \sigma_j^n \quad (Q^2, p^2) \ F_j^{\ n} \ (p^2) \tag{7}$$

and we have as usual ignored coherence and other effects of order p^2/Q^2. For this to make sense, we must assume that p^2, like $p_\perp^2$, is limited.

Now we start computing $\sigma_j^n \ (Q^2, \ p^2)$ in perturbation theory in the coupling constant $\alpha_s \ (Q_o)$ renormalized at some arbitrary momentum scale Q_o. We encounter immediately $\ln \ Q^2/p^2$ and in higher order calculation $\ln \ Q^2/Q_o^{\ 2}$ and $\ln \ p^2/Q_o^{\ 2}$ as well. There are also terms of the order of p^2/Q^2. These we will ignore because they are the same size as nonperturbative effects which we have already ignored (in field theory language, this is the restriction to twist two operators). Then the factorization takes the form

$$\sigma_j^n \ (Q^2, \ p^2) = \sum_{j'} \ \tilde{\sigma}_{j'}^{\ n} \ (\ln \frac{Q^2}{Q_o^{\ 2}} \ , \ \alpha_s \ (Qo), \ Q^2) \tag{8}$$

$$\cdot \ \Gamma_{j' \ j}^{\ n} \ (\ln \frac{p^2}{Q_o^{\ 2}} \ , \ \alpha_s(Q_o)) + 0 \ (p^2/Q^2) \quad .$$

Because Q_o is arbitrary, the Q_o dependences of $\tilde{\sigma}^n$ and Γ^n are given by renormalization group equations. If $Q^2 \simeq Q_o^{\ 2}$, then $\tilde{\sigma}_j^n$ contains no large logarithms and can be reliably calculated in perturbation theory.

Putting the factorized form back into Eq. (7) we get

$$\sigma_H^n = \sum_j \ \tilde{\sigma}_{j'}^n \ (\ln \frac{Q^2}{Q_o^{\ 2}} \ , \ \alpha_s(Q_0), \ Q^2) \quad \times$$

$$\times \sum_j \int dp^2 \, \Gamma_{j'j}^{\,n} \left(\ln \frac{\gamma^2}{Q_o^2}, \alpha_s(Q_0)\right) F_j^{\,n}(p^2) \tag{9}$$

$$= \sum_j \tilde{\sigma}_j^n \; \tilde{F}_j^n \, (Q_o^2)$$

where

$$\tilde{F}_j^{\,n} \, (Q_o^2) = \sum_{j'} \int dp^2 \; \Gamma_{jj'}^{\,n} \; F_{j'}^{\,n} \quad . \tag{10}$$

The functions $\tilde{F}_j^n \, (Q_o^2)$ are the matrix elements of the twist-two spin-n local operators renormalized at Q_o. Notice that all dependence on the small momentum p^2 has disappeared, except $0\,(p^2/Q^2)$.

Of course now the factored equation can be recast into the form of a convolution integral by inverting the moments:

$$\frac{d\sigma_H}{dx_H dy} = \sum_j \int dx_p d\xi \;\; \delta(x_H - x_p \xi) \quad \times \tag{11}$$

$$\times \quad \frac{d\tilde{\sigma}_j}{dx_p dy} \quad \tilde{F}_j \; (\xi, Q_o^2) \qquad ,$$

where

$$\tilde{F}_j \; (\xi, Q_o^2) = \sum_{j'} \int dz \; d\xi' \;\; \delta(\xi - \xi' z)$$

$$\Gamma_{jj'} \left(z, \ln \frac{p^2}{Q_o^2}, \alpha_s \, (Q_o)\right) F_{j'} \, (\xi') \; , \tag{12}$$

$$\int \frac{dz}{z} \;\; z^n \, \Gamma_{jj'} \, (Z) = \Gamma_{jj'}^{\,n} \; . \tag{13}$$

Equation (12) can be interpreted in terms of the "Partons within partons" picture of scaling. Γ_Z is the probability of finding a quark of "size $\frac{1}{Q_o}$" inside the bare quark. But factorization is the key that unlocks this entire analysis.

In the field theory language of the OPE, $\tilde{\sigma}^n_j$ are the coefficient functions. If we use our freedom to choose Q_o as we please to take $Q_o = Q$, then there are no logs in $\tilde{\sigma}^n_j$ at all. The only dependence on Q is a power dependence (given trivially by dimensional analysis) and an implicit dependence through $\alpha_s(Q)$. The zeroth-order term in α_s in $\tilde{\sigma}^n_j$ gives the standard parton model result for inclusive lepton-hadron scattering--but it has been renormalization group improved-- $\tilde{F}\ (\xi,Q^2)$ incorporates some QCD interaction corrections in the distribution functions. The scale breaking effects of the Q^2 dependence of $\tilde{F}_j$ may have been observed at SLAC and FNAL, and seems to be consistent with $\Lambda = 500$ MeV in Eq. 1.[1]

The higher-order contributions to $\tilde{\sigma}$ can also be calculated. The first order contributions can be very important because there are some components of $\tilde{\sigma}$ which vanish in lowest order. The familiar example is σ_L in electroproduction (σ_L and σ_T are characterized by different y dependence), which is explicitly of order α_S. In order α_S, it is a finite term in the effective lepton-quark cross-section which remains after the large logs have been removed by factorization. Alas, the experimental situation for σ_L is murky.[1]

GENERALIZATIONS!

In this parton language, it is easy to generalize the concept of factorization to other processes. As an example we will discuss semi-inclusive μp scattering. The parton model description of the process is depicted in Fig. 2, where k_1 (k_2) is the initial (final) momentum of the muon, P_1 is the momentum of the initial hadron (say a proton for definitness) and P_2 the momentum of the observed hadron (a pion, say). The struck parton carries momentum p_1 parallel to P_1 and satisfying $p_1^- = \xi_1 P_1^-$. The scattering produces a parton

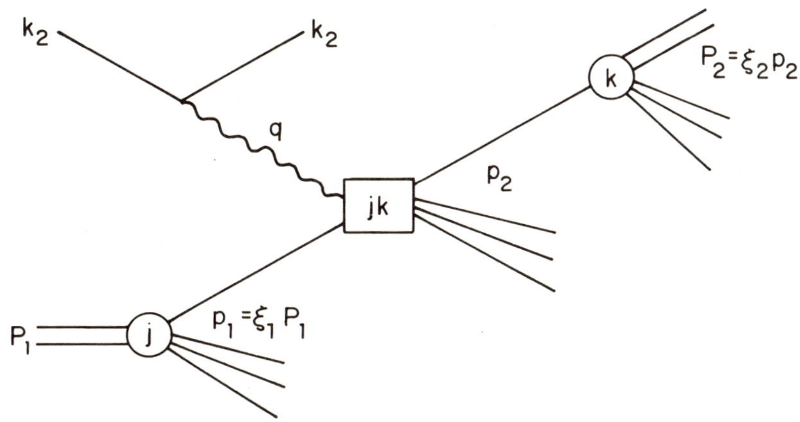

Figure 2. Semi-inclusive lepton-hadron scattering in the parton model.

of momentum p_2 which evolves into the final hadron. P_2 and p_2 are parallel and satisfy $P_2^+ = \xi_2 p_2^+$. The hadronic scattering is described by y and the two scaling variables

$$x_H = - \frac{q^-}{P_1^-} \simeq \frac{Q^2}{2P_1 q} , \tag{14a}$$

$$z_H = \frac{P_2^+}{q^+} \simeq \frac{P_1 P_2}{P_1 q} . \tag{14b}$$

The parton scattering is described by y and the parton scaling variables

$$x_p = - \frac{q^-}{p_1^-} \simeq \frac{Q^2}{2p_1 q} , \tag{15a}$$

$$z_p = \frac{p_2^+}{q^+} \simeq \frac{p_1 p_2}{p_1 q} .$$

In addition the cross-section can depend on the square of the transverse momentum $\vec{P}_{2\perp} = \xi_2 \vec{p}_{2\perp}$ of the final hadron perpendicular to $\vec{q}$ of the virtual photon and on the azimuthal angle ϕ of $\vec{P}_{2\perp}$ (or $\vec{p}_{2\perp}$) measured from $k_\perp$ of the muons. Now the hadron differential cross-sections are given by the following convolution integral

$$\frac{d\sigma_H}{dx_H dy\, dz_H\, dP_{2\perp}^2\, d\phi} = \int dx_p\, dz_p\, dp_{2\perp}^2$$

$$\times \quad d\phi\, d\xi_1\, d\xi_2\, \delta(x_H - \xi_1 x_p)\, \delta(z_H - \xi_2 z_p) \tag{16}$$

$$\times \quad \delta(p_{2\perp}^2 - \xi_2^2\, p_{2\perp}^2) \frac{d\sigma_{jk}}{dx_p dy dz_p dp_{2\perp}^2\, d\phi} F_j(\xi_1)\, D_k(\xi_2) .$$

Here $d\sigma_{jk}$ the cross-section for scattering of μ + j-parton $\to \mu$+k-parton + anything, represented by the square in Fig. 2. $F_j(\xi_1)$ is the distribution function for j partons in the proton (the j circle in Fig. 2) and $D_k(\xi_2)$ is the "decay function"- the probability that k-parton evolves into a pion carring a fraction ξ_2 of the parton momentum. (the k circle in Fig. 2).

Now as before we compute $d\sigma_{jk}$ in perturbation theory. In the best of all possible worlds, what would happen is that to all orders in perturbation theory the moments would factor (for simplicity, I will discuss only the integral over all $P_\perp^2$):

$$\sigma_{jk}^{nm} = \int dp_\perp^2 \frac{dx}{x} \frac{dz}{z} x^n z^m \frac{d\sigma_{jk}}{dx\,dy\,dz\,dp_\perp^2 d\phi} ,$$

$$D_k^{\,m} = \int \frac{d\xi_2}{\xi_2} \xi_2^{\,m} D_k(\xi_2) . \tag{17}$$

Then

$$\begin{aligned} &\sigma_{j\,k}^{n\,m}(p_1^2, p_2^2, Q^2, \ldots) \\ &= \sum_{j'k'} \tilde{\sigma}_{j'k'}^{n\,m}\left(\ell n \frac{Q^2}{Q_o^2}, \ldots\right) \times \\ &\times \Gamma_{j'j}^{n}\left(\ell n \frac{p_1^2}{Q_o^2}, \alpha_s(Q_o)\right) \times \\ &\times \Gamma_{k'k}^{m}\left(\ell n \frac{p_2^2}{Q_o^2}, \alpha_s(Q_o)\right) + 0\left(\frac{p_1^2}{Q^2}, \frac{p_2^2}{Q^2}\right) , \end{aligned} \tag{18}$$

where Γ is the same as in Eq. 8. If Eq. (18) is satisfied, then all the large logs of p_1^2 can be associated with the distribution function moments $F_j^n(p_1^2)$ as in Eq. 10, and the large logs of p_2^2 can be associated with the decay functions momenta analogously

$$\tilde{D}_k^m(Q_o^2) = \sum_{k'} \int dp^2 \, \Gamma^m_{kk'} \quad D^m_{k'} \quad . \tag{19}$$

Then all log dependence on p_1^2 and p_2^2 has disappeared, leaving distribution and decay functions which depend on Q_o according to the renormalization group. As before, we could turn the moment statements back into convolutions.

WHAT'S THE PROBLEM?

Clearly what we have to do to any order of perturbation theory is (1) check that Eq. 18 is satisfied; (2) calculate $\tilde{\sigma}_{jk}$. This has been done to first order in α_s. I will give a quick sketch of the calculation.

The lowest order contributions come from the diagram in Fig. 3 and give

$$\frac{d\sigma_{jk}}{dx_p dy dz_p dp_\perp^2 d\phi} \propto [1 + (1-y)^2] \, Q_j^2 \, \delta_{jk} \quad \times$$

$$\times \quad \delta(1-x_p) \, \delta(1-z_p) \, \delta(p_\perp^2) \quad , \tag{20}$$

which gives the standard parton result. The δ functions arise from momentum conservation ($\delta(p_1 + q - p_2)$). On general grounds there are four independent components of $d\sigma_{jk}$ (structure functions) characterized by different y and Q dependence while only one (the analog of σ_T) appears in zeroth-order. The other three, σ_L and terms linear and quadrotic in cos ϕ must be free of all large logs in first order in α_s because there are no corresponding

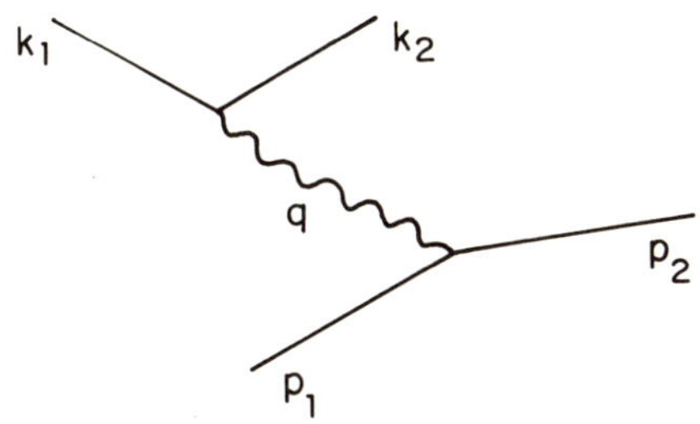

Figure 3. The zeroth-order (in αs) contribution to $d\sigma_{jk}$.

zeroth-order terms. The factorization, which in first order always takes the form

$$\sigma^{nm} = A + B\,\alpha_s\, \ln \frac{p_1^{\,2}}{Q^2} + C\,\alpha_s\, \ln \frac{p_2^{\,2}}{Q^2} + D\alpha_s + 0\,(\alpha_s^{\,2}) + 0\,\left(\frac{p_1^{\,2}}{Q^2}, \frac{p_2^{\,2}}{Q^2}\right)$$

$$= \left(A - B\alpha_s\, \ln \frac{Q^2}{Q_o^2} - C\,\alpha_s\, \ln \frac{Q^2}{Q_o^2} + D\,\alpha_s\right) \times \left(1 + \frac{B}{A}\,\alpha_s\, \ln \frac{p_1^{\,2}}{Q_o^{\,2}}\right) \left(1 + \frac{C}{A}\,\alpha_s\, \ln \frac{p_2^{\,2}}{Q_o^{\,2}}\right) + \ldots, \qquad (21)$$

cannot take place because Eq. (21) doesn't make sense for A=0 unless B=C=0 as well.

In first order in α_s, we must complete both real and virtual diagrams shown in Fig. 4, where the curly lines represent gluons. The real diagrams all have as a factor the δ function

$$\delta\left[\, p^2_{2\perp} - \frac{z_p\, Q^2}{x_p}\,(1-x_p)\,(1-z_p)\,\right] \propto \delta[(p_1 + q - p_2)^2] \quad . \qquad (22)$$

As we will see, all the logs arise from the regions in the moment integral $1 \simeq x_p$ or $1 \simeq z_p$, so $p^2_{2\perp} \simeq 0$ as we expect from Eq.(20). When we consider the $p^2_\perp$ integrated momenta (Eq. 17) the δ function of Eq. (22) disappears.

It is easy to check that σ_L and the two ϕ dependent structure

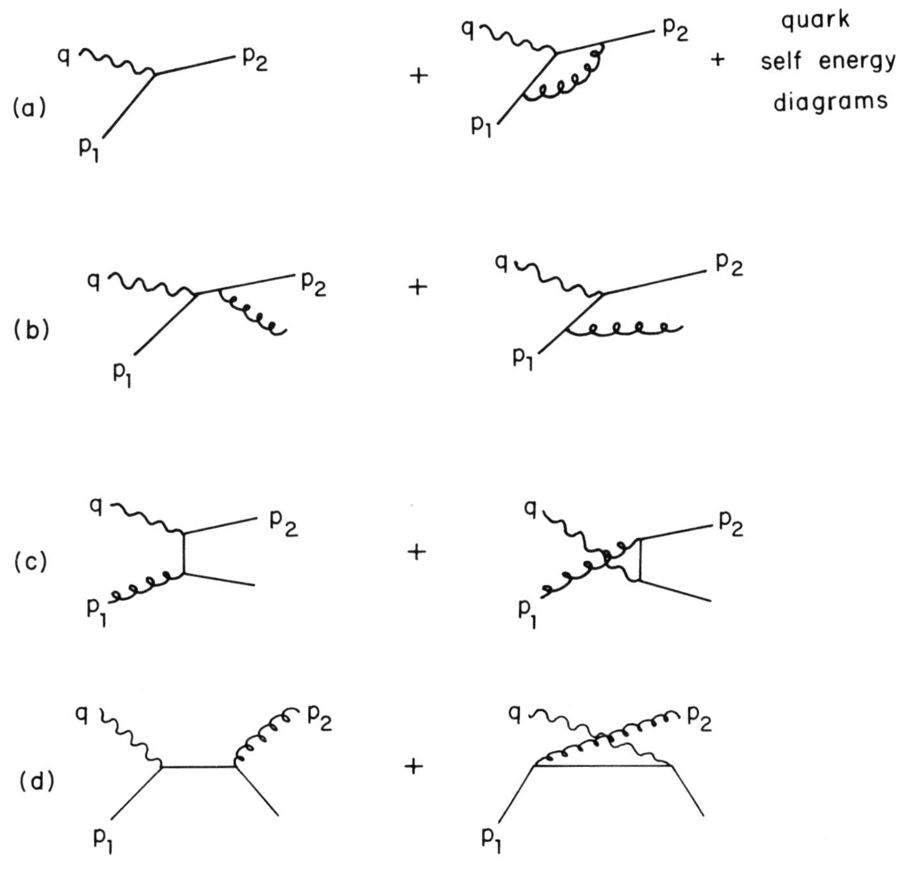

Figure 4. The first-order (in αs) contributions to $d\sigma_{jk}$.

functions are free of large logs. So they are finite predictions of QCD assuming that the rest of the picture makes sense. The predicted ϕ dependence is the following: The coefficient of $\cos^2\phi$ is always positive both in quark-quark and quark-gluon diagrams. Thus the $\vec{p}_\perp$ of the produced hadrons tends to lie in the muon scattering plane at all x_H and z_H. The magnitude of the effect is comparable to σ_L/σ_T. The coefficient of $\cos\phi$ is negative in the diagrams of Fig 4b, positive in Fig. 4d and variable in Fig. 4c. In practice, the μP data are primarily at small x_H, and the Fig. 4c contributions average to zero as $x_H \to 0$ (assuming distribution functions $\to k/\xi$ as $\xi \to 0$) so we will ignore them. It is reasonable to assume that the quark decay function into pions at large ξ is much larger than the gluon decay function. If that is the case, Fig. 4b will dominate at large z_H and the fast pions will tend to be produced with $\vec{p}_\perp$ antiparallel to $\vec{k}_\perp$. This turns out to be a fairly large effect. It has been discussed in detail by Politzer and me (3).

The factorization of σ_T is more complicated. To indicate how it goes, I will focus on Fig. 4b in which both the struck and observed partons are quarks. I will also assume for convenience that $p_1^2<0$ and $p_2^2>0$. This technical assumption makes the moment integrals much easier to deal with because it forbids the physical decay of the initial quark into on-shell quark + gluon in the second diagram in Fig. 4b and also prevents the internal quark in the first diagram in Fig. 4b from going on shell. The propagator in the first diagram is proportional to $1/(1-x_p)$ and the kinematics requires $(1-x_p) \gtrsim \dfrac{p_2^2}{Q^2}$. The propagator in the second diagram is proportional to $\sim 1/(1-z_p)$ and the kinematic restriction is

$$(1-z_p) \gtrsim -\frac{p_1^2}{Q^2}\ .$$

I calculated the real diagram in "Feynman gauge", summing over physical and unphysical polarizations of the gluons. The cross-term between the two diagrams in Fig. 4b has the form

$$\propto \alpha_s \frac{x_p z_p}{(1-x_p)\,(1-z_p)} \quad . \tag{23}$$

The moment integral is

$$\propto \alpha_s \int dx_p \;\; dz_p \;\; \frac{x_p^{\;n}}{(1-x_p)} \;\; \frac{z_p^{\;m}}{(1-z_p)} \tag{24}$$

$$= \alpha_s \int dx_p \;\; dz_p \;\; \Big\{ \frac{1}{(1-x_p)} \frac{1}{(1-z_p)}$$

$$+ \frac{(x_p^{\;n}-1)}{(1-x_p)} \cdot \frac{1}{(1-z_p)} + \frac{1}{(1-x_p)} \;\; \frac{(z_p^{\;m}-1)}{(1-z_p)}$$

$$+ \frac{(x_p^{\;n}-1)}{(1-x_p)} \;\; \frac{(z_p^{\;m}-1)}{(1-z_p)} \Big\}$$

$$= \alpha_s \; \ell n \frac{p_2^{\;2}}{Q^2} \; \ell n \frac{-p_1^{\;2}}{Q^2}$$

$$- \alpha_s \sum_{k=1}^{n} \frac{1}{k} \; \ell n \frac{-p_1^{\;2}}{Q^2}$$

$$- \alpha_s \; \ell n \frac{p_2^{\;2}}{Q^2} \sum_{k=1}^{m} \frac{1}{k}$$

$$+ \text{constants} + 0 \left(\frac{p_1^{\;2}}{Q^2} \, , \, \frac{p_2^{\;2}}{Q^2} \right) .$$

The first term could spoil factorization but it is precisely cancelled by a similar term in the virtual diagram of Fig. 4a. The other log terms factorize and reproduce the ℓn n parts of the anomalous dimensions of quark operators. When all contributions are included, Eq. 18 is satisfied with the same Γ's as in inclusive lepton-hadron scattering to first order in α_s.

PARTIAL FACTORIZATION

Even if factorization, as in Eq. 18, is not correct, the above results could be useful. If there were an additional term in Eq. 18 which doesn't factor but which is non-leading in large logs, say

$$\alpha_s (Q_o) \Sigma^{n\ m}_{j\ k} (\alpha_s \ln \frac{p_1^2}{Q_o^2}, \alpha_s \ln \frac{p_2^2}{Q_o^2}, \ldots) , \quad (25)$$

it wouldn't affect much of the analysis of σ_T, because it is the same size as the contribution of the order α_s term in $\tilde{\sigma}$ and small compared to the zeroth-order contribution. If however, the term Eq. (25) also contributes to the other structure functions, then our predictions of "σ_L/σ_T" and ϕ dependence would be useless. However, we could get lucky and find that the non-factoring contribution to these structure functions is

$$\alpha_s^2 (Q_o) \Sigma'^{n\ m}_{j\ k} (\alpha_s \ln \frac{p_1^2}{Q_o^2}, \alpha_s \ln \frac{p_2^2}{Q_o^2}, \ldots) . \quad (26)$$

Then all of our predictions for semi-inclusive μP scattering would be useful, but they would fall apart if we tried to push them to higher order. This kind of partial factorization may be much easier to demonstrate. The moral of this little fantasy is simple. The description of parton model processes obtained by assuming factorization may be wrong, but it is the only description which has a chance of being derived from QCD perturbation theory. So while we are waiting for some smart-person to prove factorization or partial factorization to all orders, the rest of us are morally justified in using these perturbative results to do interesting phenomenology. An interesting first step would be the observation of QCD effects in decay functions and ϕ dependence in μP scattering.

REFERENCES

1. For a recent study of electroproduction, see A. De Rujula, H. Georgi, and H. D. Politzer, Ann. Phys. (N. Y.) 103, 315 (1977).

2. For earlier discussions of factorization, see A. Mueller, Phys. Rev. D9 , 963 (1974); and H. D. Politzer, Phys. Lett. 70B, 430 (1977).

3. H. Georgi and H. D. Politzer, Phys. Rev. Letters, 40, 3 (1978).

4. I am grateful to S. Weinberg for interesting discussions of a related question.

EXPERIMENTAL TESTS OF QCD???

A. L. Sessoms*

Harvard University

Cambridge, MA 02138

I would like to discuss the possibility of testing Quantum Chromodynamics (QCD) in deep inelastic muon scattering from an experimental viewpoint. I will show you the instruments and quality of data that are available for such tests, and then an indication of what our very first attempts to make a comparison between these and QCD predictions in semi inclusive hadron production looks like. These results are very preliminary and in some cases merely suggestive of the possibilities that avail themselves. In all cases I think they are instructive.

I will report on work done by the Chicago-Harvard-Illinois-Oxford collaboration (CHIO) at Fermilab.[1] The participants are named in Figure 1.

A diagram of the latest version of the apparatus is given in Figure 2. An indication of the kind of events we get in this apparatus is given in Figures 3 and 4 where I have displayed a muon nucleon scatter and a muon electron scatter respectively. The scales are very compressed so sparks in the chambers are jammed together but it is clear that the muon is easily seen from its pene-

---

*Alfred P. Sloan Foundation Fellow

Supported in part by U.S. Department of Energy contract #EY-76-C-02-3064

H. L. Anderson, R. M. Fine,[1] R. H. Heisterberg,[2] H. S. Matis, L. W. Mo,[2] L. C. Myrianthopoulos, S. C. Wright

Enrico Fermi Institute[10]
University of Chicago
Chicago, Illinois 60637

B. A. Gordon, W. A. Loomis, F. M. Pipkin, S. H. Pordes,[3] A. L. Sessoms, W. D. Shambroom, L. J. Verhey[4], Richard Wilson

High Energy Physics Laboratory[11], and
Department of Physics
Harvard University
Cambridge, Massachusetts 02138

W. R. Francis,[5] R. G. Hicks, T.B.W. Kirk[6]

Department of Physics[12]
University of Illinois
Urbana, Illinois 61801

V.K. Bharadwaj,[7] N. E. Booth, G. I. Kirkbride,[8] J. Proudfoot V.W. Quirk, A. Skuja,[9] M. A. Staton, W.S.C. Williams

Nuclear Physics Laboratory[13]
Oxford University
Keble Road, Oxford, OX1 3RH, England

Figure 1

(1) Now at Nevis Laboratory, Columbia University, New York, New York 10027
(2) Now at Department of Physics, Virginia Polytechnic Institute and State University, Blacksburg, Virginia 24061
(3) Now at CERN, CH-1211, Geneva, 23, Switzerland
(4) Now at Massachusetts General Hospital, Boston, Massachusetts 02125
(5) Now at Michigan State University, Lansing, Michigan 48823
(6) Now at Fermi National Accelerator Laboratory, Batavia, Illinois 60510
(7) Now at Dept. of Physics, University of California, Santa Barbara, CA 93106
(8) Now at Hanson Laboratory, Stanford University, Stanford, Ca 94305
(9) Now at Physics Department, University of Maryland, College Park, MD 20742
(10) Supported by the National Science Foundation Contract MPS 71-03-186
(11) Supported by Energy Research and Development Administration (now Dept of Energy) Contract EY-76-C-02-3064. *000
(12) Supported by the Energy Research and Development Administration (now Dept of Energy) Contract E(11-1) 1195
(13) Supported by the Science Research Council of the United Kingdom

Figure 1 (cont'd)

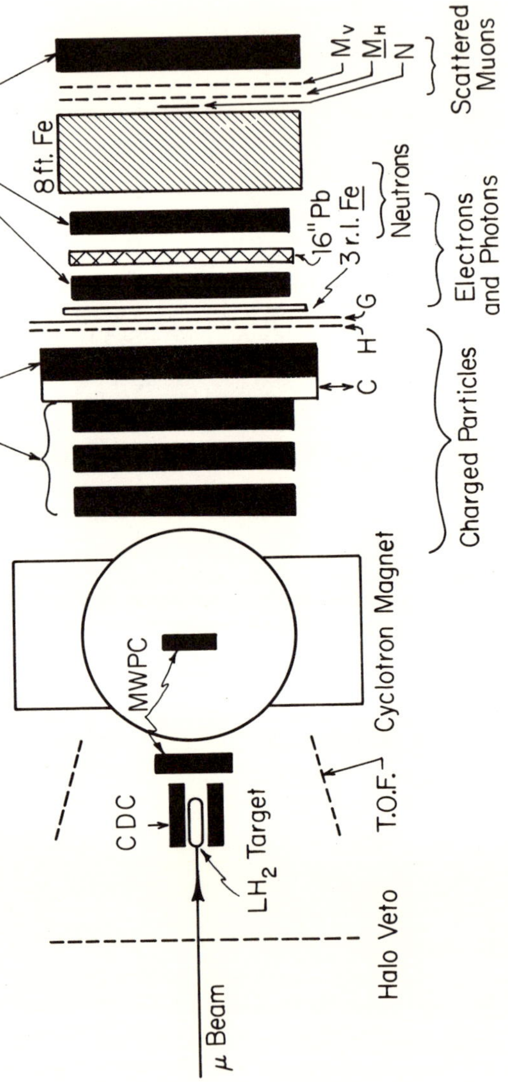
μ Beam
Halo Veto
LH2 Target
CDC
T.O.F.
MWPC
Cyclotron Magnet
Wire Spark Chambers
Charged Particles
C
H
G
3 r.l. Fe
16" Pb
Electrons and Photons
Neutrons
8 ft. Fe
N
M_H
M_V
Scattered Muons

Figure 2

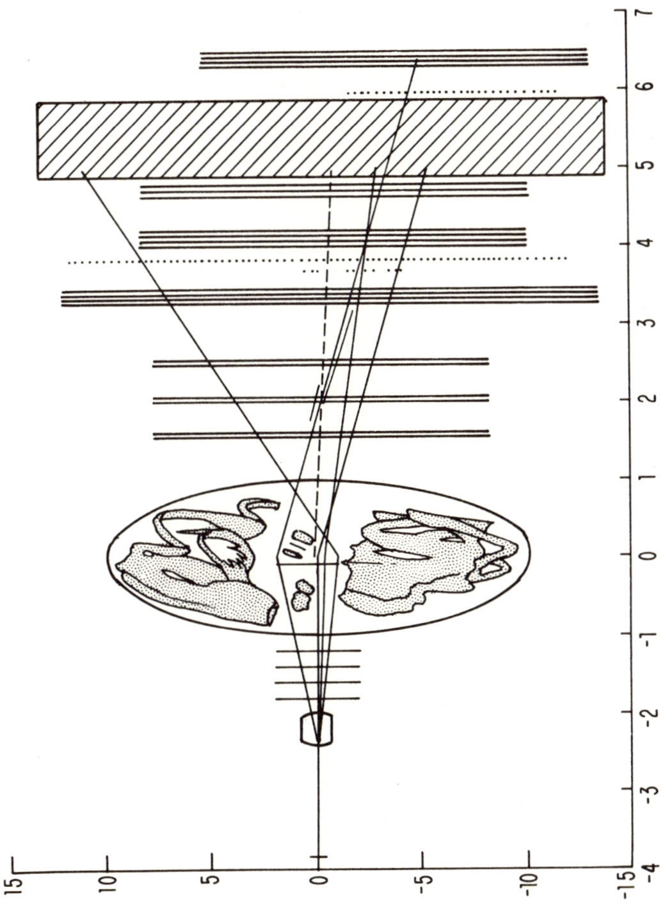

Figure 3

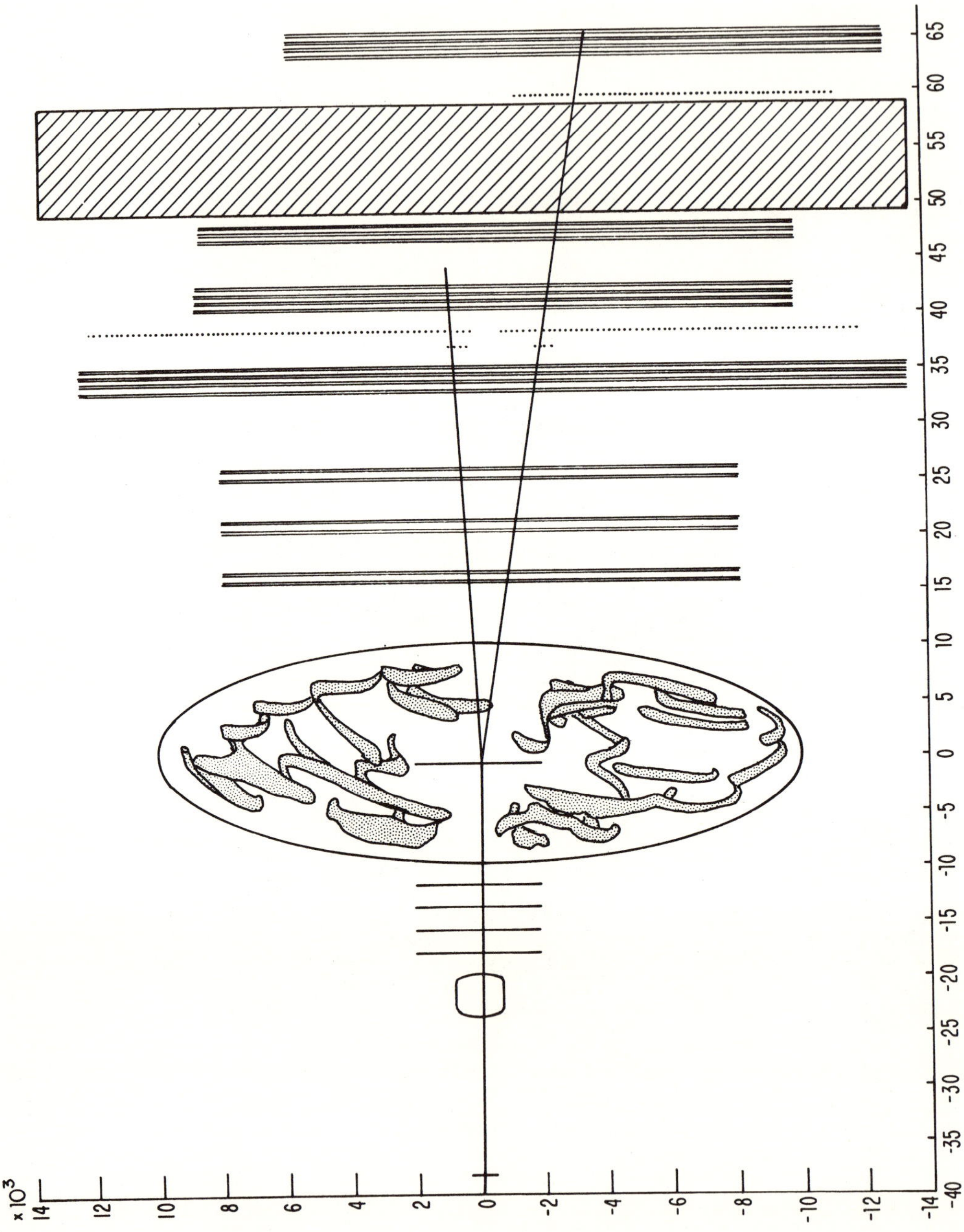
x 10³
14
12
10
8
6
4
2
0
-2
-4
-6
-8
-10
-12
-14
-40
-35
-30
-25
-20
-15
-10
-5
0
5
10
15
20
25
30
35
40
45
50
55
60
65

Figure 4

tration of 8 feet of iron and tracks in the spark chambers at the end of the spectrometer. In Figure 4 the electron is identified by its showering in the 3 radiation lengths of iron, and the subsequent spark distribution in the chambers just downstream of this point.

A close-up of an earlier arrangement of chambers around the liquid hydrogen target is shown in Figure 5. This arrangement was used for some of the data discussed below. There were wire spark chambers on two sides of the target, followed by scintillation counters used to measure the time of flight (T.O.F.) of particles leaving the target, as well as their pulse height in the plastic. The combination of these two pieces of information was often enough to separate protons from pions.

The purpose of this exercise in muon physics is to try to

1. test QCD
2. define QCD

and/or

3. destroy QCD

It is clear that the present state of the art does little to push us towards these goals, but the first halting steps are amusing. This is what I will attempt to illustrate here.

The process we are discussing is shown schematically in Figure 6, where the relevant kinematic quantities are defined. QCD makes certain predictions[2] about the behavior of the semi inclusive cross section for the reaction

$$\mu p \rightarrow \mu + \pi + \text{anything}$$

as a function of

$$Z = \frac{E_H}{\nu}, \text{ where } E_H \text{ is the energy of the pion.}$$

The cross section is expected to fall as a function of increasing q^2.

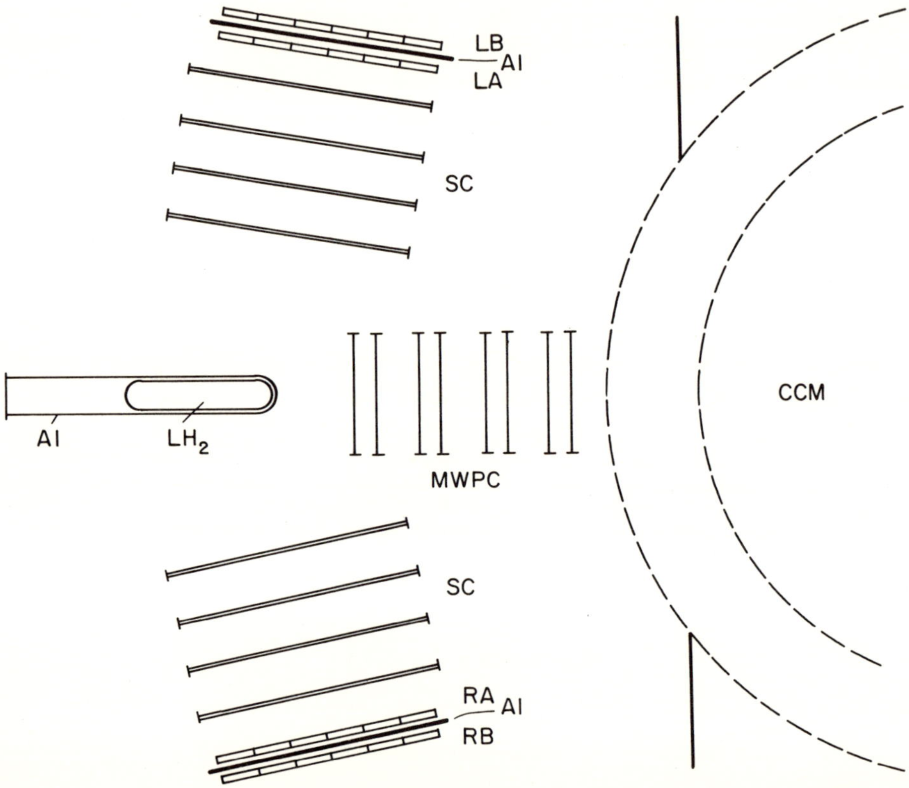

Figure 5

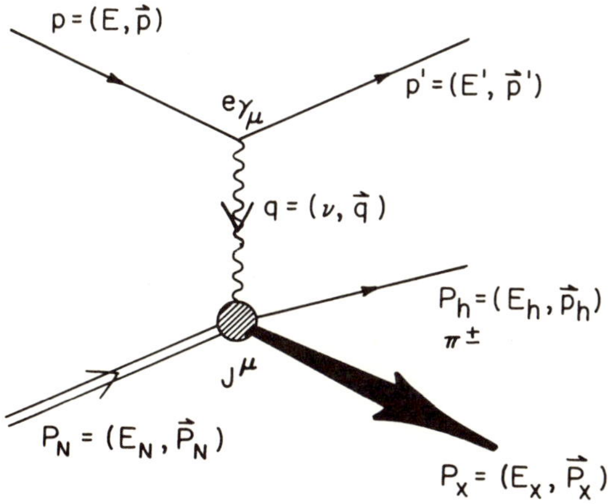

$$\nu = E - E'$$
$$\vec{q} = \vec{p} - \vec{p}\,'$$
$$Z = \frac{E_\pi}{\nu}$$

Figure 6

The results of this experiment are indicated in Figures 7 through 11. There is no indication of any trend towards a systematic decrease in $\frac{d\sigma}{dz}$ as a function of increasing q^2. A look at the predictions[2] indicates that over the range of q^2 from roughly 2 GeV^2/c^2 to approximately 10 GeV^2/c^2 the effect is small ~ 5% ; it would not show up in these data but a large effect of the order of 20% is probably ruled out, in agreement with QCD. I should note that in Figure 10 there appears to be some effect for the highest q^2 bin in $\mu p \to \mu x^-$ + anything. The possibility is ruled out by the results shown in Fig. 11 which gives $\mu d \to \mu x^-$ + anything where the statistical power of the data is greater.

In a recent paper entitled "Clean Tests of QCD in μp Scattering"[3] by Howard Georgi and H. David Politzer the authors indicated that the study of angular correlations may lead to unambiguous tests of the theory. They suggest that the value of the quantity

$$\langle\cos\phi\rangle = -\frac{\pi}{2}\,\frac{\alpha s(2-y)(1-y)^{1/2}}{\pi[1+(1-y)^2]}\,\kappa\,(1-Z_H)^{1/2} + O([1-Z_H]^{3/2})$$

where κ is a constant which is slightly model dependent, φ is the azimuthal angle between the observed hadron transverse momentum ($\vec{P}_{2_\perp}$) and the component of the initial muon momentum perpendicular to $\vec{q}$ ($\vec{K}_{1_\perp}$) and y is $P_1 \cdot q/P_1 \cdot K_1$ with P_1 the initial hadron 4-momentum and K_1 the initial muon 4-momentum, can be predicted unambiguously.

In this first, very preliminary, pass at such an analysis the quantity we measure is

$$\frac{d\sigma}{d\Phi} = \frac{1}{2\pi}\left[\sigma_T + \varepsilon\sigma_L + \varepsilon\cos 2\Phi\,\sigma_p + \sqrt{2\varepsilon(1+\varepsilon)(1+2\frac{\delta}{1+\varepsilon})}\right.$$

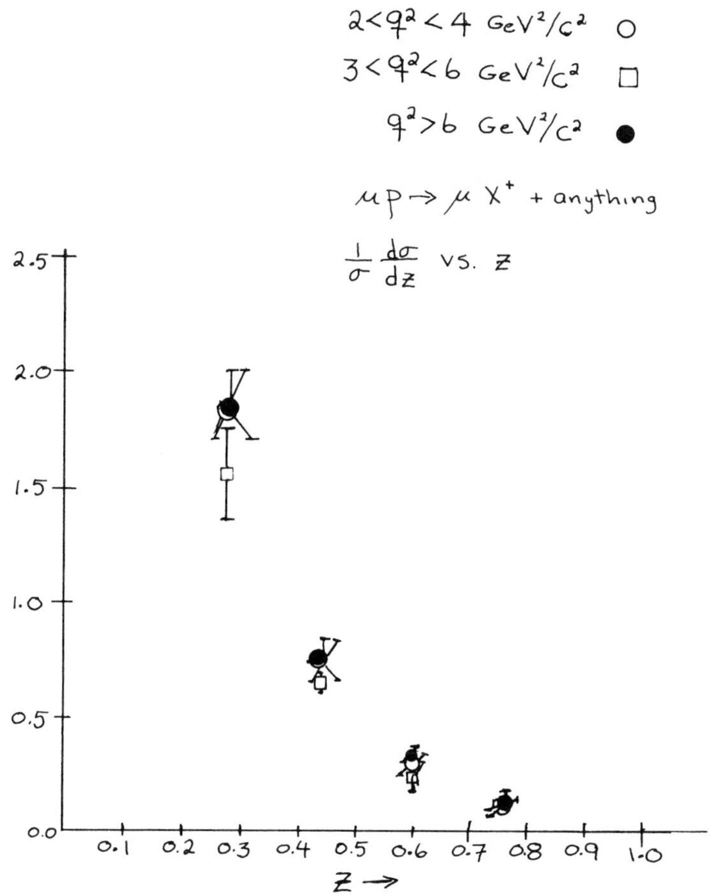

2<q²<4 GeV²/c² ○
3<q²<6 GeV²/c² □
q²>6 GeV²/c² ●
μp → μ X⁺ + anything
1/σ dσ/dz vs. z
2.5
2.0
1.5
1.0
0.5
0.0
0.1 0.2 0.3 0.4 0.5 0.6 0.7 0.8 0.9 1.0
z →

Figure 7

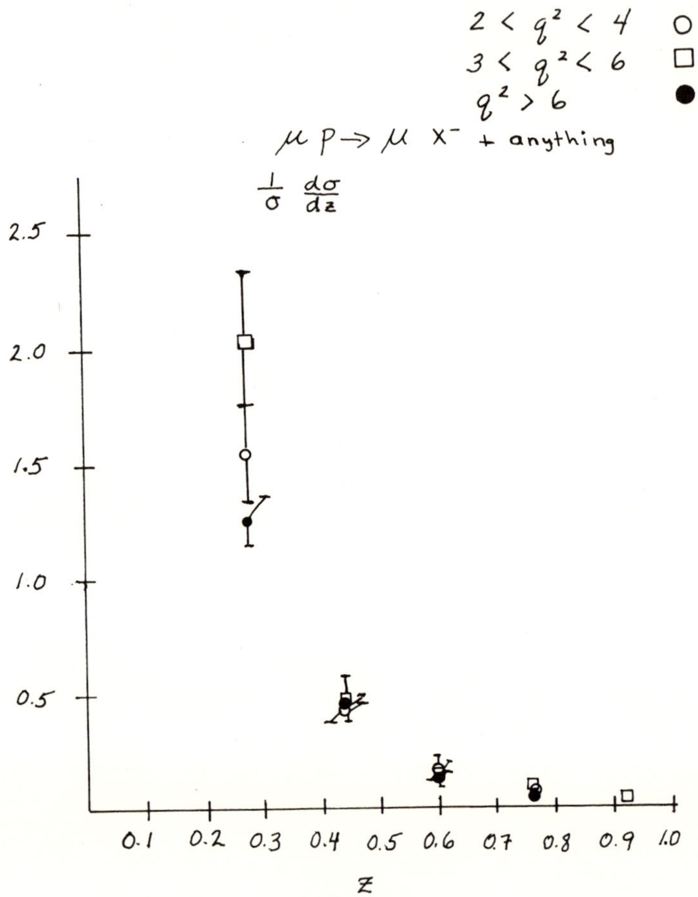

2 < q² < 4 ○
3 < q² < 6 □
q² > 6 ●
μ p → μ X⁻ + anything
1/σ dσ/dz
2.5
2.0
1.5
1.0
0.5
0.1 0.2 0.3 0.4 0.5 0.6 0.7 0.8 0.9 1.0
z

Figure 8

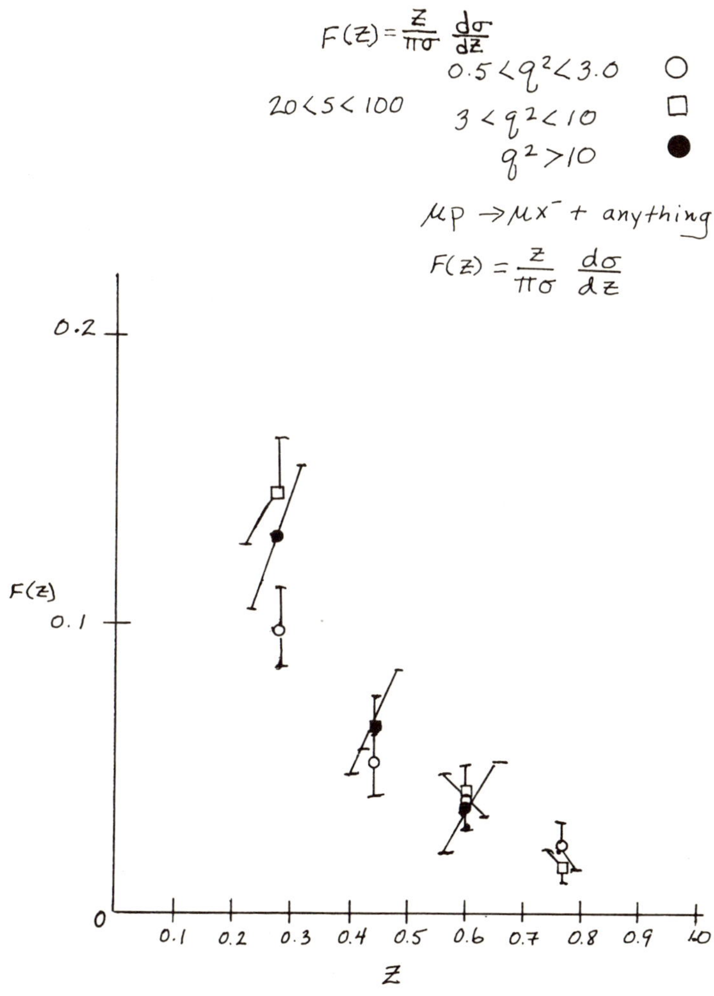
F(z) = z/πσ dσ/dz
0.5 < q² < 3.0
20 < s < 100
3 < q² < 10
q² > 10
μp → μx⁻ + anything
F(z) = z/πσ dσ/dz
0.2
F(z)
0.1
0
0.1
0.2
0.3
0.4
0.5
0.6
0.7
0.8
0.9
1.0
z

Figure 9

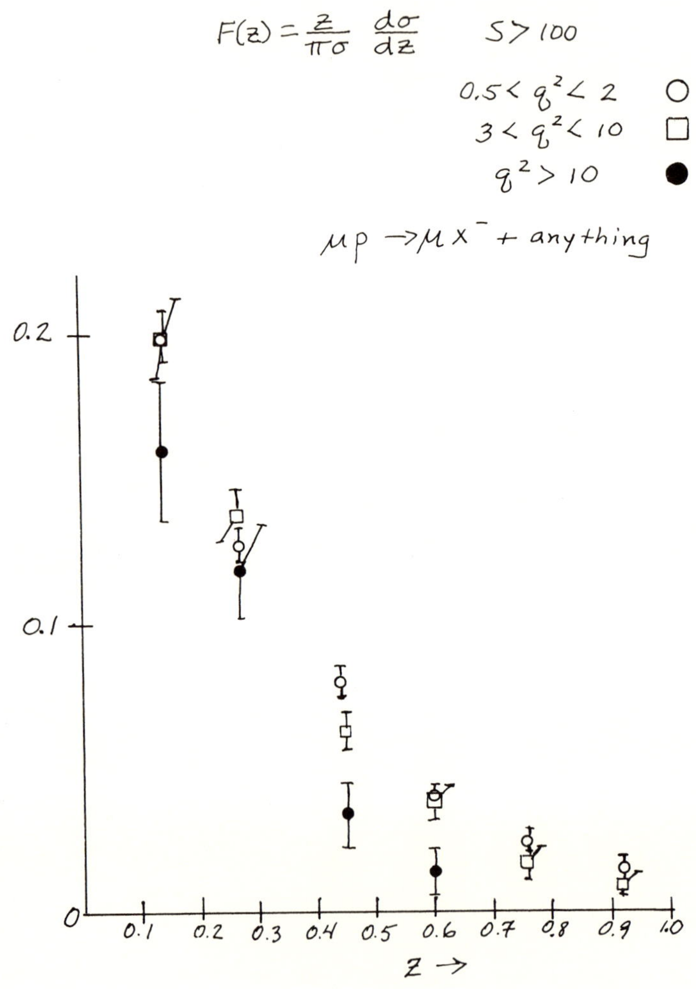

$F(z) = \frac{z}{\pi\sigma}\frac{d\sigma}{dz}$
S > 100
0.5 < q² < 2
3 < q² < 10
q² > 10
μp → μX⁻ + anything
0.2
0.1
0
0.1
0.2
0.3
0.4
0.5
0.6
0.7
0.8
0.9
1.0
Z →

Figure 10

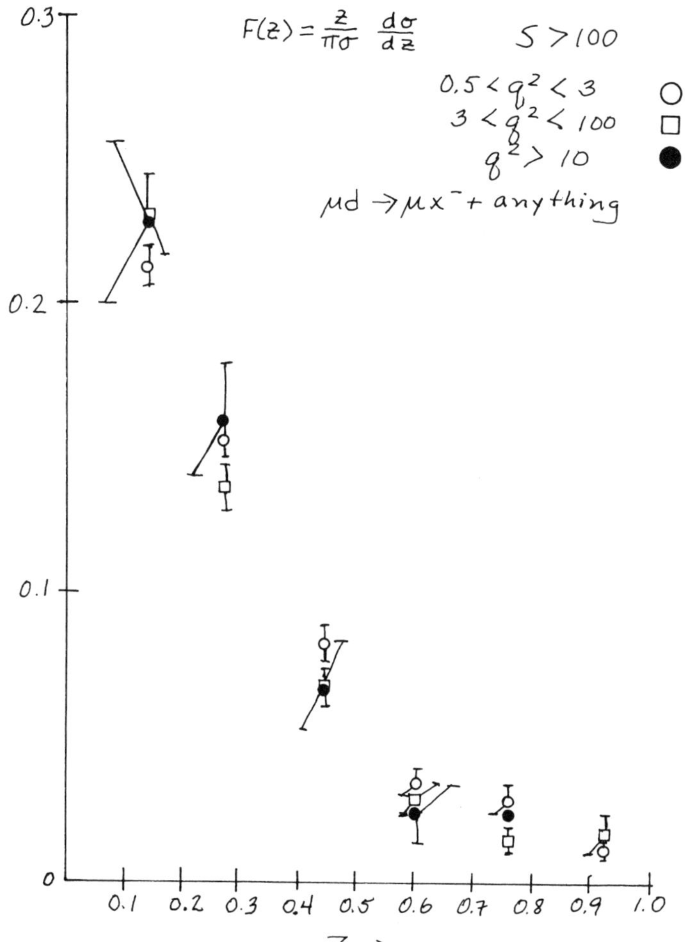

$F(z) = \frac{z}{\pi\sigma} \frac{d\sigma}{dz}$
S > 100
0.5 < q² < 3
3 < q² < 100
q² > 10
μd → μx⁻ + anything
0.3
0.2
0.1
0
0.1 0.2 0.3 0.4 0.5 0.6 0.7 0.8 0.9 1.0
Z →

Figure 11

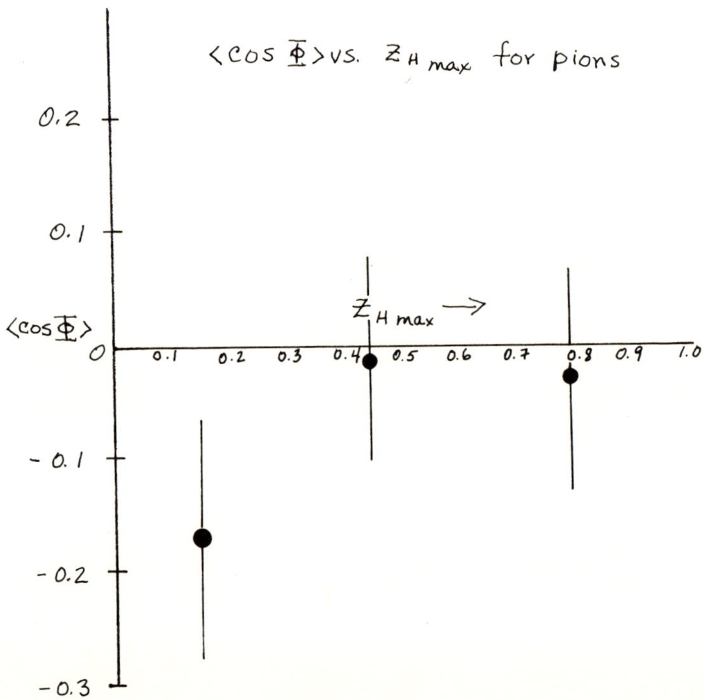

⟨cos Φ⟩ vs. Z H max for pions
0.2
0.1
⟨cos Φ⟩
0
- 0.1
- 0.2
- 0.3
Z H max →
0.1
0.2
0.3
0.4
0.5
0.6
0.7
0.8
0.9
1.0

Figure 12

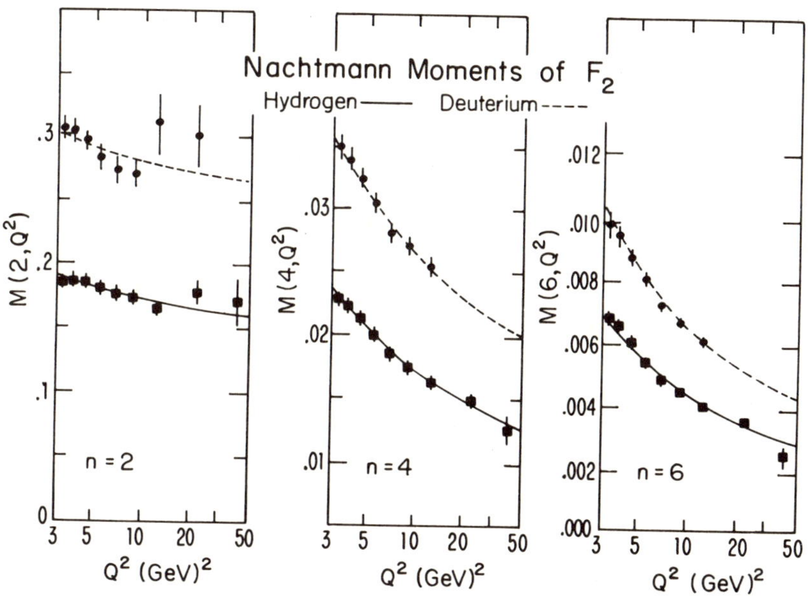
Nachtmann Moments of F2
Hydrogen ——— Deuterium - - - -
M(2,Q²)
M(4,Q²)
M(6,Q²)
n = 2
n = 4
n = 6
Q² (GeV)²
Q² (GeV)²
Q² (GeV)²

Figure 13

$$x \cos \Phi \, \sigma_{I_1} + \sqrt{\varepsilon 2(1-\varepsilon)(1 + 2\frac{\delta}{1+\varepsilon})} \, P \sin \Phi \sigma_{I_2} \Big] ,$$

where ε is the virtual photon polarizations

$$\varepsilon = [1 + 2\,(\frac{Q^2}{Q^2 + \nu^2}) \tan^2 \frac{\theta}{2}]^{1/2}$$

$$\delta = 2M_\mu^2 \,(1 - \varepsilon)/Q^2$$

Φ is the angle between the muon scattering and hadron production planes

σ_T is the unpolarized transverse cross section

σ_L is the longitudinal cross section

σ_P is the transverse interference term

σ_{I_1} and σ_{I_2} are the longitudinal-transverse interference terms

P is the muon beam polarization

A very rough correspondence between the Georgi-Politzer quantity and the measured quantity[4] is

$$\langle \cos \phi \rangle \simeq \langle \cos \Phi \rangle + 0\,(\frac{\delta}{1+\varepsilon}) \quad .$$

The quantity we measure for the Z of the hadron Z_H, is really the Z of the fastest hadron coming from the interaction, Z_{Hmax}. This is not the quantity prescribed by Georgi and Politzer in their article but it does bear some relationship to it.[4] In any case the results indicate what a detailed analysis might lead to.

Figure 12 is a plot of cos Φ versus Z_{Hmax} for pions. The first thing to note is that ⟨cos Φ⟩ does not dramatically decrease with increasing Z_{Hmax}, although one must admit that the data are rather sparse. For a given set of parameters, Georgi and Politzer predict a value for $\langle \cos \phi \rangle$ which is given along with the measured quantities in Table 1. On the basis of these numbers one cannot say much but it is clear that such an analysis is within our grasp.

QCD makes straightforward predictions[5] for the <u>moments</u>

$$M_n(Q^2) = \int_0^1 dx\, x^{n-2}\, \nu W_2(W, Q^2) \qquad n = 2, 4, 6\ldots \quad .$$

The theory predicts that, for large enough Q^2,

$$M_n(Q^2) = \sum_{i=1}^{s} C_i^{(n)} \, (\alpha_s(Q))^{\lambda_i^{(n)}} \quad ,$$

$\lambda_i^{(n)}$ are known,

$C_i^{(n)}$ come from experiment.

Using the CHIO data, Anderson, et al.[6] have made a comparison with these so-called Nachtmann moments. The results are shown in figure 13. The agreement is excellent. Note however that Λ is taken to be 0.6 GeV as opposed to the more favored 0.5 GeV.

In conclusion I would say that we are on the verge of being able to test QCD experimentally. I urge theoreticians who practice QCD to calculate and present in a reasonably clean way, predictions that can be confronted with the data experimentalists are able to generate.

I would like to acknowledge very useful discussions with my colleagues, in particular Howard Georgi, Bruce Gordon, Dave Shambroom and Charling Tao.

TABLE 1

	G-P Prediction	Data
E	200 GeV	96 GeV
x	0.05	$\langle x \rangle \sim 0.01$
y	0.3	0-0.9 $\langle y \rangle \sim 0.5$
z	0.7	0-1 $\langle z_{max} \rangle = 0.4$
ν	60 GeV	0-85 $\langle \nu \rangle \sim 50$ GeV
Q^2	6 GeV^2/c^2	0-3 $\langle Q^2 \rangle = 0.95$ GeV^2/c^2
W^2	115 GeV^2	0-175 $\langle W^2 \rangle = 95$ GeV^2
ε	0.94	0-1 $\langle G \rangle = 0.68$
$\langle \cos \phi \rangle$	$= - 0.12$	-0.07 ± 0.05

REFERENCES

1. H. L. Anderson, et al. PRL 37, 4 (1976); ibid. 38, 1450 (1977); I have used the complete set of results from Fermilab experiments 98/398 (to be published).
2. R. Field and R. Feynman, Private Communication.
   H. Georgi, Private Communication.
3. H. Georgi and H. D. Politzer, PRL 40, 3 (1978).
4. A preliminary analysis done to measure directly <cos ϕ> as a function of Z_H has gotten similar results indicating some correspondence between these two variables. The CHIO collaboration, to be published.
5. O. Nachtmann, Nucl. Phys. B63, 277 (1973); and B78, 455 (1974).
6. H. L. Anderson, et al., submitted to Phys. Rev. Lett.

APPLICATION OF QUANTUM CHROMODYNAMICS TO

HIGH TRANSVERSE MOMENTUM HADRON PRODUCTION*

Geoffrey C. Fox

California Institute of Technology

Pasadena, California

ABSTRACT

We describe the application of QCD (Quantum Chromodynamics) to high $p_\perp$ (transverse momentum) hadron production - discussing single particle and jet cross-sections and correlation measurements. We conclude that the theory is in reasonable accord with the data although uncertainties in the theory and technical ambiguities (mainly caused by the low $p_\perp$ of most of the data) make incontrovertible conclusions impossible.

INTRODUCTION - EXPERIMENTAL SUMMARY

Interest in high $p_\perp$ scattering stemmed from early ISR experiments that showed a much higher yield of hadrons at high $p_\perp$ (transverse momentum) than expected from a naive $\exp(-6\,p_\perp)$ extrapolation of the low $p_\perp$ data[1]. Further, this yield increased with energy at fixed $p_\perp$ in contrast to the scaling observed at low $p_\perp$. From the very beginning[2], it has been believed (hoped) that high $p_\perp$ hadron production was due to scattering among constituents of the

*Work supported in part by the U.S. Department of Energy under Contract No. EY76-C-03-0068.

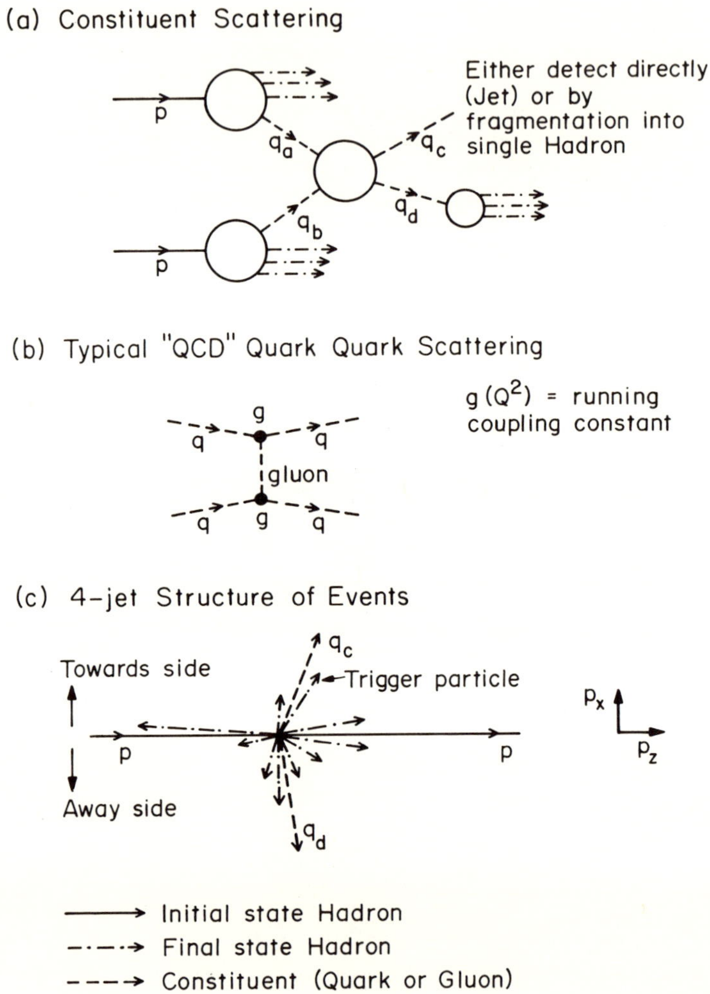

Fig. 1 Constituent scattering model for high $p_\perp$ scattering.

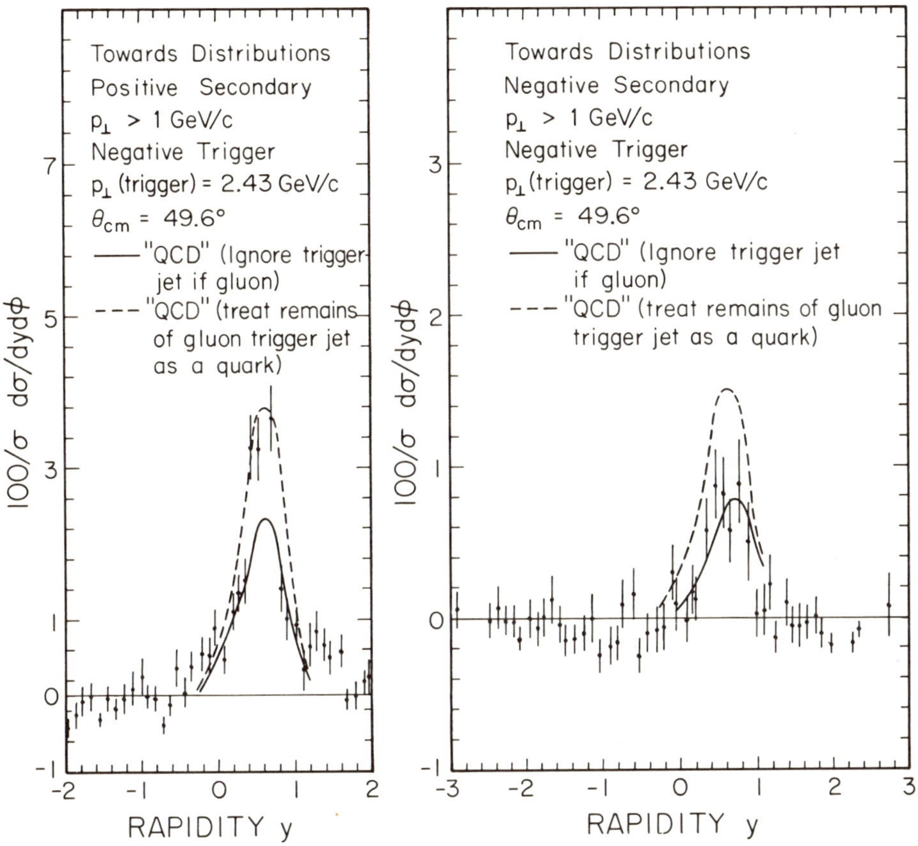

Fig. 2 Towards side correlation data from the CCHK Collaboration[9]. The data come from a negatively charged trigger at a mean $p_\perp$ of 2.43 GeV/c. Shown are the distributions in rapidity of the remaining positive and negative particles, with $p_\perp$ > 1 GeV/c and azimuth within 25° of the trigger. Background from "normal" events has been subtracted. The distribution peak near the rapidity (≈ 0.8) of the trigger. We also show the predictions of the QCD model[10]. The theoretical curves in this and the other correlation figures will be discussed later in the text. See Ref. 5 for the details of data handling and the curves from the FF quark quark scattering model[11] (which agree well with the data).

of the proton. This is illustrated in Fig. 1(a). Support for these ideas came from the power law scaling for the single particle cross-section,

$$Ed^3\sigma/d^3p = f(x_\perp = 2p_\perp/\sqrt{s})p_\perp^{-N} \tag{1}$$

Surprisingly, the observed $p_\perp$ power N was approximately 8 and not the value 4 you get from QCD with constituent quarks and gluons scattering as typified in Fig. 1(b). The power N = 8 was, in fact, predicted by the CIM model where the typical scattering is $\pi q \to \pi q$ and the π is a "constituent" of the proton[3,4]. The CIM model also predicts the correct angular (i.e., the θ_{cm} dependence not exhibited in eq. (1)) behavior[5]. It appears, at this first look, that the single particle cross-section data agree with the CIM model and disagree with quark (gluon) scattering models like QCD. On the other hand, the correlation data give exactly the opposite impression[1,5-8]. The data are in rather good agreement with the four jet structure expected in quark (gluon) scattering models and illustrated in Fig. 1(c).

A selection of some of this correlation data is:

(a) Trigger Side Correlation

Some of the best evidence that single particle triggers come from "jets" (or collection of particles at high $p_\perp$) and are not produced alone, comes from the CCHK Collaboration.[9] The data, shown in Fig. 2, indicate that a trigger particle is accompanied by particles of the same as well as opposite sign. Resonance production only naturally gives the latter correlation.

(b) Away Side Jet

Figures 3 and 4, also from the CCHK collaboration, show that:

(i) There is evidence for a clustering of particles ("jets") on the away side.

(ii) The away side particles have a broad rapidity dis-

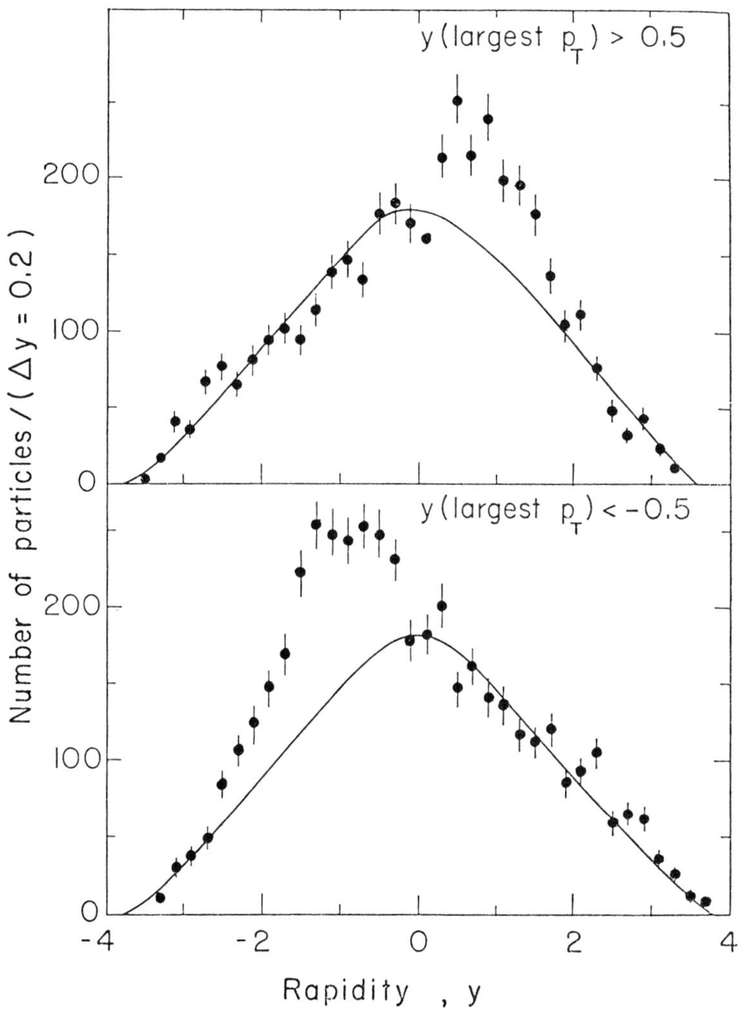

Fig. 3 Evidence from the CCHK collaboration for a jet structure on the away side. Shown are the rapidity distribution of all particles with $p_\perp > 0.8$ GeV/c on the away side <u>except</u> that one with the largest $p_\perp$. The latter is called the "pseudo trigger" and we only plot events where it has $p_\perp >$ GeV/c. We select events with pseudo trigger rapidity greater than 0.5 in (a) and less than -0.5 in (b). The smooth background estimate emphasizes the clustering of the remaining particles about the pseudo trigger rapidity.

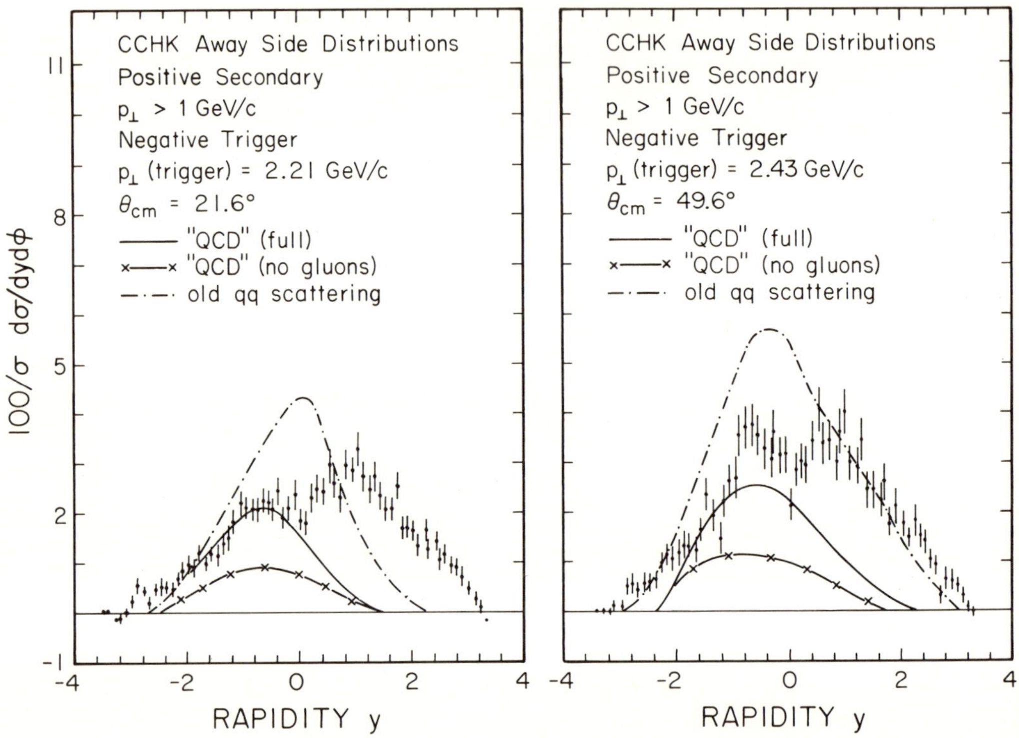

Fig. 4 Away side correlation data from the CCHK collaboration[9]. Shown are the distributions in rapidity of the remaining positive particles with $p_{\perp} > 1$ GeV/c (and within 40° in azimuth of direction opposite to trigger) in the event with the background from normal events subtracted. Data with trigger rapidities of 0.8 and 1.7 are shown. The curves come from the QCD model[10] and the FF quark quark scattering model[5,11].

tribution which is not strongly influenced by the rapidity of the trigger. We will discuss this again later because QCD appears to suggest that the away side particle tends to have the opposite sign rapidity to that of trigger ("back to back" behavior). The CIM model naturally gives a lack of correlation between the away side and trigger rapidities[5].

(c) Jet Cross-Section

The cross-section to produce a cluster of particles has been measured recently in three calorimeter triggered experiments at Fermilab[6,12,13]. There is some controversy as to the exact size of the jet cross-section which I have recently reviewed[6]. However, all the experiments agree that the jet cross-section is very large; at least five hundred times that for a single hadron. Typical results are shown in Fig. 5[13]. This observation, like that in a), supports models where the high $p_\perp$ particles are produced in clusters and not separately. It is hard to understand these observations in the CIM model but they are very natural in a quark (gluon) scattering picture. In the latter, the jet triggers are directly "observing" the quarks and gluons.

(d) Make-up of Jets

In Fig. 6, we show the distribution of momenta for charged particles making up the cluster of particles in the jet trigger[12]. The hadron data are similar to the analogous distributions in lepton processes. This is again encouraging for models based on quark (gluon) scattering. As emphasized in Ref. 6, the rather quantitative agreement between hadron and lepton processes shown in Fig. 6, should not be taken too seriously.

(e) Comparison of Jet and Single Particle Dynamics

In Fig. 7, we show that the away side hadrons have very similar distributions for jet and single particle triggers. This strongly suggests that they are controlled by the same dynamics, i.e., it is not natural to have one mechanism for the jet trigger and another one for the single particle data. In the CIM model,

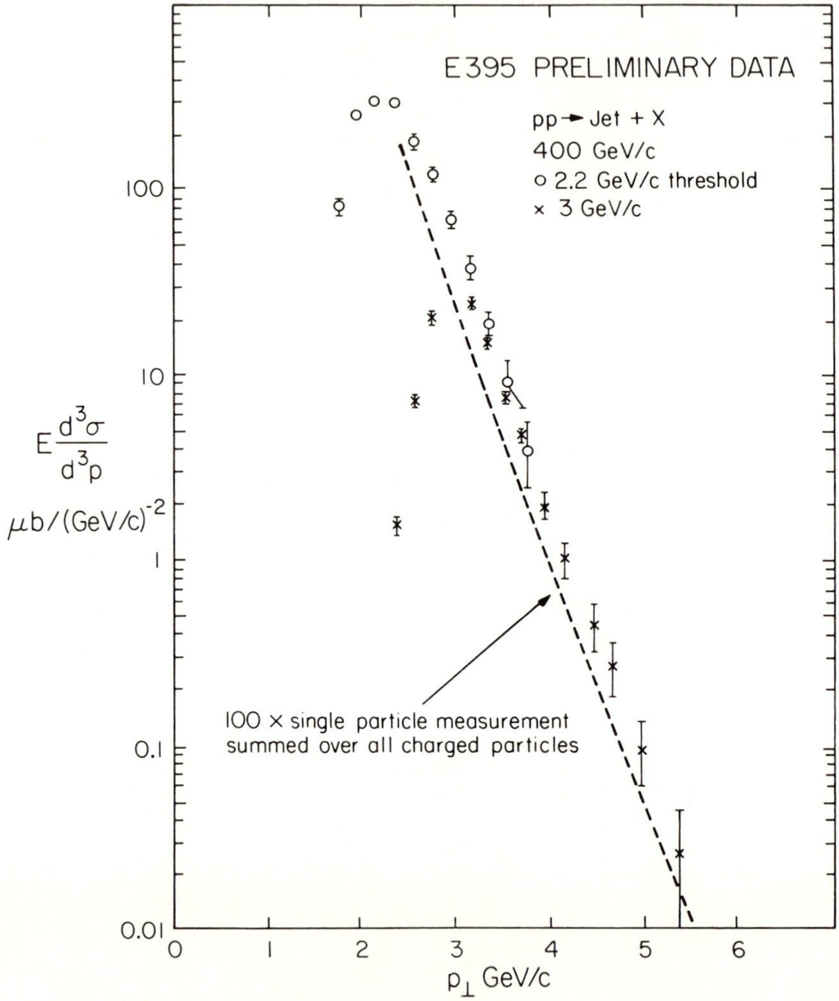

Fig. 5 The jet cross-section at 400 GeV/c from E395 (Ref. 13) compares with one hundred times the single particle cross-section[14] summed over all charged particles. (The latter is about three times the single π^o cross-section.)

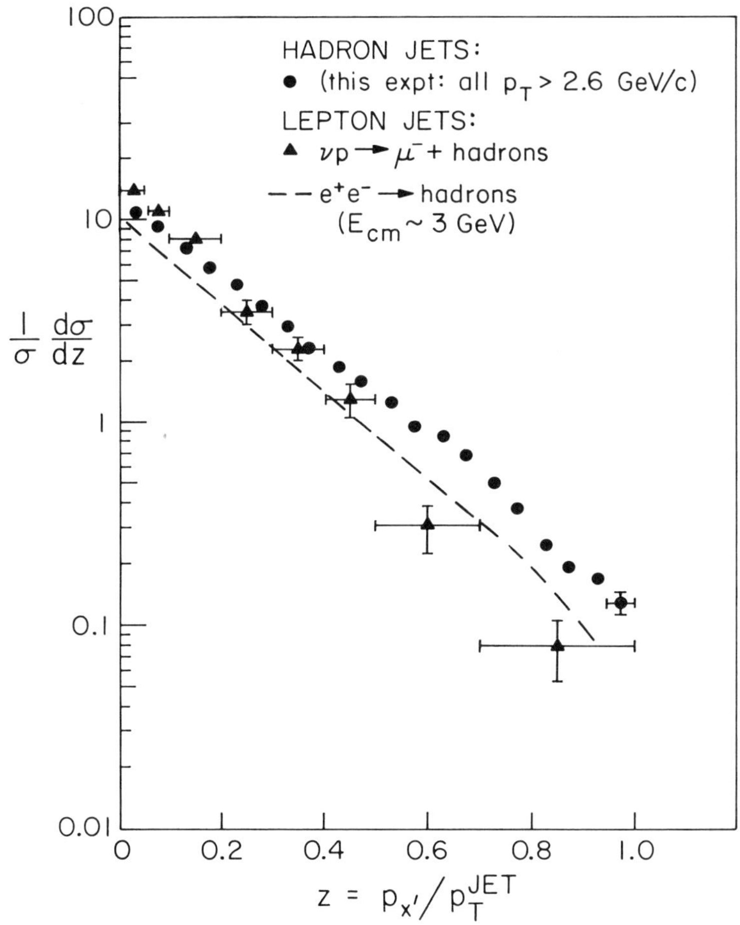

Fig. 6 Distribution of momentum fraction $z \approx p_\perp^{particle}/p_\perp^{jet}$ (the particle $p_\perp$ is projected along jet axis) for the charged particles in the cluster of particles in the jet trigger[12]. Also shown are the analogous distributions in lepton processes; e^+e^- collisions (Ref. 15) and νp interactions (Ref. 16).

one must generalize the "π" in the basic πq scattering to be a (low mass) cluster of particles with pion quantum numbers to accommodate the jet and correlation data we have already discussed. Probably the model will also describe the data in Fig. 7 when so modified.

(f) Final State Quantum Number Correlation

As $p_\perp$ increases, one sees that the yield of π^+'s increases compared to π^-'s in pp collisions. As shown in Fig. 8, the experimental results[14] are consistent with the expectation of quark fragmentation, but most models would predict this trend.

(g) Initial State Quantum Number Correlation

In Fig. 9, we compare the cross-sections for proton and π^- beams. The ratio of these two cross-sections is approximately 1.5 at low $p_\perp$ (i.e., the ratio of total cross-sections) but decreases with increasing $p_\perp$. Both jet and single π^o triggers give the same trend with the ratio becoming one for jets at $p_\perp \sim 4$ GeV/c and π^o's at 3 GeV/c. This observation is again consistent with most models; in a quark (gluon) approach, it just reflects the fact that a pion has more high momentum quarks than a proton. The displacement in $p_\perp$ between the jet and single particle triggers is predicted as the latter come from fragmentation of jets, and so the single particle properties reflect those of jets with higher $p_\perp$.

II. The Models

With one possible exception, the shape of away side rapidity spectrum in I(b), the correlation data reviewed in I have strongly supported quark (gluon) scattering ideas. How do we reconcile these data with the failure of the model for the single particle cross-section? Three approaches have been attempted.

(a). Change the Constituents

Although quarks and gluons are the "obvious" constituents, the CIM model uses pions (and other hadrons) as well[3)]. As we discussed in I, this model is then consistent with the single particle

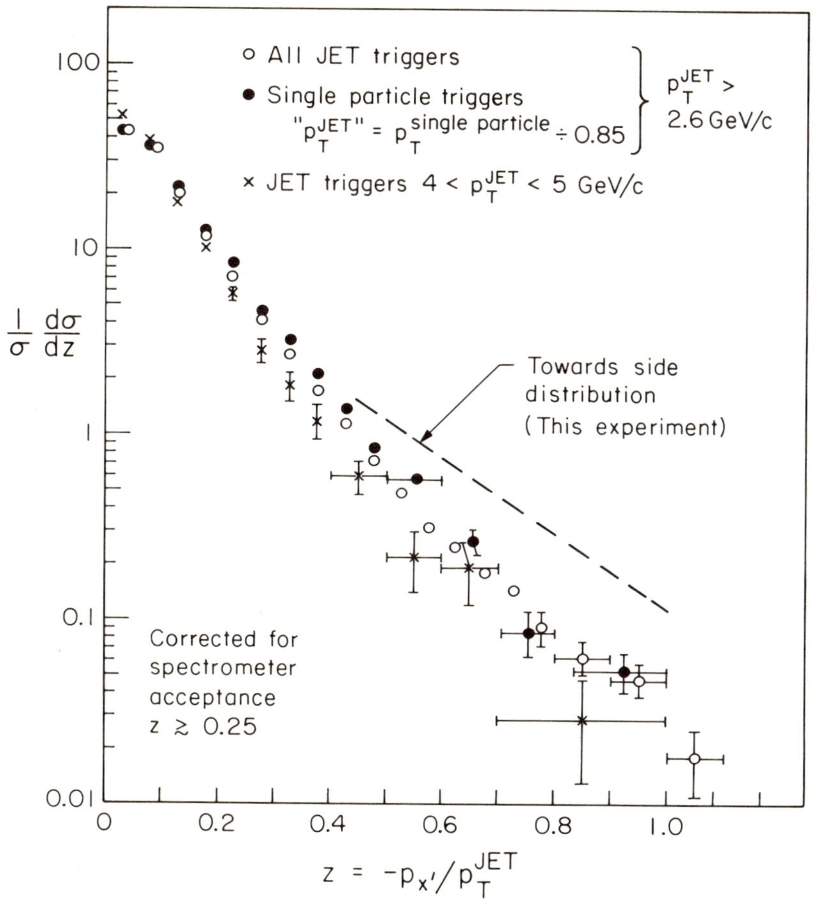

Fig. 7 The away side distribution for jet triggers compared with those for single particle triggers (Ref. 12).

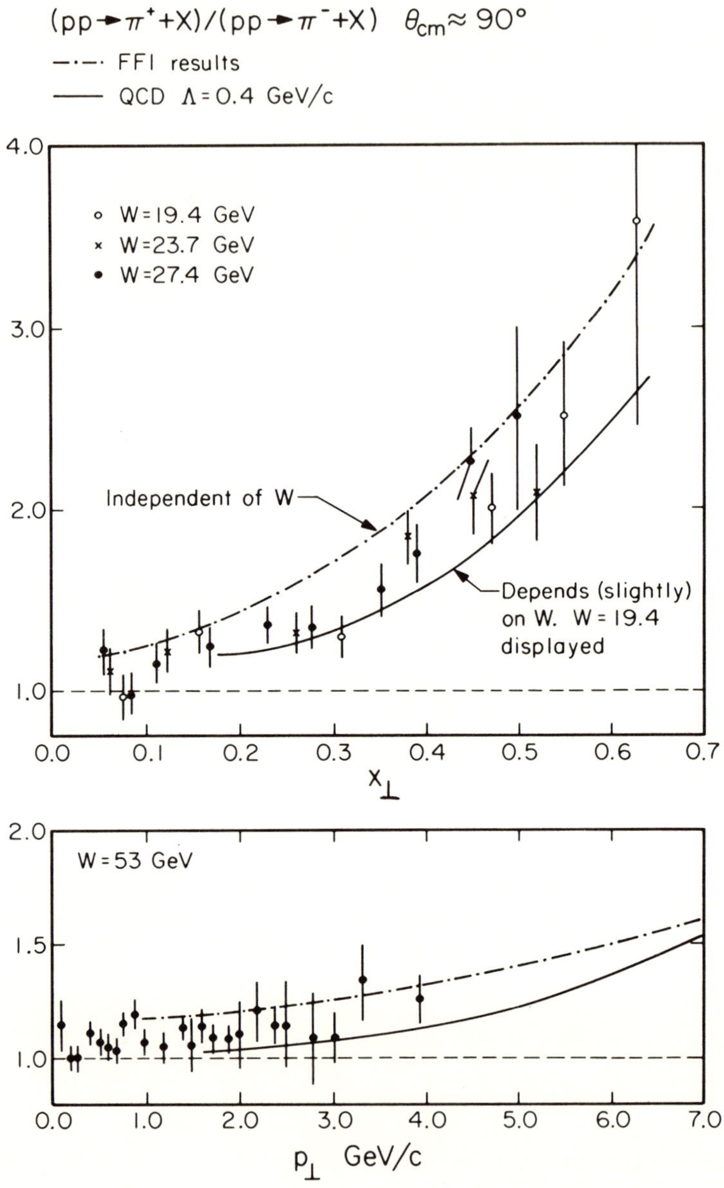

Fig. 8 The ratio π^+ and π^- single particle cross-sections for pp scattering[14,34]. Shown also are the QCD[10] and quark quark scattering model[11] (marked FF1) predictions. The presence of gluons in the QCD calculation lowers the prediction of QCD compared to the pure quark model.

cross-section but has no natural explanation for the correlation data. Theoretically, πq scattering must exist (it explains both elastic scattering and high $x_{\shortparallel}$, high $p_{\perp}$ inclusive scattering). However, the use of hadron constituents seems to me a little ambiguous; especially when one needs to produce not just single π's but the generalized π-like clusters required by the correlation data.

(b) Change the Cross-Section

The approach initiated by Field and Feynman[11] was to keep quarks as constituents but adjust the quark scattering cross-section to fit the single particle cross-section data. (This will be called the FF model hereafter.) This had many successes - in particular the towards side correlations, existence of jets and quantum number effects - which were largely independent of this cross-section. However, the model still remained theoretically unattractive because of the ad hoc cross-section choice. Further, it always predicted too many high $p_{\perp}$ particles on the away side (cf. Fig. 4 and the discussion of $1/\sigma\ d\sigma/dz_p$ later).

(c) QCD

Although the Born scattering graphs of QCD give the (incorrect) prediction of a $p_{\perp}^{-4}$ behavior for the cross-section, there are many (individually small) corrections to this in a realistic calculation[18]. I will describe an analysis of Field[10] which includes all the effects we know of at the present time. The essential features of the QCD model are:

(i) Constituents

The constituents of the hadrons are taken to be quarks and gluons. Only the charged quarks are probed in electron, muon and neutrino scattering. However, the strong QCD scattering involves both quarks and gluons. The distribution functions were taken from our recent analysis of the lepton processes[19]. This included properly the QCD corrections and came up with a coupling constant parameterized conventionally as[20]

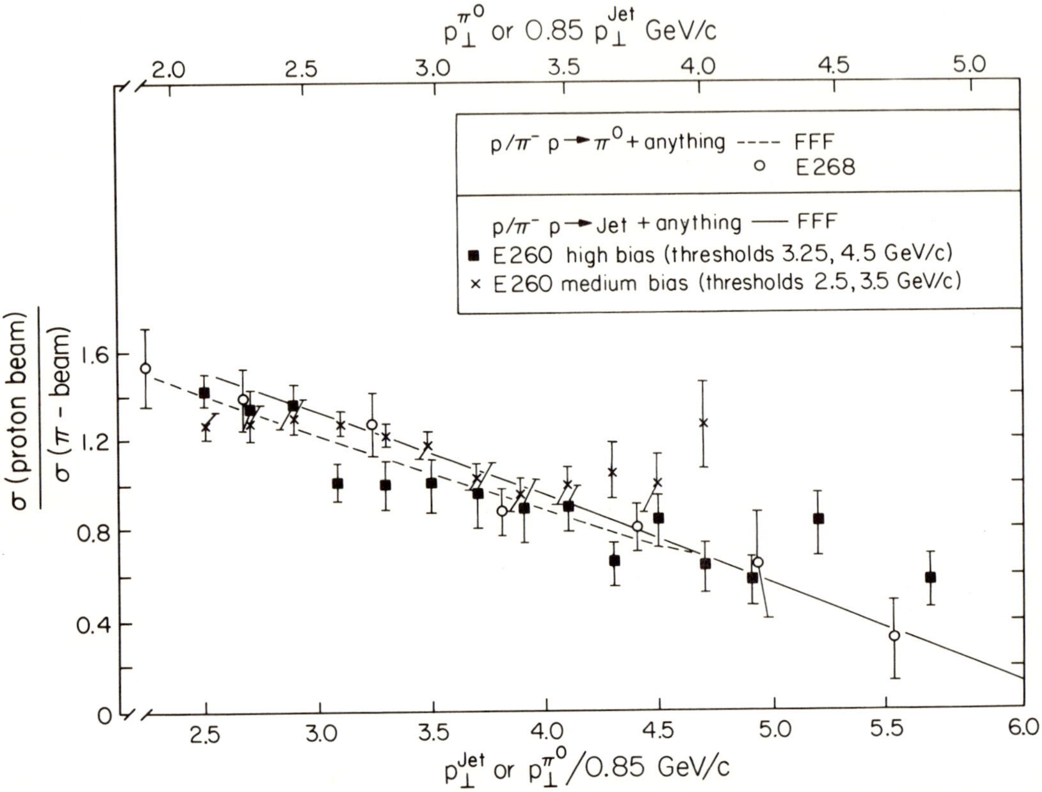

Fig. 9 The ratio of π^- and proton beam on a proton target cross-section for both single π^0's[17] and jets[6,12]. The single particle $p_\perp$ is scaled both here and in Fig. 7, because of the expectation that single particles come from the fragmentation of jets and, hence, have less momentum than the parent $p_\perp$. The explicit factor of 0.85 is partly suggested by the experiment and partly a theoretical calculation in old quark scattering model (cf. Ref. 6). The theoretical "predictions"[11] shown in this figure are sensitive to the particular distribution of quarks assumed in the π; the latter has essentially been adjusted to fit the data. The predictions in the previous figure are, however, nontrivial as the relevant distributions were fixed by lepton data.

$$g^2/4\pi \approx \frac{12\pi}{25\log(Q^2/\Lambda^2)} \quad , \qquad (2)$$

where Λ varied from 0.15 GeV/c to 0.5 GeV/c; the exact value depending on the treatment of the $1/Q^2$ (kinematic) corrections to the theory. Another analysis[21] of the electron and muon data finds Λ = 0.66 GeV/c (using the kinematic formalism where we found Λ = 0.5 GeV/c). These differences are not significant at present; the main conclusion is that the current lepton data are consistent with QCD scale violations with a coupling constant size corresponding to $\Lambda \sim$ 0.5 GeV/c. A determination of Λ and indeed a convincing argument for the validity of QCD needs data at higher Q^2 where the theory is unambiguous. A typical comparison of the scale breaking distribution functions with lepton data is shown in Fig. 10[22]. This uses Λ = 0.4 GeV/c; in the following we shall use either this value or the choice Λ = 0.6 GeV/c.

In Fig. 11, we display the Q^2 dependence for the quark ($\approx \nu W_2$) and gluon distribution functions. Note that both exhibit a rise at small x and fall at large x as Q^2 increases (x is momentum fraction of the constituents inside the proton). The effect is particularly dramatic for the gluon distribution.

(ii) Scattering Cross-Section

The scattering cross-section is straightforward[18]. A contribution to quark quark scattering is shown in Fig. 1(b). However, quark gluon and gluon gluon scattering are also important and included. Note that the coupling constant g decreases logarithmically with Q^2 as specified in eq. (2) above.

(iii) Fragmentation Function

The fragmentation functions $D(z,Q^2)$ for a scattered constituent (gluon or quark) into hadrons are needed to predict the single particle cross-section. In a naive parton model, these

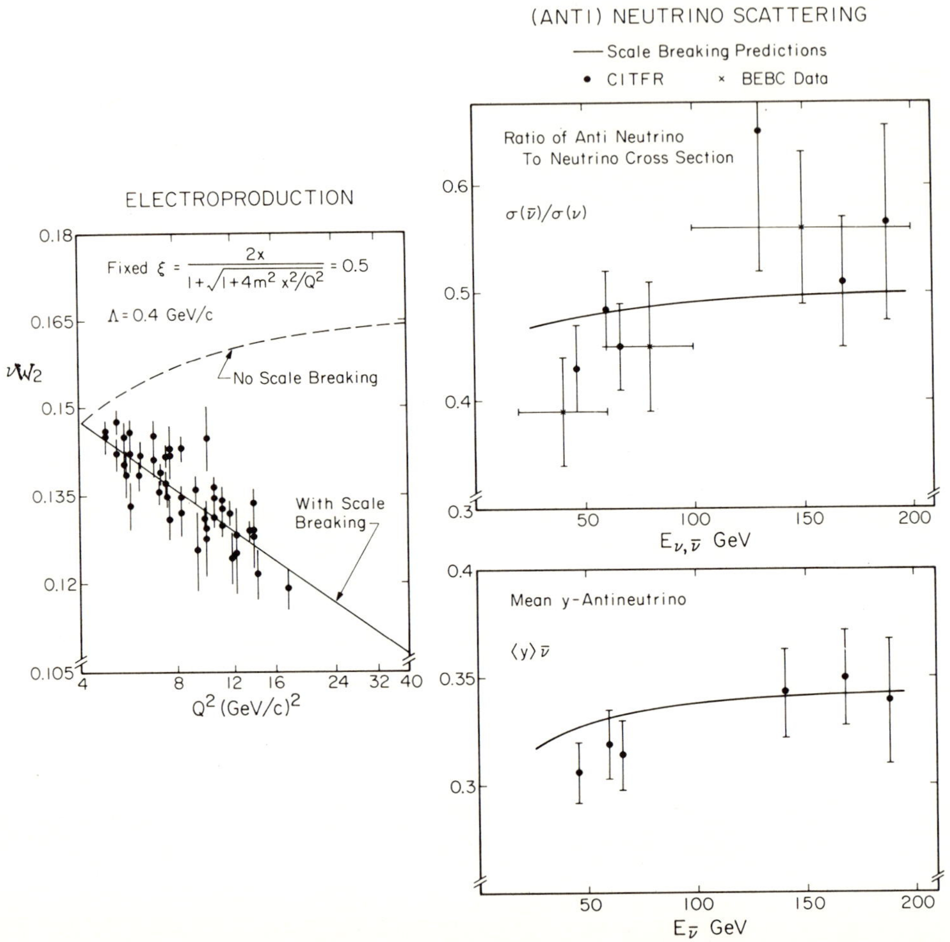

Fig. 10 Comparison of QCD scale breaking theory with electroproduction (for data, see compilation in Ref. 19) and neutrino scattering (Refs. 24 and 25). Note that as discussed in Ref. 19, the (anti) neutrino scattering results are very sensitive to low Q^2 corrections for $E_{\nu,\bar{\nu}} \lesssim 50$ GeV/c. In the electroproduction case, the theoretical distributions are functions of x and so the "no scale breaking" curve rises with Q^2 on the graph which is at fixed ξ (defined in the figure).

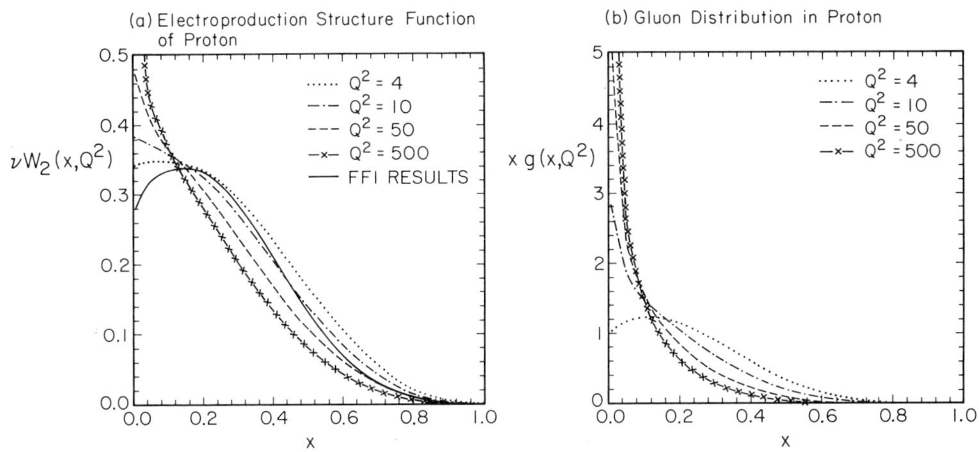

Fig. 11 x dependence of νW_2 and the gluon distributions inside a proton for various Q^2 values. The scale breaking is calculated using $\Lambda = 0.4$ GeV/c. Also shown is the old Field-Feynman distribution[11].

functions only depend on the momentum fraction z carried off by the final state hadron. However, in an asymptotically free theory, they show scale breaking (Q^2 dependence) which is governed by similar equations to those for the scale breaking of the νW_2[26]. One needs to know the fragmentation functions at one value of Q^2 and then the Mellin transform technique used in Ref. 19 enables one to predict the z dependence at any Q^2. Typical results are shown in Fig. 12 where, just as in Fig. 11, $zD(z,Q^2)$ decreases with Q^2 at large z.

Note that the mean value of z for hadrons fragmenting from the constituents is around 0.8. Thus, this fragmentation process selects the quark constituents which have a higher probability of giving large z hadrons than gluons. This is the reason that the QCD model gives very similar quantum number and towards side correlation predictions to the FF model which only included quark constituents. On the other hand, quark gluon scattering is large and whereas most of the trigger particles come from quarks, the away side particles have a large gluon component. For instance, at the ISR for low $x_\perp \sim 0.1$, over 60% of the events have an away side that comes from a gluon fragmentation, whereas this number is only 20% on the trigger side. For the jet triggers, one is no longer biased in favor of quark triggers and there is again a large gluon contribution.

(iv) Internal Transverse Momentum

There is now general agreement that the constituents show large values for their internal transverse momentum inside the proton. One must include this in a quantitative calculation because of the "trigger bias" which enhances the configuration where the internal transverse momentum points toward the trigger direction[5,6,9,11]. In QCD, one gets an effective internal momentum from the bremsstrahlung typified in Fig. 13(a). One may also have true internal transverse momenta present in the basic parton "wave-function" inside the proton. As indicated in Fig. 13(b), one might

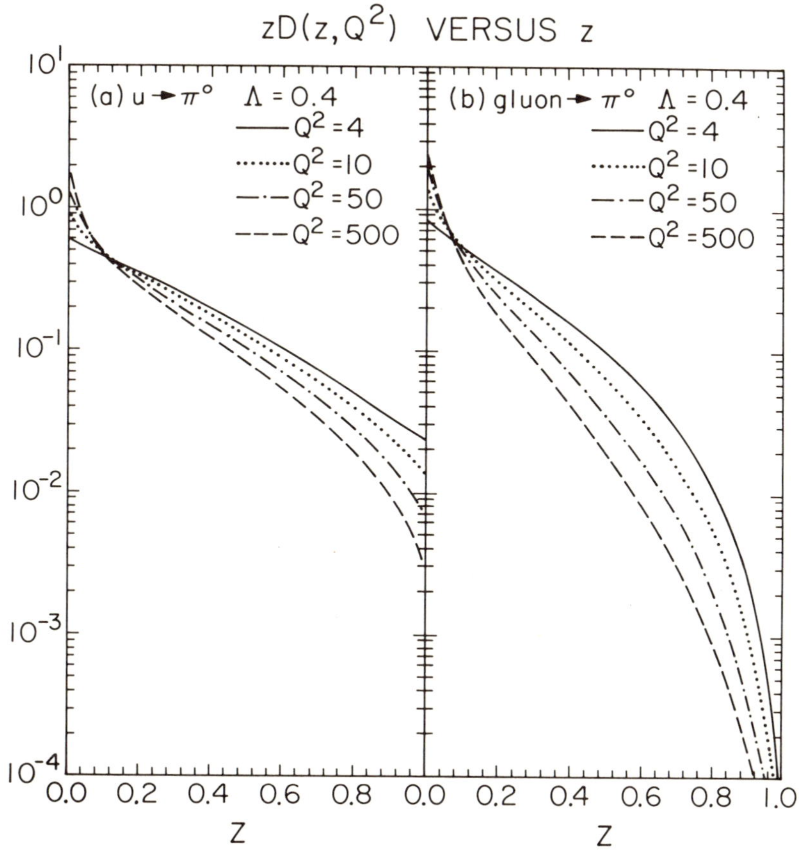

Fig. 12 z dependence of the fragmentation functions for quarks and gluons into π^0's for various Q^2 values.

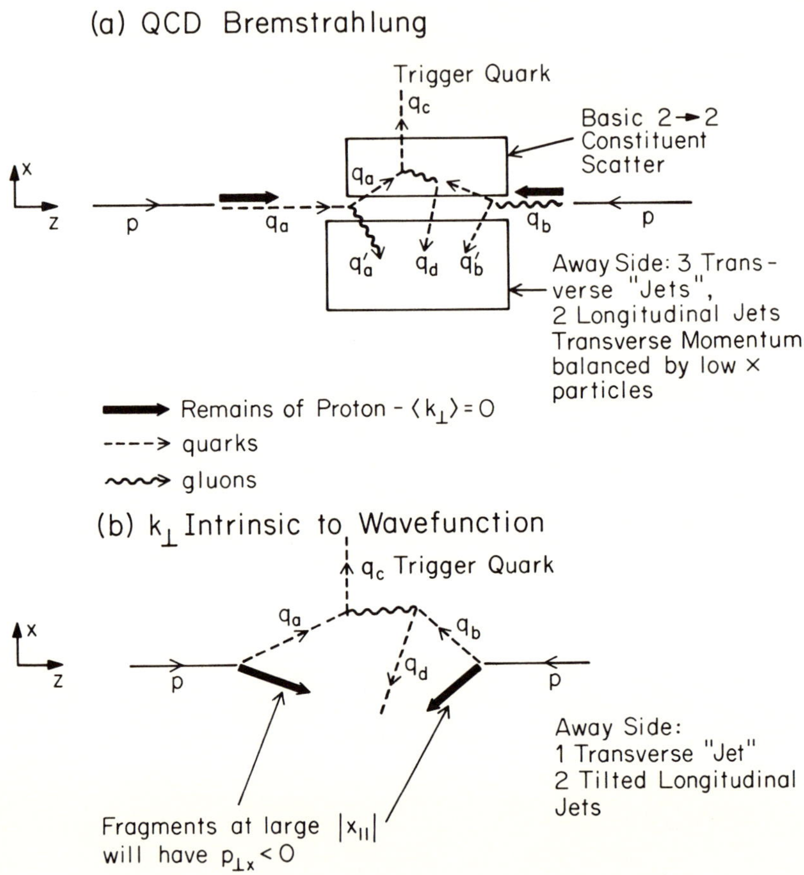

Fig. 13 Sources of internal transverse momentum discussed in the text

expect the wave function to have a term where the trigger parton $k_\perp$ is balanced by another constituent which has the opposite $k_\perp$ and most of the remaining longitudinal momentum. Now consider a trigger at 90° in the xz plane with $p_x > 0$. Typically, this comes from the fragmentation of a constituent with $k_{\perp x} > 0$ and so will be balanced by constituents with $k_{\perp x} < 0$. One expects to see this negative $k_\perp$ in the background particles with large longitudinal momentum. This effect is nicely shown in Fig. 14 in data from the British-French-Scandinavian (BFS) group at the ISR[27]. The size of the observed effect is smaller by about a factor of two than our prediction which suggests that the wave function model is not the whole story. On the other hand, the QCD bremsstrahlung model suggests that positive $k_{\perp x}$ is balanced by a host of low x particles from the fragmentation of a low momentum constituent. This model does not seem to be able to explain the data in Fig. 14. The data quite likely require the two effects: $k_\perp$ intrinsic to the wave function and that produced by bremstrahlung before the scattering.

The constituent internal momentum is also observed directly in the Drell-Yan process $pp \to \mu^+\mu^- X$, and the current data are consistent with a mean value $\langle k_\perp \rangle \approx 850$ MeV/c for constituents producing high mass μ pairs[28]. There has been much speculation that this can be explained by QCD bremsstrahlung with perhaps some additional internal momentum[29]. Many groups are trying to make this more quantitative. In view of the uncertainty in the theory and the evidence for two components in the hadron data (Fig. 14), Field chose the constant value of 850 MeV/c for the mean internal momentum. The shape of the $k_\perp$ distribution was also chosen to agree with the μ pair data[28].

(v) Uncertainties

The QCD theory, outlined above, represents an absolutely normalized prediction. The major uncertainty is the quark internal momentum discussed above. This is especially a problem for extra-

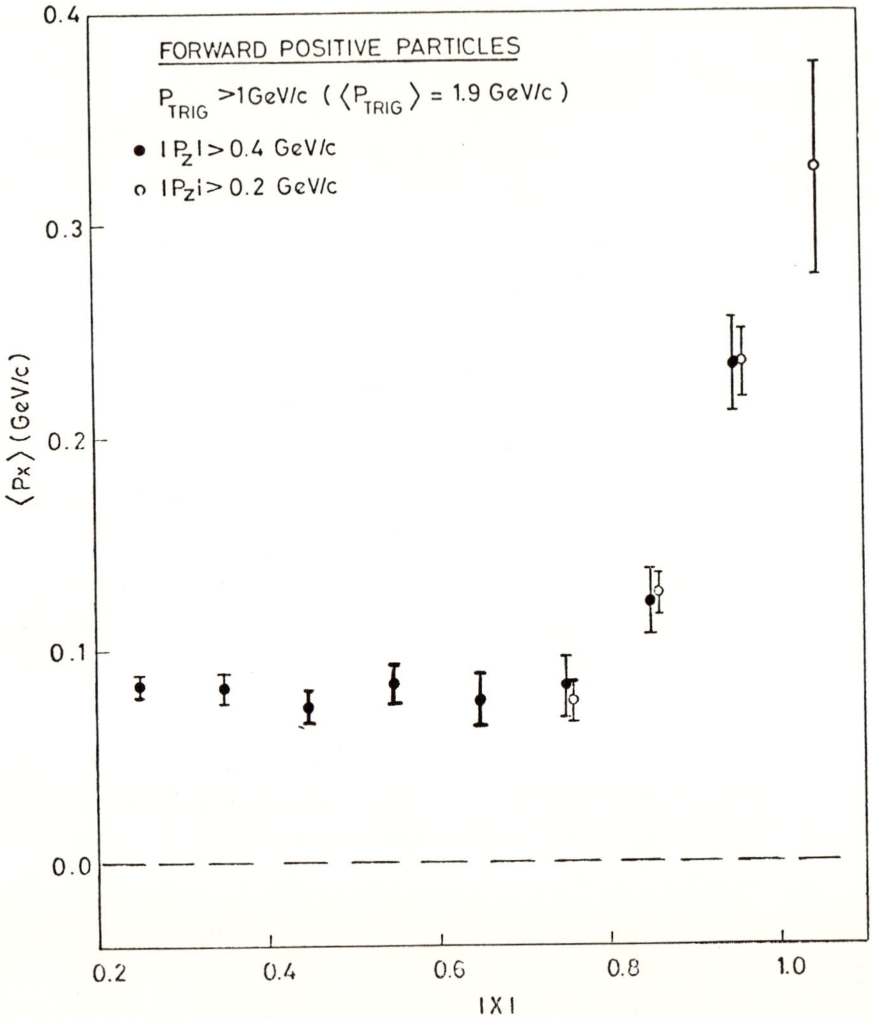

Fig. 14 Data for 90° high $p_\perp$ triggers in the xz plane (z is beam target direction) from the BFS group at the ISR[27]. Shown is the mean x component of momentum, opposite to the trigger, as a function of the longitudinal fraction $x_{\shortparallel}$ for positive particles.

polating QCD to Q^2 values higher than those probed by the current μ pair data. Thus, QCD predicts that $\langle k_\perp^2 \rangle$ has a component proportional to $Q^2/\ln Q^2$ which grows with increasing Q^2. For instance, in the region probed by the new colliding proton beams, $p_\perp = 40$ GeV/c and $x_\perp = 0.1$, the mean $k_\perp^2$ has become of order 50 $(\text{GeV/c})^2$.[30] On the other hand, there is probably not too much of a problem in the region $p_\perp \lesssim 6$ GeV/c but the predictions, using a constant $k_\perp$ for the cross-sections for higher $p_\perp$ should be regarded as lower bounds, for the increasing $\langle k_\perp \rangle$, suggested by QCD, will probably increase the prediction. A similar defect in the calculation is that the mean $k_\perp$ in the fragmentation process, constituents to hadrons, is kept Q^2 independent. This should again be specified by QCD and increase with increasing Q^2. The single particle cross-section is much less senstitive to this quantity than the internal momentum of constituents in hadrons.

Less serious uncertainties stem from both the gluon distributions inside the proton and the glue to pion fragmentation function. These were assumed to behave as $(1-x)^4$ or $(1-z)^2$ as $x,z \to 1$. These powers differ by one from the predictions of the counting laws[31]. However, as exemplified in the next section, equally good fits could be obtained with the behavior $(1-x)^5$ and $(1-z)^3$ suggested by the counting laws. QCD then predicts the evolution in Q^2 of these functions. I will discuss the sensitivity of the predictions to this assumption later.

Another problem is the specification of the variable Q^2 in the hadron process. This was chosen to be a symmetric function of the Mandelstam invariants for the basic $2 \to 2$ constituent scattering[10]. The ambiguity originating from this arbitrary choice can be estimated by considering the effect by varing Q, in the typical $\ln(Q^2/\Lambda^2)$ dependence, by a factor of two. One finds an uncertainty of around a factor 1.5 in the predictions.

III. COMPARISON OF QCD WITH EXPERIMENT

(a) SINGLE PARTICLE CROSS-SECTION

Figures 15 to 18 compare QCD with the experimental $pp \to \pi^o$ inclusive cross-section. One must remember that one is searching for a $p_\perp^{-N}$ behavior at fixed $x_\perp$ and not at fixed energy. Figure 15 plots the data at $x_\perp = 0.2$, 0.35, and 0.5. One observes that the absolutely normalized QCD predictions are within a factor of two of the published data at all $x_\perp$ and $p_\perp$. There are the substantial uncertainties mentioned at the end of Section II which could well change the prediction by a factor of two (this could either improve or worsen agreement with experiment!) One should also say that the gluon distribution inside the proton was originally[19] parameterized as $(a+bx)(1-x)^5$; this corresponds to one free parameter as one linear combination of a and b is determined by the energy momentum sum rule for the proton. This free parameter has been adjusted to improve agreement in Fig. 15. Further, the power of (1-x) has been changed from 5 to 4. This adjustment still gives distribution functions that agree with the μ and e inelastic scattering data. Using the original parameters of Ref. 19 changes the predictions by only 20%. Finally, we note that the predictions at low $p_\perp \lesssim 3$ GeV/c are sensitive to low values of the invariants of the constituent scattering. The zero mass quark gluon scattering amplitudes are formally infinite as the invariants go to zero. This infinity has been removed by introducing an m^2 equal to 1 GeV^2. Varying this cutoff by a factor of three changes the cross-section by a factor 1.5 at the lower $p_\perp$'s as shown in Fig. 18. The same theory and data are shown in Fig. 16 for fixed energy rather than fixed $x_\perp$. Note how important both the gluons and the smearing (quark internal $k_\perp$) are in increasing the cross-section. The gluons are particularly important at low $x_\perp$, i.e., at W = 53 in Fig. 16. The smearing is more important where the cross-section is steeper in $p_\perp$, i.e., at the lower W = 19.4 GeV. Figure 15 also

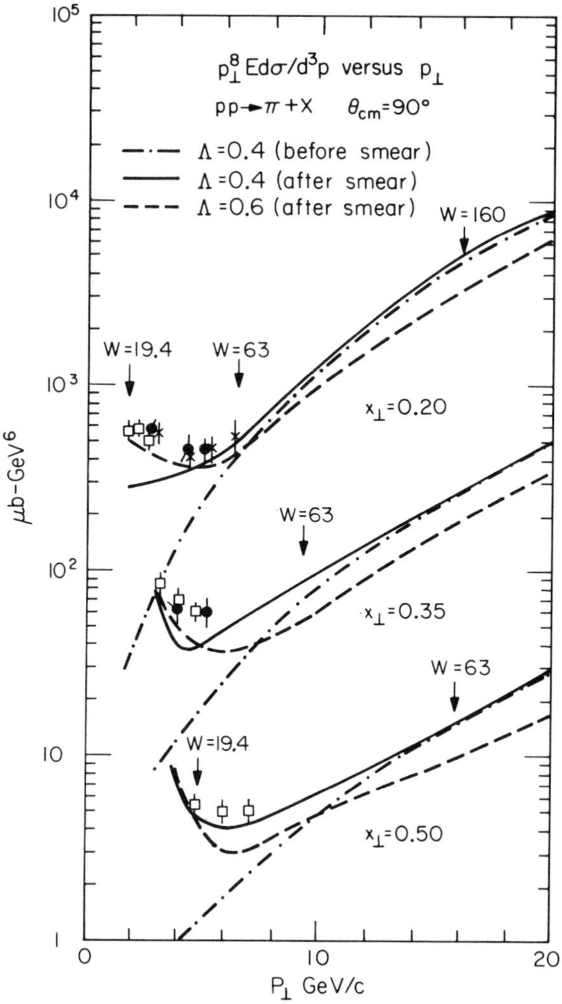

Fig. 15 Comparison of experimental values of $p_\perp^8\ Ed^3\sigma/d^3p$ with the predictions of QCD[10]. Shown are the data at fixed $x_\perp$ = 0.2, 0.35 and 0.5 (see Ref. 11 for the data sources here and in the next 3 figures). The QCD theory is shown for Λ = 0.4 and Λ = 0.6 GeV/c. The curve for Λ = 0.4 is also shown with no internal $k_\perp$ for constituents (marked "before smear").

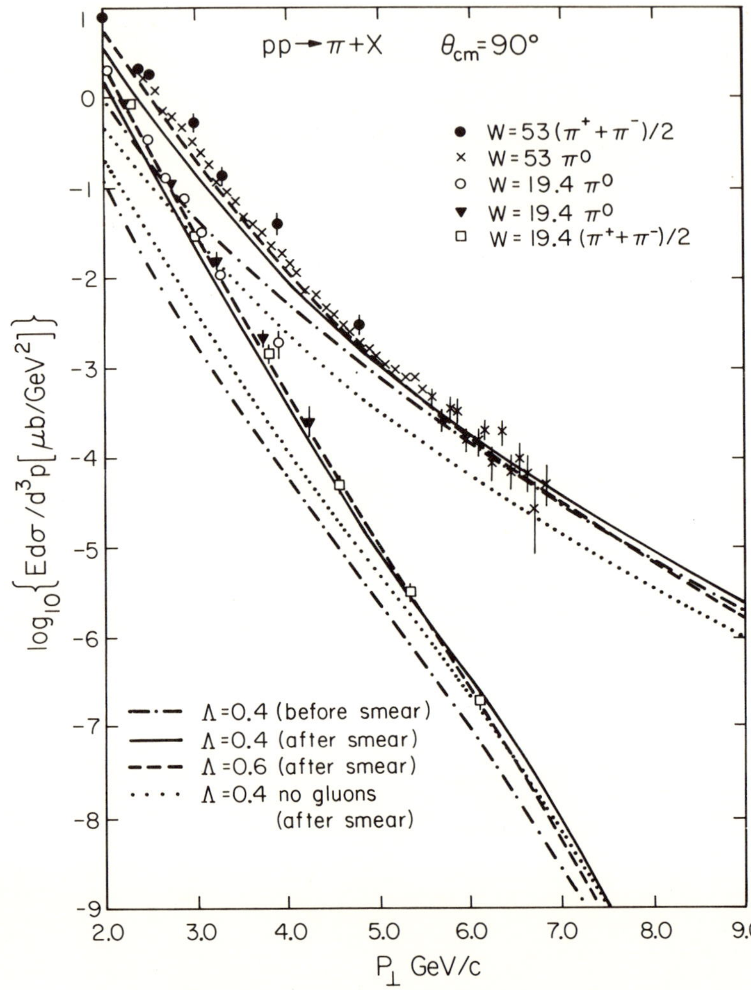

Fig. 16 Comparison of the QCD model[10] with data on $Ed^3\sigma/d^3p$ at fixed energy $\sqrt{s}$ = 19.4 and 53 GeV. The theoretical curves for Λ = 0.4 and 0.6 GeV/c are plotted. For Λ = 0.4, we show the effect of either removing the gluons or the constituent internal transverse momentum.

emphasizes that the QCD model predicts that the observed $p_\perp^{-8}$ behavior is not only a total accident but a purely temporary effect that disappears fpr $p_\perp > 6$ GeV/c. I gather that preliminary data from the ISR[32], suggest such a breakdown in the $p_\perp^{-8}$ behavior. Obviously, only experimental data at such larger $p_\perp$'s will provide unambiguous tests of QCD.

It is interesting to ask how the basic $p_\perp^{-4}$ behavior is modified to the approximate $p_\perp^{-8}$ behavior seen for $p_\perp < 6$ GeV/c. Parameterize the theory at fixed $x_\perp$ as

$$Ed^3\sigma/d^3p \;\alpha\; p_\perp^{-N_{eff}} \text{ for } 2 \leq p_\perp \leq 6 \text{ GeV/c.} \qquad (3)$$

Then Fig. 17 shows for the typical case of quark-quark scattering:

(i) Naive prediction - no scale breaking - $N_{eff} = 4$

(ii) Add Q^2 dependence of coupling constant as in Eq. (2) - $N_{eff} = 4.9$

(iii) Further, add Q^2 dependent scale breaking in structure functions - $N_{eff} = 5.2$

(iv) Further add Q^2 dependent scale breaking in fragmentation functions - $N_{eff} = 5.7$

(v) Further add the quark internal $k_\perp$ - $N_{eff} = 7.1$.

Adding the gluons increases N_{eff} to around 8 (7.6 to be precise) as shown in Fig. 15. However, it is important to realize the importance of gluons is not to change $p_\perp$ dependence at fixed $x_\perp$ but rather to raise the cross section at low $x_\perp$.

In Fig. 18, we look at the angular ($x_{\parallel}$) dependence of the data. The QCD theory is not as good as the old Field-Feynman model; however, the cross-section in the latter was adjusted precisely to fit these data.

The comparison between theory and experiment certainly does

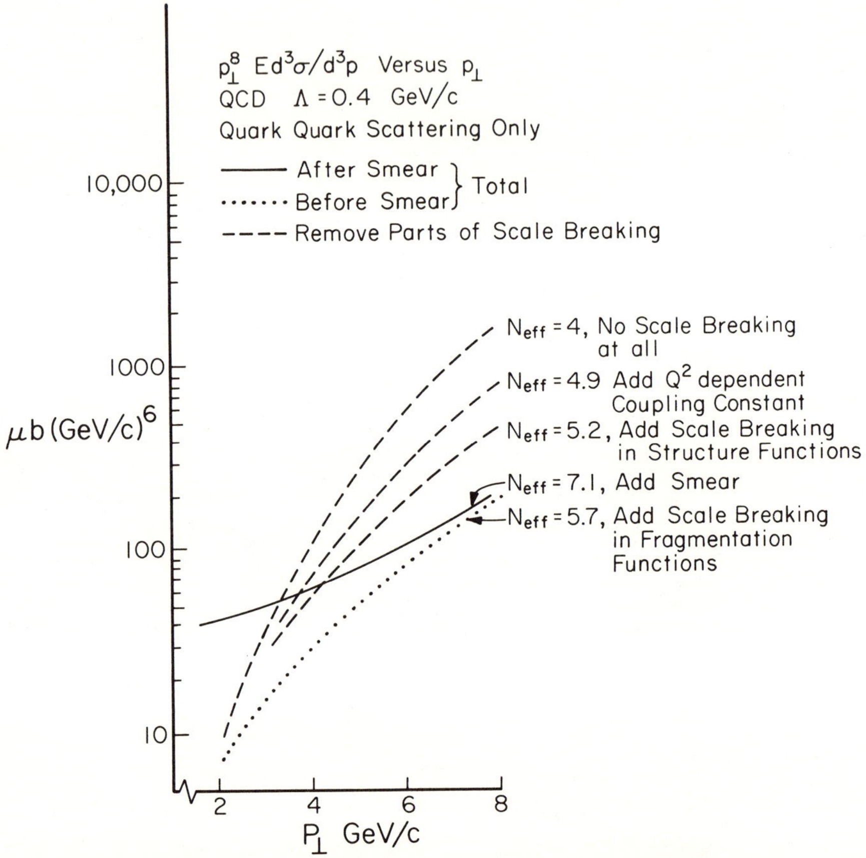

Fig. 17 $p_\perp^8\ Ed^3\sigma/d^3p$ from the quark quark scattering term of the QCD calculation with $\Lambda = 0.4$ GeV/c. Figure 15 has the cross-section with gluons included. Figure 17 shows not only the total quark-quark scattering piece but also the curves with the different sources of scale breaking removed. N_{eff} is defined in eq. (3).

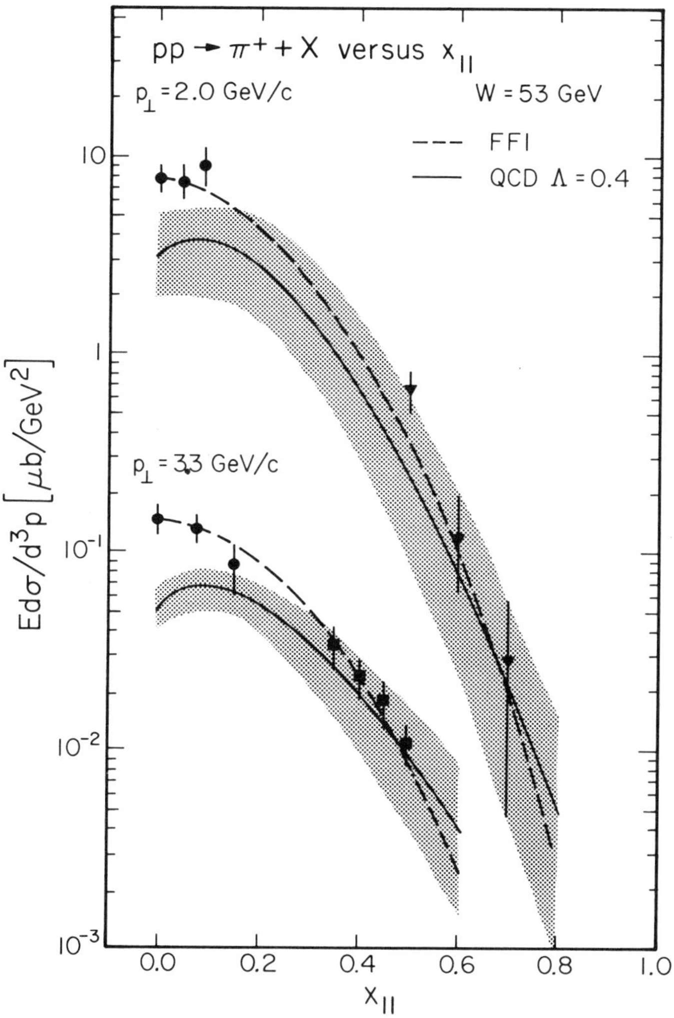

Fig. 18 The angular ($x_{\shortparallel}$) dependence of $Ed^3\sigma/d^3p$ at $\sqrt{s}$ = 53 GeV/c. Shown are the QCD model[10] with Λ = 0.4 GeV/c and the quark quark scattering model (marked FF1[11]. The band represents the uncertainty in the calculation due to the cutoff of the constituent cross-section at low $\hat{s}$, $\hat{t}$ and $\hat{u}$.

not prove that QCD describes all the high $p_\perp$ scattering for $p_\perp \gtrsim 2$ GeV/c. It does suggest that QCD is a substantial part of the observed yield and certainly cannot be ignored. (Remember the prediction was absolutely normalized with parameters taken from lepton scattering.) There is obviously the most uncertainty in the comparison at low $p_\perp$. Here one may also have the CIM diagrams; these presumably also have similar scale breaking effects to the quark gluon diagrams. Presumably, the CIM diagrams (e.g., $\pi q \to \pi q$) fall not like the naive $p_\perp^{-8}$ but rather as a higher power of $p_\perp$. Two processes that must be present are

$$q\bar{q} \to \pi\pi \quad \text{(quark fusion, Ref. 33)}$$

$$\text{and} \quad qg \to \pi q,$$

as they involve only quark and gluon constituents. It would be interesting to get a reliable normalization for these and see their effect at the lower $p_\perp$'s. Such a study should include proton and antiproton production data which have yet to be discussed in the QCD model. Finally, on a mundane level, the lower $p_\perp$ values may also have some spillover of low $p_\perp$ physics that is not simply related to our high $p_\perp$ dynamics.

(b) Quantum Number Dependence of Single Particle Cross - Section

In Fig. 8, we show the ratio of π^+/π^- production in pp collisions that we have already discussed. The QCD calculation leads to a lower value for this ratio than the FF quark scattering model. This just reflects the contribution of the gluon fragmentation. This is certainly poorly determined at present and the differences between the curves in Fig. 8 should be taken as indicative of the effects of gluons but not a solid prediction.

QCD predicts the qualitative behavior of the π^-/proton beam ratio shown in Fig. 9. Notice that QCD would predict different scale breaking for the π^- and proton structure functions. Thus the π^- should have less scale breaking than the proton but a larger value for the quark internal transverse momentum (coming from bremsstrahlung, Fig. 13(a)). Obviously, the current data would need substantial improvement before this could be tested.

(c) The Jet Cross-Section

QCD predicts a larger jet over single particle production rate than the FF quark scattering model. Jets and single particles basically differ by the fragmentation function $D(z,Q^2)$. This is reduced by scale breaking at large z and so QCD finds a larger cross-section that falls slightly slower with $p_\perp$ than the FF quark scattering model. A further effect is the presence of gluons in the QCD model which increase the jets (especially at low $x_\perp$) much more than the single particles. The QCD jet over single π^0 production ratio varies from 500 to 4000 for the range $3 < p_\perp < 6$ GeV/c, $\sqrt{s} \sim 20$ GeV/c probed by the current experiments.[12,13] This is larger than the quoted experimental value of around 500. However, this experimental number comes from a comparison of the jet and single particle transverse momentum distributions. As discussed in Ref. 6, it is more reasonable to use the jet energy which is typically about 1 GeV/c more than its $p_\perp$. This increases the experimental jet cross-section by about an order of magnitude. (We haven't quite pinned this down properly.) The experimental values are then in better agreement with the theory and even maybe rather larger than predicted. Curiously, the old FF predictions (which were a factor 3 - 6 smaller than QCD) agreed rather well with the experimental values using the $p_\perp$, not energy distributions.[6,35] We now regard this as completely fortuitous.

(d) Trigger Side Correlation

In Fig. 3, we display the predictions of QCD for the CCHK towards side correlation data. The calculation is ambiguous because although we have a reasonable model for sequential quark fragmentation[11], we have little idea how to predict the associated particles

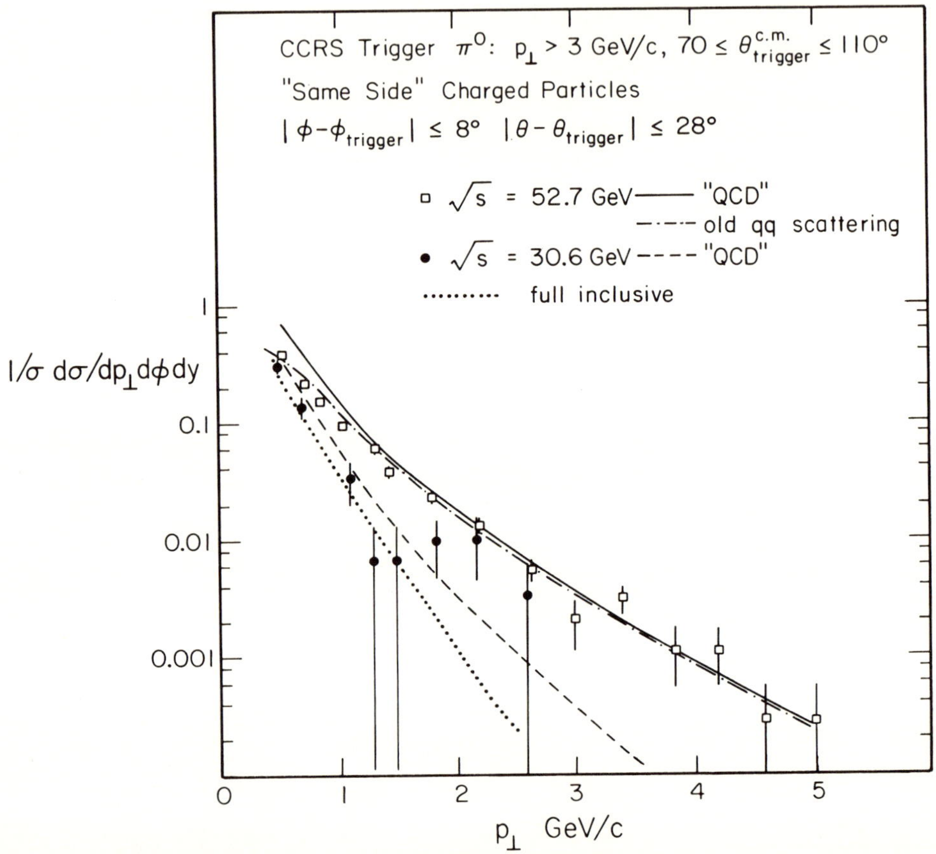

Fig. 19 Towards side correlation measurements from the CCRS collaboration[36]. We show the QCD model predictions and those of the old FF quark scattering model[5]. The QCD (but not FF) predictions have an estimate of the background (i.e., particles that are not from trigger jet) included. Note that the part due to gluons is not taken to be an upper estimate as it was in Fig. 2.

when the trigger jet was a gluon. One possible assumption is that a gluon breaks up into a $q\bar{q}$ pair and then each quark fragments independently. The dashed line in Fig. 3 used an <u>upper</u> estimate for this gluon process (see Reg. 10 for details), while the solid curve completely ignores the gluons. The current prediction lies somewhere between these two curves. I conclude that the idea and theory are in reasonable agreement although the gluon uncertainty (and the low trigger $p_\perp$ of data) rule out a strong conclusion.

In Fig. 19, we compare theory and experiment for the trigger side correlation measurement of the CCRS group[36]. Again there is excellent agreement. In this figure, we also show the FF model curves, which again agree well with the data. (The same was true for the CCHK data.)

(e) <u>Away Side Particles</u>

One of the major problems with the FF model was that it predicted too many particles at high $p_\perp$ on the away side. Figures 20 and 21 show that this has been solved in the QCD approach. In Fig. 20, we have the away side $p_\perp$ distributions from the CCRS group[36] with the QCD predictions right on the nose! This is probably slightly fortuitous as the figure also shows that gluon fragmentation makes a substantial contribution - at least for $p_\perp < 3$ GeV/c. Clearly, our gluon estimates are only qualitative; the conservative approach is just to note that the predictions of quark fragmentation alone lie below the data.

A more popular variable for studying the away side is not the single particle transverse momentum but rather this divided by the trigger $p_\perp$, which we call z_p. In simple models, the z_p distribution is independent of trigger $p_\perp$. This is only approximately true for the QCD model which is compared with data from the BFS collaboration[7] in Fig. 21. The theory lies below the experimental data at low trigger $p_\perp$. This is easy to understand qualitatively (but very hard quantitatively) in terms of additional contributions from background particles (cf. the complicated away side structure in Fig. 13(a)).

For $p_\perp^{trigger} > 3$ GeV/c, there is nice agreement between theory and experiment. Note that the FF model is typically a factor of three higher than the data in both Figs. 20 and 21. In Ref. 10, we study z_p distributions from other experiments[9,37] and show they lead to similar conclusions to the data shown in Fig. 21.

We now turn to Fig. 4 which shows not the $p_\perp$ but the rapidity distribution for away side particles from the CCHK collaboration[9]. The important feature of the data is that the away side particles are still approximately centered at zero rapidity even though the trigger had rapidity 0.8 or 1.7. This is predicted by both the FF (shown in Fig. 4) and the CIM models. However, QCD predicts a shift of the away side particles towards a rapidity opposite that of the trigger. There is no direct conflict between theory and experiment as QCD is still normalized below the data. This is especially true if you only consider the reliable quark fragmentation term. In QCD, one must assume that the background is peaked at positive rapidities. It is clearly important to either understand this background theoretically (this is quite complicated as shown in Fig. 13. There is also some dependence on the Lorentz frame used to describe fragmentation[5]) <u>or</u> test QCD experimentally by observing the away side rapidity distribution for nonzero rapidity triggers at sufficiently high $p_\perp$ that the background is negligible. In Fig. 3, we can try to isolate the away side jet as the heights above the solid line of the correlated peaks are proportional to the jet (as opposed to background) signal. There is some hint that this is indeed larger for negative rapidity as predicted by QCD. It would be interesting to see this technique developed further.

(f) Other Correlation Measurements

The other correlation data, discussed at the beginning of this article which were basically on the relation of jet and single particle dynamics, are in qualitative agreement with the QCD model. There are two other types of data that we have discussed in the past[5,11]. Firstly, there is the azimuthal and rapidity

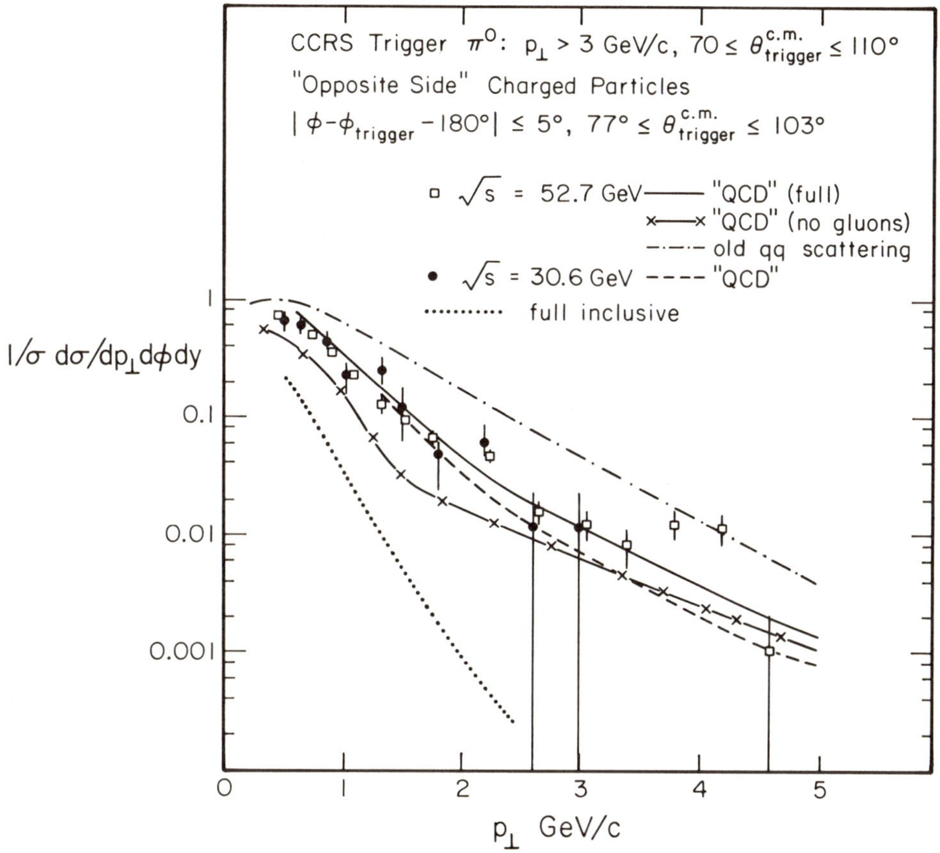

Fig. 20 Away side correlation measurements from the CCRS collaboration[36]. We show the QCD model predictions and those of the old FF quark scattering model[5]. The QCD (but not FF) predictions have an estimate of the background (i.e., particles that are not from away side jet) included.

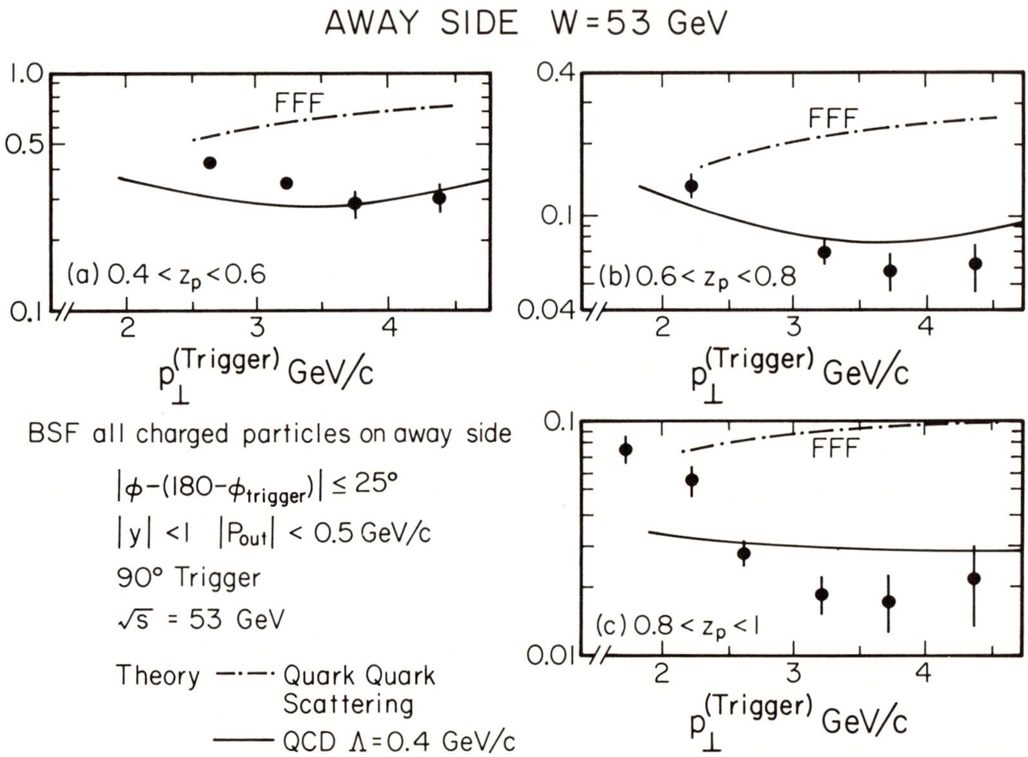

Fig. 21 The trigger $p_\perp$ dependence of $1/\sigma \; d\sigma/dz_p$ where $z_p = p_\perp^{away\ side}/p_\perp^{trigger}$ from the British-French-Scandinavian collaboration[7]. Shown are the predictions of both the QCD model and the FF quark scattering approach[6]. Neither prediction has background included.

dependence of charged particles as reported, for instance, by the ACHM group[38]. This was discussed in detail in Ref. 11 and the QCD predictions are very similar to those of the FF model which agree well with the data. Secondly, one interesting observable is "P_{out}", i.e., the component of transverse momentum for away side particles perpendicular to the trigger direction. This was always smaller in the FF model than in the CCHK data[9]. The QCD model gives a larger prediction (simply because it has a large internal constituent $k_\perp$) but this is still a little smaller than the data[10]. Strong conclusions cannot be drawn because the data have rather low trigger $p_\perp$ and further have a background contribution (cf. Figs. 4 and 21, QCD underestimates the away side magnitude for trigger $p_\perp < 3$ GeV/c) which has not been subtracted.

(c) Very High Energy

At the Q^2 values probed by colliding beams of $\sqrt{s} \sim 500$ GeV, the QCD predictions will be very easy to test. For $p_\perp > 30$ GeV/c, the QCD jet cross-section will be several microbarns and the single particle cross-section a factor of a thousand smaller. These predictions are a factor of a thousand (jet) or a hundred (single particle) larger than the $p_\perp^{-8}$ extrapolation of the FF model. This large jet cross-section will make it very hard to find the W boson (and other new particles) from its $q\bar{q}$ or jet-jet decay. Quigg showed that even the $p_\perp^{-8}$ jet extrapolation led to a W signal that was at best ten times the hadronic background. A factor of a thousand increase in the latter is obviously fatal! Further, the increasing internal $k_\perp$ of the constituents at these high energies, mentioned earlier, will make the W hard to find in its $\mu\nu$ or $e\nu$ decay mode. It is not clear to me what "jets" at these high $p_\perp$ will really look like experimentally. Large internal transverse momentum will be present in the fragmentation process and one will surely not see the collimated (mean transverse momentum of hadrons perpendicular to jet axis ~ 330 MeV/c) jets that graced, for instance, PEP proposals. This problem is now being worked on theoretically and hopefully we will soon know the answer. One prospect that intrigues me is the possibility of "observing" gluons, i.e., of seeing high $p_\perp$ jets and distinguishing quark (as seen at PEP, PETRA) from gluon induced jets. The latter will surely have more low momentum members than do quarks; further gluons are flavor singlet. However, it is difficult to put one's finger on a positive method of identifying gluons. It will be interesting to study the away side of current ISR data (where there should be a majority of gluons) to see if the events fall into two classes which could be identified with the two types of jets.

CONCLUSION

The recent developments in high $p_\perp$ phenomenology are exciting because they really suggest that we can finally observe in hadron scattering properties of the fundamental (quark, gluon) Lagrangian Clearly, the current picture is tentative and requires a great deal more theoretical and experimental work. Hopefully, a unified picture of lepton inelastic scattering, Drell-Yan production of μ pairs and hadron high $p_\perp$ scattering will emerge.

I thank R. D. Field and R. P. Feynman for a continuing fruitful collaboration.

REFERENCES

1. P. Darriulat, in Proceedings of the Eighteenth International Conference on High Energy Physics, Tbilisi, U.S.S.R. (1976), edited by N. N. Bogolubov et al.

2. S. M. Berman, J. D. Bjorken and J. B. Kogut, Phys. Rev. D4, 3388 (1971).

3. D. Sivers, S. Brodsky and R. Blankenbecler, Phys. Reports 23C, 1 (1976).

4. S. J. Brodsky, R. Blankenbecler and J. F. Gunion, "Large Transverse Momentum Processes and the Constituent Interchange Model", SLAC preprint (1977).

5. G. C. Fox, in BNL Report No. BNL-50598, edited by H. Gordon and R. F. Peierls, Proceedings of the American Physical Society Meeting, Division of Particles and Fields, Upton, New York (1976).

6. G. C. Fox, "Recent Experimental Results on High Transverse Momentum Scattering from Fermilab," CALT-68-630 preprint (1977).

7. H. Boggild, Invited talk at Eighth Symposium on Multiparticle Dynamics, Kaysersberg, France (1977).

8. H. Frisch in BNL report cited in Ref. 5.

9. M. Della Negra et al., Nucl. Phys. B127, 1 (1977). This references earlier publications by the CCHK collaboration.

10. R. D. Field, "Can Existing High Transverse Momentum Hadron Experiments be Interpreted by Contemporary Quantum Chromodynamic Ideas?," CALT-68-633 (1977). (Note that numerical values in this paper should be reduced by a factor of 2); R. P. Feynman, R. D. Field and G. C. Fox, paper in preparation.

11. R. D. Field and R. P. Feynman, Phys. Rev. D15. 2590 (1977): R. P. Feynman. R. D. Field and G. C. Fox. Nucl. Phys. B128, 1 (1977).

12. C. Bromberg et al., "Production of Jets and Single Particles at High $p_\perp$ in 200 GeV Hadron Beryllium Collisions, "Nucl. Phys., to be published. Other papers by E260 are given in Ref. 6.

13. L. Cormell et al., "Evidence for Jets, and for Correlated Jets, in High $p_\perp$ Events," preprint (1977). (Fermilab, E395).

14. J. W. Cronin et al., Phys. Rev. Letters 38, 112, 115 (1977).

15. G. Hanson et al., Phys. Rev. Letters 35, 1609 (1975).

16. J. C. VanderVelde, "Neutrino Interactions in the 15-foot Bubble Chamber and Search for Charmed Particles," preprint and invited talk at Moriond Conference (1977).

17. G. Donaldson et al., Phys. Rev. Letters 36, 1110 (1976).

18. For other, generally less complete, treatments of the QCD corrections to the $p_\perp^{-4}$ behavior, see Ref. 9 and R. Cutler and D. Sivers, Phys. Rev. D16, 679 (1977) and ANL preprint PR-77-40 (1977); B. L. Combridge, J. Kripfganz and J. Ranft, Phys. Letters B70, 234 (1977); J. F. Owens, E. Reya and M. Gluck, FSU preprint 77-09-07 (1977); R. Raitio and R. Sosnowski, University of Helsinki preprint HU-TFT-77-22; A. P. Contogouris, R. Gaskell and A. Nicolaidis, McGill University preprint (1977).

19. G. C. Fox, Nucl. Phys. B131, 107 (1977) and "Phenomenology of Asymptotically Free Theories in Deep Inelastic Scattering" II: Neutrino Scattering.

20. The coupling constant is not exactly as given in text. At low Q^2, it has $1/Q^2$ corrections proportional to charmed quark mass squared. See Ref. 19 for details.

21. H. L. Anderson, H. S. Matris and L. C. Myrianthopoulos, "A Determination of the Quark and Gluon Moments of the Nucleon", Chicago preprint (1977).

22. We use the fit in Ref. 19 which used structure functions that are functions of x. This is much simpler than the Georgi-Politzer formalism (Ref. 23) but has very similar predictions.

23. H. Georgi and H. D. Politzer, Phys. Rev. D14, 1829 (1976) and A. de Rujula, H. Georgi and H. D. Politzer, Annals of Physics 103 315 (1977).

24. (BEBC data) P. C. Bosetti et al., Phys. Letters 70B, 273 (1977).

25. (CITFR data) B. C. Barish et al., Phys. Rev. Letters 39, 1595

(1977) and other Caltech preprints (see Ref. 19).

26. H. Georgi and H. D. Politzer, Phys. Rev. Letters 40, 3 (1978).

27. M. G. Albrow et al., "Studies of Proton-Proton Collisions at the CERN ISR with an Identified Charged Hadron of High Transverse Momentum at 90° I: On Forward Particles in High $p_\perp$ Reactions", CERN preprint (1977).

28. D. M. Kaplan et al., Phys. Rev. Letters 40, 435 (1978).

29. H. D. Politzer, "Quantum Chromodynamics at High Energies", CALT-68-628 (1977); G. Altarelli, G. Parisi and R. Petronzio, "Transverse Momentum in Drell-Yan Processes," CERN-TH-2413 (1977) preprint. See also references cited in Ref. 28.

30. This estimate uses the technique outlined in Ref. 29.

31. S. Brodsky and G. Farrar, Phys. Rev. D11, 1309 (1975).

32. B. G. Pope, Preliminary data presented at the 1977 DPF meeting at Argonne.

33. P. V. Landshoff, "Large Transverse Momentum Jets: A Theoretical Review," Tutzing Conference (1976).

34. B. Alper et al., Nucl. Phys. B100, 237 (1975).

35. R. D. Field, "Implications of the Recent Large $p_\perp$ Jet Trigger data from Fermilab," preprint and invited talk at the Moriond Conference (1976).

36. F. W. Busser et al., Nucl. Phys. B106, 1 (1976).

37. Fermilab experiment E494. R. J. Englemann, private communication (1977).

38. K. Eggert et al., Nucl. Phys. B98, 49 (1975).

39. C. Quigg, Rev. Mod. Phys. 49, 297 (1977).

GAUGE FIXING DEGENERACIES, CONFINEMENT AND OTHER PROBLEMS FOR NON-ABELIAN GAUGE THEORIES*

Heinz Pagels

Rockefeller University and

Princeton University

During the last year there have been several new developments in the quantum theory of gauge fields which I will review here. First is the discovery that in non-abelian gauge theories the usual gauge fixing conditions fail to uniquely specify the gauge. There are, in general, many fields, all gauge equivalent, satisfying the same gauge condition. I will describe this observation of V.N. Gribov's[1] along with some general results of I.M. Singer. This degeneracy poses a problem for the quantum theory since one must fix the gauge to quantize the field. Consequently, at present a consistent formulation of quantized gauge fields does not exist.

The second topic of discussion will be a review of some attempts to see the conjectured confinement property of non-abelian gauge theories. The attempts that I will describe here all have to do with merons or monopoles (which are really time slices of merons). These confinement mechanisms have been studied in Landau gauge by Callan, Dashen, and Gross,[2] in axial gauge by Mandelstam[3] and in Coulomb gauge by Bender, Eguchi and Pagels.[4]

*Supported in part by the National Science Foundation under Grant No. PHY78-01221.

These approaches are probably all related (although they appear to be very different) since in the one meron (monopole) approximation they obtain an R^3 potential between color sources. It is not clear if there is any connection between the gauge fixing degeneracies and confinement.

Finally I will discuss a new set of objects, quantum solitons, that disappear in the $h \to 0$ limit and make a few remarks about integrability conditions and dual fields for the classical gauge theory.

I. GAUGE FIXING DEGENERACIES IN NON-ABELIAN GAUGE THEORIES

A. Boundary Conditions for Gauge Fields

Before discussing the gauge fixing degeneracies it is important to emphasize the role of boundary conditions on gauge fields. We will be discussing only the 3+1 dimensional SU(2) gauge theory and set the gauge coupling g=1.

What physical criteria establish the boundary conditions on gauge fields? One requirement is to have infrared finite energy. While this requirement is necessary one can be more precise by restricting oneself to fields that are asymptotically static. (Radiation fields for the gauge field can be classified into types but we do not consider this here.) Then the boundary conditions are established by the requirement that there exist electric monopoles of strength e or magnetic monopoles of strength m. The behavior of the gauge field is

$$A_o^a(x,t) \underset{|x|\to\infty}{\sim} \frac{e}{|x|} + \frac{m}{|x|^2} + O(\frac{1}{x^3}) \quad , \tag{1.1}$$

$$A_i^a(x,t) \sim \frac{m}{|x|} + \frac{e}{|x|^2} + O(\frac{1}{x^3}) \quad .$$

Of course the gauge field is not a gauge invariant object so the boundary conditions (1.1) could be altered by a gauge transformation.

However, if we restrict gauge transformations U(x,t) to satisfy

$$\begin{array}{lll} U(x,t) & & \text{non-singular} \\ U(x,t) & \rightarrow & \text{constant} \\ |x| \rightarrow \infty & & \end{array} \tag{1.2}$$

then (1.1) is left unchanged. Alternatively the boundary conditions (1.1) could instead be given a gauge-and Lorentz-invariant meaning by specifying the behavior of the invariants

$$F^a_{\mu\nu}F^a_{\mu\nu}, \quad F^a_{\mu\nu}{*F}^a_{\mu\nu}, \quad \varepsilon^{abc}F^a_{\mu\nu}{*F}^b_{\nu\lambda}F^c_{\lambda\mu}, \text{ etc.}$$

as $|x| \rightarrow \infty$.

In general we want to allow for electric (colored) charges so $e \neq 0$. Whether or not to admit monopoles is an open question. So we will consider two possibilities

$$\begin{array}{lll} \text{Weak Boundary Condition} \equiv m \neq 0 & & A_i \sim 1/|x| \\ \text{Strong Boundary Condition} \equiv m = 0 & & A_i \sim 1/|x|^2 \end{array} \tag{1.3}$$

and examine the consequence of each.

B. Gribov's Observations

In a quantum theory of gauge fields the gauge must be fixed since the gauge degrees of freedom are superfluous and are not to be quantized. In the canonical formulation this requirement is met so that if initial data, commutators at equal time etc. are specified at one time then the dynamical evolution is unique for all time. The dynamical system is then well defined. In the path integral formulation this requirement is seen by the presence of an infinity in the functional integral due to integration over gauge equivalent configurations. Gauge fixing is required to define the functional integral and restrict the integral to one member of each gauge equivalent orbit. This is the Faddeev-Popov prescription.

The most elementary example of gauge fixing is in the U(1), abelian gauge theory. For example suppose we take the Coulomb gauge and boundary condition:

$$\text{(i)}\quad \text{Gauge fixing}\quad F(A) = \partial_i A_i(x,t) = 0$$

$$\text{(ii)}\quad \text{B.C.}\quad A_o \sim 1/|x| \;,\; A_i \sim 1/|x| \quad |x|\to\infty$$

and now ask: is there another field A', gauge equivalent to the first, also satisfying conditions (i) and (ii)? So

$$A'_\mu(x,t) = A_\mu(x,t)+\partial_\mu\phi(x,t),\; \phi(x,t)\text{ nonsingular.}$$

One easily sees that (i) and (ii) imply $A'_\mu(x,t) = A_\mu(x,t)$ and this means the gauge is fixed. In the space of all functions, schematically represented in Fig. 1a, the gauge fixing condition F(A)=0 is a hypersurface. The set of all gauge equivalent potentials to A is represented as an orbit that intersects F(A)=0 only once at that unique field configuration that satisfies the gauge condition.

If we now turn for our second example to the SU(2) gauge theory the situation is dramatically altered. Again we impose a gauge fixing condition and appropriate boundary conditions:

$$\text{(i)}\quad \text{Gauge Fixing}\quad F^a(A) = \partial_i A_i^a(x,t) = 0$$

$$\text{(ii)}\quad \text{B.C.}\quad \text{weak}\quad A_o \sim 1/|x| \quad A_i \sim 1/|x|$$

$$\text{strong}\quad \text{"}\quad A_i \sim 1/|x|^2$$

and ask if there are gauge equivalent configurations

$$A'_\mu(x,t) = U(x,t)A_\mu(x,t)U^{-1}(x,t) + iU(x,t)\partial_\mu U^{-1}(x,t)$$

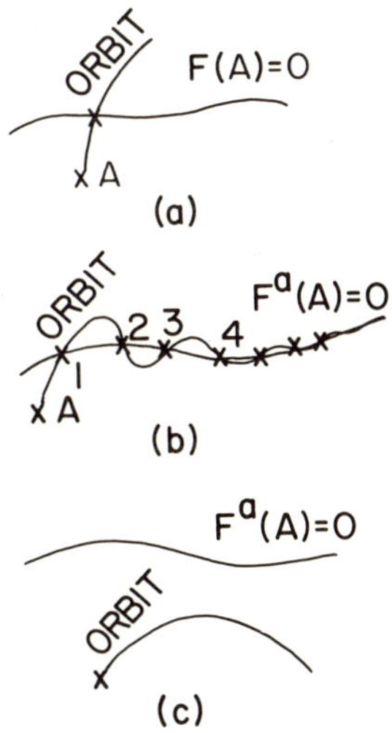

Figure 1. Trajectories in Function Space of Gauge Potentials

which also satisfy (i) and (ii). As was first emphasized by Gribov[1] the answer is yes. The requirement that A'_μ also satisfy the Coulomb gauge condition implies

$$D_i(A)(U^{-1}\partial_i U) = 0 \quad . \tag{1.4}$$

This equation has non-trivial solutions U for some A that give fields A' that satisfy the B.C. (ii). In function space the orbit of A intersects the gauge fixing hypersurface $F^a(A)=0$ many times as indicated in Fig. 1b. Conceivably there could exist configurations A that cannot be made to satisfy a given gauge condition by performing a gauge transformation, as shown in Fig. 1c.

To actually see how the gauge fixing degeneracy happens an explicit example is useful.[1] To study the solutions to the degeneracy condition (1.4) one makes the ansatz for static configurations

$$A_i^a(x) = \varepsilon^a_{ij} \frac{x_j}{r^2} f(r) \ , \quad r^2=x_1^2+x_2^2+x_3^2 \tag{1.5}$$

which automatically satisfies the condition (i), $\partial_i A_i^a(x)=0$ for arbitrary f(r). The gauge transformation is written

$$U(\alpha) = e^{i\alpha(r)\vec{x}.\vec{\tau}/r} = \cos\alpha(r) + \frac{i\vec{x}.\vec{\tau}}{r} \sin\alpha(r) \ , \tag{1.6}$$

$$\text{nonsingular if } \alpha(0)=n\pi \quad .$$

Then (1.4) is satisfied if

$$\nabla^2\alpha(r) - \frac{\sin 2\alpha(r)}{r^2}(1-f(r)) = 0 \quad ,$$

which upon changing variables $r=e^t, -\infty \le t \le +\infty$ becomes the pendulum equation with damping

$$\ddot{\alpha}(t) + \dot{\alpha}(t) - \sin 2\alpha(t)(1-f(t)) = 0$$
$$2\alpha(-\infty) = 2\pi n \quad . \tag{1.7}$$

Here $1-f(t)$ is the force of gravity (see Fig. 2a).

To analyze the pendulum problem first suppose $f(t)>1$. Then gravity points upward and the inverted pendulum does not move and the gauge is fixed. If $f(t)<1$, in particular if $f(t)=0$ so $A_i^a=0$, the pendulum swings down and comes to rest at $\alpha(-\infty)=\pi/2$ because of the damping (see Fig. 2b). This corresponds to new vacuum $A_i = i\Omega\partial_i\Omega^{-1}$ where $\Omega \to i\hat{x}.\vec{\tau}$ as $|x| \to \infty$ so $A_i \sim 1/|x|$. This vacuum is excluded if we pick strong B.C. but it is allowed with the weak B.C. Finally one can convince oneself that by picking $f(t)$ to vary with t appropriately there are non-trivial solutions $\alpha(t)$ which correspond to gauge equivalent fields satisfying the strong B.C.

Several remarks are appropriate to conclude this discussion.

(a) The static spherically symmetric vacuum configuration ($F^a_{\mu\nu} = 0$) with strong B.C. is unique in the Coulomb gauge: $A^a_\mu = 0$.

(b) The vacuum configurations with weak B.C. are $A_\mu = 0$ and $A_\mu = i\Omega(\alpha)\partial_\mu\Omega^{-1}(\alpha)$ with α a solution to the pendulum equation, $\alpha(\infty) = \pi/2$. This latter vacuum configuration is at least 4-fold degenerate corresponding to 3 translational modes and an arbitrary parameter in the solution of the pendulum equation. The homotopy of these vacuua defined by the winding number

$$n = \frac{i}{24\pi^2}\int d^3x\, \varepsilon^{ijk}\, \mathrm{Tr}(A_iA_jA_k) \quad , \tag{1.8}$$
$$A_o = 0,\ A_i = i\Omega\partial_i\Omega^{-1}$$

are

$$A_i = 0 \qquad n = 0$$

$$A_i = i\Omega\partial_i\Omega^{-1} \qquad n = 1/2 \qquad \Omega=\Omega(\alpha) \quad .$$

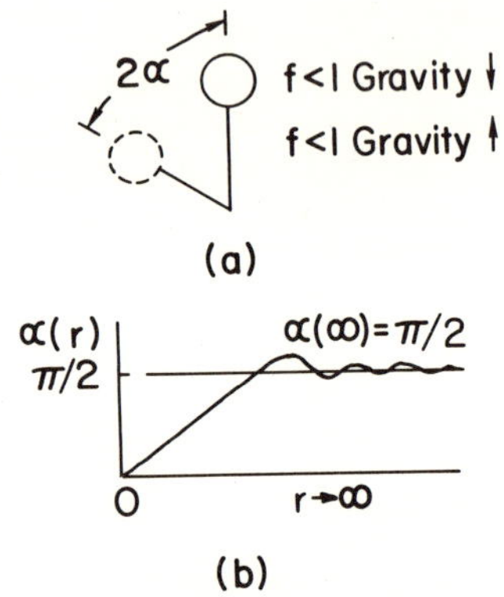

Figure 2. Analysis of Pendulum Equation

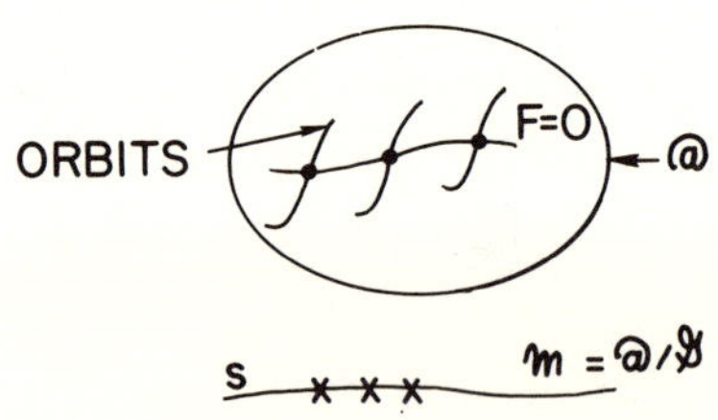

Figure 3. Manifolds in Function Space

These vacuua are interpolated by the (complex time) meron solution

$$A_i^a(x,t) = \varepsilon^a{}_{ij} \frac{x_j}{r^2}(1+\phi(x,t))$$

$$\text{Meron} = \phi(x,t) = \frac{t}{\sqrt{r^2+t^2}} \begin{cases} \xrightarrow{t=+\infty} 1 & n=1/2 \\ \xrightarrow{t=0} 0 & \text{Wu-Yang Monopole} \\ \xrightarrow{t=-\infty} -1 & n=0 \end{cases} .$$

This meron solution for finite 4-volume R^4 has an action given by $S = (3\pi^2/g^2)\ln R$.

(c) Even with strong B.C., in the Coulomb gauge, for sufficiently large curvature $|F^a_{\mu\nu}|>|F^a_{\mu\nu}|_{\text{CRITICAL}}$ there exist gauge fixing degeneracies. This was one of the main points of Gribov's work.

(d) With the strong B.C. the field configurations that give degeneracy are far from the saddle points, at least for those configurations studied so far. So it is not clear how important these configurations are in the functional integral. A point emphasized by Amati and Rouet[5] is that for small fluctuations about saddle points, like instantons, the background gauge $D_\mu(A^{\text{instanton}})A_\mu = 0$ is non-degenerate.

C. <u>I.M. Singer's Generalization</u>

Gauge field theories with their emphasis on geometry and topology have revitalized the symbiosis of mathematics and physics by providing a meeting ground for common problems. Here I will briefly report some results of I.M. Singer which I learned of through conversations with him. The detailed proofs will appear in a future publication.

The main assumptions are (i) the gauge field is defined on the compact mainfold S_4. Presumably this is equivalent to the Euclidean R_4 plus the strong B.C., $A \sim 1/|x|^2$. (ii) The gauge fixing condition $F^a(A)$ is continuous in A. The usual functional integral

$$Z = \int_{@} [DA] e^{-|F_A|^2}$$

over the full affine space @ of all potentials is divergent since the curvature invariant $|F_A|$ is gauge invariant. The question is can one remove this infinity by integrating only over a manifold M(s) parameterized by s

$$Z \to \int_{M(s)} [Ds] M(s) e^{-|F_A|^2}$$

where every element of M(s) corresponds to a single orbit in @ (see Fig. 3)? M(s) is a projection from the gauge fixing hypersurface $F^a(A) = 0$ and is given by $M = @/G$ where G is the group of automorphisms on the principle bundle sending each fiber into itself, e.g. the mapping $S^4 \to SU(2)$. G is essentially the group of gauge transformations.

Then one can show (i) that no (continuous) $F^a(A)$ exists for the non-abelian theory. For suppose, to the contrary, that there was a one-to-one relation between the intersection of the orbits on $F^a(A) = 0$ and M(s). Then one would have $@ = M \otimes G$. But @ has trivial topology and therefore so does G, contrary to the fact that the mapping $S_4 \to SU(2)$ has a nontrivial homotopy $\Pi_3(SU(2)) = Z$. Therefore one can not have a unique correspondence. This establishes the existence of gauge fixing degeneracy for general continuous gauge fixing conditions on S_4. (ii) In general the degeneracy is infinite, i.e. for some A there are infinitely many solutions to $F(A) = 0$. (iii) There is no gauge fixing for which the slopes of the intersections of the orbits with $F(A) = 0$ (given by the Faddeev-Popov determinants) are all equal. This means we can not just truncate the functional integral at one intersection and maintain invariance under finite gauge transformations.

D. Implications

Gauge fixing degeneracy is evidently a general phenomena and it is important to understand it further. Several of the implications are outlined below.

(a) The Faddeev-Popov prescription for the functional integral is incomplete since it assumes $F^a(A) = 0$ has a unique solution. In the general case there are many intercepts of the orbit with $F^a(A)=0$ (see Fig. 1b). The Faddeev-Popov determinants are a measure of the slope in the vicinity of the intercept and the determinant at the nth intercept is

$$\det M^{ad}(A^n) = \det \left(\frac{\delta F^a(A^n)}{\delta A^b_\mu} D^{db}_\mu(A^n)\right) .$$

Here A^n are the solutions of $F^a(A^n) = 0$. Then one can show that the definition of the functional integral given by Bender, Eguchi and Pagels[4]

$$Z = \int [dA_\mu] \frac{\delta(F^a(A))}{\sum_n \det^{-1} M(A^n)} e^{-s(A)} \qquad (1.9)$$

is invariant under finite as well as infintesimal gauge transformations. Unlike the case where there is only one intercept the sum on inverse determinants can not be promoted to an effective ghost field interaction in the action in an obvious way. So while in principle one can write down the fully gauge invariant generating functional (1.9) one does not have a useful tool for calculations.

A corollary of these remarks is that the Feynman rules for nonabelian gauge theories are not gauge invariant. They are only invariant for infinitesimal transformations, not finite ones.

(b) The instanton solution to the 4-D Euclidean gauge theory is usually written so it satisfies the Landau gauge condition $\partial_\mu A^a_\mu = 0$. One may ask what the behavior of the solution is in the Coulomb gauge $\partial_i A^a_i(x,t) = 0$. This configuration with strong B.C. was described by Jackiw, Muzinich and Rebbi[6] and Ademollo, Sciuto and Napolitano[7] (R. Jackiw will be discussing the solution in detail in the proceedings of this conference) and for the weak B.C. by Abbott and Eguchi[8]. The instanton has a Pontryagin index p = 1 where

$$p = \frac{g^2}{16\pi^2} \int d^4x \; \mathrm{Tr}\; F_{\mu\nu} {}^*F_{\mu\nu} \quad .$$

Since $\mathrm{Tr}(F_{\mu\nu} {}^*F_{\mu\nu})$ is a divergence we can use Gauss's law to write this formula as a surface integral:

$$1 = p = n_+ - n_- + \begin{pmatrix} \text{surface} \\ \text{terms as } |x| \to \infty \end{pmatrix} \quad , \qquad (1.10)$$

where $n_\pm$ are the $t \to \pm\infty$ limits of (1.8) and the surface terms are the contributions at spatial infinity (see **Fig.** 4). As we have seen with strong B.C. the vacuum at $t = \pm\infty$ is unique and given by $A_\mu = 0$ hence $n_\pm = 0$. Also with strong B.C. the potential behaves like $\sim 1/|x|^2$ as $|x| \to \infty$ and the surface term in (1.10) is zero. So one would have proven 1 = 0, except that formula (1.10) assumed Gauss's law which is valid only if the field is continuous. Indeed one can show that at some finite time the gauge field develops a discontinuity so that $A^a_\mu(x,t_o-\varepsilon)$ and $A^a_\mu(x,t_o+\varepsilon)$ differ by a nontrivial gauge transformation as $\varepsilon \to 0$. So the field must jump at t_o if one imposes the strong B.C. With weak B.C. the jump is displaced to the boundary and the surface term is nonzero; so in this case the fields can evolve continuously. For the instanton solution these features are pecularities of the Coulomb gauge since the Landau gauge the solution is continuous.

(c) The implication of gauge fixing degeneracy is that the dynamics of gauge fields must be defined on sections. (See Fig. 5) On each section the gauge is well defined and fixed. But such a description is not possible globally if the curvature $|F_{\mu\nu}|$ is too large even with strong B.C. Defining the gauge fields on sections is an unhappy complication since we do not know how to do quantum

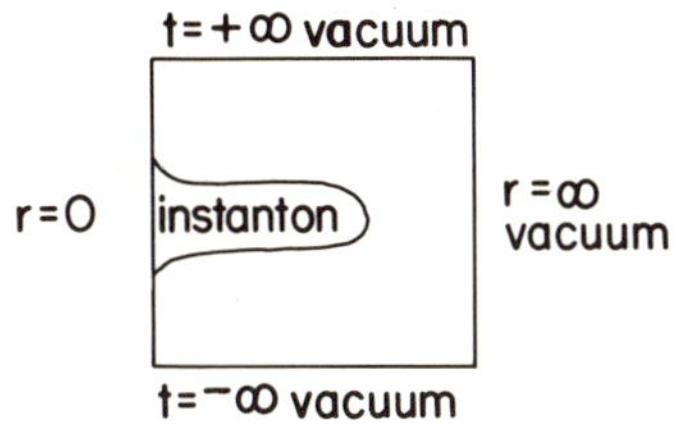

Figure 4

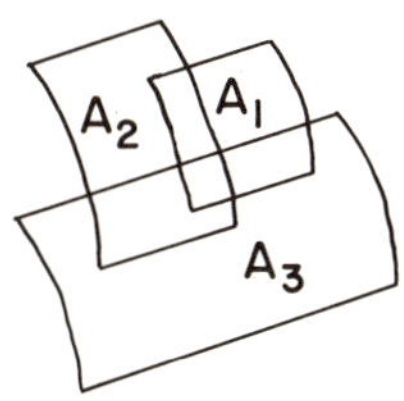

Figure 5. Patches in Function Space

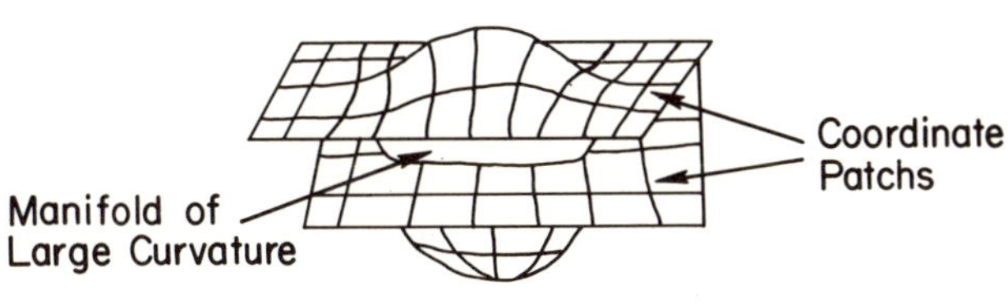

Figure 6. Coordinate Patches

theory in this case. This complication is essential as long as the dynamical variables are the gauge potentials since they jump about, require patches, etc. But it is safe to conjecture that the evolution of color singlets is smooth.

The gauge field analogy with general relativity prompts the question: are there gauge fixing degeneracies in general relativity? For the local Lorentz invariance on the vierbein, since the transformations are covariant, there is no problem. For general coordinate transformations, however, there seems to be no reason why the degeneracies are not present although I know of no explicit example having been worked out. In general relativity it is well known that if the curvature becomes large it is impossible to lay down a singularity free coordinate system. A singularity free description is possible if one allows for sections (see Fig. 6). It would be nice to have a few examples of gauge fixing degeneracy in general relativity if it really exists.

II. INTIMATIONS OF CONFINEMENT

A. Landau Gauge $\partial_\mu A_\mu^a = 0$

An ambitious program to study the phase transitions of the gauge theory has been undertaken by Callan, Dashen, and Gross.[2] R. Dashen will be reporting on recent progress in this program at this conference. The basic idea is to saturate the R_4 functional integral with saddle point configurations approximated by a dilute gas of solitons. These Euclidean solitons are instantons which behave like dipolar objects. The instanton field behaves like $F_{\mu\nu}^a \sim (\lambda^2/r^4)\eta_{\mu\nu}^a$, $r \to \infty$ which is the field of a dipole in 4 dimensions. The dipolar instanton can be viewed as made up of two constituents called merons each with 1/2 unit of topological charge. In this analogue gas model the role of temperature is played by the sliding coupling constant $g(\mu)$ where μ is the renormalization point. $g(\mu)$ becomes small for short distances and presumably grows at larger distance.

The phase transitions occur at different temperatures corresponding to the scale at which one examines the hadron. These phase transitions are depicted in Fig. 7 - a Q.C.D. thermometer. At near zero temperature $g(\mu)$ is small and the dipoles and anti-dipoles are tightly bound together. The vacuum is the ordinary vacuum and perturbation theory (asymptotic freedom) works. At a somewhat higher temperature the dipoles and anti-dipoles become liberated and this phase transition is to be associated with PCAC; the vacuum becomes γ_5 non-invariant. At a still higher temperature the dipoles themselves break apart, or almost do, into their constituent merons. This phase transition is associated with confinement.

In the one meron approximation, admittedly crude and suspicious because the dilute gas approximation is probably invalid, these authors find using the gauge invariant Wilson loop criterion for confinement that the potential between static quarks separated by a distance R grows like R^3.

B. Axial Gauge, $A_z = 0$

S. Mandelstam has attempted to see confinement in the axial gauge.[3] The hope is that the vacuum will turn out to be a superconducting fluid of magnetic monopoles. Then one traps electric flux (like magnetic flux forms vortices in an ordinary superconductor) with the string picture of confinement (see Fig. 8). The construction of the physical vacuum in the axial gauge is complicated by the requirement of a constraint equation on physical states. In a crude approximation to the physical vacuum Mandelstam estimates the vacuum value of the Wilson loop integral and finds an R^3 potential between static quarks. So this approach may be related to the others we describe.

C. Coulomb Gauge $\partial_i A_i^a(x,t) = 0$

I would now like to describe in a little more detail the work

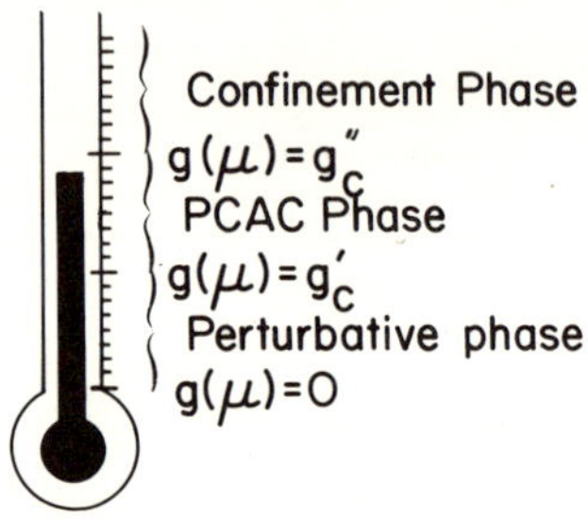

Figure 7. A QCD Thermometer

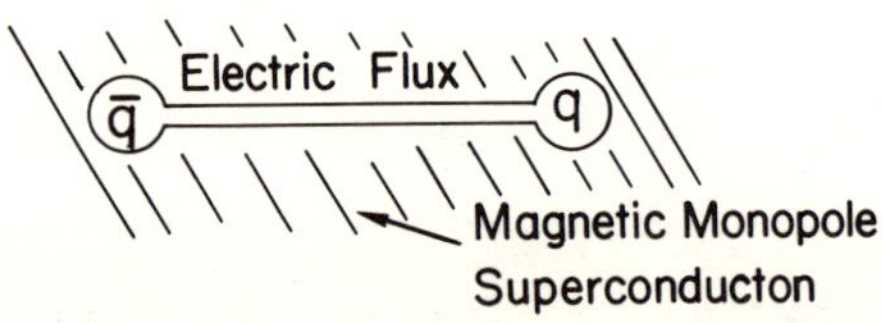

Figure 8. Electric Flux Confinement

of Bender, Eguchi, and myself[4] on the confinement problem in the canonical formalism. Crucial to our work is that we allow for magnetic monopole configurations, i.e. weak B.C. so that $A_i^a \sim 1/|x|$.[9]

In the canonical formalism the independent variables are the transverse fields ${}^TE_i^a(x,t)$ and ${}^TA_i^a(x,t)$ which satisfy

$$\partial_i {}^TE_i^a(x,t) = 0 \quad \partial_i {}^TA_i^a(x,t) = 0 \quad .$$

The Hamiltonian is

$$H = \frac{1}{2}\int d^3x[(F_{ij}^a(A))^2 + ({}^TE_i^a)^2 + (\nabla_i f^a)^2] \ ,$$

with

$$F_{ij}^a(A) = \partial_i {}^TA_j^a - \partial_j {}^TA_i^a + \varepsilon^{abc}{}^TA_i^b {}^TA_j^c \quad ,$$

$$f^a(x,t) = \int d^3y\ D^{ab}(x,y;A)\rho^b(y,t) \quad .$$

The Coulomb Green's function satisfies

$$D_i(A)\partial_i D^{ab}(x,y;A) = \delta^{ab}\delta^3(x-y)$$

and the source density is

$$\rho^a(x,t) = \varepsilon^{abc}{}^TA_i^b(x,t){}^TE_i^o(x,t) + J^a(x,t) \quad .$$

Here $J^a(x,t)$ is the colorspin density of external particles, like quarks. Notice that, unlike the abelian case, the Coulomb Green's function is a functional of A. What we will do is compute the instantaneous Coulomb interaction energy for a class of Coulomb gauge configurations between two point quark sources of colorspin. We approximate the source by

$$J^a(x,t) = T_1^a\delta^3(x-x_1) + T_2^a\,\delta^3(x-x_2)$$

and so the interaction term will be proportional to $V(R)T_1 \cdot T_2$, $R = |X_1 - X_2|$.

The field configurations we consider are of the monopole form

$$^T A_i^a(x,t) = \varepsilon^a{}_{ij}\,\frac{x_j}{r^2}\,f(r,t),\; A_o^a = 0 \tag{2.1}$$

so that $\partial_i{}^T A_i^a = 0$. The contribution of the gauge fields to the source $\rho^a(x,t)$ vanishes since $^T E_i^a = {}^T\dot{A}_i^a$ so $\varepsilon^{abc}\,{}^T A_i^b(x,t)\,{}^T E_i^c(x,t) = 0$. The interaction energy is then given by

$$H_I = -\frac{1}{2}\,T_1^b T_2^c \int d^3x D^{ab}(x,x_1;A)\nabla_x^2 D^{ac}(x,x_2;A) + (1\leftrightarrow 2)\ .$$

What we want to calculate is the expectation value

$$\langle H_I\rangle \sim T_1^b T_2^c \int d^3x\ \langle D^{ab}(x,x_1;A)\rangle\nabla_x^2\langle D^{ab}(x,x_2;A)\rangle + (1\leftrightarrow 2)\ .$$

We assume $\langle D^{ab}D^{ac}\rangle \sim \langle D^{ab}\rangle\langle D^{ac}\rangle$. The translationally invariant Green's function is gotten by averaging over all configurations according to

$$D^{ab}(x-y) \equiv \langle D^{ab}(x,y;A)\rangle = \frac{\int[d^TA][d^TE]D^{ab}(x,y;A)e^{iL(^TA,\,^TE)}}{\int[d^TA][d^TE]e^{iL(^TA,\,^TE)}}\ .$$

Here $L(^TA, {}^TE)$ is the action. The configurations we consider comprise a set obtained from our ansatz (2.1) by translation $x \to x + c$ and an O(3) rotation. Even with this ansatz the calculation of $D^{ab}(x-y)$ is very complicated. Using the representation

$$D^{ab}(x,y;A) = \sum_n \frac{V_n^a(x;A)V^{*b}(y;A)}{-E_n}\ , \tag{2.2}$$

where

$$\partial_i D_i(A) V_n^a(x;A) = - E_n V_n^a(x;A) \tag{2.2}$$

the translational and rotationally invariant Green's function we assume is approximated by averaging the configuration over a finite volume $V = L^3$ according to

$$D^{ab}(x-y) = \int_V \frac{d\Omega}{4\pi} \frac{d^3\vec{c}}{V} \sum_n \frac{V_n^a(x+c;A) V^{*b}_n(y+c;A)}{-E_n} ,$$

where $d\Omega$ is an angular average to restore rotational invariance. Here $V_n^a(x;A)$ is the eigenfunction corresponding to the configuration at $\vec{c} = 0$.

To solve the eigenvalue problem (2.2) we make the ansatz

$$V_n^a(x;A) = \frac{x^a}{r^2} y_n(r)$$

which upon using (2.1) yields from (2.2) the one dimensional Schrödinger equation

$$y''(r) + (E_n - V(r)) y_n(r) = 0 ,$$

$$V(r) = \frac{1}{r^2}(1-f(r)) ,$$

where we drop the t dependence of f. To avoid short distance singularities we will require $f(0) = 0$. Then the potential looks like Fig. 9 if $f(r)$ is monotonic. We see that if $f(\infty)$ = constant > 1 there is an accumulation of bound states at $E_n = 0$. Near $E = 0$ we can use the semi-classical WKB methods. The result of this calculation is that the accumulation phenomena for configurations with $f(\infty) > 1$ produces a strong singularity in the Fourier transform of $D^{ab}(x-y)$:

$$D^{ab}(k) = \delta^{ab} D(k) \underset{|k|\to 0}{\sim} \frac{\delta^{ab}}{L^2 |k|^4} .$$

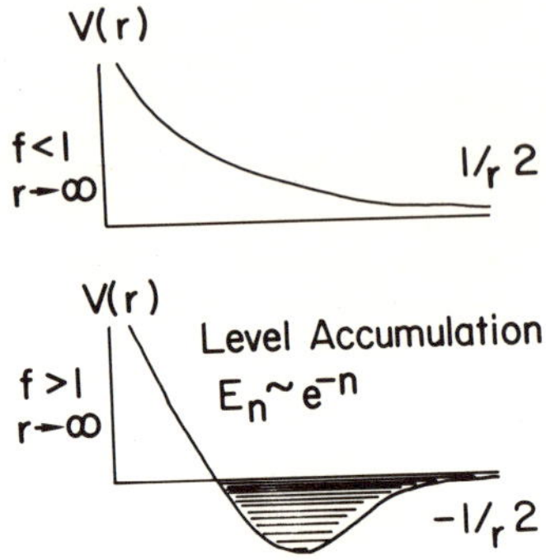

Figure 9 . Potential and Bound States

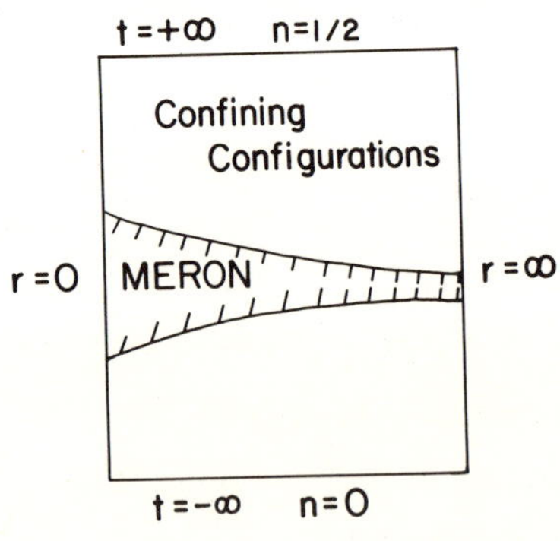

Figure 10 . Time Evolution of the Meron

Without the accumulation one obtains the usual Coulomb result $\sim 1/|k|^2$. The $|k|^{-4}$ singularity corresponds to an interaction energy

$$<H_I> \underset{R\to\infty}{\sim} T_1T_2R^3 \quad ,$$

the same result obtained in the previous approaches.

The interpretation of this calculation is that we are examining time slices of a meron (see Fig. 10), which in the Coulomb gauge is in the form of our ansatz (2.1). Admittedly our approximation to the expectation value $<H_I>$ is crude but the result is encouraging. The hope is that using multimeron configurations produces a screening effect so that the potential becomes linear.

III. QUANTUM SOLITONS

I would like to describe a few speculations that W. Marciano and I have made on solitons. The basic idea is that in some field theories, especially the gauge theories, there may exist a new set of solitons that are due to quantum effects and do not exist in the classical theory.

Consider the generating functional for a scalar field

$$Z(J) = N^{-1}\int[d\phi]\exp(i\int d^4x[\mathcal{L}(\phi(x))+J(x)\phi(x)])$$

so the generator of 1PI functions is $W(J) = i\ln Z(J)$. From the Legendre transformation we define the effective action $\Gamma(\Phi(x))$, a function of the ordinary classical field $\Phi(x)$.

$$W(J(x)) = \Gamma(\Phi(x)) + J(x)\Phi(x) \quad ,$$

$$\Phi(x) = \frac{\delta W(J)}{\delta\, J(x)} \qquad J(x) = -\frac{\delta\Gamma(\Phi)}{\delta\Phi(x)} \quad .$$

The equation of motion is simply $\delta\Gamma/\delta\Phi = 0$. The effective potential is defined by $V(\Phi) = -\Gamma(\Phi(x)) =$ constant.

In tree approximation the effective action is just the classical action

$$\Gamma^{\text{tree}}(\Phi(x)) = L(\Phi(x))$$

and one can attempt to find classical soliton modes in the classical theory. Often the classical theory can be shown to possess no soliton modes by application of a variety of no-go theorems. The virial theorems of Hobart[10] and Derrick[10] are examples. If there are no soliton solutions to the classical theory there may be solitons associated with the quantum theory, at least at one loop order,

$$\Gamma(\Phi(x)) \simeq \Gamma^{\text{tree}} + \Gamma^{\text{one loop}}$$

and these objects we call quantum solitons. One way that quantum solitons might appear is that the quantum effects stabilize an otherwise unstable classical solution. For example, the Hawking radiation evaporates a black hole. However the black hole can eventually stabilize at the Planck mass due to quantum gravity effects.[11]

An explicit example of quantum solitons is in the Coleman-E. Weinberg[12] treatment of the massless abelian Higgs model. On the classical level the effective potential for the scalar field has the minimum for unbroken symmetry (see Fig. 11) and there are no solitons. On the one loop level the scalar field develops a nontrivial minimum. This is seen to be a self consistent minimum from a renormalization group argument. Using the one loop effective action we can show there exists a vortex solution - similar to the Nielsen-Olesen vortex solution in an abelian massive Higgs model.[13] This soliton, generated by dimensional transmutation, disappears

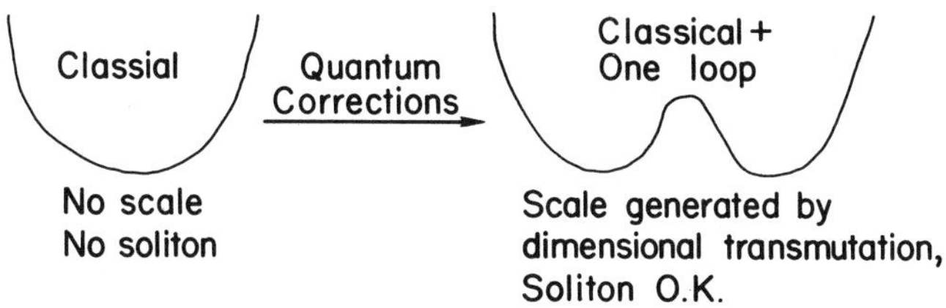

Figure 11 . Potential for Quantum Solitons

in the $\hbar \to 0$ limit - a quantum soliton. Of course this example is a bit artificial since one must impose the unnatural constraint that the physical Higgs mass in the effective action vanish.

A more natural example of quantum solitons are those theories which are classically scale invariant without constraints beyond renormalizability. The scale invariance on the classical level may forbid classical solitons. But the scale invariance must be broken by renormalization effects so one may have quantum solitons. Examples are the 1+1 Gross-Neveu model and the 3+1 nonabelian gauge theory. We consider next the latter example.

In the classical SU(2) gauge theory in 3+1 dimensions a variety of theorems[14] assure us of the absence of solitons - lump like solutions. This is because in the classical theory the trace θ^μ_μ of the energy momentum tensor vanishes and there is scale and conformal invariance. In the quantum theory, however, the trace has an anomaly[15] indicating the breakdown of scale invariance of the quantum theory. Without breaking scale invariance there is no way to tell high energy from low. So the sign of the anomaly, which on shell is a positive definite operator times the Callan-Symanzic function, determines if the theory is ultraviolet or infrared stable. For the SU(2) gauge theory

$$\theta^\mu_\mu = \frac{\beta(g)}{2g}(F^a_{\mu\nu})^2 \qquad \underset{g\to 0}{\beta(g)} = - \underset{b_o>0}{b_o g^3} \quad . \tag{4.1}$$

We now ask what effective action, as a function of the invariants of the gauge theory, produces an energy momentum tensor

$$\theta_{\mu\nu} = \frac{\delta L}{\delta g_{\mu\nu}} - \frac{1}{2} g_{\mu\nu} L \tag{4.2}$$

with the trace (4.1)? The answer is

$$L^{eff} = - \frac{1}{4}(F^a_{\mu\nu})^2(1+\gamma(g) - \frac{\beta(g)}{2g} \ln (F^a_{\mu\nu})^2/\mu^4) \quad , \tag{4.3}$$

where $\gamma(g) \sim O(g^2)$ and $\beta(g)$ are anomalous terms and μ is a renormalization mass. Now consider (4.3) as our effective action. The energy density is

$$\theta_{oo} = \frac{1}{4}(E^2+H^2)(1+\gamma(g) - \frac{\beta(g)}{2g}\,\ell n2(H^2-E^2)/\mu^4) - \frac{\beta(g)}{4g}E^2 \qquad (4.4)$$

and one finds the minimum of the energy has nonvanishing field strengths:[16]

$$E^2 = 0 \qquad H^2 \underset{g\to 0}{\sim} \frac{1}{2}\mu^4 e^{-1/2b_o g^2} \qquad . \qquad (4.5)$$

The Lorentz invariant characterization of this vacuum is $F^a_{\mu\nu}F^a_{\mu\nu} = H^2 > 0$. This suggests that the quantum vacuum is not $A_\mu = iU\partial_\mu U^{-1}$, the usual classical result, but has a background magnetic field on account of the trace anomaly. Further, the background magnetic field tends to exclude electric flux so one has the possibility of electric vortex formation. To actually construct the quantum solitons one must solve the field equations associated with L^{eff} and this has not yet been done. We view L^{eff} as a model for dimensional transmutation, and it may shed light on other phase transitions, e.g. confinement, as well.

More work remains to be done on the quantum soliton concept. But what is already clear is that quantum soliton solutions to the effective action may be of considerable importance in understanding QCD phase transitions since gauge theories have no solitons in 3+1 classical theory.

IV. FURTHER REMARKS ABOUT GAUGE FIELDS

A. Integrability Conditions for Gauge Fields

Suppose we ask: under what conditions on $F^a_{\mu\nu} = -F^a_{\nu\mu}$ does the system of equations

$$F^a_{\mu\nu}(A) = \partial_\mu A^a_\nu - \partial_\nu A^a_\mu + \varepsilon^{abc} A^b_\mu A^c_\nu \qquad (4.1)$$

admit a unique solution $A^a_\mu(F)$? To answer this we introduce the dual operation $*F^a_{\mu\nu}(A) = \frac{1}{2}\varepsilon_{\mu\nu\lambda\delta}F^a_{\lambda\delta}(A)$ with the property $** = -1$ (Minkowski) and $** = +1$ (Euclidean). It follows from (4.1) and the assumption that A^a_μ is nonsingular that

$$D_\lambda(A)*F^a_{\lambda\mu}(A) = \partial_\lambda *F^a_{\lambda\mu}(A) + \varepsilon^{abc}A^b_\lambda *F^c_{\lambda\mu}(A) = 0 \quad . \qquad (4.2)$$

As pointed out by Roskies[17] this equation has the solution

$$A^b_\nu(F) = \partial_\lambda *F^a_{\lambda\mu}(M^{-1})^{ab}_{\mu\nu} \quad , \qquad (4.3)$$

where $M^{ab}_{\mu\nu} = \varepsilon^{abc}F^c_{\mu\nu}$. Halpern[18] noticed the simplification in the case of SO(3)

$$A^b_\nu(F) = \partial_\lambda *F^a_{\lambda\mu} F^b_{\mu\alpha} *F^a_{\alpha\beta}(K^{-1})_{\beta\nu} \quad ,$$
$$K_{\mu\nu}(F) = \frac{1}{3}\,\varepsilon^{abc} *F^a_{\mu\alpha} F^b_{\alpha\beta} *F^c_{\beta\nu} \quad , \qquad (4.4)$$

so that $A^b_\nu(F)$ exists if det $K \neq 0$. Hence the answer to the question as to the condition on $F^a_{\mu\nu}$ such that A^b_ν exist is

$$F^a_{\mu\nu} = F^a_{\mu\nu}(A^b_\nu(F)) \ , \ \det K(F) \neq 0 \quad , \qquad (4.5)$$

where the function $A^b_\nu(F)$ is given by (4.4).

B. Duality for Potentials

One can introduce the concept of a dual potential. The dual potential is defined from the solution (4.4) with F replaced by *F so that

$$*A^b_\nu \equiv A^b_\nu(*F) \qquad (4.6)$$

and this exists providing that det K(*F)≠0. Clearly the dual potential is the solution to

$$D_\lambda(*A)F^a_{\lambda\mu}(A) = 0 \qquad (4.7)$$

and under a gauge transformation *A transforms just like the gauge field A. Consequently the difference A-*A transforms like the adjoint representation. We note that the dual field can be used to define electric flux if the first homotopy group is non-trivial

$$\begin{aligned} \text{magnetic flux} &\sim \operatorname{Tr} \exp \oint A_\mu dx_\mu \\ \text{electric flux} &\sim \operatorname{Tr} \exp \oint {*A_\mu} dx_\mu \quad . \end{aligned} \qquad (4.8)$$

However, if the potential A in addition satisfies the Yang-Mills equation

$$D_\lambda(A)F^a_{\lambda\mu}(A) = 0 \qquad (4.9)$$

it follows from subtracting (4.7) from (4.9) that det K(*F)≠0 implies *A=A. The interested reader can work out the consequences of the equations *F(A)=F(*A).

I don't know how useful this concept of the dual potential is. For many interesting solutions the det K≠0 condition is not met.

ACKNOWLEDGEMENT

It is a pleasure to thank Carl Bender and Thoru Eguchi for help in preparing my remarks on the work we did together on confinement in the Coulomb gauge. It's also a pleasure to thank William Marciano for discussions about quantum solitons.

REFERENCES

1. V.N. Gribov, Materials for the XII LNPI Winter School V. 1, p. 147 (1977); preprint "Quantization of Non-Abelian Gauge Theories" #367 LNPI, Leningrad 1977.
2. C. Callan, R. Dashen, and D. Gross, Phys. Lett. 63B (1976) 334; preprint "Towards a Theory of the Strong Interactions" Princeton University, August 1977.
3. S. Mandelstam, "Confinement in Non-Abelian Gauge Theories" Talk given at Washington Meeting of the APS, April 1977.
4. C. Bender, T. Eguchi and H. Pagels, preprint "Gauge Fixing Degeneracies and Confinement in Non-Abelian Gauge Theories" September 1977, to be published in Phys. Rev. D.
5. D. Amati and A. Rouet, CERN preprint "Gauge Unambiguous Quantization of Yang-Mills Theory Around Classical Solutions" (1978).
6. R. Jackiw, I. Muzinich and C. Rebbi "Coulomb Gauge Description of Lange Yang-Mills Fields" preprint October 1977.
7. M. Ademollo, S. Sciuto, and E. Napolitano, CERN preprint, "A Further Pathology of the Coulomb Gauge in Non-Abelian Yang-Mills Theories".
8. L. Abbott and T. Eguchi, Phys. Lett. 72B, (1977), 215.
9. It is difficult to see confinement if one restricts oneself to strong B.C. A recent preprint of D.H.I. Abarbanel and J. Bartels conclude that confinement does not occur for configurations with strong B.C. For weak B.C. the integral equations they use do not exist so their criticism does not apply to Ref. 4.
10. R. Hobart, Proc. Phys. Soc. Lond. 82 (1963) 201; G.H. Derrick, J. Math. Phy. 5 (1964) 1252.
11. S.W. Hawking, preprint (to be published Comm. Math. Phys., 1978).
12. S. Coleman and E. Weinberg, Phys. Rev. D7 (1973) 1888; see also M.R. Brown and M.J. Duff, Phys. Rev. D11 (1975) 2124.

13. H.B. Nielsen and P. Olesen, Nucl. Phys. B61 (1973) 45.

14. S. Coleman, Lectures at 1975 International School of Subnuclear Physics, "Ettore Majorana"; S. Deser, Phys. Lett. 64B (1976) 463; H. Pagels, Phys. Lett. 68B (1977) 466.

15. M.S. Chanowitz and J. Ellis, Phys. Lett. 40B (1972) 397; Phys. Rev. D7 (1973) 2490; R.J. Crewther, Phys. Rev. Lett. 28 (1972) 1421; R. Jackiw, in "Lectures on Current Algebra and Its Applications," S. Treiman, R. Jackiw and D. Gross, Princeton Univ. Press 1972, p. 97; J.C. Collins, A. Duncan and S.D. Joglekar, Phys. Rev. D16 (1977) 438.

16. This result is also obtained by G.H. Savvidy, Phys. Lett. 71B, (1977), 133.

17. R. Roskies, Phys. Rev. D15 (1977), 1731.

18. M. Halpern, Phys. Rev. D16 (1977), 1798.

GAUGE SPECIFICATION IN A NON-ABELIAN GAUGE THEORY

R. Jackiw[+]

Rockefeller University

New York, New York

INTRODUCTION

When Yang-Mills theory was invented a quarter of a century ago, the principle which guided its construction was local gauge invariance - the example of Maxwell theory was extended to non-Abelian symmetry. In the ensuing years that the model has been studied, we have recognized the electromagnetic analogy to be inadequate; the non-Abelian gauge theory differs importantly from the Abelian one, for example, in the topological structure of the gauge fields. In this lecture I describe how also gauge specification in the non-Abelian theory works quite differently from Maxwell theory. Additionally, I present a particular gauge - the unconstrained temporal gauge - which seems especially appropriate for Yang-Mills theory.

By "gauge specification" we mean the imposition of some condition on the gauge potentials A^{μ}. (We use an anti-Hermitian matrix description, with the coupling constant scaled out, and we discuss the SU(2) theory.)

$$f(A^{\mu}) = 0 \quad . \tag{1}$$

[+] J. S. Guggenheim Fellow

Two questions naturally arise about any gauge specification. Firstly, we may wonder whether it is attainable, i.e., given an arbitrary potential A^μ not satisfying (1), does there exist a gauge function g, such that by a gauge transformation

$$A^\mu \to A'^\mu = g^{-1}A^\mu g + g^{-1}\partial^\mu g \qquad (2)$$

one can attain the gauge potential A'^μ which does satisfy (1)? (Here g is a unitary 2 x 2 matrix, with unit determinant.) Secondly, we can ask whether the gauge specification is unique, i.e., given one potential A^μ which does satisfy (1), can one find a gauge transform of it, A'^μ, which continues to satisfy (1)? If not, the gauge specification is unique; otherwise (1) must be supplemented with further conditions in order to attain uniqueness.

A gauge condition need not be specified if one merely wants to solve the classical Yang-Mills equations. However, if a Hamiltonian formulation of the theory is desired, for instance as a pre-requisite to quantization, some gauge specification is necessary. Also an enumeration of physical degrees of freedom requires the elimination of unphysical, gauge degrees of freedom; this too is achieved by imposing a gauge condition.

All of us know how these considerations work for electrodynamics, and here we examine them in the context of SU(2) Yang-Mills theory. Specifically we shall focus on the Coulomb gauge condition, there being several reasons for this choice: the Coulomb gauge is employed most frequently and most naturally in electrodynamics; consequently it has also been used in the non-Abelian models. Indeed the first attenpts at quantization of Yang-Mills theory were in this gauge.[1] More recently Mandelstam[2] and Gribov[3] have argued that there are difficulties associated with the Coulomb gauge condition, which may affect in a non-trivial way the physical content of the theory. To clarify these problems, my colleagues and I decided to study carefully the Coulomb gauge, and the results of our investigation are the following.[4]

COULOMB GAUGE IN YANG-MILLS THEORY

Specification of the Coulomb gauge begins by imposing a transversality condition on the gauge potentials.

$$\vec{\nabla}\cdot\vec{A} = 0 \ .$$

But we know that even in the Maxwell theory the transversality condition is insufficient to fix the gauge completely; Eq. (3) must be supplemented by boundary conditions. (In the Abelian theory, a gauge transformation takes $\vec{A}$ into $\vec{A}+\vec{\nabla}\theta$; hence the latter remains transverse provided $\nabla^2\theta=0$. Since non-constant harmonic functions are never regular everywhere, this additional gauge freedom is removed by demanding that $\vec{A}$ be everywhere regular). We must expect therefore that similarly in the non-Abelian theory boundary conditions are needed.

In order to determine which boundary conditions should be imposed, let us consider first vacuum gauge configurations, i.e. configurations with vanishing field strength. Clearly a vanishing vector potential is transverse; it will be the unique transverse vacuum potential, provided its gauge transforms are disallowed. Thus we need to impose conditions such that the differential equation $\vec{\nabla}\cdot(g^{-1}\vec{\nabla}g) = 0$ has no admissible solutions for g. A study of this equation reveals that the transverse vacuum is uniquely the vanishing potential when the following boundary conditions are imposed on $\vec{A}$,[3,4,5]

$$\vec{A} \quad \text{everywhere regular} \ , \tag{4a}$$

$$\lim_{r\to\infty} r\vec{A} = 0 \ . \tag{4b}$$

Eq. (4a), familiar from electrodynamics, must be augmented by

Eq. (4b) to achieve a unique vacuum. Note that the latter can be satisfied in the absence of magnetic monopoles; however if magnetic monopoles are present, physical effects are associated with singular, long-range, $O(1/r)$ vector potentials, and Eq.(4) cannot be imposed on the theory. Since at the present time there is no conclusive evidence that magnetic monopoles play any role in the description of physical reality, we adopt the boundary condition (4) and achieve a unique Coulomb gauge for the vacuum.

What happens with non-vacuum configurations? By continuity, we expect those close to the vanishing potential also to be described by a unique Coulomb gauge. However, as we now show, for sufficiently large vector potentials this is no longer possible, where "sufficiently large" will mean that the configuration is topologically non-trivial and has non-zero Pontryagin index. Recall that the Pontryagin index, given by

$$q = -\frac{1}{16\pi^2}\int d\vec{r}dt \; \mathrm{tr}\, {}^*F^{\mu\nu}F_{\mu\nu} ,$$
$$F^{\mu\nu} = \partial^\mu A^\nu - \partial^\nu A^\mu + [A^\mu, A^\nu] ,$$
$$^*F^{\mu\nu} = \tfrac{1}{2}\varepsilon^{\mu\nu\alpha\beta}F_{\alpha\beta} , \tag{5}$$

takes on integer values for sufficiently well-behaved gauge fields. Since the integrand may be written as a total divergence,

$$\mathrm{tr}\, {}^*F^{\mu\nu}F_{\mu\nu} = \partial_\mu X^\mu ,$$
$$X^\mu = 4\varepsilon^{\mu\alpha\beta\gamma}\mathrm{tr}\left(\tfrac{1}{2}A_\alpha \partial_\beta A_\gamma + \tfrac{1}{3}A_\alpha A_\beta A_\gamma\right) \tag{6}$$

an application of Gauss' theorem allows q to be expressed in terms of surface integrals, provided X^μ, i.e. A^μ, is continuous ,

$$q = -\frac{1}{16\pi^2}\{\lim_{t\to+\infty} \int d\vec{r}X^0 - \lim_{t\to-\infty} \int d\vec{r}X^0 + \int_{-\infty}^{\infty} dt \lim_{r\to\infty} \int d\vec{S}\cdot\vec{X}\} \quad . \tag{7}$$

Let us consider a gauge potential, not necessarily transverse, for which $g\neq 0$, for example one given by

$$\begin{aligned} A^0 &= 0 \quad , \\ A^i &= \eta(t) g^{-1} \partial^i g \quad . \end{aligned} \tag{8}$$

Here η interpolates between 0 and 1 as t ranges from $-\infty$ to $+\infty$: $\eta(-\infty)=0$;$\eta(+\infty)=1$. Consequently at $t=-\infty$ A^μ vanishes, while it is a pure gauge at $t=+\infty$. For g we take a time-independent, unitary, 2x2 matrix of unit determinant sufficiently well-behaved at $r\to\infty$ so that $\vec{A}$ vanishes faster than $1/r$ at large r. As is well-known, such matrices fall into homotopy classes labelled by the integers, and in particular one can choose g so that q=1.[7] It is easily seen that the entire contribution to q comes from the first term in (7); the second vanishes since A^μ vanishes at $t=-\infty$; the third gives no contribution from the spatial surface at infinity due to the rapid decrease of $\vec{A}$ at large r.

Suppose now the configuration (8) is gauge-transformed to the Coulomb gauge, where $\vec{A}$ is tranverse and satisfies (4). The Pontryagin index, given by the gauge invariant formula (5), remains 1. However, the expression (7) cannot yield a non-vanishing value in the Coulomb gauge, since each term vanishes: the last one because of the rapid decrease at spatial infinity due to (4); the first two because at $t=\pm\infty$ the configuration is a vacuum, but the Coulomb vacuum is the null potential. (Of course by dropping the condition (4) at spatial infinity we could describe a non-vanishing q by Eq. (7), but then, as we have stated, even the vacuum is not unique.) Only continuity of the gauge potential is needed to establish (7) from (5); thus we must conclude that sufficiently large gauge fields, carrying non-zero Pontryagin index, do not admit a Coulomb gauge description with continuous gauge potentials.

Detailed investigation[4] shows that the discontinuity is in time; for any fixed time the configuration (8) can be transformed into the Coulomb gauge, but the transformation is not continuous in time. At some time t_o, the Coulomb gauge transverse potential involves a discontinuity of the type

$$\lim_{\varepsilon\to 0} \vec{A}(\vec{r},t_o+\varepsilon) = A_1(\vec{r}) ,$$

$$\lim_{\varepsilon\to 0} \vec{A}(\vec{r},t_o-\varepsilon) = \vec{A}_2(\vec{r}) , \tag{9}$$

with $\vec{A}_1$ and $\vec{A}_2$ transverse and not equal. However, since both are gauge transformations of $-\eta(t_o)g^{-1}\vec{\nabla}g$, they must be gauge transforms of each other. This establishes that the Coulomb gauge is not attainable in terms of continuous gauge potentials, nor is it unique.[8]

When the potentials are discontinuous, the right-hand-side of (7) acquires an additional contribution, and the Pontryagin index is given by

$$q = -\frac{1}{16\pi^2} \int d\vec{r} \text{ discontinuity } \{ X^o\} . \tag{10}$$

Since the discontinuity in the potentials is given by a gauge transformation U

$$A_2^i = U^{-1} A_1^i U + U^{-1} \partial^i U \tag{11}$$

the discontinuity of X^o may be computed from (6). One finds

$$\text{discontinuity } \{ X^o\} =$$
$$-\frac{2}{3}\varepsilon^{ijk}\text{tr}(U^{-1}\partial_i U)(U^{-1}\partial_j U)(U^{-1}\partial_k U) - 2\varepsilon^{ijk}\text{tr}\,\partial_i(\partial_j U U^{-1} A_k) . \tag{12}$$

Thus the Pontryagin index is given precisely by the winding number of the gauge transformation which relates the two different transverse descriptions of the same field configuration.

$$q = \frac{1}{24\pi^2} \int d\vec{r} \varepsilon^{ijk} \mathrm{tr}(U^{-1}\partial_i U)(U^{-1}\partial_j U)(U^{-1}\partial_k U) . \tag{13}$$

The above considerations are general and formal. We have also constructed explicitly, by numerical, machine integration, the gauge transformation which takes a specific vector potential in the temporal gauge (where A^0 vanishes) to the Coulomb gauge.[4] The result is presented schematically in Fig. 1, which exhibits the qualitative relationship between a vector potential in the temporal gauge, $\vec{A}_T$, (plotted on the vertical axis) and the corresponding vector potential in the Coulomb gauge, $\vec{A}_C$,(plotted on the horizontal axis). Vanishing $\vec{A}_T$ leads to a unique, vanishing $\vec{A}_C$. When $\vec{A}_T$ increases in magnitude, it continues to determine a unique Coulomb potential, until the critical value (a) is reached, when two Coulomb potentials are gauge equivalent to the same temporal potential - one on the branch continuously connected to the origin, the other on the discontinuous branch. For the range of $\vec{A}_T$ between (a) and (b) one has three Coulomb gauge potentials, and the multiplicity keeps increasing with increasing $\vec{A}_T$. (It even becomes infinite for definite values of $\vec{A}_T$.) This multivaluedness may be eliminated by requiring that only the unique Coulomb potential which is continuously connected with the vacuum be kept. However, beyond the value (c) in the temporal gauge, there are no solutions for the Coulomb gauge on the branch through the origin, and one must pass discontinuously to the second branch. The situation now repeats itself as $\vec{A}_T$ is increased further. (The number of branches is infinite; only three are plotted in Fig. 1.) Thus the explicit calculation confirms entirely the general discussion.

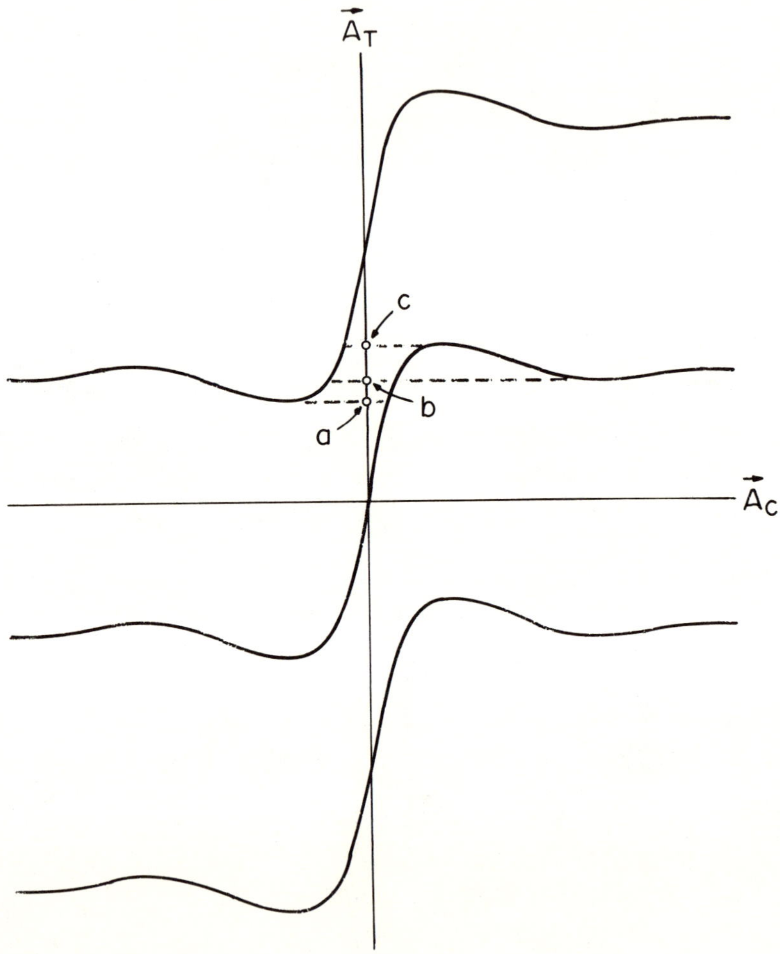

Fig. 1 Schematic portrayal of the gauge equivalence between a vector potential in the temporal gauge ($\vec{A}_T$) and one in the Coulomb gauge ($\vec{A}_C$).

PHYSICAL IMPLICATION

One may hope that the pecularities of the Coulomb gauge in the non-Abelian theory have something to do with confinement.[9] We have not been able to establish such a connection. The only apparent physical implication of the necessary discontinuity and non-uniqueness of the Coulomb gauge - consequences of non-trivial topological structures in a non-Abelian theory - is that the Yang-Mills theory has a degree of freedom not encountered in an Abelian model. This is just the angle, whose existence was previously established in the temporal gauge.[7,10,11]

The angle arises in the present context when the quantal Yang-Mills theory is formulated, in the Schrödinger picture, in terms of a wave-functional $\psi(\vec{A})$, with $\vec{A}$ transverse. For sufficiently small $\vec{A}$, the description is conventional. However, when $\vec{A}$ reaches a definite magnitude, non-unique, gauge-equivalent $\vec{A}$'s describe the same physics. The wave-functional for two such potentials must be equal in modulus, but a difference in phase can be accommodated. This phase is then the genesis of the angle.

BETTER GAUGE CHOICE FOR YANG-MILLS THEORY

We have seen that in a non-Abelian gauge theory the attempt to specify completely the Coulomb gauge fails. We do not find this failure signalling the presence of unexpected features in the theory; rather the pathologies (non-unique, discontinuous potentials) serve to reproduce phenomena (non-trivial Pontryagin index, non-trivial angle) that are easily seen in other gauges, for example the temporal gauge. Admittedly, in the Coulomb gauge, these physically important phenomena appear in an obscure and complicated way. Consequently it is natural to seek a gauge which combines the virtues of the temporal gauge (simplicity of the canonical formalism) with those of Coulomb gauge (unconstrained canonical variables). However, such a search must take into account

Singer's result: on S_4 attempts to specify a gauge completely and uniquely must fail.[12]

I conclude this lecture with a discussion of a gauge which has recently been constructed, and which seems to be particularly appropriate for non-Abelian theories.[13] (In the following the coupling constant e and Planck's constant $\hbar$ are retained, so that dependence on them is readily appreciated.) We begin by recalling the Hamiltonian formulation in the temporal gauge. The Hamiltonian is

$$H = \int d\vec{r}\, \tfrac{1}{2}\vec{E}_a \cdot \vec{E}_a + V(A) ,$$

$$V(A) = \int d\vec{r}\, \tfrac{1}{2}\vec{B}_a \cdot \vec{B}_a ,$$

$$E_a^i = F_a^{io} = -\dot{A}_a^i ,$$

$$B_a^i = -\tfrac{1}{2}\varepsilon^{ijk} F_a^{jk} = \varepsilon^{ijk}\left(\partial_j A_a^k - \tfrac{1}{2}e\,\varepsilon_{abc} A_b^j A_c^k\right) ,$$

$$A_a^o = 0 , \tag{14}$$

and we may identify the canonical variables to be A_a^i and $\Pi_a^i = -E_a^i = \dot{A}_a^i$. The canonical commutation relations between these imply Hamiltonian equations of motion which reproduce the spatial component of the Yang-Mills equation,

$$\dot{\vec{E}}_a - \vec{\mathcal{D}}_{ac}(A) \times \vec{B}_c = 0 ,$$

$$\vec{\mathcal{D}}_{ac}(A) \equiv \vec{\nabla}\, \delta_{ac} - e\varepsilon_{abc} \vec{A}_b . \tag{15}$$

However, the time-component of the field equation – the non-Abelian version of Gauss' law –

$$G_a \equiv \vec{\mathcal{D}}_{ac}(A) \cdot \vec{E}_c = 0 \tag{16}$$

does not emerge as a Hamiltonian equation motion, in the temporal gauge. It is incorporated in the following way. Observe that the Hamiltonian (14) possess a Noether symmetry as a consequence of the invariance of V(A) under spatial gauge transformations, which preserve $A_a^0=0$,

$$\vec{A}_a \to \vec{A}'_a = T_{ab}\vec{A}_b - \frac{1}{2e}\varepsilon_{abc}(\vec{\nabla}T\tilde{T})_{bc} \quad ,$$

$$\vec{E}_a \to \vec{E}'_a = T_{ab}\vec{E}_b \quad . \tag{17}$$

Here T is a time-independent, 3x3 rotation matrix, $\tilde{T}$ is its transpose. The generators of this transformation

$$G_a = \vec{\mathcal{D}}_{ac}(A)\cdot\vec{E}_c \quad ,$$

$$[G_a(\vec{x}),\, G_b(\vec{y})] = i\ \varepsilon_{abc}G_c(\vec{x})\ \delta\ (\vec{x}-\vec{y}) \quad , \tag{18}$$

commute with H. Therefore, we regain the physical Yang-Mills theory as that subset of the Hamiltonian theory for which the constants of motion G_a vanish. Specifically in the quantum theory, a constraint on the states is imposed

$$G_a\psi = 0 \tag{19}$$

which requires that the physical states be invariant under the gauge transformation (17).[14]

Precisely because of the subsidiary condition (19), the temporal gauge does not make explicit the canonical structure of a gauge theory and one wants to solve the constraint. The procedure is to recognize that V(A) possesses "ignorable coordinates" - that is why a Noether symmetry is present in the problem. One therefore passes to a coordinate system in which the ignorable coordinate is one of the canonical variables, and G_a is simply related to the (conserved) conjugate momentum. In that coordinate system, the Hamiltonian will depend explicitly on G_a, but not on the ignorable variable, and Gauss' law can be simply satisfied by setting G_a to zero. (The situation is very analogous to the way s waves in a Schrödinger problem with radial symmetry are studied: one passes to a radial coordinate system and sets the angular momentum to zero).

It is obvious how to effect the above in an Abelian theory: the ignorable coordinate is the longitudinal component of $\vec{A}$, and elimination of the corresponding canonical momentum is equivalent to the use of the Coulomb gauge, which is a "natural" gauge for the problem. It is equally clear, that the Coulomb gauge is not in any way "natural" for the non-Abelian problem; V(A) is not independent of $(\vec{A})_{\text{longitudinal}}$. It is therefore not surprising that various obscurities arise when the problem is forced into the Coulomb gauge.

In order to identify the ignorable coordinate and solve the constraint (19) in a systematic fashion for the Yang-Mills theory, we work in a Schrödinger picture, where $\vec{E}_a$ is diagonalized and $\vec{A}_a$ is realized by functional differentiation. Next, observe that the $\vec{E}_a$'s may be considered as components of a 3x3 matrix E: $E^i_a=(E)_{ia}$. Since any 3x3 matrix may be decomposed into the product of an orthogonal matrix, diagonal matrix and another orthogonal matrix, we may write

$$E^i_a=R_{i\alpha}(\theta^J)\hat{E}_{\alpha\beta}\tilde{R}_{\beta a}(\theta^T) \ ,$$

$$\hat{E}_{\alpha\beta} = - \delta_{\alpha\beta}\pi_\alpha \quad \text{(no summation on } \alpha). \tag{20}$$

Here the orthogonal matrices have been parametrized by three angles θ_n,

$$R_{nm}(\theta) = \cos\theta(\delta_{nm} - \hat{\theta}_n\hat{\theta}_m) + \hat{\theta}_n\hat{\theta}_m + \varepsilon_{nmo}\hat{\theta}_o \sin\theta \quad ,$$

$$\theta = \sqrt{\theta_n\theta_n}, \quad \hat{\theta}_n = \theta_n / \theta \; . \tag{21}$$

A gauge transformation multiplies $R(\theta^T)$ by another orthogonal matrix; the remaining quantities, θ^J and π, are gauge-invariant.[15] By the chain rule it follows that

$$A^i_a = - i\hbar \frac{\delta}{\delta E^i_a} = R_{i\alpha}(\theta^J) R_{a\beta}(\theta^T) \hat{A}_{\alpha\beta} \; , \tag{22}$$

where the operator $A_{\alpha\beta}$ is

$$\hat{A}_{\alpha\beta} = \delta_{\alpha\beta} i\hbar \frac{\delta}{\delta\pi_\alpha} + X^n_{\alpha\beta} M^{-1}_{nm}(\theta^J) i\hbar \frac{\delta}{\delta\theta^J_m} + Y^n_{\alpha\beta} M^{-1}_{nm}(\theta^T) i\hbar \frac{\delta}{\delta\theta^T_m} \; . \tag{23}$$

Here we have defined the matrix $M(\theta)$ by

$$\tfrac{1}{2}\varepsilon_{cab}\,(\hat{R}\delta\hat{R})_{ab} = \delta\theta_d M_{dc} \quad . \tag{24}$$

The other quantities appearing in (23) are constructed as follows,

$$X^n_{\alpha\beta} = S^n_{\alpha\beta}\pi_\beta , \; Y^n_{\alpha\beta} = \pi_\alpha S^n_{\alpha\beta} \text{ (no summation on } \alpha \text{ or } \beta). \tag{25a}$$

The only non-vanishing components of S are

$$S^3_{12} = S^3_{12} = \frac{1}{-\pi_1^2 + \pi_2^2} ,$$

$$S^1_{23} = S^1_{32} = \frac{1}{-\pi_2^2 + \pi_3^2} ,$$

$$S^2_{31} = S^2_{13} = \frac{1}{-\pi_3^2 + \pi_1^2} . \qquad (25b)$$

A more intrinsic formulation emerges when we consider the combinations which occur in the formula for A^i_a ,

$$\mathcal{J}_n = M^{-1}_{nm} (\theta^J) \, i\hbar \frac{\delta}{\delta \theta^J_m} , \qquad (26)$$

$$\mathcal{T}_n = M^{-1}_{nm} (\theta^T) \, i\hbar \frac{\delta}{\delta \theta^T_m} . \qquad (27)$$

The pairs $R^J \equiv R(\theta^J), \mathcal{J}$ and $R^T \equiv R^T(\theta^T), \mathcal{T}$ are not canonically conjugate, but their commutators exhibit the rotational action of these quantities ,

$$[\mathcal{J}_n(\vec{x}), R^J_{ab}(\vec{y})] = i\hbar \, R^J_{ac}(\vec{x}) \varepsilon_{cbn} \, \delta(\vec{x}-\vec{y}) , \qquad (28a)$$

$$[\mathcal{J}_n(\vec{x}), \mathcal{J}_m(\vec{y})] = -i\hbar \, \varepsilon_{nmp} \mathcal{J}_p(\vec{x}) \, \delta(\vec{x}-\vec{y}) , \qquad (28b)$$

$$[\mathcal{T}_n(\vec{x}), R^T_{ab}(\vec{y})] = i\hbar \, R^T_{ac}(\vec{x}) \, \varepsilon_{cbn} \delta(\vec{x}-\vec{y}) , \qquad (29a)$$

$$[\mathcal{J}_n(\vec{x}),\mathcal{J}_m(\vec{y})] = -i\hbar\,\varepsilon_{nmp}\mathcal{J}_p(\vec{x})\,\delta(\vec{x}-\vec{y}) . \tag{29b}$$

We thus see that the vector potential is described at each point of space by a body-fixed spin and iso-spin angular momentum. Furthermore the spin density $-\varepsilon_{ijk}A^j_a E^k_a$ is readily shown to be $R^J_{ij}\mathcal{S}_j$; while similarly the iso-spin density $-\varepsilon_{abc}A^i_b E^i_c$ is $R^T_{ab}\mathcal{J}_b$. One verifies that these quantities satisfy, as they should, commutation relations of space-fixed angular momentum.

Let us now turn to the constraint equation (19), which becomes a functional differential equation.

$$\vec{\mathcal{D}}_{ac}\left(-i\hbar\frac{\delta}{\delta E}\right)\cdot\vec{E}_c\,\psi(E) = 0 . \tag{30a}$$

The most general solution for ψ (E) is readily obtained;[13,14] it is

$$\psi(E) = \Phi\,(R^J E)\exp - \frac{i}{e\hbar}\,\Omega\,(E) ,$$

$$\Omega(E) = \tfrac{1}{2}\int d\vec{x}\;\;\varepsilon_{abc}\vec{E}_a\cdot(\vec{\nabla}R^T\tilde{R}^T)_{bc} , \tag{30b}$$

where we have used the decomposition (20) for $\vec{E}_a$, and Φ is any functional depending on $\hat{R}^J E$, but not on R^T. In other words, it is possible to make explicit the dependence of the functional on R^T; that dependence is contained in the phase (30b). (In an Abelian theory the result is quite different: one learns that

$$\psi(E)=0 \text{ unles } \vec{\nabla}\cdot\vec{E}=0.]$$

Since all physical states carry the same phase, it may be removed provided all operators 0 of the theory are replaced by $\bar{0}$ = $\exp\{+\frac{i}{e\hbar}\Omega(E)\}\, 0\, \exp\{-\frac{i}{e\hbar}\Omega(E)\}$. Clearly this affects only the gauge potential ,

$$\bar{E}^i_a = E^i_a \quad ,$$

$$\bar{A}^i_a = A^i_a + \frac{i}{e\hbar}[\Omega(E), A^i_a]$$

$$= A^i_a - \frac{1}{e}\frac{\delta}{\delta E^i_a}\Omega(E) \, ,$$

$$= R^T_{ab} R^J_{i\alpha}[\hat{A}_{\alpha b} - \frac{1}{e} Y^m_{\alpha b}\,\partial_j (R^J\hat{E})_{jm}] - \frac{\varepsilon_{abc}}{2e}(\partial_i R^T \tilde{R}^T)_{bc} \tag{31}$$

This formula shows that $\bar{A}^i_a$ is a gauge transformation of another vector potential $\mathcal{a}^i_a$.

$$\mathcal{a}^i_a = R^J_{i\alpha}\hat{\mathcal{a}}_{\alpha a} \, , \tag{32}$$

where $\hat{\mathcal{a}}_{\alpha\beta}$ is just the same as $\hat{A}_{\alpha\beta}$, except that $\mathcal{Y}_n$ is decreased by $\frac{1}{e}\partial_j(R^J\hat{E})_{jn}$. The matrix which effects the gauge transformation is precisely R^T , which also occurs in the definition of $\bar{E}^i_a = E^i_a$. Hence we arrive at the important result that all operators of the theory, when operating on the physical states $\phi\,(R^J\hat{E})$, are gauge transforms with R^T. Gauge invariant quantities, like the Hamiltonian, are independent of R^T, which therefore is recognized as the ignorable coordinate. (This statement is obviously true for the classical theory. In the quantum theory R^T does not commute with $\vec{\mathcal{a}}_a$, and one must keep track of reordering terms which arise when various R^T's are combined to form $R^T\tilde{R}^T=I$. It is easy to show that R^T does disappear in the quantum theory, but non-classical additions are present in the Hamiltonian. These additions involve the delta function and its derivatives, evaluated at the origin; the reordering terms are infinite. We do not present here the calculation of these objects; we believe that they are inessential, since in a properly renormalized theory all infinities vanish. Consequently it should be possible to regulate the theory so that the infinite reordering terms are absent).

The Hamiltonian is independent of R^T; it depends on $\mathcal{T}_n$ through $\vec{\mathcal{a}}_a$. The functional $\phi\ (R^J\hat{E})$ also is independent of R^T and θ^T, hence it is annihilated by $\mathcal{T}_n$, which consequently may be set to zero. This is equivalent to satisfying Gauss' law since

$$\bar{G}_a = R^T_{ab}\mathcal{T}_b \ . \tag{33}$$

In summary, we find that an unconstrained formulation of Yang-Mills theory involves states depending on $R^J\hat{E}$. The Hamiltonian is

$$H = \int d\vec{x}\ \tfrac{1}{2}[e^2\pi_a^2 + \frac{1}{e^2}\ \vec{B}_a\cdot\vec{B}_a]\ , \tag{34a}$$

$$B^i_a = \varepsilon_{ijk}(\partial_j\mathcal{a}^k_a - \tfrac{1}{2}\varepsilon_{abc}\mathcal{a}^j_b\mathcal{a}^k_c)\ , \tag{34b}$$

$$\mathcal{a}^i_a = R^J_{i\alpha}\hat{\mathcal{a}}_{\alpha a}\ , \tag{34c}$$

$$\hat{\mathcal{a}}_{\alpha\beta} = \delta_{\alpha\beta}\ i\hbar\frac{\delta}{\delta\pi_\alpha} + X^n_{\alpha\beta}\mathcal{J}_n - Y^n_{\alpha\beta}\ \partial_j(R^J\hat{E})_{jn}\ ,$$

$$\text{(no summation on } \alpha) \tag{34d}$$

$$\hat{E}_{\alpha\beta} = -\ \delta_{\alpha\beta}\pi_\alpha\ . \qquad \text{(no summation on } \alpha) \tag{34e}$$

We have rescaled π to $e\pi$ to make the dependence on the coupling constant explicit and simple. Note that $\hat{a}_{\alpha\beta}$ is not symmetrized between the multiplicative operators and the derivative ones. The expression is nevertheless Hermitian with respect to the new measure

$$dE^i_a = dR^J \tilde{R}^J \quad dR^T \tilde{R}^T \quad d\,\pi_a$$

$$(\pi_1^2 - \pi_2^2)\,(\pi_2^2 - \pi_3^2)\,(\pi_3^2 - \pi_1^2) \qquad . \qquad (35)$$

The structure of the gauge-invariant Hamiltonian (34) bears no resemblance to the one which arises in the lowest order of perturbative expansions in the conventional approach, the latter being simply three non-interacting Maxwell theories. Also the exact gauge dependence of states (30b) involves the phase $\frac{1}{e}\,\Omega(E)$ exhibiting a non-perturbative 1/e factor. These states also are quite different from the perturbative ones, which in lowest order are proportional to $\delta\,(\vec{\nabla}\cdot\vec{E}_a)$.

We do not know at present whether our Hamiltonian can serve as a useful starting point for practical calculations. We note that the transformation to angular variables, which we have carried out, has singular points, a circumstance already familiar from the radial Schrödinger equation. Here the singular points correspond to a degenerate electric field for which two or more eigenvalues of $E_{i\alpha}$ are equal in magnitude. In the effective Hamiltonian $\mathcal{J}_n$ and $\partial_j(R^J\hat{E})_{jn}$ serve as a "centrifugal barrier" which in general prevents the dynamical system from achieving degenerate configurations. Only when $\mathcal{J}_n$ and $\partial_j(R^J\hat{E})_{jn}$ are in special relationships with each other in a degenerate configuration possible. These singular points appear to be, in the present context, a manifestation of the generalized gauge ambiguities, discussed by Singer.[12] However, they do not seem to produce any novel effects in the theory.

REFERENCES

1. J. Schwinger, Phys. Rev. 125, 1043 (1962).
2. S. Mandelstam, Lecture at American Physical Society Meeting, Washington DC, April 1977.
3. V. Gribov, Lecture at 12th Winter School, Leningrad, 1977.
4. R. Jackiw, I. Muzinich and C. Rebbi, Phys. Rev. D 17, 1576 (1978) For a related investigation, see Reg. 5; difficulties with the Coulomb gauge were also noted by the authors of Ref. 6.
5. M. Ademollo, E. Napolitano and S. Sciuto, CERN preprint.
6. N. Christ, A. Guth and E. Weinberg, Nucl. Phys. B114, 61 (1976)
7. R. Jackiw and C. Rebbi, Phys. Rev. Lett. 37, 172 (1976).
8. Further analysis (see Ref. 4) demonstrates that configurations with zero Pontryagin index also have time discontinuities in a Coulomb description, provided the winding number, defined as $n(t) = -\frac{1}{16\pi^2}\int d\vec{r}X^0(\vec{r},t)$, changes with time by at least one unit.
9. Such a hope is implicit in Gribov's discussion; Ref. 3. Explicit investigations, which follow Gribov's ideas, are by C. Bender, T. Eguchi and H. Pagels, Phys. Rev. D (in press). It is difficult to assess the relevance of these calculations. They explore properties of the Coulomb propagator, which, however, in a non-Abelian theory, is neither gauge invariant nor directly related to the interaction energy of test changes. differences between the two in perturbation theory have been identified by S. Davis and F. Feinberg (in preparation).
10. G. 't Hooft, Phys. Rev. D14, 3432 (1976).
11. C. Callan, R. Dashen and D. Gross, Phys. Lett. B63, 334 (1976)
12. I. Singer, Berkeley preprint. Because Singer works on S_4, rather than in Minkowski space, it is not clear how widely relevant is his result to physical considerations.
13. J. Goldstone and R. Jackiw. Phys. Lett. B (in press).
14. We ignore the phase discussed in Refs. 7, 10 and all. For a more careful treatment, which takes the phase into account, see

Ref. 13.

15. Parametrization of the electric field as a gauge choice has also been advocated by M. Halpern, Phys. Rev. D16, 1978 (1977).

QUANTIZED ELECTRIC FLUX TUBES IN QCD

Itzhak Bars*

Yale University

New Haven, Connecticut

INTRODUCTION

I am going to present a formulation [1] of QCD which leads to the much speculated about string-like quantized electric flux tubes. My approach is heavily based on a new set of quantized variables. The new variables satisfy commutation rules which are equivalent to the canonical commutation rules of the ordinary gauge potential and electric field in an axial gauge. The latter variables are explicitly constructed in terms of the new variables.

The method consists of finding the complete set of eigenstates of the electric field term in the Hamiltonian. This provides a Hilbert space in which the remaining part of the Hamiltonian can, at least in principle, be computed. The new Hilbert space is quite different from the ordinary Fock space. The Fock space is commonly used in perturbative calculations, in powers of the coupling constant g, at short distances. The new Hilbert space is suited to investigate large distance behavior and may be taken as the starting point for perturbative calculations in powers of (1/g) when

*Alfred P. Sloan Foundation Fellow

g is large. This is consistent with the results of the renormalization group for non-Abelian theories which indicate that the effective coupling constant grows with distance.

As a result of the diagonalization, I find that the electric field is quantized and forms string-like flux tubes, as will be described shortly. It should be emphasized that this is not a classical result. It follows from the quantum theory as a direct consequence of the canonical commutation rules.

The role played by the new variables is reminiscent of the O(4) treatment of the Hydrogen atom. The new variables bring out an algebraic structure corresponding to 3 local commuting SU(N) groups. It is then possible to use the algebraic properties of Lie algebras and their representations to find the eigenstates.

The computation has to be done with a short-distance cutoff which is usually needed to define a field theory, as one multiplies singular field operators at the same point. The cutoff then becomes a measure of the thickness of the flux tube. It is a very interesting and open question whether it is possible to remove the cutoff, after renormalization, to all orders in (1/g). It is, of course, known that for QCD the same short distance cutoff is removable by renormalization in ordinary perturbation theory.

My results, but not my methods, are analagous to those of lattice theories [2], at least in lowest order in (1/g).

II. DEGREES OF FREEDOM

The theory is defined by the familiar QCD Lagrangian

$$L = -\frac{1}{4} F_{\mu\nu}^{a} F^{\mu\nu}_{a} + \bar{\psi} (i \not{D} - m) \psi \quad . \tag{2.1}$$

The symmetric energy-momentum tensor $\theta_{\mu\nu}$ leads to the gauge invariant Hamiltonian density

$$H = \frac{1}{2} (\vec{E}^2 + \vec{B}^2) + \psi^+ (-i \vec{\alpha} \cdot \vec{D} + \beta m) \psi. \tag{2.2}$$

The boundary conditions, $F^a_{\mu\nu} \to 0$, $\psi^i_a \to 0$ as $|\vec{x}| \to \infty$, must be supplied to complete the definition of the theory. However, since there is a constraint

$$\frac{1}{g} \vec{D}.\vec{E}^a = \psi^+ \frac{\lambda^a}{2} \psi \quad , \tag{2.3}$$

one must choose a gauge and identify the canonical degrees of freedom.

I will use the axial gauge $A^a_3 = 0$ which is very convenient for solving the constraint (2.3). This gauge has been discussed in the literature many times [3]. It has been reanalyzed recently with particular attention to boundary terms [4]. Classically, there are no Gribov ambiguities [5] in the axial gauge once the boundaries are specified by exhausting all gauge freedoms [4]. Quantum mechanically there are peculiar problems as was noticed before [6]. These are discussed and resolved in refs. (4,1) and (7).

In the $A^a_3 = 0$ gauge the canonical variables are (A^a_1, E^a_1), (A^a_2, E^a_2), (ψ, ψ^+) while A^a_o and $E^a_3 = -\partial_3 A^a_o$ are dependent operators. E^a_3 can be solved from the constraint equation (2.3)

$$\frac{1}{g} E^a_3 (x_1 x_2 x_3 t) = \int_{x_3}^{\infty} dx'_3 \; G^a_3 (x_1 x_2 x'_3 t) \tag{2.4}$$

where

$$G^a_3 = -\frac{1}{g} D_1 E^a_1 + \frac{1}{g} D_2 E^a_2 + \psi^+ \frac{\lambda^a}{a} \psi \quad . \tag{2.5}$$

The boundary condition $E^a_3 (x_1, x_2, x_3 = -\infty, t) = 0$ is the source of the problem mentioned above. Classically this condition can be imposed without difficulty. However, quantum mechanically the operator

$$\frac{1}{g} E^a_3 (x_3 = -\infty) = \int_{-\infty}^{\infty} dx'_3 \; G_3 (x'_3) \tag{2.6}$$

has non-trivial commutation rules with the canonical variables, indicating that the boundary condition $\int_{-\infty}^{\infty} dx_3' \, G_3^a \, (x_3') = 0$ should be considered as a constraint. Therefore (A_i^a, E_i^a) etc., are not all independent variables. One may resort to two possible solutions: (1) One can work in a larger Hilbert space by considering all (A_i^a, E_i^a) independent, but select the physical sector by imposing the constraint on the states [6,4,1], or (2) One can specify the gauge further, and solve explicitly this remaining constraint[7]. I choose the first alternative as it turns out to be rather simple to implement [1]. Actually, had one ignored the problem completely, but used the gauge invariant Hamiltonian in Eq. (2), the unphysical states would be removed automatically from the spectrum since their energy would be ∞ (see below). This observation is further elaborated upon in ref. (4).

One is now in a position to quantize the theory canonically in the standard way ($i = 1,2$; $a = 1,2\ldots,N^2-1$; $\alpha = 1,2,3,4$; $k=1,2,\ldots N$):

$$[A_i^a\,(\vec{x}),\; E_j^b\,(\vec{x}\,')] = i\,\delta^{ab}\delta_{ij}\;\delta\,(\vec{x}-\vec{x}\,') \tag{2.7}$$

$$\{\psi_\alpha^K\,(\vec{x}),\; \psi_\beta^\ell\,(\vec{x}\,')\} = \delta^{K\ell}\,\delta_{\alpha\beta}\;\delta(\vec{x}-\vec{x}\,') \tag{2.8}$$

while all other canonical commutators are set equal to 0. To do computations one must work in a Hilbert space in which a complete set of observables are diagonalized simultaneously. Because of asymptotic freedom, perturbative calculations in powers of a small effective coupling constant g may be done at short distances. Then the appropriate Hilbert space is provided by the familiar Fock space. However, at large distances the effective coupling constant is large according to the renormalization group. Therefore, it is extremely inconvenient to attempt to do large distance calculations in Fock space. A more convenient Hilbert space may be provided by diagonalizing the electric field term in the Hamiltonian

$$H_o = \int d^3x \, \mathcal{H}_o = \int d^3x \frac{1}{2} \left(E_1^2 + E_2^2 + E_3^2 \right) . \tag{2.9}$$

In such non-perturbative Hilbert space the full $H = H_o + H_1$ may be computed at least in principle. I will argue shortly that H_o dominates in the strong coupling limit. Thus this may be taken as the starting point for a strong coupling expansion of the theory. I will exhibit the eigenstates of H_o in the large distance limit, and demonstrate that the electric field forms quantized flux tubes in the form of closed and open strings with quarks at the ends.

First, it is important to identify some commuting observables. The canonical rules lead to the algebra

$$[G_3^a(\vec{x}), G_3^b(\vec{x}')] = if^{abc} G_3^c(\vec{x}) \, \delta(\vec{x}-\vec{x}') , \tag{2.10}$$

$$[E_3^a(\vec{x}), E_3^b(\vec{x}')] = igf^{abc} \delta(x_1'-x_1) \, \delta(x_2'-x_2) \{\theta(x_3'-x_3) \, E_3^c(\vec{x}') + \theta(x_3-x_3') \, E_3^c(\vec{x})\} , \tag{2.11}$$

$$[E_3^a(\vec{x}), E_i^b(\vec{x}')] = igf^{abc} \delta(x_1'-x_1) \, \delta(x_2'-x_2) \, \theta(x_3'-x_3) E_i^c(\vec{x}') . \tag{2.12}$$

From these equations it is straightforward to derive that E_1^2, E_2^2, E_3^2 and G_3^2 commute with each other at all points in space. Therefore, $H_o = \int d^3x \, \mathcal{H}_o$ consists of commuting pieces which can be diagonalized at each space point independently. Hence the eigenstates of H_o can be written in the form of a direct product

$$\prod_{\vec{x}_n} |e_1^2(\vec{x}_n), e_2^2(\vec{x}_n), e_3^2(\vec{x}_n), g_3^2(\vec{x}_n), \ldots> \tag{2.13}$$

where the dimensionless numbers $(e_I^2, g_3^2, \ldots)$ are, up to a scale (to be specified below), the eigenvalues of the operators $(E_I^2, G_3^2, \ldots)$ at the point $\vec{x}_n$. The complete set of commuting observables is larger as will be seen soon. My task is to insure that the states (13) form a basis for the canonical commutation rules. This will give restrictions on the allowable values for the labels $(e_I^2, G_3^2, \ldots)$. For this purpose it is extremely useful and illuminating to introduce a new set of variables which convert the problem to a group theoretical one. It will then be sufficient to use properties of Lie algebras to arrive at the desired result.

III. New Variables

In fig. 1 the three possible gauge conditions $A_3 = 0$ or $A_1' = 0$ or $A_2'' = 0$ are considered. The gauge transformation from $A_1' = 0$ to $A_3 = 0$ is denoted by the unitary operator $B_{13}^{k\ell}$ $(B_{13}^+ B_{13} = 1)$. Similarly B_{23} is the transformation from $A_2'' = 0$ to $A_3 = 0$ and $B_{12} = B_{13}^+ B_{23}$ from $A_1' = 0$ to $A_2'' = 0$.

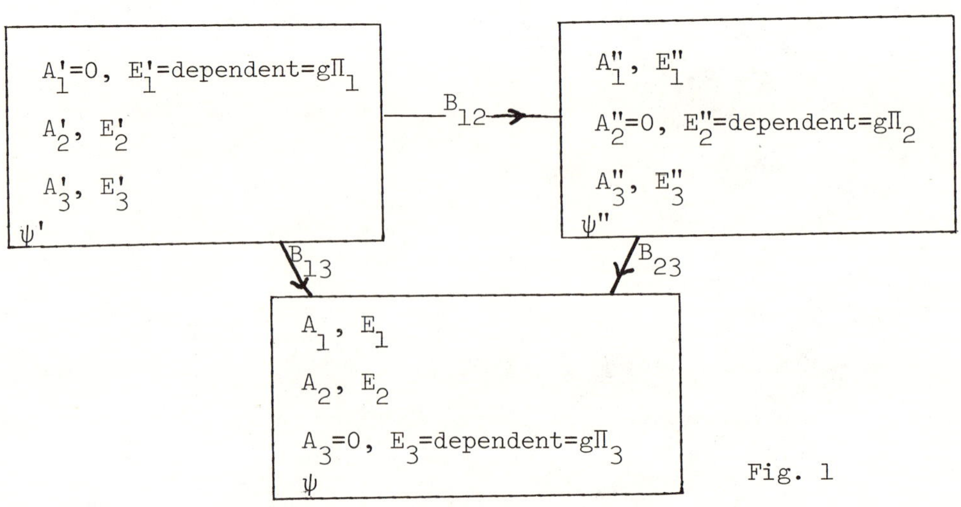

Fig. 1

In the $A_3 = 0$ gauge (A_1, E_1), (A_2, E_2) satisfy canonical commutation rules while E_3 is a dependent operator. If we write $E_3=g\Pi_3$ it can be seen from Eqs. (2.11) and (2.12) that Π_3 satisfies g-independent commutation rules. Similarly, in the $A_1' = 0$ (or $A_2'' = 0$) gauge $E_1' = g\Pi_1$ (or $E_2'' = g\Pi_2$) is a dependent operator and Π_1^a (or Π_2^a) is expected to have commutation rules similar to those of Π_3^a except for the replacement $\theta(x_1'-x_1)$(or $\theta(x_2'-x_2)$) instead of $\theta(x_3'-x_3)$ etc. This expectation will be confirmed shortly.

For a general gauge transformation U one can write

$$A_\mu = U^+ (A_\mu' + \frac{i}{g}\partial_\mu)U \quad , \quad F_{\mu\nu} = U^+ F'_{\mu\nu} U \tag{3.1}$$

where $A_\mu = \frac{1}{2}\lambda^a A_\mu^a$, $F_{\mu\nu} = \frac{1}{2}\lambda^a F_{\mu\nu}^a$ etc. Thus,

when $U = B_{13}$ or B_{23} eq. (3.1) leads to

$$A_1 = \frac{i}{g} B_{13}^+ \partial_1 B_{13} \qquad A_2 = \frac{i}{g} B_{23}^+ \partial_2 B_{23} \tag{3.2}$$

$$E_1 = g\!:\! B_{23}^+ \Pi_1 B_{13}\!: \qquad E_2 = g\!:\! B_{23}^+ \Pi_2 B_{23}\!: \tag{3.3a}$$

The notation :(...): in eq. (3.3a) means

$$:B^+\Pi B: = \Pi^a B^+ \frac{\lambda^a}{2} B = B^+ \frac{\lambda^a}{2} B\Pi^a = B^+\Pi B + \frac{1}{2}(N-\frac{1}{N})\theta(o)\delta^{(2)}(0) \,. \tag{3.3b}$$

This insures that this is a traceless operator and is in accordance with the commutation rules specified below.

My purpose is to use the unitary group variables B_{13} and B_{23} as the independent variables (they have the same number of degrees of freedom as A_1 and A_2). It can be checked that the canonical commutation rules $[A_1^a, A_1^b] = 0 = [E_1^a, E_1^b]$, $[A_1^a, E_1^b] = i\delta(\vec{x}-\vec{x}')$ etc., are reproduced as follows: a) All components of $B_{13}^{k\ell}, B_{23}^{k\ell}$, $B_{13}^{+k\ell}$, $B_{23}^{+k\ell}$ commute with each other, b) All variables with label "1" commute with all variables with label "2", c) B and Π commute with the quark variables ψ, d) The non-trivial commutators are

$$[\Pi_1^a(\vec{x}) , B_{13}^{k\ell} (\vec{x}')] = - (\frac{\lambda^\alpha}{2} B_{13}(\vec{x}'))^{k\ell} \theta(x_1'-x_1) \delta(x_2'-x_2) \delta (x_3'-x_3) \tag{3.4}$$

and the hermitian conjugate equation

$$[\Pi_1^a(\vec{x}), B_{13}^{+\,k\ell} (\vec{x}')] = (B_{13}^+(\vec{x}') \frac{\lambda^a}{2})^{k\ell} \theta(x_1'-x_1) \delta(x_2'-x_2) \delta(x_3'-x_3) . \tag{3.5}$$

Using Jacobi identities one arrives at

$$[\Pi_1^a(\vec{x}), \Pi_1^b(\vec{x}')] = if^{abc}\delta(x_2'-x_2) \delta(x_3'-x_3) \{\theta(x_1'-x_1)\Pi_1^c (\vec{x}')+\theta(x_1-x_1') \Pi_1^c (\vec{x})\} . \tag{3.6}$$

Similar equations hold for the variables (Π_2, B_{23}) for which $\theta(x_2'-x_2)$ appears. For the commutation rules to work out the order of the operators must be kept as in eq. (3.3). One also must use the identity.

$$\frac{1}{2} \lambda^a_{k\ell} \lambda^a_{mn} = \delta_{kn}\delta_{m\ell} - \frac{1}{N} \delta_{k\ell}\delta_{mn} = \frac{1}{2} (B^+ \lambda^a B)_{k\ell} (B^+ \lambda^a B)_{mn} . \tag{3.7}$$

With the new variables one obtains $D_1E_1 =:gB_{13}^+ \partial_1\Pi_{13}:$ and $D_2E_2 =:gB_{23}^+ \partial_2\Pi_2B_{23}:$ so that G_3 and E_3 in eqs. (2.4, 2.5) can be written in terms of B and Π. The notation: ( ): means

$$:B^+ \partial \Pi B: = \partial \Pi^a B^+ \frac{\lambda^a}{2} B = B^+\frac{\lambda^a}{2} B \partial\Pi^a = B^+\partial \Pi B - \delta (\vec{o}) \frac{1}{2} (N- \frac{1}{N}) .$$

This is necessary to guarantee the tracelessness of D_1E_1 etc. as it should be.

It is convenient to define G_1^a and G_2^a analogous to G_3^a as

$$G_1^a = -\partial_1 \Pi_1^a \ , \ G_2^a = -\partial_2 \Pi_2^a \ , \ G_3^a = -\partial_3 \Pi_3^a \tag{3.8}$$

so that the constraint equation (2.4) takes the form

$$\lambda^a G_3^a + :B_{31}\lambda^a G_1^a B_{13}: + :B_{32}\lambda^a G_2^a B_{23}: = \rho_3 \quad , \tag{3.9}$$

where the notation $B_{31} = B_{13}^+$ and $B_{32} = B_{23}^+$ is introduced for clarity. The quark color density ρ_3^{ij} can be written in 3 equivalent ways by using eq. (3.7)

$$\rho_3^{ij} = \psi^+ \frac{\lambda^\alpha}{2}\psi (\lambda^a)^{ij} = \psi^+ B_{31} \frac{\lambda^a}{2} B_{13}\psi \, (B_{31}\lambda^a B_{13})^{ij} = \psi^+ B_{32} \frac{\lambda^a}{2} B_{23}\psi$$
$$\times (B_{32}\, \lambda^a \, B_{23})^{ij} \quad , \tag{3.10}$$

exhibiting transformations of the quark field into the gauges $A_1'=0$ and $A_2''=0$. Thus, through eqs. (3.8-3.10) one can solve for Π_3 (using 2.4) in terms of Π_1, Π_2 etc., and vice versa. It is then possible to calculate the commutation rule

$$[\Pi_3^a(\vec{x}) \ , \ B_{13}^{k\ell}(\vec{x}')] = (B_{13}(\vec{x}')\frac{\lambda^a}{2})^{k\ell}\, \delta(x_1'-x_1)\, \delta(x_2'-x_2)\, \theta(x_3'-x_3) \tag{3.11}$$

and similarly for $[\Pi_3^a, B_{23}]$. It is interesting to compare this result to eq. (3.4) and note the symmetry between Π_1^a and Π_3^a in relation to the labels 1,3 on B_{13}. A similar relationship can be noticed between the variables B_{12}, B_{13}, B_{23} on one hand and Π_1, Π_2, Π_3, on the other.

A complete symmetry between the labels 1,2,3 can be exhibited by writing <u>local</u> commutation rules in a compact way for I=1,2,3 (using $B_{JI}=B_{IJ}^+$ and $B_{IJ} = 1$ for I = J)

$$[G_I^a(\vec{x}),\ B_{JK}^{k\ell}(\vec{x}')] = \{\delta_{IJ}(\frac{\lambda^\alpha}{2} B_{JK}(\vec{x}'))^{k\ell} - \delta_{IK}(B_{JK}(\vec{x}')\frac{\lambda^a}{2})^{k\ell}\}\delta(\vec{x}-\vec{x}') \tag{3.12}$$

$$[G_I^a(\vec{x}),\ G_J^b(\vec{x}')] = if^{abc}G_I^c(\vec{x})\ \delta(\vec{x}-\vec{x}')\ \delta_{IJ}\ ; \tag{3.13}$$

all other commutation rules follow from these! I emphasize now the remarkable property that (3.12 - 3.13) are the type of commutation rules found in the non-linear sigma model. Here the B_{IJ} correspond to three "sigma models" written in the form of unitary matrices. The G_I act as the local currents of left or right-handed unitary transformations on these matrices. In the standard SU(N)xSU(N) sigma model denoted by U there are only two generators G_L and G_R which are constrained by $\lambda^a G_L^a = :U^+\lambda^a G_R^a U:$ so that $TrG_L^2 = TrG_R^2$. In this case there are 3 commuting <u>local</u> groups $SU(N)_1 \times SU(N)_2 \times SU(N)_3$ generated locally at each $\vec{x}$ by $G_I^a(\vec{x})$ which obey the constraints eqs. (3.9, 3.10)

$$\sum_{J=1}^{3} B_{IJ}\ \lambda^a G_J^a\ B_{JI} = \lambda^a \rho_I^a \qquad \text{any } I\ , \tag{3.14}$$

$$\rho_I^a = \psi^+ B_{3I}\frac{\lambda^a}{2}\ B_{I3}\psi\ , \tag{3.15}$$

$$B_{IJ} = B_{IK}\ B_{KJ} \qquad \text{(no sum)}\ . \tag{3.16}$$

In this light the electric fields Π_I^a may be interpreted as sums of generators (like sums of angular momenta) along straight lines embedded in 3-dimensions since one can write

$$\Pi_1^a\ (x_1x_2x_3) = \int_{x_1}^{\infty} dx_1'\ G_1^a\ (x_1'x_2x_3)\ ,$$

$$\Pi_2^a\ (x_1x_2x_3) = \int_{x_2}^{\infty} dx_2'\ G_2^a\ (x_1x_2'x_3)\ , \tag{3.17}$$

$$\Pi_3^a\ (x_1x_2x_3) = \int_{x_3}^{\infty} dx_3'\ G_3^a\ (x_1x_2x_3').$$

The boundary conditions $F^a_{\mu\nu}=0$ demand

$$\Pi_1^a\ (-\infty, x_2, x_3) = \Pi_2^a(x_1, -\infty, x_3) = \Pi_3^a(x_1, x_2, -\infty) = 0 \tag{3.18}$$

on physical states.

Having noticed these properties it is important to check self-consistency by computing the commutation rules of the remaining variables in the other gauges. For example by writing A_3', E_3' (see Fig. 1) in the form $A_3' = \frac{i}{g} B_{13}\ \partial_3 B_{31}$ and $E_3' = :gB_{13}\Pi_3 B_{31}:$ one finds $[A_3'^a, E_3'^b] = i\ \delta(\vec{x}-\vec{x}')\ \delta^{ab}$ as it should be, etc.

Finally the Hamiltonian can be expressed in terms of the new variables in the form $(\mathrm{Tr}\ \frac{\lambda^a}{2}\ \frac{\lambda^b}{2} = \frac{1}{2}\delta^{ab})$.

$$\begin{aligned} H = {} & g^2 \sum_{I=1}^{3} \mathrm{Tr}\ \Pi_I^2 + \frac{1}{g^2} \sum_{I>J}^{3} \mathrm{Tr}\ [\partial_I(B_{IJ}\ \partial_j B_{JI})]^2 \\ & + \psi^+\ \beta m \psi + \sum_{I=1}^{3} (\psi^+ B_{3I})\ (-i\,\alpha_I\,\partial_I)\ (B_{I3}\psi)\ . \end{aligned} \tag{3.19}$$

The second term may be rewritten by using the property

$$\mathrm{Tr}\ [\partial_1(B_{12}\,\partial_2 B_{21})]^2 = \mathrm{Tr}\ [\partial_2(B_{21}\,\partial_1 B_{12})]^2 \quad \text{etc.}$$

I can now discuss the dependence of H on the coupling constant.

The operators Π, B are normalized so that their commutation rules are independent of g. Thus, the only coupling constant dependence in the quantum theory can come from the explicit dependence of H on g as in eq. (3.19). Since B_{IJ} is a compact operator bounded by its unitarity property $B^{+}_{IJ}B_{IJ} = 1$, it cannot become large in any quantum state. The derivatives $\partial_I B_{IJ}$ etc. could become large; however, if one introduces a short distance cutoff Δ and investigates phenomena at distances larger than Δ the derivatives would be smaller than $\lesssim 1/\Delta$. Thus, if g is defined at this cutoff distance and is taken large, it is clear that the first term of the Hamiltonian would dominate. According to the renormalization group for QCD, it is consistent to take $g(\Delta)$ = large if Δ is large enough. Since I have not altered the theory merely by rewriting it in the new variables, the renormalization group argument should apply. Thus, I conclude that for large distance phenomena, the electric field term would give an approximate qualitative description of the energy dependence of the states. Therefore the eigenstates of H_o may be taken as a starting point for a strong coupling expansion of the theory. I emphasize that the arguments given above must break down for QED since for a Abelian theory $g(\Delta) \to$ small as $\Delta \to$ large, according to the renormalization group.

IV. RESTRICTIONS ON THE SPECTRUM AND EXPLICIT EIGENSTATES

Eqns. 3.13 and 3.17 give $[\Pi^a_I, G^2_J] = 0$ which shows one can diagonalize simultaneously the larger set of operators G^2_1, G^2_2, G^2_3, $\Pi^2_1, \Pi^2_2, \Pi^2_3$ at each space time point $\vec{x}$ independently. Therefore the eigenstates take the form $\prod_{\vec{x}_n} \prod_{I=1}^{3} |e^2_I(\vec{x}_n), g^2_I(\vec{x}_n), \ldots\rangle$ where the dimensionless $(e^2_I, g^2_I, \ldots)$ are the eigenvalues of $(\Pi^2_I, G^2_I, \ldots)$ up to a scale. For an SU(2) pure gauge theory this is a <u>complete</u> set of labels (6 degrees of freedom (A^a_i, E^a_i)). For SU(N) one must use additional labels corresponding to e.g. cubic, quartic, etc. Casimir operators and other operators as is well known. Finally in a theory with quarks the states must have additional labels corresponding to those variables, as will be shown below after defining the vacuum.

The algebraic relations which I have presented amount to certain restrictions on the spectrum that follow directly from the properties of the $SU(N)_1 \times SU(N)_2 \times SU(N)_3$ local Lie algebra. These are: 1. One should expect that $g_I^2(\vec{x}_n)$ are determined by quantized pure numbers since $G_I^2(\vec{x})$ are Casimir operators of 3 independent non-Abelian Lie algebras at each $\vec{x}$ (like $J^2 = j(j+1)$ for angular momentum).

2. Since Π_I^a are sums of generators (eqn. 3.17) (like total angular momentum) Π_I^2 must be determined by quantized numbers as well.

3. Furthermore, given a state with a configuration of points $(\vec{x}_n)$ at which $g_I^2(\vec{x}_n) \neq 0$ are specified, then the Clebsch-Gordon series puts a restriction, in terms of $g_I^2(\vec{x}_n)$, on the maximum and minimum quantum numbers allowed for $e_I^2(\vec{x}_n)$ (like $|j_1 - j_2| \leq j \leq j_1 + j_2$ for the sum of angular momenta).

4. The boundary conditions eq. (3.18) mean that the total quantum number corresponding to the sum of generators along any line from $(-\infty)$ to $(+\infty)$ must be 0 on physical states.

5. The constraint eqn. (3.14) puts a restriction among (g_1^2, g_2^2, g_3^2) at a given point $\vec{x}_n$, such that they should add up to the quantum number specified by the quark color density $\rho^2(\vec{x}_n)$ (like adding 3 angular momenta).

Let me now be more specific on the eigenvalues of G_I^2, say G_1^2. Mathematically, this is ill-defined since there is a $\delta(\vec{x}-\vec{x}')$ in the algebra (3.13) satisfied by the distribution G_1^a. I will introduce a short distance cutoff to define the singular operator product $G_1^2(\vec{x})$ (and Π_1^2, etc.). This can be done in a number of ways:

1. One can smear the delta function in eq. (3.12) $\delta(\vec{x}-\vec{x}) \to \delta_\Delta(\vec{x}-\vec{x}')$, by taking e.g. $\delta_\Delta(x_1) = \frac{\sqrt{\pi}}{\Delta} e^{-x_1^2/\Delta^2}$, so that

at the origin $\delta_\Delta(\vec{0})=(\sqrt{\pi}/\Delta)^3$. All other $\vec{x}$, $\vec{x}'$ dependence in eq. (3.12) is left intact. This prescription is sufficient to do calculations as I will demonstrate.

2. An alternative approach is to introduce a lattice with a spacing of order $\frac{\Delta}{\sqrt{\pi}}$, and allow the operators $G_I^a(\vec{x})$ and $B_{JI}(\vec{x})$ to be non-zero only at lattice sites. I emphasize that this would be a different lattice theory than the ones of Wilson and Kogut and Susskind[2]. I have no bilocal link operators but only local corner operators. This is possible because I am working in a specific gauge and need not worry about constructing a gauge invariant lattice theory. I will choose the first cutoff method since it is a continuum approach and is simple to manipulate. Δ will be fixed at a finite value. It is an open question whether it can be removed by renormalization to all orders in $1/g$. However, there must be dimensional transmutation since a dimensionful parameter must appear. The value of Δ determines the thickness of the string. Eventually one expects it to be related to the Regge slope and of course to the first massive state.

One can now write

$$[G_1^a(\vec{x}) \ , \ G_1^b(\vec{x})] = (\sqrt{\pi}/\Delta)^3 \ i \ f^{abc} \ G_1^c(\vec{x}) \ , \qquad (4.1)$$

which shows that the dimensionless operator $(\Delta/\sqrt{\pi})^3 G_1^a(x)=g_1^a(x)$ leads to pure quantum numbers determined by the Lie algebra SU(N). For example, the Casimir operator is $g_1^2(x)=1/2(N-1/N)$ for the fundamental representation, and other quantum nos. for higher representations. One may, of course, have different quantum numbers at different points $\vec{x}_n$. With similar arguments I conclude that

$(\Delta/\sqrt{\pi})^2\Pi_1^a(\vec{x})= e_1^a(\vec{x})$ leads to pure quantum numbers for $e_1^2(x)$. Thus, the electric field $\frac{1}{g^2}E_1^2=\Pi_1^2=(\sqrt{\pi}/\Delta)^4 e_1^2$ is quantized, and so is E_2^2 and E_3^2.

An explicit construction of states can now be given. The

simplest state has zero labels $g_I^2 = e_I^2 = 0$ at all points $\vec{x}$. This state trivially satisfies all the constraints, including the boundary conditions. This is the vacuum in the $g\to\infty$ limit. It is characterized by

$$G_I^a(\vec{x})\ |o\rangle = 0 = \Pi_I^a(\vec{x})\ |o\rangle \tag{4.2}$$

and has energy $H_o = 0$.

Next, consider the state $|\vec{y}\rangle$ in which g_I^2 are non-vanishing only at one point $\vec{y}$. It is possible to satisfy the constraint eq. (3.14) by making e.g. $G_3^a = 0 = \rho_3^a$ for all $\vec{x}$ and $G_1^2 = G_2^2 \neq 0$ only at $\vec{y}$. Therefore only Π_1^a and Π_2^a are non-zero in this state. From eq. (3.17) one concludes that $\Pi_1^a\ (\vec{x}) \sim \theta(y_1-x_1)\ \delta(x_2-y_2)\ \delta(x_3-y_3)$ and $\Pi_2^a\ (\vec{x}) \sim \theta(y_2-x_2)\ \delta(x_1-y_1)\ \delta(x_3-y_3)$ thus violating the boundary condition (3.18), and leading to an infinite energy. More explicitly the state can be constructed by applying the $B_{12}(\vec{y})$ operator on the vacuum.

$$|\vec{y}\rangle = B_{12}^{ij}\ (\vec{y})\ |o\rangle \quad . \tag{4.3}$$

It can be verified that this is non-trivial eigenstate of $\Pi_I^2(\vec{x})$:

$$\Pi_1^2(\vec{x})|\vec{y}\rangle = [\Pi_1^a(\vec{x}),[\Pi_1^a(\vec{x}),\ B_{12}^{ij}\ (\vec{y})]]\ |o\rangle$$

$$= |\vec{y}\rangle\ \frac{1}{2}\ (N-\frac{1}{N})\ [\theta_\Delta(y_1-x_1)\delta_\Delta(x_2-y_2)\delta_\Delta(x_3-y_3)]^2 \tag{4.4}$$

and similarly for $\Pi_2^2\ (\vec{x})$ for which $\theta_\Delta\ (y_2-x_2)$ appears, while $\Pi_3^2(\vec{x}) = 0$ on this state. It is now clear why smeared δ functions were necessary. The state describes a stringlike quantized flux tube of thickness $\sim \Delta^2$ and of ∞ length as shown in Fig. 2.

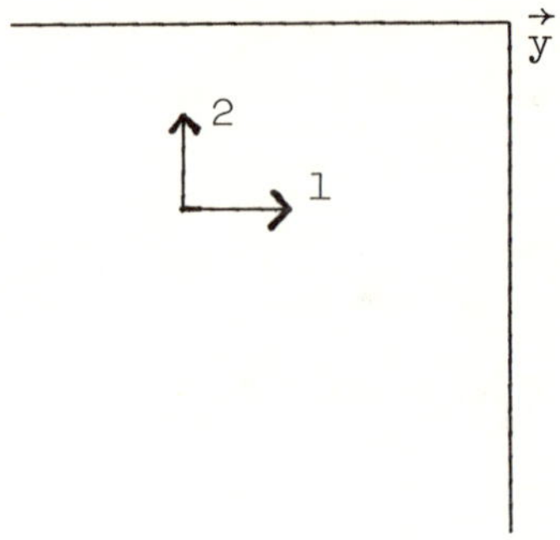

Fig. 2

The kink at $\vec{y}$ is due to the presence of charge carried by the "gluon" $B_{12}^{ij}(\vec{y})$, which has index i associated with the "1" direction and index j associated with the "2" direction as seen from eq. (3.12). This state has ∞ energy $H_o = \sum_{I=1}^{3} \int d\vec{x}^3\, \Pi_I^2(\vec{x})$ due to the the length of the string. Thus, it is automatically removed from the spectrum, as required by the boundary conditions (eq. 3.18).

It is now clear that states with finite energies will be either closed strings or strings with quarks at the ends. The simplest closed string requires 4 corners at which charges are located as in Fig. 3. Strings with arbitrary shapes require many more charges.

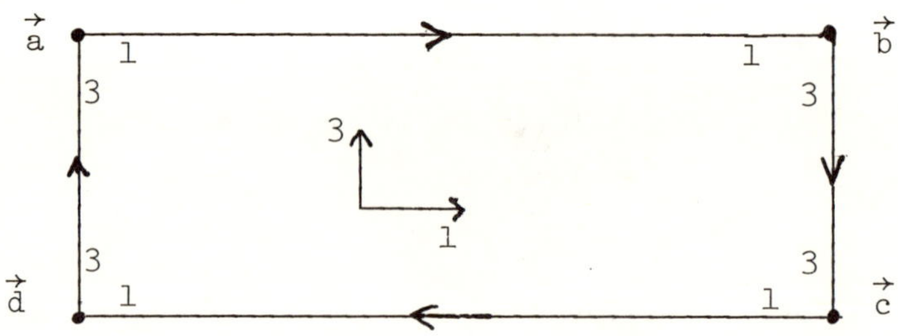

Fig. 3

The state corresponding to Fig (3) is constructed as follows:

$$|\vec{a}, \vec{b}, \vec{c}, \vec{d}> = \mathrm{Tr}\,[B_{13}(\vec{a})\, B_{31}(\vec{d})\, B_{13}(\vec{c})\, B_{31}(\vec{b})]\,|o> . \quad (4.5)$$

Only Π_1^2 and Π_3^2 are non-zero on this state, since it is constructed only with the operator B_{13}. The value of Π_1^2 (x) can be calculated as in eq. (4.4):

$$\Pi_1^2 (\vec{x}) \, |\vec{a}, \vec{b}, \vec{c}, \vec{d} > \cong |\vec{a}, \vec{b}, \vec{c}, \vec{d}> (\sqrt{\pi}/\Delta)^2 \, e_1^2 \, (\vec{x};\vec{a},\vec{b},\vec{c},\vec{d})$$

$$+ 0\{ \delta_\Delta(x_3-d_3)\ldots\} , \qquad (4.6a)$$

where the eigenvalue e_1^2 is

$$e_1^2 = \frac{1}{2} (N-\frac{1}{N}) \, \delta_\Delta(a_1-x_1)\theta_\Delta(x_1-b_1) \, \{\delta_\Delta(x_2-a_2) \, \delta_\Delta(x_3-a_3) + \delta_\Delta (x_2-d_2)$$

$$\times \, \delta_\Delta (x_3-d_3)\} . \qquad (4.6b)$$

I emphasize that this state is an eigenstate to the extent that the product of two smeared delta functions can be neglected: $|\vec{a} - \vec{d} \,| \gtrsim \Delta \rightarrow \delta_\Delta(x_3-a_3).\delta_\Delta(x_3-d_3) \approx 0$, as in eq. (4.6a). Therefore provided charges are separated by distances larger than Δ the above calculation demonstrates that Π_1^2 (x) has the shape of flux tubes of thickness $(\Delta/\sqrt{\pi}\,)^2$ along the top and bottom sides of the rectangle in Fig. (3). Similar arguments indicate that $\Pi_3^2(x)$ behaves similarly along the vertical sides of the rectangle. For shorter distances one could attempt, in principle, to construct the exact eigenstates of H_o, but this would not lead to a meaningful description of the physics without including the remaining part of the Hamiltonian. Thus, it is interesting that the formalism hints of a different behavior of the electric field at shorter distances, as expected by asymptotic freedom.

The energy H_o can now be easily computed. I obtain $H_o = \frac{1}{2}(N- \frac{1}{2}) \frac{1}{2}g^2 \, (\sqrt{\pi}/\Delta)^2$ L, where L is the perimeter of the rectangle. Note that this has the form of (length/(thickness). This indicates a linear "potential energy", which would confine color charges.

Turning the attention to quarks, one needs to further specify the properties of the vacuum with respect to them. A prime requirement is that

$$\rho_I^a(\vec{x})\ |0\rangle = 0 \tag{4.7}$$

so that eq. (4.2) is consistent with (3.14). The only basis for the fermions that would satisfy this property is

$$\psi_a^i(\vec{x}) = \begin{pmatrix} b_\alpha^i(\vec{x}) \\ d_\alpha^{+i}(\vec{x}) \end{pmatrix}, \tag{4.8}$$

where $b_\alpha^i(\vec{x})$ and $d_\alpha^{+i}(\vec{x})$ are two-component spinors corresponding to the upper and lower components of a Dirac fermion. The canonical anticommutation rules correspond to

$$\{b_\alpha^i(\vec{x}),\ b_\beta^{+j}(\vec{y})\} = \delta^{ij}\delta_{\alpha\beta}\ \delta(\vec{x}-\vec{y}) = \{d_\alpha^i(\vec{x}), d_\beta^{+j}(\vec{y})\} \tag{4.9}$$

and all other anticommutators taken to be 0. These operators are related to the perturbative Fock space creation and annihilation operators by a Bogoliubov type transformation.

The fermion color charge takes the form

$$\rho_3^a(\vec{x}) = \psi^+ \frac{\lambda^a}{2} b + d^+\left(-\frac{\lambda^a}{2}\right)^T d \ . \tag{4.10}$$

Therefore, if the vacuum is now defined by

$$b_\alpha^i(\vec{x})|0\rangle = 0 = d_\alpha^i(\vec{x})\ |0\rangle \tag{4.11}$$

then eq. (4.7) is satisfied. The perturbative operators would violate this property. If one works in the basis in which

$\beta = \left(\begin{array}{c|c} 1 & 0 \\ \hline o & -0 \end{array}\right)$, $\vec{\alpha} = \left(\begin{array}{c|c} o & \vec{\sigma} \\ \hline \vec{\sigma} & o \end{array}\right)$ then the mass operator is diagonal

$$:\psi^+\beta m\psi: \; = \; b^+mb + d^+md \tag{4.12}$$

while the kinetic term is not diagonal. For heavy quarks the kinetic term is small, so that this basis is convenient for nearly static calculations.

It is now possible to construct open strings with quarks at the ends. I will only write down a few states as examples. They correspond to Fig. 5

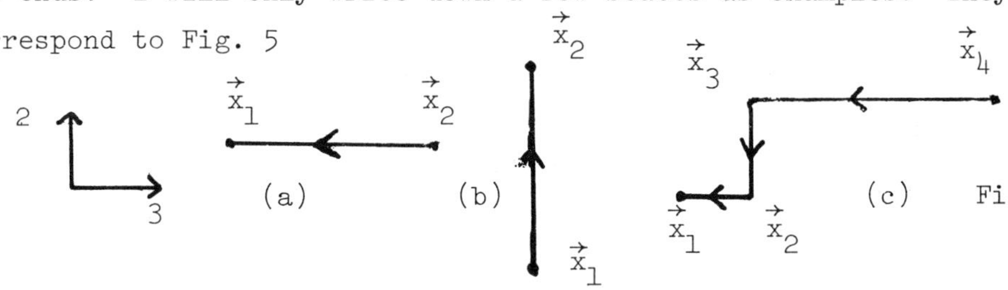

Fig. 5

(a) $\quad |\vec{x}_1,\vec{x}_2> \; = b_a^{+i}(\vec{x}_1)\Gamma_{\alpha\beta}d_\beta^{+i}\;(\vec{x}_2)\;|o> \;;\; \Gamma=1,\vec{\sigma}$

(b) $\quad |\vec{x}_1,\vec{x}_2> \; = \Gamma_{\alpha\beta}\; b_\alpha^+(\vec{x}_2)\; B_{32}(\vec{x}_2)B_{23}(\vec{x}_1)d_\beta^+\;(\vec{x}_1)\;|o>$

(c) $\quad |\vec{x}_1,\vec{x}_2,\vec{x}_3,\vec{x}_4> \; = \Gamma_{\alpha\beta}\; b_\alpha^+(\vec{x}_1)\; B_{32}(\vec{x}_2)\; B_{23}(\vec{x}_3)\; d_\beta^+\;(\vec{x}_4)\;|o>$

In case (a) the string is along the "3" direction and is constructed directly in terms of the fermion operators defined in the $A_3^a = 0$ gauge. In case (b) the string is in the "2" direction. Accordingly it was necessary to first transform the fermions to the $A_2'' = 0$ gauge (see Fig. 1) by using the operator B_{32} or B_{23}. Finally in case (c) the appropriate strings are produced by gluon as well as fermion charges. It is seen that the role played by the gluonic operators is to make transformations to gauges that are "tangent" to the string along the appropriate string bit.

The electric fields Π_I^2 can be computed on these states as in eq. (4.4). Then the energy can be calculated with the result that

$$\text{Energy} = \left(\frac{1}{2} g^2\right) \frac{1}{2} \left(N - \frac{1}{N}\right) \times (\text{length})/(\text{thickness}).$$

This shows once again linear confinement.

DISCUSSION

Obviously there is some global similarity between the present formulation and the lattice gauge theory of Wilson and Kogut and Suskind(2). However, they are different in detail. In particular I note that for, $\Delta = 0$, the Hamiltonian written in terms of either the old (A,E) or the new (Π,B) variables, is Poincaré covariant(4). The theory is not as drastically changed as the usual lattice theories by the introduction of Δ. The B operators are local, unlike the link operators on the lattice which are bilocal. Furthermore at the location of B the "electric" flux changes direction. Despite the differences in details the similarity of the results induces one to think that my theory corresponds to a continuum version of the lattice formulation.

I have shown that, <u>in the strong coupling limit</u>, QCD leads to closed strings as well as to open strings with quarks at the ends. To make further progress one must develop methods of calculation to include the neglected terms of the Hamiltonian. One must also obtain a better understanding of the deep question of the scale parameter Δ, and find out if it can be removed by the renormalization process.

FOOTNOTES AND REFERENCES

(1) I. Bars, Phys. Rev. Letters, March 1978.

(2) K. Wilson, Phys. Rev. D.10, 2445, (1974)
J. Kogut and L. Suskind, Phys. Rev. D11, 395, (1975)

(3) R. L. Arnowitt and S. I. Fickler, Phys. Rev. 127, 1821, (1962)
J. Schwinger, Phys. Rev. 130, 402, (1963)
W. Kummer, Acta Physica Austriaca 14, 149, (1961)
R. N. Mohapatra, Phys. Rev. D4, 2215, (1971)
W. Konetschny and W. Kummer, Nucl. Phys. B100, 106, (1975) and references contained therein.

(4) I. Bars and M. B. Green, Phys. Rev. D. Jan (1978).
I. Bars and Frederic Green, in preparation.

(5) V. N. Gribov, Lecture at the 12th Winter School, Leningrad Nuclear Physics Institute (1977). English translations can be found in J. Bartels and W. Nahm, CERN preprint, and in SLAC-Trans. 176. See also the lectures by H. Pagels and R. Jackiw in this conference and references contained therein.

(6) J. Schwinger in ref. (3), and more recently S. Mandelstam in notes from APS Washington meeting (1977). See also W. A. Bardeen and R. Pearson Phys. Rev. D14, 547, (1976), who arrive at similar conditions from different arguments.

(7) Alan Chodos, Canonical Quantization of Non-Abelian Gauge Theories in Axial Gauge, Yale Report COO-3075-183.

REMARKS ON THE TOPOLOGY OF GAUGE FIELDS*

Y. Nambu+

Stanford Linear Accelerator Center

In contrast to the short distance behavior of quantum chromodynamics (QCD), which is within the scope of perturbation theory and thus can be subjected to quantitative tests, the large distance or strong coupling regime of QCD is not well understood yet even qualitatively. An overriding problem of interest in quark confinement, and various theoretical schemes have been put forward to show that the quarks can indeed be confined. These schemes, though varying from one to another in detail, rely on the idea that the growing coupling constant at large distances plays a key role. It is not clear yet, however, whether confinement is a natural consequence of QCD alone, or it requires some independent and extraneous assumptions.

Another important aspect of gauge fields is the existence of topologically nontrivial configurations such as vortices, monopoles and instantons. They must be regarded as natural consequences of gauge theories, and in fact there are some familiar examples in condensed matter physics. Most likely these configurations also

---

*Amplified version of introductory remarks made at the 1978 Coral Gables Conference.

+On leave from the University of Chicago

are crucial to the understanding of confinement as is claimed by a number of recent papers. For this reason I have planned this session to be organized mainly around the topological problems. My own remarks on these questions will bery brief:

1. Action versus Free Action

There is a formal analogy between statistical mechanics and the Feynman formulation of quantum field theory, which seems to become especially relevant in gauge theories. This analogy was emphasized by the Princeton group[1] who made use of the concept of entropy. Let us write down the Feynman integral (in the Euclidean form) for a gauge field

$$Z = \int \exp \left[- \frac{1}{4g^2} \int F_{\mu\nu} F_{\mu\nu} d^4 X\right] D(A_\mu) \quad , \tag{1}$$

where the coupling constant does not enter the definition of $F_{\mu\nu}$. The functional measure is not clearly defined, but we intend to integrate over the A_μ without imposing special gauge conditions. For a given value of the action, there is a corresponding phase space volume which we write symbolically as exp S:

$$Z = \int \exp \left[- \frac{1}{g^2} I + S(I)\right] dI$$

$$= \exp \left[-Ⓗ/g^2\right] \quad , \tag{2}$$

$$Ⓗ \approx (I - g^2 S)_{min.}$$

Let us call Ⓗ free action in contrast to the action I. In principle, we should minimize Ⓗ rather than I, and this difference could become important for large g^2, which plays the role of temperature.

This statement takes on a real significance when the topology of gauge fields is considered. Suppose we start from the lowest value of I, i.e. the pure gauges: $F_{\mu\nu} = 0$, $I = 0$. This part of phase space can be parametrized by a unitary matrix field u,

$$A_\mu = i\, u^+ \partial_\mu u, \quad u^+ u = 1 \quad . \tag{2a}$$

The field u, however, may be topologically non-trivial. For example, choose

$$u = \tau_\mu x_\mu / |x| \ , \quad \tau_\mu = (\vec{\tau}, i) \tag{3}$$

in the case of SU(2) theory. Then u becomes singular at the origin, and $F_{\mu\nu}$ and I cannot be identically zero since the Pontrjagin index

$$\frac{1}{16\pi^2} \operatorname{Tr} \int F_{\mu\nu} \tilde{F}_{\mu\nu} d^4x = \int Q_\mu dS_\mu ,$$

$$Q_\mu = - \frac{1}{3} T_r\, u^+ \partial_\nu u\, \partial_\lambda u^+ \partial_\rho u\, \varepsilon^{\mu\nu\lambda\rho} \tag{4}$$

is computed to be nonzero. Thus we might say that pure gauges are not a well defined concept, or the entropy cannot be defined for given I, at least not for I = 0. There seems to be a sort of uncertainty principle between I and S due to topology. Clearly this is related to the Gribov problem[2] and Singer's observation[3] of its generality. At any rate we are forced to enlarge pure gauges to their neighborhood which I will call almost pure gauges. It is conceivable that for large enough g^2 the gain in S by including various topologies can outweigh the cost of larger I, thus actually lowering the free action. Then the topology-averaged field configurations which minimize the free action may deviate substantially from the naive classical configurations which minimize only I.

2. A class of almost pure gauges.[4]

It is both natural and convenient to consider the following class of almost pure gauges

$$A_\mu = ifu^+ \partial_\mu u \quad , \tag{5}$$

where f is a scalar function which vanishes at the singularities of u. The topological characterization of a singularity will not be altered by this if $f \to 1$ in a region surrounding it. From Eq. (5) we obtain

$$F_{\mu\nu} = f(1-f)\ \partial_\mu u^+ \partial_\nu u + \partial_\mu f u^+ \partial_\nu u - (\mu \leftrightarrow \nu). \tag{6}$$

It is not surprising that this class of configurations cover various known examples of nontrivial topology such as instanton, meron, monopole, and string.

A more interesting point is that we can generalize Eq. (5) further as

$$A_\mu = i \sum_i f_i \ u_i^+ \partial_\mu u_i \quad . \tag{7}$$

The many-instanton solution of 't Hooft (in the singular gauge) is indeed of this form. Furthermore, Eq. (7) can be written as a London relation[5]

$$A_\mu = \lambda\, J_\mu = \lambda \sum_{i=1}^{N} J_{i\mu} \quad ,$$

$$J_{i\mu} = ih_i u_i^+ D_\mu u_i \quad ,$$

where

$$D_\mu u = \partial_u u + iu A_\mu \quad , \tag{8}$$

f_i and h_i are related by

$$f_i = \lambda h_i / (1+\lambda\Sigma h_i). \tag{9}$$

Then a simple superposition principle holds for the currents $J_{i\mu}$ to generate many instantons, and the total current is conserved: $D_\mu J_\mu = 0$. It might also be instructive to observe that Eq. (8) can be further simplified if J_μ is defined in terms of a 2N x 2 rectangular matrix U:

$$U = \Sigma \oplus h_i^{\frac{1}{2}} u_i \quad (\text{ assuming } h_i \geq 0\ ),\ U^+U = (\Sigma h_i)1$$

$$J_\mu = \frac{i}{2}\,[U^+(D_\mu U) - (D_\mu U^+)\ U]. \tag{10}$$

U has a gauged SU(2) symmetry acting from the right, and a global SU(2) x S_N symmetry acting from the left which may be generalized to SU(2N). These two are independent, like color and flavor.

3. A remark concerning the Wilson criterion.

The Wilson criterion is widely used to test the confinement property of a theory. In the context of functional integration, one evaluates

$$< W > = \Sigma <W_i> \rho_i \quad , \tag{11}$$

$$W = \mathrm{Tr}\ \exp\ [i\oint A_\mu dx_\mu] \quad ,$$

where ρ_i is the weight of a configuration i. In general Eq. (11) is expected to yield an asymptotic form

$$<W> \sim a\ \exp\ [-\lambda L] + b\ \exp\ [-\mu L^n] + \cdot\cdot,\ \tilde{n} > 1 \tag{12}$$

where L is the linear dimension of the Wilson loop. What I would like to emphasize is that to prove confinement, one must first show the absence of configurations contributing to the first term rather than the presence of configurations contributing to the second term. I suppose that the former task is more demanding and difficult than the latter.

REFERENCES

1. C. Callan et al., Phys. Lett 66B, 375 (1977).
2. V. M. Gribov, Lectures at the 12th Winter School of the Leningrad Physics Institute (1977).
3. L. M. Singer, Berkeley preprint (Math. Department), 1977.
4. See also Y. Nambu, Cal Tech preprint CALT 68-634, 1977.
5. Y. Nambu, Annals of the New York Academy of Sciences 294 (Five Decades of Weak Interactions) p. 74 (1977).

DYNAMICAL EFFECTS OF INSTANTONS AND MERONS*

Roger Dashen

Institute for Advanced Study

Princeton, New Jersey

ABSTRACT

The dynamical effects of instantons and merons in QCD are reviewed. Instantons act like four dimensional color magnetic dipoles and make the coupling increase very rapidly. At a certain scale they break up into meron pairs and the coupling diverges.

I like to think of instantons as just another kind of vacuum fluctuation.[1] For years we have been studying the small Gaussian zero point oscillations around the classical vacuum $A^{\mu} = 0$. They are fluctuations of strength $A^{\mu} \sim g$ (g is the coupling) and can get very wiggly producing the ultraviolet phenomena which we know so well. The fluctuations associated with instantons are tunnellings from the classical vacuum $A^{\mu} = 0$ to a (topologically distinct) gauge rotated vacuum.[2] The most efficient way to tunnel from one vacuum to another is to follow a path which satisfies the imaginary time equations of motion. This is the instanton. It is a large, smooth fluctuation with fields of order unity. In QCD this new class of fluctuations has <u>very</u> large effects. Ordinary zero point oscillations

* Research sponsored by the Department of Energy under Grant no. EY-76-S-02-2220.

renormalize masses and coupling constants. The instantons produce similar mass and coupling shifts, the only real differences being the size of the effect and the absence of ultraviolet divergences.

I am going to talk about these dynamical effects of instantons and merons (a half instanton--more about this later) in QCD. Actually, I am going to spend most of my time on pure Yang-Mills theory with heavy "test quarks" represented by the Wilson loop. Real, light quarks bring in some fascinating new physics associated with chiral symmetry breaking. I will mention this briefly at the end.

Defining fields so that the action is $\frac{1}{4}g^2\int \mathrm{tr}F^2$, the BPST instanton[3] in singular gauge is

$$A^\mu_a(x) = 2R_{aa'}\bar{\eta}^{\mu\nu}_{a'} \frac{(x-x_I)^\nu}{(x-x_I)^2} \frac{\rho^2}{\rho^2+(x-x_I)^2} , \qquad (1)$$

where $\bar{\eta}^{\mu\nu}_a$ is 't Hooft's tensor,[4] $(\bar{\eta}^{\mu\nu}_a = \varepsilon_{oa\mu\nu} - \frac{1}{2}\varepsilon_{abc}\,\varepsilon_{bc\mu\nu})$, and ρ, x_I and R are an arbitrary scale, location and global gauge. For an anti-instanton, replace $\bar{\eta}$ by $\eta^{\mu\nu}_a = \varepsilon_{oa\mu\nu} + \frac{1}{2}\varepsilon_{abc}\,\varepsilon_{bc\mu\nu}$. A minor extension[1] of 't Hooft's[4] calculation for SU(2) shows that for SU(3) the probability of finding an instanton in a space time volume VT is

$$VT\int \frac{d\rho}{\rho^5} D(\rho) , \qquad (2)$$

where

$$D(\rho) = .10X^6 e^{-X} ,$$
$$X(\rho) = 8\pi^2/\bar{g}^2(\rho) \qquad (3)$$

and $\bar{g}$ is the running coupling as defined by 't Hooft.[4]

Actually, for a reason that will become apparent below, single or even multiple instanton solutions are not very interesting. What is interesting are configurations containing many instantons and many anti-instantons. In singular gauge, configurations of this kind which approximately satisfy the equations of motion can be obtained by simple superposition. If the distance between instantons or anti-instantons is large compared to their scale size, the action is additive and the determinant which is needed to find the analog of Eq. 2 is multiplicative. In this approximation one can sum over configurations containing n_+ instantons and n_- anti-instantons.

The answer is

$$\sum_{n_+,n_-=1} \left[\frac{1}{n_+!} \frac{1}{n_-!} \left(VT \int \frac{d\rho}{\rho^5} D(\rho) \right)^{n_+ + n_-} \right]$$

$$= \exp\left[2VT \int \frac{d\rho}{\rho^5} D(\rho) \right] \qquad (4)$$

and the dominant term in the series is the one for which $n_+ = n_- = TV \int \frac{d\rho}{\rho^5} D$. Thus the physics lies in configurations containing many instantons and anti-instantons.

Eq. 4 is just the grand partition function for a perfect gas of instantons and anti-instantons with equal densities $\int \frac{d\rho}{\rho^5} D$. This gas analogy is extremely useful in practical calculations: it enables one to use all the machinery of classical statistical mechanics. Knowing the density of instantons, one can now go back and check whether or not the assumption that they are far apart is valid. It is not, unless we cut off the integration over ρ. Using the lowest order relation $d\,X/d\ln\rho = 11$, one finds[1] that the gas becomes dense at a scale corresponding to $X \sim 12$. For the purpose of this talk I will suppose that the integral has been cut off at some $X \gtrsim 12$.

It is quite feasible to go one step beyond the perfect gas and consider instanton interactions. For the long distance part of the interaction we need only the asymptotic instanton field

$$A_a^\mu = 2\rho^2 \bar{\eta}_a^{\mu\nu} x^\nu / x^4 \quad . \tag{5}$$

This is the field of a <u>four dimensional color magnetic dipole</u>. For fields that fall faster than x^{-1}, the Yang-Mills equations linearize asymptotically and solutions can be classified as dipoles x^{-3}, quadrupoles x^{-4}, etc., in the same way as in an Abelian theory. A magnetic dipole moment is an antisymmetric tensor $\int [j^\mu x^\nu - x^\mu j^\nu] d^4x$ (equivalent to a vector in three dimensions) and the general magnetic dipole is asymptotically

$$A_a^\mu = M_a^{\mu\nu} \frac{x^\nu}{x^4} \quad , \tag{6}$$

where M is an antisymmetric $\left(M_a^{\mu\nu} = -M_a^{\nu\mu}\right)$ numerical matrix. Evidently an instanton has a moment $M_a^{\mu\nu} = 2\rho^2 \bar{\eta}_a^{\mu\nu}$ and an anti-instanton, a moment $M_a^{\mu\nu} = 2\rho^2 \eta_a^{\mu\nu}$. The instanton is a magnetic (as opposed to electric) dipole because the Euclidean Yang-Mills equations correspond to static color magnetism in four dimensions. The interaction between two four-dimensional dipoles with moments $M_a^{\mu\nu}$ and $N_a^{\mu\nu}$ is

$$S_I = \frac{\pi^2}{g^2 R^4} \operatorname{tr} [M_a \{T, N_a\}] \quad , \tag{7}$$

where

$$T^{\mu\nu} = \delta^{\mu\nu} - 4\hat{R}^\mu \hat{R}^\nu \quad . \tag{8}$$

R^μ is the separation and the trace is over Lorentz indices. Since two instantons do not interact, the dipole picture cannot make sense unless S_I vanishes when both M and N are (anti) self dual. In fact, it does! If N is self dual, it belongs to the (1,0) representation of O(4). Then, since T belongs to (1,1), the antisymmetric matrix {T,N} belongs to (0,1) and the trace vanishes if M is self dual. For M self dual and N anti-self dual S_I does not vanish and one can show[1] that it does give the correct instanton anti-instanton interaction at large distances.

Since well separated instantons and anti-instantons act just like color magnetic dipoles, our gas is a gas of dipoles. A gas of magnetic dipoles has one particularly important property: it is paramagnetic giving rise to a positive magnetic susceptibility which is equivalent to an upward renormalization of the coupling constant. The interaction of a four-dimensional dipole with a weak constant external field is

$$S_I = \frac{\pi^2}{g^2} M_a^{\mu\nu} F_a^{\mu\nu} \text{ (ext)} \quad . \tag{9}$$

A textbook calculation of the color magnetic susceptibility χ then gives[1]

$$\chi = \frac{\pi^2}{2} \int \frac{d\rho}{\rho} \; X(\rho) D(\rho) \quad , \tag{10}$$

where as usual $X = 8\pi^2/g^2$. The effect on the static ($q^2 = 0$) coupling is a renormalization by $1+\chi$, i.e., $g^2 \rightarrow (1+\chi)g^2$. Changing variables to $X = -11 \ln(\rho\mu)$ in the integral for χ we have

$$\chi = (.10)\frac{\pi^2}{22} \int_{X_c}^{\infty} X^7 e^{-X} dX \quad , \tag{11}$$

where as mentioned before, $X_c \gtrsim 12$. A plot of χ versus X_c is shown in Fig. 1. Note that χ is very big, even in a coupling regime $g^2/8\pi^2 \sim .05$ to $.07$ where one would have thought that perturbation

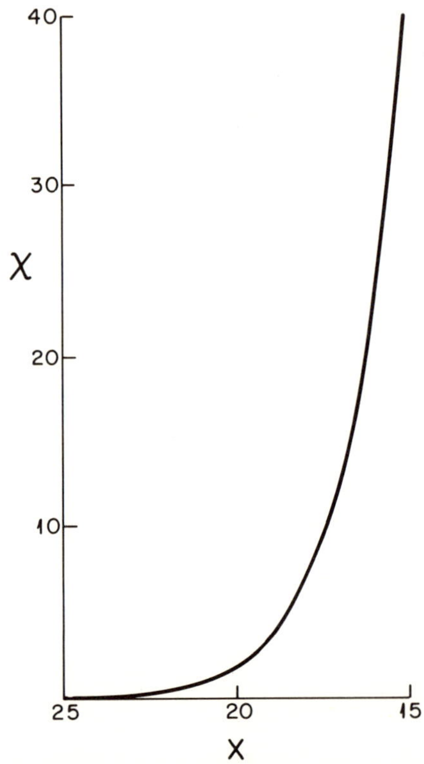

Fig. 1 The color magnetic susceptibility of an instanton gas. The coupling g is renormalized by $g^2 \to (1 + x)g^2$. The parameter X_c is the asymptotic freedom coupling constant $X_c = -11 \ln \rho_c \mu$ corresponding to the scale size ρ_c at which instantons are cut off.

theory is still valid. It is expected that this rapid rise in the coupling at a sharply defined scale size sets the stage for dimensional transmutation, i.e., the trade off of a coupling for an arbitrary mass scale. If so, the strong interaction coupling is $g^2/8\pi^2 \sim .07$, a small number. In any case, the new vacuum fluctuations have big effects.

The effects of the new fluctuations on the Wilson loop have also been calculated[1]. According to Wilson, the energy of a pair of heavy test quarks separated by $\vec{r}$ is

$$\exp\,[-E(\vec{r})T] = \left\langle P\exp\,[i\oint A\cdot\, dx]\right\rangle \quad , \qquad (12)$$

where P denotes path ordering, the average is over all vacuum fluctuations and the loop integral runs around a Euclidean rectangle which is long $\sim$ T in the time direction and of spacial extent $|\vec{r}|$. Averaging over a gas of instantons gives

$$E(\vec{r}) = 2\Delta M + V(r) \quad , \qquad (13)$$

where ΔM is a (finite) positive mass shift and V is a potential with the properties that

$$V(0) = -2\Delta M \quad , \qquad (14)$$

i.e., the net effect is zero when the quarks are on top of each other, and for large r

$$V(r) \rightarrow \chi\, V_{coul}\,(r) \quad , \qquad (15)$$

where $V_{coul}(r)$ is the lowest order coulomb interaction and χ is the renormalization constant discussed above. For finite r, V varies smoothly between $-2\Delta M$ and its asymptotic value. The value of ΔM depends on the cutoff: for $X_c \sim 12$ it is about 10 GeV. This is not confinement, but a 20 GeV potential well at a coupling $g^2/8\pi^2 \sim 1/12$ is no small effect.

The fact that instantons are dipoles suggests that they might be made out of something simpler; two monopoles perhaps? Monopoles don't work. The instanton should share its unit of topological charge $Q = \frac{1}{8\pi^2}\int \mathrm{tr}\,(F\tilde{F})d^4x$ between its constituents. A monopole field falling like x^{-2} with at most gauge singularities must have integral topological charge. However, in a non-Abelian theory like QCD, there is another possibility. It is possible to have "null poles" which fall like r^{-1} (one unit slower than monopoles) and for which the Yang-Mills equations do not linearize asymptotically. An example of a null pole is the meron

$$A^{\mu}_{a} = \eta^{\mu\nu}_{a} \frac{x^{\nu}}{x^{2}} \quad . \tag{16}$$

As was shown originally by de Alfaro, Fubini and Furlan[5] the meron satisfies the Yang-Mills equations and has a half unit of topological charge concentrated at the origin and another half unit at infinity. An inversion brings in the topological charge from infinity and produces the two meron solution,

$$A^{\mu}_{a} = \eta^{\mu\nu}_{a} \frac{(x^{\nu} - x^{\nu}_{I})}{(x-x_{I})^{2}} + \eta^{\mu\nu}_{a} \frac{(x^{\nu} - x^{\nu}_{II})}{(x-x_{II})^{2}} \quad . \tag{17}$$

The asymptotic tail of the two meron solution $2\eta^{\mu\nu}_{a}x^{\nu}/x^{2}$ is a pure gauge and can be removed by a singular gauge transformation leaving a field that falls like x^{-3}. A meron pair therefore forms a dipole. The dipole moment of the pair is[1]

$$M^{\mu\nu}_{a} = - \frac{\Delta^{2}}{2} [\eta^{\mu\nu}_{a} - \hat{\Delta}^{\mu}\hat{\Delta}^{\lambda}\eta^{\lambda\nu}_{a}\ \eta^{\mu\lambda}_{a}\hat{\Delta}^{\lambda}\hat{\Delta}^{\nu}] \quad , \tag{18}$$

where $\Delta^{\nu} = (x_{I} - x_{II})^{\nu}$ and $\hat{\Delta}^{\nu} = \Delta^{\nu}/|\Delta|$. Since merons carry half units of topological charge and pairwise form dipoles, they are candidates.

for the constituents of instantons. It has in fact been shown that after smearing the singularities at x_I and x_{II} an instanton can be continuously deformed into two merons.[1] (To avoid a potential source of confusion, it is the intrinsically non-Abelian character of null poles (merons) which allows two of them to form a dipole rather than, as one might have thought, a monopole.)

A year ago when Callan, Gross and myself[6] suggested merons as a mechanism for confinement, they were rather speculative objects. Their status has improved considerably. The singularities at x_I and x_{II} need to be smoothed out. This can be done in a way which does not introduce an arbitrary scale by inserting constraints (and their associated Jacobians) into the functional integral and integrating over the radii of the cores.[1] The Gaussian functional integral around the (constrained) solution is then perfectly well defined and, up to a constant, has been done.[1] In a gas of meron pairs the density of pairs is, of core radii r_1 and r_2 small compared to the separation Δ

$$C\int \frac{dr_1}{r_1^5} \frac{dr_2}{r_2^5} d^4\Delta \exp\left[-\frac{1}{2} X(r_1) - \frac{1}{2} X(r_2) - S_m(r_1,r_2,\Delta)\right] \tag{19}$$

where C is a constant containing a few powers of $\bar{g}$ but otherwise independent of r_1,r_2 and Δ. The terms $\frac{1}{2}X(r_1)$ and $\frac{1}{2}X(r_2)$ correspond to half an instanton in each core and S_m is the renormalized interaction of two merons; its dependence on r_1,r_2 and Δ is given by

$$\begin{aligned} \frac{\partial}{\partial \ln\Delta} S_m &= \frac{3}{4} \hat{X}(\Delta) , \\ \frac{\partial}{\partial \ln r_1} S_m &= \frac{3}{8} \hat{X}(r_1) , \\ \frac{\partial}{\partial \ln r_2} S_m &= \frac{3}{8} \hat{X}(r_2) , \end{aligned} \tag{20}$$

where $\hat{X} = X + 6.55$. The derivatives are the same as those of the classical actions $S = \frac{6\pi^2}{g^2} \ln \frac{\Delta}{\sqrt{r_1 r_2}}$ with the bare coupling replaced by the indicated running couplings. Note that the dimensional factors in the density are 10 powers of core radii and do not contain Δ. With the Gaussian integral now done merons are on the same footing as instantons: they are not optional.

The color magnetic susceptibility χ_m due to meron pairs is[1]

$$\chi_m = C' \int \frac{dr_1 dr_2 d^4\Delta}{r_1^5 \, r_2^5} \quad [\Delta^4] \exp\left[-\tfrac{1}{2}X(r_1) - \tfrac{1}{2}X(r_2) - S_m\right] \tag{21}$$

where C' is a numerical constant. The integrals over r_1 and r_2 are harmless and the integral over Δ has the same convergence properties as

$$\int d\ln\Delta \exp\left[- S_m(\Delta) + 8\ln\Delta\right] \quad . \tag{22}$$

As a function of $\ln\Delta$ the argument of the exponent decreases for small Δ, has a minimum when $\hat{X}(\Delta) = {}^{32}/_3$ and then begins to grow. In other words, if $\hat{X}(\Delta)$ can exceed ${}^{32}/_3$, χ is divergent. This is the phase transition suggested a year ago.[6] At a scale corresponding to $\hat{X} = {}^{32}/_3$ the static coupling blows up and (for reasons not apparent from this discussion) free merons appear. It is difficult to be quantitative about what is on the other side of the transition, but in all likelihood it is a strong coupling phase where quarks are confined. The coupling at which this happens $\hat{X} = {}^{32}/_3$ or $X = 17.22$ ($g^2/8\pi^2 = 0.58$) is remarkably small. At this scale the density of instantons is still small and they are just beginning to produce a large χ.

To summarize, the picture so far is one of a rapid rise in the effective coupling followed by a transition to confinement at $X \sim 17$. In real QCD this is only half the story: there are (almost) massless fermions and chiral symmetry breaking to contend with. As pointed out by 't Hooft,[4] if there are N_f light flavors, the instanton is replaced by a $2N_f$ fermion interaction. In nature there are at least 2 very light quarks and for simplicity I will stick to the case $N_f = 2$: an instanton then becomes a four fermion interaction. Instantons no longer exist in isolation and one must begin the discussion with instanton-anti-instanton pairs bound together by massless fermion exchange. The fermion propagators provide a multiplicative factor of R^{-12} (R is the separation) equivalent to an interaction $S_1 = 12 \ln R$.

This logarithmic interaction between instantons and anti-instantons plays a role similar to the logarithmic attraction between merons: one expects a phase transition when the pairs break up, chiral symmetry is spontaneously broken and quarks acquire a dynamical mass.[1,7] Approximate calculations indicate that this takes place at about the same scale $X \sim 17$ where everything else happens. This is a story in itself but I don't have time to present any details.

REFERENCES

1. This talk is an abstract of and/or guide to C. Callan, R. Dashen and D. Gross, "Toward a Theory of the Strong Interactions," IAS preprint COO-2220-115, to be published in Phys. Rev.

2. C. Callan, R. Dashen and D. Gross, Phys. Lett 63B, 334 (1976) G. 't Hooft, Phys. Rev. Lett. 37 8 (1976), R. Jackiw and C. Rebbi, Phys. Rev. Lett. 37, 172 (1976).

3. A. Belavin, A. Polyakov, A. Schwartz and Y. Tyupkin, Phys. Lett. 59B, 85 (1975).

4. G. 't Hooft, Phys. Rev. D14, 3432 (1976).

5. V. deAlfaro, S. Fubini and G. Furlan, Phys. Lett. 65B, 1631 (1977).

6. C. Callan, R. Dashen and D. Gross, Phys. Lett. 65B, 375 (1977); D. Gross, talk at Orbis Scientiae, 1977.

7. C. Callan, R. Dashen and D. Gross, Phys. Rev. D16, 2526 (1978).

PARTICIPANTS

Richard Arnowitt
Northeastern University

Marshall Baker
University of Washington

Itzhak Bars
Yale University

M.A.B. Beg
The Rockefeller University

Carl M. Bender
Washington University

Edmond L. Berger
Argonne National Laboratory

S. Blaha
Williams College

Arthur A. Broyles
University of Florida

R. Budny
Princeton University

Kevin Cahill
Louisiana State University

Robert Cahn
University of Michigan

Peter Carruthers
Los Alamos Scientific Laboratory

Mou-shan Chen
Center for Theoretical Studies

Rodney L. Cool
The Rockefeller University

Bruce Cork
University of California LBL
Berkeley

John M. Cornwall
University of California
Los Angeles

Ernest D. Courant
Brookhaven National Laboratory
and State University of New York

D.G. Crabb
University of Michigan

Roger Dashen
Institute for Advanced Studies

Hans G. Dehmelt
University of Washington

M. Derrick
Argonne National Laboratory

Jean O. Dickey
California Institute of Technology

P.A.M. Dirac
Florida State University

Tohru Eguchi
Stanford Linear Accelerator Center

Martin B. Einhorn
University of Michigan

Victor Elias
University of Maryland

Pierre Fayet
California Institute of Technology
and LPTENs, Paris

George W. Ford
University of Michigan

G.C. Fox
California Institute of Technology

Paul H. Frampton
Ohio State University

Murray Gell-Mann
California Institute of Technology

Howard Georgi
Harvard University

O.W. Greenberg
University of Maryland

M.T. Grisaru
CERN, Switzerland

Eugene P. Gross
Brandeis University

Feza Gürsey
Yale University

Francis Halzen
University of Wisconsin

Barry J. Harrington
University of New Hampshire

Richard Haymaker
Los Alamos Scientific Laboratory

Jarmo Hietarinta
University of Maryland

David Horn
Cornell University

Roman Jackiw
Massachusetts Institute of Technology

Osman Kadiroglu
Center for Theoretical Studies

Michio Kaku
City College of the City University of New York

G. Karl
University of Guelph, Canada

Stuart H. Kasdan
Princeton University

Boris Kayser
National Science Foundation

Y.S. Kim
University of Maryland

Toichiro Kinoshita
Cornell University

T. Kitazoe
Ohio State University

Konrad Kleinknecht
Universität Dortmund, Germany

Frank Krausz
Center for Theoretical Studies

A.D. Krisch
University of Michigan

Behram Kursunoglu
Center for Theoretical Studies

H.C.S. Lam
McGill University, Canada

Herbert Lashinsky
University of Maryland

Leon Lederman Group
Columbia University

Frank J. Loeffler
Purdue University

Jerzy Lukierski
State University of New York at Stony Brook

K.T. Mahanthappa
University of Colorado

W. Marciano
The Rockefeller University

Marvin L. Marshak
University of Minnesota

Sydney Meshkov
California Institute of Technology and National Bureau of Standards

Hannu I. Miettinen
Fermi Laboratory

Peter Minkowski
University of Bern, Switzerland

Rabindra N. Mohapatra
The City College of the City University of New York

Yoichiro Nambu
University of Chicago

D.V. Nanopoulos
Harvard University

Pran Nath
Northeastern University

Homer A. Neal
Indiana University

Jerzy Nowakowski
Center for Theoretical Studies

Patrick J. O'Donnell
University of Toronto, Canada

Reinhard Oehme
University of Chicago

John O'Fallen
Argonne Universities Association

Martin Olsson
University of Wisconsin

Oliver Overseth
University of Michigan

Heinz R. Pagels
The Rockefeller University

Sandip Pakvasa
University of Hawaii

William F. Palmer
Ohio State University

D.C. Peaslee
Department of Energy
Washington, D.C.

Arnold Perlmutter
Center for Theoretical Studies

Stephen Pinsky
Ohio State University

Enrico C. Poggio
Brandeis University

Chris Quigg
Fermi Laboratory

Helen Quinn
Stanford Linear Accelerator Center

L.G. Ratner
Argonne National Laboratory

Arthur Rich
University of Michigan

Rudolf Rodenberg
Technische Hochschule Aachen Germany

S.P. Rosen
Purdue University

Ralph Roskies
University of Pittsburgh

Howard J. Schnitzer
Brandeis University

Jonathan F. Schonfeld
California Institute of Technology

John H. Schwarz
California Institute of Technology

Gino Segre
University of Pennsylvania

Allen Lee Sessoms
Harvard University

Sun-Sheng Shei
New York University

L.M. Simmons, Jr.
Los Alamos Scientific Laboratory

A. Sirlin
New York University

Stuart Smith
Princeton University

Harold Spinka
Argonne National Laboratory

V.L. Teplitz
Virginia Polytechnic Institute and State University

Kent M. Terwilliger
University of Michigan

Gerald H. Thomas
Argonne National Laboratory

T.L. Trueman
Brookhaven National Laboratory

Hung-Sheng Tsao
The Rockefeller University

P. van Nieuwenhuizen
State University of New York at Stony Brook

Walter W. Wada
Ohio State University

Kameshwar C. Wali
Syracuse University

Yasushi Watanabe
Argonne National Laboratory

Richard Weiner
University of Marburg and
Los Alamos Scientific Laboratory

Arthur B. Wicklund
Argonne National Laboratory

R.W. Williams
University of Washington

G.B. Yodh
University of Maryland

Norman J. Zabusky
University of Pittsburgh

F. Zachariasen
California Institute of Technology

Paul F. Zweifel
Virginia Polytechnic Institute and State University

PROGRAM

ORBIS SCIENTIAE 1978

MONDAY, JANUARY 16, 1978

LAW SCHOOL AUDITORIUM

PROLOGUE

Henry King Stanford, University of Miami
WELCOMING ADDRESS (5 min.)

Behram Kursunoglu, University of Miami
15th Anniversary
"HOPES AND EXPECTATIONS" (15 min.)

P.A.M. Dirac, Florida State University
"A NEW APPROACH TO COSMOLOGICAL THEORY" (30 min.)

SESSION I: "SPIN EFFECTS AT HIGH ENERGIES"

Moderator: Alan D. Krisch, University of Michigan

Dissertators: Alan D. Krisch, University of Michigan
"SPIN EFFECTS IN LARGE $p_{\perp}^2$ PROTON--PROTON SCATTERING" (50 min.)

Homer A. Neal, Indiana University
"SPIN EFFECTS IN SMALL $p_{\perp}^2$ HIGH ENERGY SCATTERING" (50 min.)

Ernest Courant, Brookhaven National Laboratory
"PROSPECTS FOR POLARIZED BEAMS FOR STORAGE RINGS AT VERY HIGH ENERGIES" (50 min.)

SESSION II: "LEPTONIC MAGNETIC MOMENTS"

Moderator: Hans Dehmelt, University of Washington

Dissertators: Toichiro Kinoshita, Cornell University
"WHAT CAN ONE LEARN FROM VERY ACCURATE MEASUREMENTS OF LEPTON MAGNETIC MOMENTS?" (30 min.)

G. William Ford, University of Michigan
"LEPTON SPIN MOTION IN WEAK MAGNETIC MIRROR TRAP" (30 min.)

Hans Dehmelt, University of Washington
"ELECTRON MAGNETIC MOMENT FROM GEONIUM SPECTRA" (30 min.)

Robert W. Williams, University of Washington
"MUON ANOMALOUS MAGNETIC MOMENT: THE LAST WORD" (30 min.)

Arthur Rich, University of Michigan
"ELECTRON AND POSITRON G - 2 EXPERIMENTS - PAST, PRESENT, AND FUTURE" (30 min.)

TUESDAY, JANUARY 17, 1978

SESSION III: "NEW LEPTONS, NEW QUARKS, CURRENTS: EXPERIMENTS AND THEORIES"

Moderator: Sydney Meshkov, California Institute of Technology and National Bureau of Standards

Dissertators: Sydney Meshkov, California Institute of Technology and National Bureau of Standards
"OPENING REMARKS"

Konrad Kleinknecht, Universität Dortmund
"RECENT RESULTS FROM THE CERN-DORTMUND-HEIDELBERG-SACLAY COLLABORATION"

Chris Quigg, Fermi Laboratory
"NEW FLAVORS"

Robert Cahn, University of Michigan
"NEW ABSOLUTELY STABLE HADRONS"

Feza Gürsey, Yale University
"OCTONIONIC QUANTUM MECHANICS AND APPLICATIONS"

SESSION IV: "QUANTUM FLAVORDYNAMICS (QFD)"

Moderator: Mirza A. Beg, Rockefeller University

Dissertators: M.A. Beg, Rockefeller University
"OPENING REMARKS"

W. Marciano, Rockefeller University
"PRESENT STATUS OF MUON-NUMBER CONSERVATION"

R. Mohapatra, CUNY
"SPONTANEOUSLY BROKEN LEFT-RIGHT SYMMETRY"

Helen Quinn, Stanford Linear Accelerator Center
"NATURAL SUPPRESSION OF STRONG CP-VIOLATION AND THE 'AXION' ALTERNATIVE"

WEDNESDAY, JANUARY 18, 1978

SESSION V: "ATTEMPTS AT UNIFICATION INCLUDING SUPERSYMMETRY AND SUPERGRAVITY"

Moderator: Murray Gell-Mann, California Institute of Technology

Dissertators: Murray Gell-Mann, California Institute of Technology
"STATUS OF SUPERTHEORIES"

Pierre Fayet, Ecole Normale Superieure, Paris and California Institute of Technology
"PHYSICAL CONSEQUENCES OF SUPERSYMMETRIES"

John Schwarz, California Institute of Technology
"SPINNING STRING THEORY FROM A MODERN PERSPECTIVE"

Peter van Nieuwenhuizen, SUNY, Stony Brook
"COVARIANT QUANTIZATION OF SUPERGRAVITY"

THURSDAY, JANUARY 19, 1978

SESSION VI: "THEORY AND PHENOMENOLOGY OF THE WEAK COUPLING REGIME OF QUANTUM CHROMODYNAMICS (QCD)"

Moderator: Howard Georgi, Harvard University

Dissertators: Howard Georgi, Harvard University
"USE AND MISUSE OF THE PARTON MODEL"

Hannu Miettinen, Fermi Laboratory
"QUARKS AND GLUONS IN SOFT HADRONIC COLLISIONS"

A.L. Sessoms, Harvard University
"EXPERIMENTAL TESTS OF QCD?"

Geoffrey C. Fox, California Institute of Technology
"APPLICATION TO QUANTUM CHROMODYNAMICS TO HIGH TRANSVERSE MOMENTUM HADRON PRODUCTION"

SESSION VII: "ATTEMPTS TO UNDERSTAND THE STRONG COUPLING REGIME OF QUANTUM CHROMODYNAMICS (QCD)"

Moderator: Yoichiro Nambu, University of Chicago
"INTRODUCTORY REMARKS"

Dissertators: Heinz Pagels, Rockefeller University
"GAUGE FIXING DEGENERACIES AND CONFINEMENT IN NON-ABELIAN GAUGE THEORIES"

R. Jackiw, Massachusetts Institute of Technology and Rockefeller University
"GAUGE DETERMINATION AND AMBIGUITY YANG-MILLS THEORY"

Itzhak Bars, Yale University
"QUANTIZED ELECTRIC FLUX TUBES IN QCD"

Roger Dashen, Institute for Advanced Study, Princeton
"DYNAMICAL EFFECTS OF MERONS AND INSTANTONS"

SUBJECT INDEX